Improvement of Crops in the Era of Climatic Changes

Parvaiz Ahmad • Mohd Rafiq Wani
Mohamed Mahgoub Azooz • Lam-Son Phan Tran
Editors

Improvement of Crops in the Era of Climatic Changes

Volume 2

Springer

Editors
Parvaiz Ahmad
Department of Botany
Sri Pratap College
Srinagar
Jammu and Kashmir
India

Mohd Rafiq Wani
Department of Botany
Govt. Degree College (Boys)
Anantnag
Jammu and Kashmir
India

Mohamed Mahgoub Azooz
Department of Botany
South Valley University
Qena
Egypt

Lam-Son Phan Tran
Signaling Pathway Research Unit
RIKEN Center for Sustainable Resource Science
Yokohama
Japan

ISBN 978-1-4614-8823-1
ISBN 978-1-4614-8824-8 (eBook)
DOI 10.1007/978-1-4614-8824-8
Springer New York Heidelberg Dordrecht London

Library of Congress Control Number: 2013950910

Printed on acid-free paper

Springer is part of Springer Science+Business Media (www.springer.com)

Preface

"Improvement of Crops in the Era of Climatic Changes" Volume 2

Climate change is an unprecedented threat to the food security for hundreds of millions of people who depend on small-scale agriculture for their livelihoods. Abiotic stress is the prime cause for deteriorating the average yield of major crops by more than 50 %, which cause losses worth hundreds of millions of dollars each year. Plants are exposed to rapid, unpredicted, and diverse environmental disturbances, resulting in stressful conditions. Stress signal is first perceived by the receptors present on the membrane of plant cells. The signal information is then transduced downstream, resulting in an activation of various stress-responsive genes. The products of these stress genes, ultimately lead to stress tolerance response and enable the plant to survive and surpass the unfavourable conditions. Recent trends in population growth suggest that global food production is unlikely to gratify the future demands under predicted climate change scenarios, unless rates of crop improvement are accelerated or sweeping changes occur in patterns of human food consumption. The situation is generally more staid in less developed countries, where agro-ecosystems are already fragile, investment in agriculture is limited, and climate change is predicted to have its most devastating effects. Global climate change is likely to increase the problems of food insecurity, hunger and malnutrition for millions of people, particularly in south Asia and sub-Saharan Africa, and further exacerbate the problem by remarkably restricting the plant growth and development.

The potential yield of economically important crops is drastically coming down every year just because of abiotic stresses. It has been projected that global food production must increase by 70 % by 2050 to meet the ever-increasing demand caused by burgeoning human population, increasing incomes and consumption. Several factors are contributing to high plant performance under different environmental conditions, therefore an effective and complementary use of all the available technological tools and resources is needed to meet the challenge. The mechanisms underlying endurance to environmental stress factors have long been the focus of intense research. The progress in biotechnology, genomic research, and molecular marker applications have brought to the forefront an interdisciplinary science that is revolutionizing twenty-first century crop improvement. Many novel genomics technologies like next generation sequencing and omics have emerged as powerful tools for understanding the genomic variation among crop species at different molecular levels. Climate change is no more an illusion, its ruinous impact is globally witnessed and interventions must be highly addressed at international, regional and national levels. In this context, the book "***Improvement of Crops in the Era of Climatic Changes" Volume 2*** will serve as avant-garde resource for researchers and students who are immersed in developing the improved crop cultivars and management methods. Written by a varied group of internationally distinguished experts, "***Improvement of Crops in the Era of Climatic Changes" Volume 2*** is a concise, yet comprehensive resource for researchers, students and others seeking knowledge expansion in this burning area of research and will lead them to new pondering on the subjects of climate changes and crop improvement.

In this book, we present a collection of 14 chapters written by 51 reputed experts in the fields of plant abiotic stress tolerance, induced mutagenesis and crop improvement. It is a well-timed and painstakingly compiled contribution of the topics that are of vast scientific eminence. Chapter one (1) throws light on Brassicas: responses and tolerance to heavy metal stress. In this chapter, the authors stated that there is a great scope for understanding methodically the genetics and genomics of *Brassica* species and the mechanisms of actions underlying the metal-induced toxicities and the tolerances developed therein. Chapter two (2) addresses the recent advances in rapid and sensitive screening for abiotic stress tolerance, wherein the authors have elaborated the development and identification of molecular markers associated with tolerance response and their value in sensitive indirect selection among few crop species. Chapter three (3) is about transcriptomics of heat stress in plants. This chapter primarily provides the current understanding on the role of regulatory genes (transcription factors), heat shock protein genes, metabolic genes, signaling compounds, osmolytes, reactive oxygen species and role of miRNAs as well as small RNAs of plants under high temperature. In addition, it gives a brief account of various transcriptome approaches to study the expression profiling of genes during the heat stress. Chapter four (4) is about biotic stress and crop improvement, wherein the authors fussily tried to identify the most widespread plant viruses in Azerbaijan, using different molecular techniques and precisely evaluated some characteristics of plant responses to viral stress. Chapter five (5) is regarding the salt stress and sugar beet improvement: challenges and opportunities. In this chapter, the authors present the comprehensive discussion on the challenges and opportunities for improvement of salt tolerance in sugar beet and emphasized that future research should chiefly focus on physiological, molecular and metabolic dimensions to facilitate the development of such crops with inherent stress tolerance capacity.

Chapter six (6) describes the genotypic variation for drought tolerance in wheat plants. In this chapter, a rich gene pool, comprising of thousands of wheat genotypes with contrasting photosynthetic traits, productivity and tolerance to drought stress, and introduced from world gene pool, particularly, CIMMYT and ICARDA was created in Azerbaijan which could be of great help for monitoring the environmental stresses in field grown plants and in the selection of stress-resistant varieties. Chapter seven (7) deals with soil contaminants: sources, effects and approaches for remediation. Here, the authors authoritatively stated that the remediation of heavy metal contaminated soils is necessary to reduce the associated risks, make the land resource available for agricultural production, enhance the food security, and scale down the land tenure problems arising from the changes in land use pattern. Chapter eight (8) describes the role of macronutrients in plant growth and acclimation: recent advances and future prospective. This chapter deals with the recent progress made in finding out the roles of macronutrients in plant growth and acclimation processes as well as future prospective of elemental research in plants. Chapter nine (9) is concerned with mutation breeding: a novel technique for genetic improvement of pulse crops particularly chickpea (*Cicer arietinum* L.). This chapter compresses various facets of contemporary knowledge for pulse crop varietal improvement, particularly chickpea, through induced mutagenesis with special thrust on qualitative as well as yield attributing traits. Chapter ten (10) deals with organic farming: the return to nature. In this chapter, the authors enumerated that organic foods have more plant secondary metabolites, higher micronutrient content and more conjugated fatty acids for better human health, including lower incidences of non-communicable diseases. Additionally, they stated that organic agriculture merges modernism, custom and science to manage the shared surroundings by encouraging the fair relationship and high quality of life for everyone involved. Chapter eleven (11) is about the role of cytological aberrations in crop improvement through induced mutagenesis. In this chapter, the authors scrupulously revealed the impact of mutagens on cytological behaviour and their overall role in crop improvement. Chapter twelve (12) deals with the wheat improvement: historical perspective and mutational approach-a review. This review enfolds various historical aspects, in addition to contemporary knowledge of wheat crop improvement programs through induced mutagenesis. Chapter thirteen (13) is about the cotton leaf curl virus disease predictive model based on environmental variables. This chapter was initiated to develop a disease predictive model to characterize the epidemiological factors conducive for disease spread and severity. The authors also envisaged that such models would be highly helpful in forecasting the diseases and subsequently help to decide the correct timing of pesticide applications.

Chapter fourteen (14) deals with transcription factors in abiotic stress responses -their potentials in crop improvement. In this chapter, the authors summarized contemporary understanding about TF activities in plants under adverse stress conditions and their use in crop improvement.

Chapters contributed in this book have been published keeping intact author's justifications, however suitable editorial changes were made, wherever considered necessary. In spite of our best efforts, there is a chance of some errors still creeping in the book, for which we seek reader's feedback. We wish to express abstemious appreciation to our well versed contributors, who readily accepted our invitation to write their chapters. Moreover, we would like to thank Springer Science+Business Media, LLC, New York, particularly Eric Stannard (Editor Botany, Springer), Andy Kwan (Assistant Editor, Springer), Kevin Wright (Developmental Editor, Springer), Flora Kim (Developmental Editor, Springer) and all the other staff members of Springer, New York who were directly or indirectly associated with us in the current project for their constant support and help in bringing out the timely publication of this volume.

Parvaiz Ahmad
Mohd Rafiq Wani
Mohamed Mahgoub Azooz
Lam-Son Phan Tran

Government Sri Pratap College, Srinagar, Jammu and Kashmir, India
Government Degree College (Boys), Anantnag, Jammu and Kashmir, India
Department of Botany, South Valley University, Qena, Egypt
Signaling Pathway Research Unit, RIKEN Center for Sustainable Resource Science, Yokohama, Japan

Contents

Contributors

Zahra Abbasi Department of Sugar Beet Research, Isfahan Agriculture and Natural Resources Research Center, Isfahan, Iran

Parvaiz Ahmad Department of Botany, S.P. College, Srinagar, Jammu and Kashmir, India

Safdar Ali Department of Plant Pathology, University of Agriculture, Faisalabad, Pakistan

Jalal A. Aliyev Institute of Botany, Azerbaijan National Academy of Sciences, Baku, Azerbaijan

Department of Plant Physiology and Biotechnology, Research Institute of Crop Husbandry, Ministry of Agriculture of the Republic of Azerbaijan, Baku, Azerbaijan

M. Atiq Department of Plant Pathology, University of Agriculture, Faisalabad, Pakistan

MM Azooz Department of Botany, Faculty of Science, South Valley University, Qena, Egypt

Devendra Kumar Chauhan Department of Botany, University of Allahabad, Allahabad, India

Nawal Kishor Dubey Department of Botany, Center of Advanced Study, Banaras Hindu University, Varanasi, India

Asiya Hameed Faculty of Science, Department of Botany, Jamia Hamdard, New Delhi, India

Madeeha Hamid Atta-ur-Rahman School of Applied Biosciences, National University of Sciences and Technology, Islamabad, Pakistan

A. Hannan Irrigation Department, Directorate of Land Reclamation, Sargodha, Pakistan

Mirza Hasannuzaman Faculty of Agriculture, Department of Agronomy, Sher-e-Bangla Agricultural University, Dhaka, Bangladesh

Xuan Lan Thi Hoang School of Biotechnology, International University, Vietnam National University HCMC, Ho Chi Minh City, Vietnam

Irada Huseynova Institute of Botany, Azerbaijan National Academy of Sciences, Baku, Azerbaijan

Irada M. Huseynova Institute of Botany, Azerbaijan National Academy of Sciences, Baku, Azerbaijan

Shaheen Hussain Department of Biosciences, Jamia Millia Islamia, New Delhi, India

Alvina Gul Kazi Atta-ur-Rahman School of Applied Biosciences, National University of Sciences and Technology, Islamabad, Pakistan

M. Aslam Khan Department of Plant Pathology, University of Agriculture, Faisalabad, Pakistan

Samiullah Khan Mutation Breeding Laboratory, Department of Botany, Aligarh Muslim University, Aligarh, Uttar Pradesh, India

Samar Khayamim Department of Agronomy, Sugar Beet Seed Institute, Karaj, Iran

Mohd Imran Kozgar KERB Biotech, Srinagar, Jammu and Kashmir, India

Alamdar Mammadov Institute of Botany, Azerbaijan National Academy of Sciences, Baku, Azerbaijan

Satendra K. Mangrauthia Plant Physiology Section, Directorate of Rice Research, Hyderabad, India

Nitin Mantri Health Innovations Research Institute, School of Applied Sciences, RMIT University, Melbourne, VIC, Australia

Behzad Murtaza Institute of Soil and Environmental Sciences, University of Agriculture, Faisalabad-38040, Pakistan

Ghulam Murtaza Institute of Soil and Environmental Sciences, University of Agriculture, Faisalabad-38040, Pakistan

Nabeel Khan Niazi Institute of Soil and Environmental Sciences, University of Agriculture, Faisalabad-38040, Pakistan

NahidaTun Nisa Department of Botany, Govt. Degree College for Women, Srinagar, Jammu and Kashmir, India

Eric Ober The John Bingham Laboratory, National Institute of Agricultural Botany (NIAB), Cambridge, UK

Edwin Pang Health Innovations Research Institute, School of Applied Sciences, RMIT University, Melbourne, VIC, Australia

Vikas Patade Defence Institute of Bio-Energy Research, Defence Research and Development Organisation, Nainital, Uttarakhand, India

Sheo Mohan Prasad Department of Botany, Ranjan Plant Physiology and Biochemistry Laboratory, University of Allahabad, Allahabad, India

Shaista Qadir Department of Botany, Govt. Degree College for Women, Srinagar, Jammu and Kashmir, India

Abazar Rajabi Department of Plant Breeding, Sugar Beet Seed Institute, Karaj, Iran

Muhammad Sabir Institute of Soil and Environmental Sciences, University of Agriculture, Faisalabad-38040, Pakistan

Boghireddy Sailaja Plant Physiology Section, Directorate of Rice Research, Hyderabad, India

Neelamraju Sarla Plant Physiology Section, Directorate of Rice Research, Hyderabad, India

Subzar Ahmad Sheikh Department of Botany, Government Degree College (Boys), Anantnag, Jammu and Kashmir, India

Saima Siddique Atta-ur-Rahman School of Applied Biosciences, National University of Sciences and Technology, Islamabad, Pakistan

Vijay Pratap Singh Govt. Ramanuj Pratap Singhdev Post Graduate College, Korea, Chhattisgarh, India

Saftar Suleymanov Institute of Botany, Azerbaijan National Academy of Sciences, Baku, Azerbaijan

Nargiz Sultanova Institute of Botany, Azerbaijan National Academy of Sciences, Baku, Azerbaijan

Shahbaz Talib Sahi Department of Plant Pathology, University of Agriculture, Faisalabad, Pakistan

Ameema Tariq Atta-ur-Rahman School of Applied Biosciences, National University of Sciences and Technology, Islamabad, Pakistan

Nguyen Phuong Thao School of Biotechnology, International University, Vietnam National University HCMC, Ho Chi Minh City, Vietnam

Nguyen Binh Anh Thu School of Biotechnology, International University, Vietnam National University HCMC, Ho Chi Minh City, Vietnam

Nasya Tomlekova Molecular Biology Laboratory, Department of Breeding, Variety Maintenance and Introduction, Maritsa Vegetable Crops Research Institute, Plovdiv, Bulgaria

Lam-Son Phan Tran Signaling Pathway Research Unit, RIKEN Center for Sustainable Resource Science, Tsurumi, Yokohama, Japan

Durgesh Kumar Tripathi Department of Botany, Center of Advanced Study, Banaras Hindu University, Varanasi, India

Sitapati Rao Voleti Plant Physiology Section, Directorate of Rice Research, Hyderabad, India

Mohd Rafiq Wani Department of Botany, Government Degree College (Boys), Anantnag, Jammu and Kashmir, India

About the Editors

Dr. Parvaiz Ahmad is senior Assistant Professor in Department of Botany at Sri Pratap College, Srinagar, Jammu and Kashmir, India. He has completed his post graduation in Botany in 2000 from Jamia Hamdard, New Delhi, India. After receiving a Doctorate degree from the Indian Institute of Technology (IIT), Delhi, India, he joined the International Centre for Genetic Engineering and Biotechnology, New Delhi, in 2007. His main research area is Stress Physiology and Molecular Biology. He has published more than 35 research papers in peer-reviewed journals and 29 book chapters. He is also an Editor of 12 volumes (1 with Studium Press Pvt. India Ltd., New Delhi, India, 8 with Springer NY, USA and 3 with Elsevier USA). He is a recipient of Junior Research Fellowship and Senior Research Fellowship by CSIR, New Delhi, India. Dr. Parvaiz has been awarded the Young Scientist Award under Fast Track Scheme in 2007 by the Department of Science and Technology (DST), Govt. of India. Dr. Parvaiz is actively engaged in studying the molecular and physio-biochemical responses of different agricultural and horticultural plants under environmental stress.

Dr. Mohd Rafiq Wani is currently Assistant Professor in the Department of Botany at Government Degree College (Boys), Anantnag, Jammu and Kashmir, India. Dr. Wani did his Masters in Botany in 2003 with specialization in "Genetics and Plant Breeding" from Aligarh Muslim University (AMU), Aligarh, Uttar Pradesh, India. After obtaining the Degree of Doctorate in 2008 for his research work on "Chemical Mutagenesis in Mungbean" from the same University, joined Department of Higher Education, Jammu and Kashmir Government, in 2009. He teaches a range of bioscience related subjects at undergraduate and post graduate levels. At present, his research interests are mainly focused on the improvement of pulses through induced mutations and exploring the physiological and biochemical responses of crop plants to a range of biotic and abiotic stresses. As a part of his research endeavour, Dr. Wani has extensively researched and written on the issues of induced chemo-mutagenesis among the food crops, with special reference to pulses. He has published around 30 research papers in peer reviewed journals and 12 book chapters in various research oriented volumes. Dr. Wani, while constantly working for his academic and research interests, has so far edited five (05) volumes of books on the subjects of Plant Stress Physiology, Induced Plant Mutagenesis, and Crop Improvement with reputed international publishers like Springer New York, USA and Wageningen Academic Publishers, Netherlands. In addition, he is an editorial member and reviewer of few online journals pertaining to plant sciences, besides being the life member of various scientific societies like Indian Society of Pulses Research and Development (ISPRD) and Indian Society of Genetics and Plant Breeding (ISGPB).

Dr. Mohamed Mahgoub Azooz is a Professor of Plant Physiology in Department of Botany, Faculty of Science, South Valley University, Qena, Egypt. He received his B.Sc. in 1984 and M. Sc. in 1990 from Assiut University, Egypt and Ph.D. in 1997 from South Valley University, Qena, Egypt in collaboration with Tübingen University, Germany. He has been deputed to King Faisal University, Saudi Arabia as Professor (from 2005 to present) in Department of Biological Sciences, College of Science. Prof. Azooz has published more than 85 research articles in peer-reviewed journals and 10 book chapters in international books. He has also authored 3 books and edited 4 volumes with international publishers like springer NY. He has supervised many M.Sc. students and Ph.D. scholars till date and many are in progress. He is serving as a member of editorial board and reviewer of many international journals. Prof. Azooz has been the principal investigator, co-investigator, consultant and collaborator in more than 20 research projects. He has been awarded scientific publication prize 2012, 2013 and 2014 from both King Faisal University and South Valley University for his scientific articles published in international scientific journals with high impact factors. Prof. Azooz has been included in Marquis Who's in the World (Biography, 28th Edition, 2011), and in the "Top 100 Scientists" list, International Biographical Centre, Cambridge-UK, 2011. Major focus of his research has been on the physiological, biochemical and molecular responses of plants against various biotic and abiotic stresses and the mechanisms of their tolerance.

Dr. Lam-Son Phan Tran is Head of the Signaling Pathway Research Unit at RIKEN Center for Sustainable Resource Science, Japan. He obtained his M.Sc. in Biotechnology in 1994 and Ph.D. in Biological Sciences in 1997 from Szent Istvan University, Hungary. After doing his postdoctoral research at National Food Research Institute and Nara Institute of Science and Technology of Japan, from 10/2001, he joined Japan International

Research Center for Agricultural Sciences, to work on functional analysis of transcription factors and osmosensors in stress responses in Arabidopsis. In 08/2007, he moved to University of Missouri-Columbia, USA as a Senior Research Scientist to coordinate a research team working on discovery of soybean genes for genetic engineering of drought-tolerant soybean plants. His current research interests are elucidation of the roles of phytohormones and their interactions in abiotic stress responses, and translational genomics, with the main focus on soybean, aiming to enhance crop productivity under adverse environmental conditions. He has published over 70 peer-reviewed papers with more than 50 research and 15 review articles.

Chapter 1
Brassicas: Responses and Tolerance to Heavy Metal Stress

Shaista Qadir, Asiya Hameed, NahidaTun Nisa, MM Azooz, Mohd Rafiq Wani, Mirza Hasannuzaman, Alvina Gul Kazi and Parvaiz Ahmad

Abstract *Brassica* is considered as an important crop all over the world owing to its economically important products. *B.juncea* and *B.napus* are cultivated as oilseed crops globally. Heavy metal (HM) stress is one of the abiotic stresses that limit plant growth and development. Root and shoot lengths and fresh and dry weights have been observed to act as accumulators as well as indicators of metal toxicity in crops. *Brassica* has a potential to combat the metal-induced stress, thereby reducing

P. Ahmad (✉)
Department of Botany, S.P. College, Srinagar 190001, Jammu and Kashmir, India
e-mail: parvaizbot@yahoo.com

S. Qadir · N. T. Nisa
Department of Botany, Govt. Degree College for Women, Maulana Azad Road, Srinagar 190001, Jammu and Kashmir, India
e-mail: sqadirs@yahoo.co.uk

N. T. Nisa
e-mail: nahida123@yahoo.com

A. Hameed
Faculty of Science, Department of Botany, Jamia Hamdard, New Delhi, India
e-mail: hameed901@gmail.com

M. Azooz
Department of Botany, Faculty of Science, South Valley University, 83523 Qena, Egypt
e-mail: azzozm@yahoo.com

M. R. Wani
Department of Botany, Government Degree College (Boys), Anantnag 192102, Jammu and Kashmir, India
e-mail: botanyrafiq@gmail.com

M. Hasannuzaman
Faculty of Agriculture, Department of Agronomy, Sher-e-Bangla Agricultural University, Sher-e-Bangla Nagar, Dhaka-1207, Bangladesh
e-mail: mhzsauag@yahoo.com

A. G. Kazi
Atta Ur Rehman School of Applied Biosciences, National University of Sciences and Technology, Islamabad, Pakistan
e-mail: alvina_gul@yahoo.com

P. Ahmad et al. (eds.), *Improvement of Crops in the Era of Climatic Changes,* Volume 2,
DOI 10.1007/978-1-4614-8824-8_1,

the damage by undergoing various types of adaptations. However, cost-effective techniques are available in order to minimize the toxicity and to protect the surroundings from HM stress. Decrease in chlorophyll content confers to weak uptake of mineral ions due to the interference of HMs in plants. Nevertheless, low concentrations of some HMs demonstrate an efficient yield in some species. HMs disturb the composition of fatty acids and as a result lead to tremendous changes in lipid membrane that may ultimately cause lipid peroxidation. Proline accumulation enhances the tolerance level under osmotic stress and is known to regulate the water balance in crop plants. Increased glutathione (GSH) in *B. napus*and *B. juncea,* on exposure to HMs, has shown its active involvement in detoxification of free radicals either directly or through certain enzymes. Phytochelations are one of the important methods to reduce the phytotoxicity by binding complexes with high-affinity ligands in the vacuole, thereby keeping the released toxins away from the metal-sensitive metabolic centers in the cytoplasm. Ascorbate–GSH cycle plays an efficient role in reactive oxygen species (ROS) detoxification released through abiotic stress. Besides, ROS shows release of new isozymes of peroxidases. Genetic engineering has been established to enhance the plant's ability to endure and mitigate the environmental stress. This involves the insertion of foreign DNA into nuclear genome and genomic chloroplast. However, gene expression can be regulated by various promoters. Several transgenic approaches have been carried out successfully with enhanced accumulation of HMs in *B. juncea* cultivars. There is lot of scope to understand the mechanism of HM uptake as well as the capacity of plants to withstand the environmental stresses.

Keywords *Brassica* · Heavy metals · Growth · Osmotic stress · Oxidative stress · Antioxidants · Oil content · Metal uptake

1 Introduction

Brassica is an important genus of the family Brassicaceae, also called as mustard family. The family includes 375 genera and 3,200 species, along with a large number of cultivars and hybrids of cultivated origin. It is widely distributed throughout the world. Most species of the genus originated in Western Europe, the Mediterranean, and temperate regions of Asia. *Brassica* species are broadly considered as valuable source of dietary fiber. The oileferous Brassicas are found within the species like *B. juncea, B. carinata, B. rapa, B. campestris,* and *B. napus,* and are jointly called as rapeseed–mustard oil. The plant is thus associated with great agricultural and horticultural importance because of the economically important products which the genus provides in the form of edible roots, leaves, stems, buds, flowers, seed, and oil (Hasanuzzaman 2008).

Among the family Brassicaceae, several wild relatives of *Brassica,* including *B. adpressa, B. fruticulosa, B. pinescens, B. oxyrrhina, B. barrelieri,*and *B. tournefortii,*are collectively referred as *Brassica*coenospecies. These have useful agronomic traits that can be introgressed into the cultivated Brassicas using the wide hybridization programs (Warwick 1993). In fact, Brassicashave been successfully crossed with members of *Brassica*coenospecies and other wild relatives like *Sinapsis, Diplotaxis, Erucastrum,* and *Raphanus* to obtain the wide hybrids (Nandakumar et al. 1988). There are mainly six species of *Brassica* that merit attention for their economic importance. Among the six species, three are diploid: *B. campestris* (AA, $2n=20$), *B. nigra* (BB, $2n=16$), and *B. oleracea* (CC, $2n=18$),and the other three are amphidiploids: *B. juncea* (AABB, $2n=4x=36$), *B. napus* (AACC, $2n=4x=34$),and *B. carinata* (BBCC, $2n=4x=34$). The amphidiploid species likely originated in nature from diploid ancestors through unidirectional hybridization followed by spontaneous chromosome doubling (Murphy 1994). The botanical or genomic relationship between these six species was established by UN (1935) and is represented in the form of a triangle, usually known as U's triangle. Three diploid species are orientated at the corners of the triangle. In the world, India ranks fourth in oil seeds economy. Mustard contributes 28.6% to the total oil seed production in India (Shekhawat et al. 2012). Heavy metal (HM) causes many direct and indirect effects on plants and animals. Exposure to HM stress leads to decrease in crop yield worldwide (Ahmad K et al. 2011; Ahmad et al. 2011a, 2012; Hasanuzzaman and Fujita 2012; Hasanuzzaman et al. 2013). Excessive amount of HMs causes disturbances in the mineral nutrition and carbohydrate metabolism of plants, thus strongly reducing the biomass production (John et al. 2009). HMs are reported to inhibit the metabolic processes such as nitrogen assimilation, photosynthesis, respiration, water uptake, and transcription (Boussama et al. 1999; John et al. 2009). HMs are responsible for oxidative stress in *Brassica* sp. because: (1) they generate reactive oxygen species (ROS), (2) they inhibit or stimulate the activities of antioxidant enzymes, and (3) HM stress also leads to lipid peroxidation (John et al. 2009; Ahmad et al. 2011a, 2012). This chapter focuses on the effect of HMs on growth, physiological, and biochemical aspects of *B. juncea,* besides discussing the metal tolerance through transgenic approaches.

2 Production in India

India is one of the major rapeseed–mustard-growing countries in the world, holding the top position in area under cultivation and second in production after China. From the past several decades, there has been a continuous increasing demand for edible oilseeds and other by-products in India as well as in other countries, because of which there has been a rapid increase in its production. *Brassica* contributes to 28.6% of the total oilseed crops in India, and occupies the second place among the most important edible oils after groundnut. The rapeseed/mustard oil has the lowest amount of saturated fatty acids as compared to other edible oils. However, the two essential fatty acids linoleic and linolenic acids are present in adequate amount as compared to other edible oils. Globally, 59.93 million t of *Brassica* oilseeds are produced from an area of 30.74 ha with an average yield of 1.9 t ha^{-1} (USDA 2010).

However, India accounts for 10.7 and 21.7% of the yield and area, respectively. Owing to extensive population growth rate, the percapita oil expenditure is expected to increase from the present 13.4 to 23.1 kg/annum by 2030. This may be due to the change in the way of living, and around 102.3 mt of oil seeds will be needed to fulfill this demand. Regarding the present scenario of the rapeseed–mustard as compared to oilseed production in India, it has been predicted that more rapeseed–mustard need to be produced to fill the gap between demand and supply.

3 Effects of HMs on *Brassica*

HM toxicity is one of the major environmental health problems in modern era (Hasan et al. 2009; Alvarez-Ayuso 2008; Suzukiet al. 2001; Ahmad et al. 2011a, 2012). The term HMs is usually used for any element that has metallic properties (ductility, conductivity, density, stability,ligand specificity, etc.), atomic number greater than 20, and density five times higher than water. The release of HMs in nature is a consequence of anthropogenic activities, including industrial processes and damages to both man-made and natural ecosystems (Karimi and Zulkifili 2010; Nagajyoti et al. 2008; Tyler et al. 1989; Hasanuzzaman and Fujita 2012). These HMs do not vanish but in turn get accumulated in soil (Gisbert et al. 2003; Tangahu et al. 2011), which acts as a sink (Karami and Zulkifili 2010). Although a few of them have nutritional importance, e.g., iron and copper, most of them are cytotoxic. Some are toxic even at very low concentrations (Pehlivan et al. 2009; John et al. 2009; Ahmad et al. 2011a, 2012), while all are toxic at higher concentrations (Wuana and Okieimen 2010) and posea threat to plants as well as human beings (Rauser and Meuwly 1995). As the concentration of metals is increasing day by day, techniques are required to reduce the release of toxins into the soil and water bodies. The HMs have great impact on morphological, physiological, and biochemical aspects of *Brassica* and are discussed below.

3.1 Growth

Plants, being essential component of our ecosystem, have contributed a lot for serving as biological monitors at the cost of forest dieback and decline, and have thus have become a debatable topic. Deficiencies as well as surplus accessibility of an element also exhibit its negative effects. Exposure to HMs results in so many physiological breakdowns that it becomes nearly impossible to determine which effects are primary and which are secondary (Prasad 1995). Despite recent progress in understanding the individual aspects of metal toxicity and resistance mechanisms, little is known about the coordination of cellular sequestration mechanisms with adaptation of plant growth. Plant uptake of metals is dependent on a system that is metabolically mediated and also competitive with Zn and other metals.

Development of plant is considered as an essential mechanism that is formed as a result of coordination of the main processes (Vassilev et al. 1998). During this process, they are not found resistant to various stresses including those of HMs. These metals adversely affect the overall plant development, among which the most common are stunted growth (Fariduddin et al. 2009), leaf chlorosis, and variations in the activity of several essential enzymes of various metabolic pathways (Arduini et al.1996; Godbold and Hutterman 1985). Parameters such as fresh weight, dry weight, and shoot as well as root lengths have been used as indicators of metal toxicity in plants (Baker and Walker 1989; Ahmad et al. 2011a, 2012; Mohamed et al. 2012; Shanmugaraj et al. 2013). The HMs obstruct the various plant physiological and growth processes.This association of HMs with different developmental processes of plant represents a dose–response curve. Various plants have been observed to act in different ways such as excluders, indicators, and accumulators (Ouzounidou et al. 1998). The same HM, if excessively present, demonstrates the decline in growth and yield (Raziuddin et al. 2011; Shafiq and Iqbal 2005; Shafi et al. 2009, 2010), blocks cell division and development and eventually causes death of the plant.

Raziuddin et al. (2011) observed a significant decrease in all growth parameters of *B. juncea* cv. NIFA-Raya and *B. napus* cv.Abasin on exposure to metal stress, although the former was less susceptible to toxic effects. Similarly, lead (Pb), one of the most important HMs, is frequently available in the environment (Shahid et al. 2011; Grover et al. 2010) and its most common sources are vehicles and automobiles (Wierzbicka and Antosiewiez 1993; Nicholson et al. 2003; Sezgin et al. 2003). This metal accumulates in the roots, stems, leaves, and seeds of the plant (Singh et al. 1998; Sekara et al. 2005; Yilmaz et al. 2009; John et al. 2009) and damages whole uptake system including chlorophyll formation and cell division (Gupta et al. 2010; Krzesłowska et al. 2010; Liu et al. 2008; Sharma and Dubey 2005a; John et al. 2009; Ahmad et al. 2011a, 2012). The severe reduction in root growth among *Brassica* (Canola) cultivars was attributed to Pb-induced impaired nutrient uptake and altered metabolism in plants (Ashraf et al. 2011). The transport of various ions like N, P, K, Ca, Mg, Zn,etc. was reported to get reduced in both roots and shoots of all the cultivars of *Brassica* due to the influence of Pb (Wensheng et al. 1997; Panda and Choudhary 2005; Fodor et al. 1998; Sinha et al. 2006; Gopal and Rizvi 2008), thereby hampering the various physiological processes. The reduction in absorption of nutrients in the presence of lead may result from their competition (e.g., those with atomic size similar to lead). According to Sharma and Dubey (2005a), the strong interaction of K^+ ions with Pb could result from their similar radii ($Pb^{2+}=1.29$ Å and $K^+=1.33$ Å). These two ions may compete for entry into the plant through the same potassium channels. Similarly, lead affects K^+ -ATPase and -SH groups of cell membrane proteins and causes an efflux of K^+ from roots. However, lead does not cause nitrogen efflux. Nitrate reductase acts as the rate-limiting step in the overall assimilation of nitrate (Xiong et al. 2006; Sengar et al. 2009). Xiong et al. (2006) demonstrated that lead induces a significant reduction in shoot nitrate content (70 and 80%), nitrate reductase activity (100 and 50%), and free amino acid content (81 and 82%) in *B. pekinensis*.

Lead also decreases the concentration of divalent cations (Zn, Mn, Mg, Ca, and Fe) and has been reported in *B. oleracea* (Sinha et al. 2006). Peško et al. (2011) also investigated the effect of seven HM ions (Cd(II), Cr(VI), Cu(II), Hg(II), Ni(II), Pb(II), and Zn(II)) on root growth of five cultivars of *B. napus*and observed that the toxicity of metal ions decreased in the following order Cu > Cr > Hg > Cd > Pb > Ni > Zn. Prasad et al. (1999) observed curtailed growth in *B. juncea* under Zn stress.

Mercury, another HM of concern, exhibits significant phytotoxicity in two cultivars of Indian mustard at elevated concentrations, and its uptake induces a significant reduction in both biomass and leaf relative water content (Shiyab et al. 2009). Different concentrations of HMs also influence the root volume. For example, in response to 20 ppm $CdCl_2$ in sand cultures, roots of saplings were 50 % smaller in volume (Smith and Brennan 1984). Maize seedlings were found to be decreased at 25 µg Cd/g (Hasset et al. 1976). Root growth and weight also show a considerable reduction in American sycamore by Cd treatment (Carlsson and Bazzaz 1977). Sandalio et al. (2001) observed 20 and 90 % decrease in growth rates on exposure to 5 and 50 µM Cd concentration, respectively. Root elongation has been observed to be totally stopped after 5 days of Cd treatment in *Platanus occidentalis* (Kahle 1993). Shoot growth, leaf length, as well as leaf area also reveal the same results against the stress except low doses of cadmium that show comparatively better growth and development of plant (Setia et al. 1993; John et al. 2009). Setia et al. (1993) demonstrated that 8 mM Cd^{2+} stress is responsible for decrease in diameter of new stems by 23 % in wheat. Carlson and Bazaaz (1977) also reported reduced growth in American sycamore (*P. occidentalis*) on exposure to Pb–Cd interaction.

A gradual decrease of plant dry matter has been observed in response to HMs. Reduction of plant dry matter (root) was 10–30 % in seedlings of American sycamore (*P. occidentalis*) at 10–100 ppm Cd in the nutrient media (Kahle 1993). Likewise, shoot dry weight shows 37–48 % decrease with increasing concentrationsof Cd and Pb in soybean (Haung et al. 1974). Bhattacharya and Choudhuri (1994) also demonstrated a marked 33.6 % reduction in biomass of *Vigna* seedlings with Cd concentration of 10^{-5} M. Vassilev et al. (1998) observed a decrease in dry mass accumulation in barley plants grown in pots. The inhibitory effect was 32–35 % at the first harvest (the stage of tillering in control), decreasing to 10–73 % at the fourth harvest (the stage of full ripeness in control), with 45 mg Cd/Kg soil. Malan and Farrant (1998) observed a significant decrease in number of pods (83 %) and seed mass (16 %) with 0.05 mg/L Cd in soybean. HMs (Cd, Zn, and Hg) have been found to reduce the root and shoot lengths of *B. oleracea* var. Botrytis (Theriappan et al. 2011) and 10 different cultivars of *B. juncea* grown under different concentrations of Cd (Qadir 2003). The overall decrease arises due to the suppression of growth of plants owing to an irreversible inhibition by Cd on proton pump responsible for the process (Aidid and Okamotto 1993, 1992). Nevertheless, low concentration of Cd shows a good yield in tomato and eggplants (Khan and Khan 1983), tobacco cells (Hirt et al. 1989), and seedlings of alfalfa (Peralta et al. 2000).

3.2 Yield

The productivity of any crop is dependent on its inherent capacity for photosynthesis, photosynthetic area developed, and availability of photosynthetically active radiations within the canopy. Reduction in photosynthesis accompanied by biomass decrease against the metal stress appears to be an almost universal finding (Piotrowska et al. 2009; Islam et al. 2008; Ouzounidou et al. 1997; Dudka et al. 1996; John et al. 2009). While few studies report yield enhancement with very low concentration of metals (Breckle et al. 1991), a significant decrease in biomass after exposure to metal stress was observed by Anjum et al. (2008) in *B. napus* (rapeseed) and *B. juncea* (John et al. 2009) plants at different stages of growth. A marked reduction was observedin the number of siliquae per plant, number of seeds per siliqua, seeds per plant, and seed weight per plant in *B. napus* due to sewage water treatment containing Pb, Cd, and Cr (Ahmad K et al. 2011). Similar findings have been reported earlier by different workers (Bazai and Achakzai 2006; Farid 2006; Kang et al. 2007; Khan et al. 2009). Tamoutsidis et al. (2002) reported that the application of increasing doses of municipal wastewater reduces the overall yield of some vegetable crops, e.g., lettuce, endive, spinach, radish, carrot, and sugar beet. Sharma and Dubey (2005b) reported flower and pod senescence as a consequence of metal toxicity, which leads to the production of less number of viable pods and seeds and reduced yield under metal stress in rice crop. Kakar et al. (2010) and Raziuddin et al. (2011) reported that *B. napus* grown under the influence of HMs turns the soil saline, and as a result of which the plants are unable to take essential elements needed for their vegetative growth, which ultimately results in yield reduction. Ahmad K et al. (2011) proved that the wastewater having higher concentrations of HMs adversely affects the plant growth and development as well as yield. Prins et al. (2011) observed that *B. juncea* growing on seleniferous soils shows decreases in biomass, pollen germination, individual seed and total seed weight, number of seeds produced, and seed germination.

3.3 Photosynthesis

Photosynthetic inhibition is a well-known HM toxicity and has been reported by a number of workers (Xiong et al. 2006; Cenkci et al. 2010; Singh et al. 2010; Touiserkani and Haddad 2012). The decrease in photosynthetic rate may be explained on the basis of number of changes that occur in normal metabolic pathway on exposure to HMs. HMs have been found to inactivate the Photosystem II (PSII) activity along with gradual loss of granal stacks, thereby damaging the thylakoid acyl lipids, with subsequent formation of some polypeptides associated with oxygen-evolving complex and disorganization of light-harvesting complex II (LHC II) antenna system (Islam et al. 2007; Tukendorf and Baszynski 1991). PSII regulates the process of photosynthesis, because it undergoes water oxidation and sustains the process of electron transport. It is formed of three parts, i.e., a core composed of reaction center proteins D1 and D2, cytochrome CP47, and 33-kDa Mn stabilizers, including various low molecular weight proteins, with an oxygen-evolving system (Barber et al. 1997). Chlorophyll as well as prosthetic groups are essential for charge separation and stabilization of proteins (i.e., both D1 and D2) (Nanba and Satoh 1987). HM involves the breakdown of PSII core proteins like D1 (Oquist and Hunner 1993). The root problem may be that the photosynthesis is coupled with a series of electron transport systems (ETS), and metal ions are suggested to block the electron transport pathway (Qufei and Fashui 2009), thus causing inactivation of PSII as a result of metal instability involved in sequences of reactions. Photosynthetic carbon fixation is the primary target of metal toxicity. Most of the products of this fixation produced by primary light reaction are utilized by carbon metabolism. Oxidized nicotinamide adenine dinucleotide phosphate (NADPH) constantly reinstates the terminal electron acceptor in order to balance the photochemical reaction centers. This also provides the coupling of electron transport withadenosine triphosphate (ATP) synthesis where proton motive forces carry out the feedback inhibition of electron flow,therebyproviding a dissipation mechanism for excessive excitation energy, as the products of primary light reaction exceed its synthesis and simultaneously modify the rate of electron transport through PSII. It also affects the quantum yield of linear electron transport. Further, HM tends to accumulate in the chloroplastsand substitutes for essential divalent ions (Cenkci et al. 2010; Gupta et al. 2009) leading to ion imbalance, thereby affecting the chloroplast activity which is believed to be the cause of PSII disassembly (Geiken et al. 1998). Decreased photosynthetic rate in *B. juncea* cv. NIFA-Raya and *B. napus* cv.Abasin, under the influence of HMs, can also be due to the interaction of metal ions with pigment systems of the plants (Raziuddin et al. 2011).

3.3.1 Influence on Pigments

Pigments are the most important part in plants (Hall and Rao 1999), and nutrient efficiency, plant development, and yield are the result of these pigments (Seyyedi et al. 1999). The chlorophyll consists of tetrapyrrole ring with Mg^{2+} as the central atom. Mn is a key element for photosynthesis as well as for regulation of enzyme synthesis (Doganlar et al. 2012). A decrease in chlorophyll as a mark of metal toxicity has been reported by many workers (Kang et al. 2007; Sepehr and Ghorbanli 2006; Larsson et al. 1998; Skorzynska-Polit and Baszynski 1997; Rascio et al. 1993; Schlegel et al. 1987; John et al. 2009; Ahmad K et al. 2011; Ahmad et al. 2011a, 2012). Touiserkani and Haddad (2012) reported that the concentration of chlorophyll shows a gradual decrease on exposure to HMs, followed by chlorosis and necrosis (Skorzynska-Polit and Baszynski 1997). A significant decrease in chlorophyll content in *B. chinensis* and *B. pekinensis* (Liu et al. 2004), *B. juncea* cv. Vitasso and *B. napus* cv. Atlantic (Vatehova et al. 2012), and *B. juncea* (Mohamed et al. 2012) was also reported in response to

HM stress. Similarly, Raziuddin et al. (2011) observed a significant decrease in chlorophyll content of *B. juncea* cv. NIFA-Raya and *B. napus* cv. Abasin on exposure to metal stress. Touiserkani and Haddad (2012) observed a significant decrease in chlorophyll content of different *B. napus* cultivars grown under the influence of HMs.

δ-Aminolevulinic acid (ALA) is the first specific precursor in chlorophyll synthesis for ALA-dehydratase (a metal-sensitive enzyme) that is directly affected by HMs (Sasa and Sugahara 1976). The enzyme for the condensation of two molecules of ALA involves the presence of thiol group at the binding sites for both biosynthesis (Nandi and Shamim 1968) and formation of tetrapyrrole protochlorophyllide, and hence prevents the production of chlorophyll (Dahlin et al. 2000; Boddi et al. 1996; Masuda et al. 1996; McEven et al. 1996). The HM stresses have been reported to increase the chlorophyllase activity and subsequent enzymatic degradation, which leads to the deletion of chlorophyll molecules. Reduction may also confer to poor uptake of mineral ions owing to the presence of HM in growth medium. Horvath et al. (1996) reported a reduction in chlorophyll content on exposure to Cd. Ouzounidou et al. (1997) observed the same in leaves of wheat when subjected to 1 mM Cd, which reflects the indirect effect of cadmium on the content of essential nutrients. Erdei et al. (2002) also demonstrated a gradual reduction in barley seedlings against varying concentrations of cadmium. Chatterjee and Chatterjee (2000) reported a decline in chlorophyll content in *B. oleracea* var. Botrytis and cv. Maghi on exposure to copper, cobalt. and chromium.

Apart from inhibition of biosynthetic enzymes for chlorophyll formation, the increased levels of free radicals of fatty acids produced from polyunsaturated fatty acids due to the higher activity of lipoxygenase may also contribute to the decreased level of chlorophyll with HM treatments (Somashekaraiah et al. 1992; Klein et al. 1984; Hildebrand and Hymowitz 1982). The decrease in chlorophyll content may also be correlated with the adverse toxicity of metals on their uptake and accumulation of essential nutrients in plants, viz., Fe, Mg, Ca, and K as reported in wheat (Ouzounidou et al. 1997; Greger and Ogrer 1991; Greger and Lindberg 1987). Earlier reports suggest that the change in Fe to Zn ratio may be responsible for the reduced chlorophyll content in plants (Root et al. 1975). Recent studies, however, suggest that the formation of LHC is disturbed in HM-treated leaves (Horvath et al. 1996), and as a result, the LHC protein synthesis stops at the transcriptional level (Tzivelka et al. 1999). The product formed during stress also causes photoxidative breakdown, as demonstrated in barley/oat leaves (Luna et al. 1994). Similar results were reported by Hegedus et al. (2001) in barley, where the high quantity of chlorophyll was disintegrated after the initial Cd treatment. Mg, being an important component of chlorophyll molecule, gets removed by an enzyme called Mg dechelatase; subsequently, a ring opens, and as a result of dioxygenase activity, the binding protein is set free for degradation. The rest of the chlorophyll catabolites are carried to the vacuole where further metabolism occurs (Buchanan-Wollaston 1997). Liu et al. (2008) reported an increase in chlorophyllase activity upon Pb stress in *Sedum alfredii*.

3.3.2 Carotenoids

Carotenoids (accessory pigments) are present in the thylakoid membrane of chloroplasts (Collins 2001). They form the diverse set of pigments found in nature and are produced by all photosynthetic and many non-photosynthetic organisms (Nishio 2000). Due to chlorophyll degradation, leaves appear yellow, but this is seen merely when the plant reaches to senescence naturally or under the stress conditions. Carotenoids have an essential role in photoprotection and in scavenging ROS (Young and Britton 1990). During this mechanism, the transfer of excited chlorophyll and singlet oxygen to carotenoid takes place,which quenches as well as dissipates them without undergoing any chemical change (Young and Britton 1990; Bartley and Scolnik 1994). They accumulate primarily in photosynthetic membranes in association with LHC and reaction center complex. Carotenoid content is least affected (Clijsters and Van Assche 1985) or usually enhanced under the stress conditions (Ralph and Burchelt 1998; Foyer and Harbinson 1994). The plant's potential to adjust the levels of accessory pigments is significantly essential to endure the abiotic stress and, thereby, enhance the tolerance, as they shield the photosynthetic tissues against photosensitization. The carotenoids are responsible for de-epoxidation of singlet oxygen and oxygenated state of carotenoids. The xanthophylls undergo interconversion from one form to another in a cyclic way, leading to the non-photochemical quenching of the excessive excitation energy (Demming-Adams and Adams III 1992). While an enhancement in non-photochemical absorption of this excitation energy has been observed in land plants against HMs (Krupa et al. 1993a), the use of xanthophyll cycle in such dissipation has not been reported so far. The ability of HMs to stimulate the production of carotenoids is supposed to restore the damaged pigments during their interaction with excited chlorophyll molecules.

3.4 Oil Content

In the global oilseeds context, rapeseed–mustard occupies an important place, and in India these are next to groundnut, contributing about 32% of the total oilseed production. Rapeseed–mustard showed a remarkable progress in terms of production and productivity during the last decade, but the data from National Productivity reflect that despite a major gain in realized yield, substantial breakthrough in terms of potential yield remains to be achieved. Abiotic stresses, including drought, heat, cold, salinity, and HMs, are the increasing problems not only affecting the plant yield but also altering the composition of fatty acids of membrane lipids.

Fatty acids are free carboxylic groups with a biological chain of commonly occurring 16–18 carbon atoms. Some of the chains are either saturated or unbranched and few are unsaturated and branched containing three carbon rings or hydroxyl groups. Components of phospholipids, glycolipids, hormones, and intracellular messengers are essential groups of fatty acids synthesized in cytosol. Synthesis and breakdown of citrate carrying an acetyl group occur from mitochondria to cytoplasm. NADPH is required for this process, which is produced by pentose phosphate pathway as well as during the movement of reducing equivalents from mitochondria by malate–pyruvate shuttle. Fatty acids are triacylglycerols (neutral fat) present in adipose tissue, which can be activated to acyl CoA by the hydroxylic action of lipases.They are then translocated to inner mitochondrial membrane by carnitinine, and simultaneously destroyed in the mitochondrial matrix. They are elongated and desaturated by enzyme systems in the endoplasmic reticulum membrane. Besides nicotinamide adenine dinucleotide (NADH) and O_2, the complex consists of a flavoprotein, a cytochrome, and a nonheme iron protein required for carrying out this process.

Sinha et al. (2010) showed that Brassica juncea cultivated on Cu, Cr (VI), As(VIII), As(V) contaminated soils, non-significant decrease in oil yield was observed, except for Cr and higher concentrations of As(V). Both As (V) and Cr showed a decrease in oil content, but the maximum reduction has been found in Cr-treated plants (Sinha et al. 2010). Adverse climatic changes like temperature (Pleines et al.1987; Tremolieras et al. 1982), salinity (Allakhverdiev et al. 1999; Elenkov et al. 1996), and HMs (Howlett and Avery 1997; Fodor et al. 1995; Frostegard et al. 1993; Krupa and Baszynski 1989; Hameed et al. 2012) change the composition of fatty acids in plants. High temperature leads to tremendous enhancement of $C_{18:1}$ content and a decrease in $C_{18:3}$ content. Low temperature increases $C_{18:1}$ and $C_{18:2}$ desaturation, resulting in higher $C_{18:3}$ content (Pleines et al. 1987; Tremolieres et al. 1982). In a study carried out by Ouarti et al. (1997), cadmium enhances the proportion of $C_{16:0}$ and lowers the proportion of $C_{18:2}$ and $C_{18:3}$ in 17-day-old tomato seedlings. The result of this study suggests that metal stress induces changes in the unsaturated fatty acids. In addition, the buildup of $C_{16:0}$ rather than $C_{18:0}$ signifies a change in the ratio of products from fatty acid synthase. Likewise, Krupa and Baszynski (1989) reported the similar findings with thylakoids and demonstrated the lowering of all individual glycolipids and phospholipids, and the maximum reduction was found in phosphatidylcholine content. Composition of acyl lipids removed from thylakoids was illustrated by a considerable decrease in the trans-δ-3-hexadecanoic acid with an exception of linolenic acid that tends to fall among all lipids.

4 Responses of *Brassica* Towards HM *Stress*

4.1 Osmotic Stress

Water is considered as a major factor in the regulation of plant growth. Plant water relation is strongly disrupted by metal ion toxicity (Brunet et al. 2009). This effect is primarily dependent on the chemical property of metals like its valence state, ionic radii, and potential of forming organic system. Beyond the limits, all metal ions in plants cause alteration of water relations (Haug and Caldwell 1985), alteration in turgor pressure (Qureshi et al. 2007), and alteration in plasma membrane properties, along with inhibition of root growth (Hagemeyer and Breckle 2002). These actions in turn give rise to secondary impacts like hormonal imbalances, deficiency of essential nutrients, changes in photoassimilate translocation, and alteration of water relations. Increased interest in plant water relations under metal toxicity stress also derives from the fact that plants in metal-enriched soil frequently suffer from osmotic stress, mainly because of poor physical soil conditions as well as shallow root system. Osmotic stress involves some alterations in the solute concentration around a cell, causing rapid modifications in the water across its cell membrane. When the concentration of salts is high outside the cell, exosmosis of water takes place. Salt stress also prevents the mobilization of essential ions, substrates, and cofactors from the cell and results in "cell shock." Likewise, at low concentrations, water is drawn into the cell in greater quantity, thus causing it to enlarge, rupture, or undergo apoptosis. HM-induced drought conditions threaten the cells with dehydration, and the productivity and quality of many commercially grown agronomical and horticultural crops are often adversely affected. To overcome this stress, plants are equipped with many beneficial molecules like osmolytes and osmoprotectants.

4.2 Osmolytes and Osmoprotectants

4.2.1 Proline

For a long time, proline was considered as an inert compatible osmolyte that protects the subcellular structures and macromolecules under osmotic stress (Mehta and Gaur 1999; Kavi-Kishor et al. 2005; Ahmad et al. 2008, 2010c, 2011a, 2014a, b; Rasool et al. 2013; Katare et al. 2012). However, proline accumulation can influence the stress tolerance in multiple ways (Szabados and Savoure 2010). Other than plants, proline accumulations have been widely distributed in eubacteria, protozoa, and algae (Rasool et al. 2013). A common response of plants to HM stress is the accumulation of proline (Muneer et al. 2011; John et al. 2009; Qureshi et al. 2007; Ahmad et al. 2011a, 2012; Parmar et al. 2013), other organic acids, etc. So far, research has been carried out to study the impact of HM-induced ion toxicities on different plants (Zhang et al. 2000; Schat et al. 1997; Bassi and Sharma 1993), including cultivars of *B. juncea* (Buddh and Singh 2012; Sharma et al. 2010; Qadir 2003; Saradhi et al. 1993; Ahmad et al. 2011a, 2012), and maximum studies revealed the accumulation of proline contents under HM stress. An experiment conducted on *B. oleracea*var. Botrytis under the influence of Cd, Zn, and Hg reported an increase in proline content (Theriappan et al. 2011). Proline gets accumulated in plants under different stresses like mineral nutrient deficiencies (Possingharri 1956; Ahmad et al. 2014a), high salinity (Arshi et al. 2002; Thomas and Bohnert 1993; Katare et al. 2012; Rasool et al. 2013), osmotic stress (Voetberg and Sharp 1991), and oxidative stress (Ahmad et al. 2010b, 2011b; Ahmad and Umar 2011; Ahmad 2014). Proline accumulation has been shown to increase the resistance of plants to various stresses (Aspinall and Paleg 1981) by acting as a solute that protects them against enzyme denaturation (Sharma and Dubey (2005a, b); Paleg et al. 1984, 1981; Nikolopoulos and Manettas 1991), a reservoir of energy sources like carbon (C) and nitrogen (N), a solute that alleviates the protein synthesis mechanism, a means of reducing the cell acidity, a sink for energy to regulate the redox potentials, and a shuttle for reducing the power among cell components (Forlani et al. 2000). Kishor et al. (1995) also demonstrated that accumulated proline showed the tolerance capacity against the osmotic stress. Accumulation of proline played a great role in maintenance of water balance in plant tissues (Costa and Morel 1994) and scavenging of free radicals from the cells (Smirnoff and Cumbes 1989). It also acts as a part of nonenzymatic free radical detoxification similar to the mechanism of glutathione (GSH), α-tocopherol, and glucose (Alia et al. 1995). It is very difficult to speculate the mechanism that accounts for the accumulation of proline under extremely varying conditions (Bassi and Sharma 1993; Alia and Saradhi 1991; Saradhi and Saradhi 1991). The movement of ETS has been reported to reduce in plants or its parts against the stress. As a result, there is a buildup of $NADH^{+}H^{+}$. In addition, the enhancement of NADH to NAD^{+} ratio under water stress has been already demonstrated. Such an increase in NADH might affect the substrate despite metabolic reactions requiring NAD^{+}. Furthermore, buildup of organic acids causes a reduction in cytosolic pH. The accumulation of citrate is attributed to declining NAD^{+} as well as synthesis of malate and lactate (by oxidizing NADH). Hence, proline synthesis from glutamic acid is regarded as an adaptive mechanism in order to decrease the buildup of NADH and also to lessen the acidity, whereby two of $NADH^{+}2H^{+}$ are used for combining each molecule of proline with glutamic acid. Boussama et al. (1999) suggested that the induction of NADH glutamate-dehydrogenase activity under Cd stress might provide the glutamate required for enhancing the synthesis of proline. These findings strikingly show that proline has a key role in nonenzymatic free radical detoxification processes.

4.3 Oxidative Stress and Antioxidants

Aerobic metabolism provides marvelous energy benefits to the organisms, but at the same time a constant hazard is the oxidative damage arising from by-products of respiration and photosynthesis, such as superoxide and hydrogen peroxidetocellular macromolecules such as proteins, lipids, and DNA. The phenomenon is known as oxidative stress. HMs were found to induce oxidative stress in plants (Yadav 2010; Singh et al. 2010; Grover et al. 2010; Liu et al. 2008; Pourret et al. 2008; Okamoto et al. 2001; Piqueras et al. 1999; Hendry et al. 1992; Somashekaraiah et al. 1992; John et al. 2009; Ahmad et al. 2011a, 2012; Shanmugaraj et al. 2013; Parmar et al. 2013). Oxidative stress is induced by the generation of ROS in the cells against the environmental stress (Ahmad et al. 2010a, b, 2011a, b, 2012; Ahmad and Umar 2011). These ROSs are deleterious to plants, especially for biomolecules like proteins, nucleic acids, carbohydrates, lipids, etc. ROS formed as a result of HMs (Foyer et al. 1997) reveals the new isozymes of peroxidases in roots as well as leaves of *Phaseolusvulgaris* (Van Assche and Clijsters 1990). Further, evidence of the metal-induced oxidative stress comes from the detection of lipid peroxidation, increased lipogenase activity, chlorophyll breakdown, as well as various antioxidant activities, both nonenzymatic and enzymatic, viz., superoxide dismutase (SOD), glutathione reductase (GR), dehydroascorbate reductase (DHAR), catalase (CAT), ascorbate peroxidase (APX), and guaiacol peroxidase (GP; Schutzendubel et al. 2001; Dalurzo et al. 1997; Chaoui et al. 1997; Lozano-Rodriguez et al. 1997; Gallego et al. 1996; Ahmad et al. 2011a, 2012). Antioxidant enzymes are good indicators of metal stress, and enhancement of their activities could be of great help for immediate remediation (Morsy et al. 2012).

4.4 Nonenzymatic Antioxidants

4.4.1 Ascorbic Acid

Ascorbic acid (AsA) is a water-soluble nonenzymatic antioxidant and a defense component present in plants, where it can accumulate to millimolar concentrations in both photosynthetic and non-photosynthetic tissues (Foyer et al. 1983). Its ability to show antioxidant properties is related to the fact that the dehydrogenase radical is much less reactive than other radicals (Rose and Bode 1993). Enzymatic systems exist in vivo to reduce this radical to ascorbate using NADH or GSH as a source of reducing power. Leaves frequently consist of about 10 % of soluble carbohydrates of ascorbate. AsA, being the key antioxidant (Nijs and Kelley 1991), reacts with reactive oxygen radicals (Buettner and Jurkiewiez 1996),besidessignifying the role in photoprotection and regulation of photosynthesis (Forti and Elli 1995; Foyer and Harbinson 1994). Ascorbate plays a key role in the protection of enzymes containing prosthetic transition metal ions (Padh 1990). Also, AsA is considered as a powerful antioxidant derivative that reduces α-tocopherol (oxidized) in an aqueous phase (Padh 1990).

Recent years have witnessed a large number of reports correlating increase in one or more of antioxidant systems while combating with HM stress. Several workers have observed the fate of AsA under HM stress, and differential responses have frequently been observed (Gupta et al. 2009; Qureshi et al. 2007; Zengin and Munzuroglu 2005; Qadir et al. 2004; Schutzendubel et al. 2001; Prasad et al. 1999; Kubo et al. 1995; Malan et al. 1990). When *B. juncea* L. cultivars were exposed to toxic concentration of Zn (Prasad et al. 1999) and Cd (Qadir et al. 2004), a significant increase in accumulation of AsA occurs. Schutzendubel et al. (2001) observed that the total ascorbate increased significantly after 12 h when root tip was treated with 5 and 50 μM of Cd in scots pine roots.

4.4.2 Glutathione

GSH is a widely distributed cellular tripeptide that plays an important role in scavenging cellular ROS by providing reducing equivalents for antioxidant enzyme defense systems such as APX and glutathione peroxidases (GPXs). Among the responses of plants to HM stress, the GSH enhancement plays an important role in conferring the metaltolerance to *B. juncea* (Szollosi et al. 2009; Qadir et al. 2004; Bogs et al. 2003) and other plant species (Qureshi et al. 2007). Concerted efforts in recent years have given indications of physiological significance of HMs in exerting variations in the magnitude of GSH and mechanisms of GSH functions that help in balancing of redox potential during the stress. Increasing GSH levels have also been found in extracts of pea, tobacco, maize, and tomato treated with Cd (Ruegsegger and Brunold 1992; Ruegsegger et al. 1990). Haribabu and Sudha (2011) observed an increase in GSH content in *B. juncea* after exposure to copper and cadmium.

Enhanced levels of GSH in *B. napus* (Mendoza-Cozatl et al. 2008), *B. juncea* genotypes (Seth et al. 2012; Mobin and Khan 2007; Qadir et al. 2004 under Cd and Zn toxicity (Prasad et al. 1999) suggest its active participation in detoxification of oxygen species and free radicals directly as well as through certain enzymes (Asada and Takahashi 1987). It is presumed that GSH (Wingate et al. 1988) or glutathione disulfide (GSSG; Wingsle and Karpinski 1996) or variations among GSH and GSSG (Foyer et al. 1997) may serve as signals for gene expression to get adjusted to stress. The demand-driven increase in cellular GSH levels shows its implications in stress sensors and signal transduction systems that stimulate GSH biosynthesis pathway. Boussama et al. (1999) also observed an increase in GSH under metal-enriched conditions required for the biosynthesis of HM-binding peptides. Zhu et al. (1999a) reported a fivefold increase in GSH levels in roots of Cd-treated transgenic plants as compared to control. Both the GSH- and Cd-induced production of phytochelatins (PCs) for metal-sensitive genotypes is low, suggesting a lower capacity of Cd-sensitive genotypes forthe synthesis of GSH to compensate for the metal-induced stress. PCs and GSH play a major role in mitigating the toxicity of plants against HMs (Jozefizak et al. 2012).

4.4.3 Tocopherol

The tocopherols, specifically α-tocopherol (vitamin E),area well-established part of chloroplastic membrane present in photosynthetic as well as non-photosynthetic tissues, and act as a membrane-stabilizing agent. Essential features of tocopherols is to protect the important components (biomolecules) from ROS damage. At the same time, the product α-tocopherolquinone formed during quenching is helpful in cyclic electron transport and, hence, provides photoprotection for chloroplasts (Zengin and Munzuroglu 2005). α-Tocopherol maintains the fluidity to regulate the membrane function (Munné-Bosch and Alegre 2002). Furthermore, α-tocopherol balances the oxygen radicals as well as hormones like jasmonic acid within the cell that helps in the normal functioning of plants by inducing responses against the stress (Munné-Bosch and Alegre 2002). The α-tocopherol levels have been shown to increase under various stressful conditions including HMs (Artetxe et al. 2002) in duckweed and *B. juncea* (Yousuf et al. 2010; Kumar et al. 2013). α-Tocopherol and AsA content was increased in primary leaves of *P. vulgaris* when exposed to Cd, Cu, Pb, and Hg stresses after 10 days of growth (Zengin and Munzuroglu 2005). Overexpression of γ-tocopherol methyl transferase (γ-TMT) gene from *Arabidopsis thaliana* in transgenic *B. juncea* shows manyfold increase in α-tocopherol.

4.4.4 Nonprotein Thiols

Nonprotein thiols (NP-SH) are important antioxidant components whose level increases under the metal stress (Seth et al. 2012; Mendoza-Cozatl et al. 2008; Leustek et al. 2000; Heiss et al. 1999). Vogeli-Lange and Wagner (1990) observed a marked increase in the total NP-SH contents from 0.19 to 1.23 μmol SHg^{-1} fw in the leaves of tobacco seedlings treated with 20 μM $CdCl_2$ for 1 week. Cadmium with thiol group-induced enhancement results from de novo synthesis (Howe and Merchant 1992). The increase in NP-SH is accompanied by the increase in PCs content. This increase is important because PCs undergo detoxification not only due to their function as chelators but also because they act as transporters from cytoplasm to the vacuole. Heiss et al. (1999) also observed 8.4-fold increases of thiol groups in leaves and 6.5-fold increases in roots under the stressed conditions. Zhu et al. (1999a) observed a twofold increase in thiol levels in roots of Cd-treated transgenic plants than the wild types.

The increased level of NP-SH under HM (Cd) is particularly obvious because HMs stimulate the various levels in biosynthetic pathway of thiol (cys), together with adenosine triphosphatesulfurylase (APS)-sulfotransferase (Leustek et al. 2000; Heiss et al. 1999; Leustek and Saito 1999; Lee and Leustek 1998) and *O*-acetylserlyase (Schäfer et al. 1998). An increase in NP-SH content under Cd stress in *B. juncea* (Seth et al. 2012) and *B. napus* (Mendoza-Cozatl et al. 2008) could be attributed to an increased level of γ-glutamyl-cysteine (γ-EC) synthetase proteins that lead to cys formation. In the resistant varieties, the activity of γ-EC synthetase may be high, thus resulting in an increased γ-EC formation. Increased γ-EC levels among the resistant plants result in high NP-SH as well as GSH content when compared with those found in sensitive plants.

4.4.5 Phytochelatins

No doubt the HMs exert adverse effects on all living cells, but at the same time, they are also essential for the normal growth and development. The excess of metal ions causes death of cells, and thus can be overcome by homeostasis (regulate the metals within the cell). Plant cells have developed one general mechanism to achieve this goal, i.e., through synthesis of PCs. The process involves the chelation of metal ions by specifichigh-affinity ligands that reduce the concentration of free metal ions, and consequently decrease their phytotoxicity. A number of metal binding ligands have now been recognized in plants (Mejare and Bulow 2001; Cobbett 2000). PC formation is selected as biomarkers for cellular metal sequestration (Sylwia et al. 2010; Jiang and Liu 2010; Yadav 2010; Brunet et al. 2009), because the genetic analysis has provided the direct evidence for PC involvement in metal detoxification (Pal and Rai 2010: Ha et al. 1999; Cobbett et al. 1998). PCs found in cytosol possess great affinity for binding with different metals, especially Cd. These complexes formed in the vacuole undergo sequestration, i.e.,they keep the metals away from enzymes that are susceptible (Rauser 1990). This system provides an environmentally well-conserved mechanism to deal with metal toxicity (Cobbett 2000). The major detoxification mechanism(s) in plants is based on vacuolar compartmentalization and ligand complexation. Prolonged cadmium exposures cause a significant increase in PC synthetase (enzyme for synthesis of PC in leaves of *B. juncea*) (Heiss et al. 2003). High levels of PCs were identified in the phloem sap of *B. napus* within 24h of Cd exposure using combined mass spectrometry and fluorescencehigh-performance liquid chromatography (HPLC) analysis (Mendoza-Cozatl et al. 2008).

In general, while vacuolar compartmentalization keeps HMs away from metal-sensitive metabolic centers in the cytoplasm, sequestration ligands seem to safeguard them from readily moving by reducing their chemical selectivity. For an effective and efficient internal metal tolerance, both roles are important. While the mechanism of low molecular weight endogenous or induced organic acids, particularly citrate, may be employed by plants as a strategy to detoxify the low-level exposure of HMs, an additional mechanism of producing large molecules and more specific compounds such as PCs may be employed by plants to combat high-level exposure to HMs. Although production of PCs may be a mechanism that plants employ in internal detoxification of HM, yet it is unlikely that such is the only mechanism in HM tolerance. This is because induction of PCs is associated with exposure to high external metal concentrations. Apparently, at relatively low metal exposure, plants employ mechanism(s) other than PC production to tolerate the metal stress internally.

4.4.6 Metallothioneins

Metallothioneins (MTs) were first discovered in horse (equine) kidney and have since been broadly studied among the animals, eukaryotic microorganisms, certain prokaryotes, and plants (Kawashima et al. 1992, 1991; Lane et al. 1987). These are small gene-coded cys-rich polypeptides, generally lacking aromatic amino acids (Kagi 1991) and have a high metal content in coordination with metal thiolate clusters. Their molecular weight varies from 8 to14 kDa (Robinson et al. 1993) and is thought to be the aggregates of PCs. MTs behave similarly as

PCs and oftenmetal complexation duties are shared between MTs and PCs, as seen in *Datura* and *Zea mays* (Rivai et al. 1990). In fact, PCs were originally classified as class III MTs, until they were deemed sufficiently different in structure and synthesis pathway to be classified as PCs. The classification of MTs places all animal MTs into class I and all other MTs (including plants) into class II. Class II MTs, those identified in plants, are further classified into two types. Both types of MTs are characterized by having patterns of 12 cysteine residues, but the amino acid composition and the arrangement of cysteine residues varied among the proteins.

Type I MTs have 12 cysteine residues, and they are arranged in such a way in the protein that there is 6cys–*X*aa–6cys, with *X*aa comprising about 40 amino acids. Type II MTs are configured either as 6cys–6cys or 6cys–*X*aa–*X*aa–6cys (Murphy et al. 1997). It is thought that these variations lying in their sequences tell us the feature of MT, i.e., whether it provides the power of tolerance or is able to detoxify the toxicity of various metals. MTs have highest affinity for Cu and are induced by exposure to it (Murphy et al. 1997). Various MT genes, Mouse MT I, Human MTIA, MT II, Chinese hamster MTII, yeast cupl, and pea PsMTA, have been transferred to *Nicotiana* spp., *Brassica* spp., and *A. thaliana* (Hasegawa et al. 1997; Hattori et al. 1994; Maiti et al. 1989; Misra and Gedamu 1989; Lefebvre et al. 1987) and have shown to exhibit high capability to endure stress and show 20-fold greater enhancement against control, thereby revealing that the MT geneis an essential tool to improve the resistance (Kärenlampi et al. 2000). The actual function of MTs is illustrated through various theories and thoughts. Onetheory states that MTs provide a reservoir for essential metal ions that undergo chelation till plants utilize them when needed. A second school of thought is that MTs are carriers (mobile proteins) that are supposed to move excess HMs from the point where they get accumulated to toxic levels to such areas of the plant wherever needed, or at least where the ion levels are not toxic.

4.5 Enzymatic Antioxidants

Plants are armed with a defensive mechanism that helps them to scavenge the ROS that are generated in different cellular compartments (Singh et al. 2010; Gupta et al. 2010; Brunet et al. 2009; Ahmad et al. 2010a, b, 2011a, b, 2012; Ahmad and Umar 2011, Ahmad 2014). Enzymatic antioxidants include SOD, CAT, and a number of enzymes involved in AsA–Glu cycle. HM-induced toxicity may inhibit the activity of these enzymes. Inhibition of enzyme activity may occur because of the fact that most of the HMs, viz., Cd, Pb, and Hg, have –SH groups present on enzymes (Prasad and Prasad 1987a, b; Gupta et al. 2009, Sharma and Dubey 2005a). HM ions also interact with metalloid enzymes and, thus, cause disruption in plant absorption of metals like Zn, Fe, and Mg, which are essential for the activity of most of the enzymes. Induction of antioxidant enzymes by overexpression of genes responsible has also been reported by Li and Luo (2012) in ryegrass.

4.5.1 Superoxide Dismutase

This enzyme was recognized by different names like erythrocuprein, indophenol oxidase, and tetrazolium oxidase till its catalytic function was discovered by McCord and Fridovitch (1969). SOD catalyzes a reaction in which two identical substrates have different metabolic fates. In this case, one molecule of superoxide is oxidized and the other is reduced, thus resulting in the formation of hydrogen peroxide and oxygen:

$$O_2^{\bullet -} + O_2^{\bullet -} + 2H^+ \rightarrow H_2O_2 + O_2$$

Since SOD is found in all aerobic organisms and plays a key role to protect the organisms against ROS damage (Beyer et al. 1991; Scandalias 1993; Ahmad et al. 2010a, b, 2011a, b, 2012), a lot of work reveals the specialty of this enzyme, i.e., the cloned complementary DNA (cDNA) sequences and genes, and their effects of overexpression in transgenic plants (Bowler et al. 1994; Foyer et al. 1994; Scandalios 1993). SOD activity enhances against the various toxicities (Ahmad et al. 2010a, b, 2011a, b, 2012, Ahmad 2014; Ahmad and Umar 2011). SOD enhancement is attributed to de novo synthesis of enzymatic proteins (Touiserkani and Haddad 2012). Transgenic expression of SOD activity has been shown to improve the protection against oxidative stress. Studies proposed that SOD activity regulates the balance of ions and maintains the integrity of plants when exposed to oxidative damage, except for barley species where SOD demonstrates the decreasing trend under Zn and Cd stress (Patra and Panda 1998). Luna et al. (1994) reported that increased H_2O_2 and its derivative active oxygen species (AOS) are shown to be responsible for splitting the enzyme activity with increasing concentration and have been reported in oat leaves, owing to Cu stress.

4.5.2 CAT Activity

Metabolism of peroxide is one of the main functions of CAT activity following the conversion of glycolate during photorespiration (Foyer et al. 1994). Similar findings have been observed in the metabolism of fatty acids in germinating seeds (Holtman et al. 1994). The CAT was observed to reduce when grown in high levels of CO_2 (Azevedo et al. 1998). Vitoria et al. (2001) reported an increase in CAT activity upon the Cd stress and presented it as a proof to support the theory that ROS is always formed when exposed to HM stress, especially Cd. They observed that enzyme CAT activity in the leaves gets enhanced by decreasing the concentration of Cd in the roots (Noctor and Foyer 1998). There may also be the possibility of oxidative stress when carried from roots to leaves (Vitoria et al. 2001).

A decrease in CAT activity in ten different genotypes of *B. juncea* L. was observed under different concentrations of Cd (Qadir et al. 2003). Very few of them show contradictory results. Decrease in CAT activity is also reported in *B. napus* (Touiserkani and Haddad 2012), *P. vulgaris* (Somashekariah et al. 1992), *P. aureus* (Shaw 1995), *Pisum sativum* (Dalurzo et al. 1997), *Lemna minor* (Mohan and Hossetti 1997), and *Amaranthus lividus* (Bhattacharjee 1998), following the application of HMs to growth medium. However, in *P. vulgaris,* CAT activity was decreased in roots and leaves, and there was no effect on stem (Chaoui et al. 1997). Tremendous changes in CAT activity were found in *Helianthus annuus* (Gallego et al. 1999) and *Solanum tuberosum* (Stroinski and Kozlowska 1997) on exposure to cadmium. Shah et al. (2001); Vaglio and Landriscina (1999); Ahmad et al. (2011a, 2012) also observed a general reduction of CAT activities in *B. juncea* on Cd exposure.

4.5.3 Ascorbate–GSH Cycle

This cycle has an essential role in sequestration of reactive oxygen radicals associated with AsA and GSH (Asada 1994). H_2O_2 produced as a result of SOD activity as well as photorespiration (Hernandez et al. 1995; Gille and Singler 1995; Corpas et al. 1993) gets dispersed across the membrane and is harmful because it acts as an oxidant as well as reductant (Gille and Singler 1995; Ahmad et al. 2010a, b, 2011a, b, Ahmad and Umar 2011, Ahmad 2014). APX reduces H_2O_2 into water using ascorbate as electron donor; the resulting monodehydroascorbate (MDAsA) radical can dismutate spontaneously to AsA and dehydroascorbate (DAsA) or may be enzymatically reduced to AsA by NADPH-dependent monodehydroascorbate reductase (MDAR); DAsA is also reduced to AsA enzymatically in a reaction mediated by GSH-dependent dehydroreductase. GSH is an electron donor and the GSSG formed is converted back to GSH by NAD (P) H^--dependent GR (Foyer et al. 1994). HMs induce an increased GR activity (Singh et al. 2010; Brunet et al. 2009; Qureshi et al. 2007; Reddy et al. 2005; Verma and Dubey 2003; Ahmad et al. 2011a, 2012). Reduced GSSG produced in As–GSH pathway helps to enhance the metal tolerance by increasing the GR. Furthermore, it plays an essential role in the formation of cysteine, i.e., the prosynthesis of GSH (Suter et al. 2000). The higher GR activity in metal-tolerant varieties, as compared to sensitive ones, is demonstrated in two ways: (1) In resistant varieties, the ascorbate–GSH cycle may be operating at a high rate in order to detoxify the ROSand (2) it is essential to keep the GSH in a reduced form prior to its incorporation into PCs (Cobbett 2000). Various reports have detected an enhancement under Cd stress (Chaoui et al. 1997), although the effect is dose dependent and varies with time. Schutzendubel et al. (2001) observed that the GR activity decreased initially, but with the increase of incubation time, the GR activity gets increased. Previous reports show a decline in GR activity process when subjected to toxic levels of Fe, Cu, and Cd (Mishra et al. 2006; Verma and Dubey 2003; Patra and Panda 1998;

Gallego et al. 1996). Similarly, HMs have been found to increase (Gupta et al. 2010; Piotrowska et al. 2009; Qureshi et al. 2007; Verma and Dubey 2003; John et al. 2009; Ahmad et al. 2011a, 2012) or decrease the activity of APX (Mishra et al. 2006) in *Brassica* sps.

5 Metal Uptake by *Brassica*

Metal uptake by plants and their mechanism have been reported by several researchers. Uptake of HMs depends on following conditions: (a) part of metal ions that are present in root system and (b) the pH that drastically affect the cation exchange capacity (CEC) by restricting the ready form of exchange sites. Decrease in pH causes H^+ ions to bind tightly to soil particles in order to get booted off in the presence of excess H^+ (Garcia-Miragaya and Page 1978). At high pH, cations are less bioavailable because of the less competition of H^+ binding sites. Low soil pH increases the metal bioavailability, and hence enhances the uptake until they become harmful to plants. Plants respond to HMs differently, such as excluders, indicators, or accumulators (Ouzounidou et al. 1998). Accumulators thrive well even though the concentration of metals remains high in their shoots after transforming or biodegrading the metal ions to inert forms. Plants develop a highly specific and well-organized mechanism in order to acquire essential micronutrients from the surroundings. Plant roots are provided with mechanisms like chelation, pH changes, and oxidation–reduction reactions, which help to solubilize and hence utilize these essential elements with low concentrations even from nearly insoluble precipitates. It is demonstrated that xylem is required for the translocation of various metals from roots to stems. Further, a number of transport mechanisms as well as specialized proteins present in the plasma membrane of the plant cells are engaged in movement and translocation of ions. These include: (1) proton pumps, (2) cotransporters/anti-transporters, and (3) channel proteins. Transporters for Cu, Zn, and Fe were cloned from *A. thaliana* (Salt et al. 1998). Some of the HMs like Cd show passive absorption and movement of ions occurs freely. Plants generally require about 10–15 ppm of trace elements and do not show any accumulation (Report by US Department of Energy 2004). Hyperaccumulators can concentrate the toxic ions above the normal range. Evaporation of water through leaves acts as driving force to take up nutrients and other soil particles into plant roots. This method regulates the soil substances into the shoots as well. This technique is known as phytoremediation and the plants used therein for the purpose are known as hyperaccumulators. Most members of the genus *Brassica* are hyperaccumulators and are used in such strategy. Hyperaccumulators can accumulate HMs to nearly 100 or 1,000 times as compared to non-accumulator plants (Tangahu et al. 2011). Microorganisms, bacteria and fungi, present in root zone colony help to solubilize metal ions, and thereby enhance their bioavailability (Erdei et al. 2005).

6 HM Stress Tolerances Through Transgenic Approach

Advances in science and technology offer a remarkable potential for improvement of crop plants. Modern techniques in tissue culture like somatic cell fusion, somaclonal variation, marker-guided breeding, mapping and sequences of plant genomes, and transformation are of great help to scientists for designing a crop with desired traits. Genetic engineering has been modifying the plants (*Brassica* cultivars) to increase their abilities to tolerate and remediate the environmental pollutants. In genetic engineering, plants are stimulated to take up a piece of DNA-containing specific gene(s) derived from either the same or different species, together with bacteria and animals. The foreign piece of DNA is not only inserted into the nuclear genome but also used into the genome. The expression of that gene can be regulated by using various promoters. The gene protein may show expression constitutively at all times, in all tissues or only in certain tissues (e.g., in roots or in shoots) or at certain times under the presence of specific factors (e.g., light or chemical). Moreover, using different targeting sequences as "address labels," the protein may be subjected to various compartments, such as chloroplast, the vacuole, or the cell wall. In addition, a marker gene is usually included in the gene construct so that transgenics can be selected after the transformation event. As a rule, these marker genes show resistance towards the herbicides as well as antibiotics. The introduced genes integrate into the host DNA and are consequently inherited by the offsprings. This approach has been followed while designing the transgenic *Brassica*cultivars so as to overexpress the genes encoding protein carriers and enzymes that are rate limiting to trace the element uptake and sequestration. Here, the aim is to gain high biomass plants along with the potential to improve their tolerance and also to eliminate the HMs from the soil. For example, *B. juncea* cv. 173874 has been transformed by the insertion of a foreign gene *Arabidopsis APSI* encoding ATP-sulfurylase activity using *Agrobacterium tumefaciens* as vector and CMV 35S as promoter. This resulted in overexpression of plastidic ATP sulfurylase; twofold to threefold higher Se accumulation in shoots and 1.5-fold higher Se in roots as compared to wild-type plants (Pilon- Smits et al. 1999). The same cultivar has been transformed for higher accumulation of Cd (threefold) by the insertion of foreign genes like *Escherichia coli gshII* that codes for glutathione synthetase (GS) and *E. coli gshII* encoding ECS, causing an overexpression of cytosolic GS as well as overexpression of ECS, respectively (Zhu et al. 1999a). Pilon-Smits et al. (2000) also transformed Indian mustard by insertion of foreign *E. coli gor* gene encoding GR, which resulted in overexpression of GR targeted to plastids (cpGR) as well as cytosol (cystGR). cpGr-transformed plants show two times higher root glutathione levels as compared to wild types. The transgenic *B. napus,* in association with plant growth-promoting bacteria, has also been used for remediation of Ni-contaminated soils (Farwell et al. 2006). The three transgenic plants of *B. juncea* have been examined in vitro for their ability to eliminate the selenium (Se) from Se- and Boron-contaminated saline sediments. The transgenic lines that are overexpressed code for respective enzymes, viz., APS, ECS, and

GS. The APS, ECS, and GS transgenic plants accumulated 4.3-, 2.8-, and 2.3-fold Se in their leaves than wild types, respectively. GS plants significantly thrive on contaminated soils better than wild types, reaching to an aboveground biomass/ area to almost 80 % of that of GS plants grown on unpolluted ones (Banuelos et al. 2005). Transgenic Indian mustard overexpressing selenocysteinelyase and selenocysteinemethyl transferase has been shown to exhibit enhanced potential for selenium phytoremediation under field conditions (Banuelos et al. 2007). Experiments have been carried out successfully with *B. juncea* cultivars using several transgenic approaches to further enhance the plant selenium accumulation (Banuelos 2007). These transgenic plants have yielded promising results, showing upto ninefold higher levels of selenium accumulation and threefold faster volatilization rates even under the field conditions (Pilon-Smits and Le Duc 2009; Pilon-Smits et al. 2010).

7 Conclusions and Future Perspective

*Brassica*members have remained wonderful crops to mankind since time immemorial. They have not only served the mankind in the form of their seeds which are used for extraction of edible oils, spices, and condiments but also provided other products in the form of food (vegetables) and fodder. Considerable research has been carried out in plant breeding and genetics fields to develop hybrids or cultivars of great economic importance. Now the thrust of research should be on exploring new technologies for producing better quality of oil by improving various traits like pungency of crop, activity, its shelflife, and medicinal properties. HM pollution is emerging as a major environmental threat and numerous studies have revealed their impacts on morphological, biochemical, and physiochemical aspects of plant life. Starting with the inhibition of seed germination to stunted growth, they interfere with every metabolic pathway of plants, thus leading to curtailed root growth, decreased pigment content, altered antioxidant systems, and decreased biomass and yield. However, in recent past, many members of the genus *Brassica,*e.g., Indian mustard, canola cultivars, etc., have been identified as plant species with superior ability to tolerate and accumulate different HMs. These plants have the ability to combat metal-induced ROS, and thus decrease the extent of damage caused by disruption of redox system of the cell. This is mainly achieved by reduced metal uptake, their sequestration in the cell vacuoles, formation of metal-binding chelators, viz., phytochelatins, and through the induction of enzymatic and nonenzymatic detoxification systems upon the exposure to metal stresses. Further, a greater tolerance level has been achieved through the transgenic technology, as the mustard plant provides the striking model for many experimental purposes owing to its small genome size. There is a great scope for understanding comprehensively the genetics and genomics of *Brassica* species and the mechanism of actions underlying the metal-induced toxicities and tolerance developed therein.

References

Ahmad K, Ejaz A, Azam M, Khan MI, Ashraf M, Al-Qurainy F, Fardous A, Gondal S, Bayat R, Valeem EE (2011) Lead, cadmium and chromium contents of canola irrigated with sewage water. Pak J Bot 43(2):1403–1410

Ahmad P (2014) Oxidative damage to plants, antioxidant networks and signaling. Academic, USA

Ahmad P, Umar S (2011) Antioxidants: oxidative stress management in plants. Studium Press Pvt. Ltd., New Delhi

Ahmad P, Jhon R, Sarwat M, Umar S (2008) Responses of proline, lipid peroxidation and antioxidative enzymes in two varieties of *Pisum sativum* L. under salt stress. Int J Plant Produc 2(4):353–366

Ahmad P, Jaleel CA, Salem MA, Nabi G, Sharma S (2010a) Roles of enzymatic and nonenzymatic antioxidants in plants during abiotic stress. Crit Rev Biotechnol 30(3):161–175

Ahmad P, Umar S, Sharma S (2010b) Mechanism of free radical scavenging and role of phytohormones during abiotic stress in plants. In: Ashraf M, Ozturk M, Ahmad MSA (eds) Plant adaptation and phytoremediation. Springer, Dordrecht, pp 99–108

Ahmad P, Jaleel CA, Sharma S (2010c) Antioxidative defence system, lipid peroxidation, proline metabolizing enzymes and Biochemical activity in two genotypes of *Morus alba* L. subjected to NaCl stress. Russ J Plant Physiol 57(4):509–517

Ahmad P, Nabi G, Ashraf M (2011a) Cadmium-induced oxidative damage in mustard [*Brassica juncea* (L.) Czern. & Coss.] plants can be alleviated by salicylic acid. South Afr J Bot 77:36–44

Ahmad P, Nabi G, Jeleel CA, Umar S (2011b) Free radical production, oxidative damage and antioxidant defense mechanisms in plants under abiotic stress. In: Ahmad P, Umar S (eds) Oxidative stress: role of antioxidants in plants. Studium Press Pvt. Ltd., New Delhi, pp 19–53

Ahmad P, Ozturk M, Gucel S (2012) Oxidative damage and antioxidants induced by heavy metal stress in two cultivars of mustard (L) plants. Fres Environ Bull 21(10):2953–2961

Ahmad P, Ashraf M, Azooz MM, Rasool S, Akram NA (2014a) Potassium starvation-induced oxidative stress and antioxidant defense responses in *Brassica juncea*. J Plant Inter 9(1):1–9

Ahmad P, Ozturk M, Sharma S, Gucel S (2014b) Effect of sodium carbonate-induced salinity-alkalinity on some key osmoprotectants, protein profile, antioxidant enzymes, and lipid peroxidation in two mulberry (*Morus alba* l.) cultivars. J Plant Inter 9(1):460–467

Aidid SB, Okamoto H (1992) Effects of lead, cadmium, zinc on the electrical membrane potential at the xylem/symplast interface and cell elongation of *Impatiens balsamina*. Environ Exp Bot 32:439–448

Aidid SB, Okamoto H (1993) Responses of elongation rate, turgor pressure and cell wall extensibility of stem cells of *Impatiens balsamina* to lead, cadmium, zinc. Bio Metals 6:245–249

Alia K, Saradhi PP (1991) Proline accumulation under heavy metal stress. J Plant Physiol 138:554–558

Alia K, Prasad VSK, Saradhi PP (1995) Effect of Zn on free radicals and proline in *Brassica* and *Cajanus*. Phytochemistry 39(1):45–47

Allakhverdiev SI, Nishiyama Y, Suzuki I, Tasaka Y, Murata N (1999) Genetic engineering of the unsaturation of fatty acids in membrane lipids alters the tolerance of synechacystis to salt stress. Proc Natl Acad Sci U S A 96:5862–5867

Alvarez-Ayuso E (2008) Cadmium in soil-plant systems: an overview. Intl J Environ Pollut 33(2–3):275–291

Anjum NA, Umar S, Ahmad A, Iqbal M, Khan NA (2008) Ontogenic variation in response of *Brassica campestris* L. to cadmium toxicity. J Plant Interact 3(3):189–198

Arduini I, Godbild DL, Onnis A (1996) Cadmium and copper uptake and distribution in Mediterranean tree seedlings. Physiol Plant 97:111–117

Arshi A, Abdin MZ, Iqbal M (2002) Growth and metabolism of senna as affected by salt stress. Biol Plant 45:595–298

Artetxe U, Gorae-Plazola JI, Hernandez A, Becerril JM (2002) Low light grown duckweed plants are more protected against the toxicity induced by Zn and Cd. Plant Physiol Biochem 40(10):859–863

Asada K (1994) Chloroplast: formation of active oxygen and its scavenging. Methods Enzymol 105:422–442

Asada K, Takahashi M (1987) Production and scavenging of active oxygen in photosynthesis. In: Kyle DJ, Osmond CJ, Arntzen CJ (eds) Photoinhibition: topics in photosynthesis. Elsevier, Amsterdam, pp 227–287

Ashraf MY, Azhar N, Ashraf M, Hussain M (2011) Influence of lead on growth and nutrient accumulation in canola (*Brassica napus* L.) cultivars. J Environ Biol 32:659–666

Aspinall D, Paleg LG (1981) Proline accumulation: physiological aspects. In: Paleg LG, Aspinall D (eds) Physiology and biochemistry of drought resistance in plants. Academic, New York, pp 206–241

Azevedo RA, Alas RM, Smith RJ, Lea PJ (1998) Response of antioxidant enzymes to transfer from elevated carbon dioxide to air and ozone fumigation, in the leaves and roots of wild-type and a catalase-deficient mutant of barley. Physiol Plant 104:280–292

Baker AJM, Walker PL (1989) Physiological responses of plant to heavy metals and the quantification of tolerance and toxicity. Chem Speciation Bioavail 1:7–17

Banuelos G, Terry Leduc D, Pilon-Smits EH, Mackey B (2005) Field trial of transgenic Indian mustard plants shows enhanced phytoremediation of selenium-contaminated sediment. Environ Sci Technol 39:1771–1777

Banuelos G, LeDuc DL, Pilon-Smits EAH, Tagmount A, Terry N (2007) Transgenic Indian mustard overexpressing selenocysteine lyase, selenocysteine methyltransferase, or methionine methyltransferase exhibit enhanced potential for selenium phytoremediation under field conditions. Environ Sci Technol 41:599–605

Barber J, Neild J, Morris EP, Zhelera C, Hankamer B (1997) The structure, function and dynamics of PSII. Physiol Plant 100:817–827

Bartley GE, Scolnik PA (1994) Molecular biology of carotenoid biosynthesis in plants. Annu Rev Plant Physiol Plant Mol Biol 45:287–301

Bassi R, Sharma S (1993) Proline accumulation in wheat seedling exposed to zinc and copper. Phytochemistry 33 (6):1339–1342

Bazai ZA, Achakzai KK (2006) Effect of wastewater from Quetta city on germination and seedling growth of lettuce (*Lactuca sativa* L.). App Sci J 6(2):380–382

Beyer W, Imlay J, Fridovich I (1991) Superoxide dismutases. PNAS 40:221–253

Bhattacharya M, Chaudhuri MA (1994) Effect of lead and cadmium on the biochemical changes in the leaves of terrestrial (*Vigna*) and aquatic (*Hydrilla*) plants under solution culture. Indian J Plant Physi 37:99–103

Bhattacharjee S (1998) Membrane lipid peroxidation, free radical scavengers and ethylene evolution in *Amaranthus* as affected by lead and cadmium. Biol Plant 40:131–135

Boddi B, Evertson I, Ryberg M, Sundquist C (1996) Protochlorophyllide transformation and chlorophyll accumulation in epicotyl of pea (*Pisum sativum*). Physiol Plant 96:706–713

Bogs J, Bourboulaux A, Cagnac O, Wachter A, Rausch T, Delrot S (2003) Functional characterization and expression analysis of a glutathione transporter BjGT1 from *Brassica juncea*: evidence for regulation by heavy metal exposure. Plant cell Environ 26(10):1703–1711

Boussama N, Ouariti O, Suzuki A, Ghorbal MH (1999) Cadmium stress on nitrogen assimilation. J Plant Physiol 155(3):310–317

Bowler C, Van Camp W, Van Montagu M, Inze D (1994) Superoxide dismutase in plants. Crit Rev Plant Sci 13:199–218

Breckle SW (1991) Growth under stress: heavy metals. In: Waiser Y, Eshel A, Kafkafi U (eds) Plant roots: the hidden half. Marcel Dekker, New York, pp 351–373

Brunet J, Varrault G, Zuily-Fodil Y, Repellin A (2009) Accumulation of lead in the roots of grass pea (*Lathyrus sativus* L.) plants triggers systemic variation in gene expression in the shoots. Chemosphere 77(8):113–1120

Buchanan-Wollaston V (1997) The molecular biology of leaf senescence. J Exp Bot 48(307):181–199

Buddh K, Singh RP (2012) Cadmium tolerance and its phytoremediation by two oil yielding plants *Ricinus communis* (L.) and *Brassica juncea* (L.) from the contaminated soil. Int J Phytorem 14(8):772–785

Buettner GR, Jurkiewiez BA (1996) Chemistry and biochemistry of ascorbic acid. In: Cadenas E, Packer L (eds) Handbook of antioxidants. Marcel Dekker, New York, pp 91–115

Carlson RW, Bazzaz FA (1977) Growth reduction in American sycamore (*Plantanus occidantalis*) caused by Pb-Cd interaction. Environ Pollut 12:243–253

Cenkci S, Cigerci IH, Yildiz M, Özay C, Bozdag A, Terzi H (2010) Lead contamination reduces chlorophyll biosynthesis and genomic template stability in *Brassica rapa* L. Environ Exp Bot 67(3):467–473

Chaoui A, Mazhaudi S, Ghorbal MH, Ferjoni EE (1997) Effect of cadmium and zinc interaction on hydroponically grown bean (*Phaseolus vulgaris*). Plant Sci 126:21–28

Chatterjee J, Chatterjee C (2000) Phototoxicity of cobalt chromium and copper in cauliflower. Environ Pollut 109:69–74

Clijsters H, Van-Assche F (1985) Inhibition of photosynthesis by heavy metals. Photosynth Res 7:31–40

Cobbett CS (2000) Phytochelatins and their roles in heavy metal detoxification. Plant Physiol 123:825–832

Cobbett CS, May MJ, Howden R, Rolls B (1998) The glutathione-deficient, cadmium sensitive mutant, cad 2-1 of *Arabidopsis thaliana* is deficient in γ-glutamyl cysteine synthetase. Plant J 16:73–78

Collins RA (2001) Carotenoids and genomic stability. Mutat Res 2081:21–28

Corpas JF, Gomez M, Hernendez JA, del Rio LA (1993) Metabolism of activated oxygen in peroxisomes from *Pisum sativum* cultivars with different sensitivity to sodium chloride. Plant Physiol 144:160–165

Costa G, Morel JL (1994) Water relations, gas exchange and amino acid content in cadmium treated lettuce. Plant Physiol Biochem 32 (4):561–570

Dahlin C, Aronsson H, Almkvist J, Sundquist C (2000) Protochlorophyllide-independent import of two NADPH: pchlide oxidoreductase proteins (PORA and PORB) from barley into isolated plastids. Physiol Plant 109:298–303

Dalurzo HC, Sandalio LM, Gomez M, del Rio LA (1997) Cadmium infiltration of detached pea leaves: effect on its activated oxygen metabolism. Phyton- Annalea Rei Botanicae 37:59–64

Demmig-Adams B, Adams W III (1992) Photoprotection and other response of plants to high light stress. Ann Rev Plant Physiol Plant Mol Biol 43:599–626

Doganlar ZB, Cakmak S, Yanik T (2012) Metal uptake and physiological changes in *Lemna gibba* exposed to manganese and nickel. Int J Biol 4(3):148–157

Dudka S, Piotrowska M, Terlak H (1996) Transfer of cadmium, lead, and zinc from industrially contaminated soil to crop plant: a field study. Environ Pollut 94:181–188

Elenkov I, Stefanv K, Dimitrova Konaklieve S, Popov S (1996) Effect of salinity on lipid composition of *Cladophora vagabunda*. Phytochemistry 42(1):39–44

Erdei S, Hegadus A, Hauptmann G, Szalai J, Horvath G (2002) Heavy metal induced physiological changes in the antioxidant response system. Proceedings of the 7th Hungarian Congress on Plant Physiol, pp 89–90

Erdei L, Mezôsi G, Mécs I, Vass I, Fôglein F, Bulik L (2005) Phytoremediation as a program for decontamination of heavy-metal polluted environment. Acta Biol Szeged 49(1–2):75–76

Farid S (2006) Status of cadmium concentration in soil and vegetable irrigated with city effluents. China J App Env Biol 12(3):414–419

Fariduddin Q, Yusuf M, Hayat S, Ahmad A (2009) Effect of 28-homobrassinolide on antioxidant capacity and photosynthesis in *Brassica juncea* plants exposed to different levels of copper. Environ Exp Bot 66:418–424

Farwell AJ, Vasely S, Nero V, Rodriguez H, Shah S, Dixon DG, Glick BR (2006) The use of transgenic canola (*Brassica napus*) and plant growth-promoting bacteria to enhance plant biomass at a nickel-contaminated field site. Plant Soil 288:309–318

Fodor E, Szabo-Nagy A, Erdei L (1995) The effect of cadmium on the fluidity and H + ATPase activity of plasma membrane from sunflower and wheat roots. J Plant Physiol 147(1):87–92
Fodor F, Sarvari E, Lang F, Sizigeti Z (1998) Effect of Pb and Cd on cucumber depending on the Fe-complex in the culture solution. J Plant Physiol 148:434 -439
Forlani G, Mangiagalli A, Pinter C, Nielson E (2000) Expression of δ pyrroline 5′carboxylate dehydrogenase and proline/arginine homeostasis in solanum tuberosum. Physiol Plant 110:22–27
Forti G, Elli G (1995) The function of ascorbic acid in photosynthetic phosphorylation. Plant Physiol 109:1207–1211
Foyer CH, Harbinson J (1994) Oxygen metabolism and the regulation of photosynthetic electron transport. In: Foyer CH, Mullineaux PM (eds) Causes of photooxidative stress and amelioration of defense systems in plants. CRC Press, London, pp 2–42
Foyer CH, Rowell J, Walker D (1983) Measurements of ascorbate content of spinach leaf protoplasts and chloroplasts during illumination. Planta 157:239–244
Foyer CH, Lopez Delgado H, Dat JF, Scott IM (1997) Hydrogen peroxide and glutathione associated mechanism of acclamatory stress tolerance and signaling. Physiol Plant 100:241–254
Frostegard A, Tunlid A, Baath E (1993) Phospholipid fatty acid composition, biomass and activity of microbial communities from two soil types experimentally exposed to different heavy metals. Appl Environ Microbiol 59(11):3605–3617
Gallego SM, Benavides MP, Tomaro ML (1996) Effect of heavy metal ion excess on sunflower leaves: evidence for involvement of oxidative stress. Plant Sci 121:151–159
Gallego SM, Benavides MP, Tomaro ML (1999) Effect of cadmium ions on antioxidant defense system in sunflower cotyledons. Biol Plant 42:49–55
García Miragaya J, Page A (1978) Sorption of trace quantities of cadmium by soils with different chemical and mineralogical composition. Water Air Soil Pollu 9:289–299
Geiken B, Masojid EJ, Rizzuto M, Pompili ML, Giardi MT (1998) Incorporation of [S35] methionene in higher plants reveals that stimulation of the D1 reaction center protein turnover accompanied tolerance to heavy stress. Plant Cell Environ 21:1265–1273
Gille G, Singler K (1995) Oxidative stress in living cells. Folia Microbiol 2:131–152
Gisbert C, Ros R, Deharo A, Walker DJ, Pilarbernal M, Serrano R, Navarro-Avino J (2003) A plant genetically modified that accumulates Pb is especially promising for phytoremediation. Biochem Biophys Res Commun 303:440–445
Godbold DL, Huttermann A (1985) Effect of zinc, cadmium and mercury on root elongation of *P. abies* (Karst) seedlings and the significance to forest dieback. Environ Pollut 38:375–381
Gopal R, Rizvi AH (2008) Excess lead alters growth, metabolism and translocation of certain nutrients in radish. Chemosphere 70:1539–1544
Greger M, Lindberg S (1987) Effects of cadmium and EDTA on young sugarbeat (*Beta vulgaris*) II Net uptake and distribution of Mg^{2+}, Ca^{2+} and Fe^{2+}/ Fe^{3+}. Physiol Plant 69:81–86
Greger M, Ogrer E (1991) Direct and indirect effects of cadmium on photosynthesis in sugerbeet (*Beta vulgaris*). Physiol Plant 83:129–135
Grover P, Rekhadevi P, Danadevi K, Vuyyuri S, Mahboob M, Rahman M (2010) Genotoxicity evaluation in workers occupationally exposed to lead. Int J Hyg Environ Health 213(2):99–106
Gupta D, Nicoloso F, Schetinger M, Rossato L, Pereira L, Castro G, Srivastava S, Tripathi R (2009) Antioxidant defense mechanism in hydroponically grown *Zea mays* seedlings under moderate lead stress. J Hazard Mater 172(1):479–484
Gupta D, Huang H, Yang X, Razafindrabe B, Inouhe M (2010) The detoxification of lead in *Sedum alfredii* H. is not related to phytochelatins but the glutathione. J Hazard Mater 177(1–3):437–444
Ha SB, Smith AP, Howden R, Dieetrich WH, Bugg S, O'Conell MJ, Goldsbrough PB (1999) Phytochelatin synthase genes from *Arabidopsis* and the yeast (*Schizosaccharomyces pombe*). Plant Cell 11:1153–1164
Hagemeyer J, Breckle S (2002) Trace element stress in roots. In: Waisel Y, Eshel A, Kafkafi U (eds) Plant roots: the hidden half, 3rd ed. Marcel Dekker, Inc., New York, pp 763–785
Hall DO, Rao KK (1999) Photosynthesis, 6th Edition. Cambridge University Press, UK

Hameed A, Qadri TN, Mahmooduzzafar Siddiqi TO (2012) Plant tolerance and fatty acid Profile in responses to heavy metals. In: Ahmad P, Prasad MNV (eds) Abiotic stress responses in plants: metabolism, productivity and sustainability. Springer, New York, pp 369–386

Haribabu TE, Sudha PN (2011) Effect of heavy metals copper and cadmium exposure on the antioxidant properties of the plant *Brassica juncea*. Int J Chem Res 1(5):22–33

Hasan S, Aiman Q, Fariduddin B, Hayat S, Ahmad A (2009) A cadmium toxicity and tolerance in plants. J Environ Biol 30:165–174

Hasanuzzaman M (2008) Siliqua and seed development in rapeseed (*Brassica campestris* L.) as affected by different irrigation levels and row spacings. Agric Conspectus Scientificus 73(4):221–226

Hasanuzzaman M, Fujita M (2012) Heavy metals in the environment: current status, toxic effects on plants and possible phytoremediation. In: Anjum NA, Pereira MA, Ahmad I, Duarte AC, Umar S, Khan NA (eds) Phytotechnologies: remediation of environmental contaminants. Taylor and Francis/CRC Press, Boca Raton, pp 7–73

Hasanuzzaman M, Nahar K, Fujita M (2013) Adverse effects of cadmium on plants and possible mitigation of Cd-induced damages. In: Hasanuzzaman M, Fujita M (eds) Cadmium: characteristics, sources of exposure, health and environmental effects. Nova Science, New York, pp 1–48

Hasegawa I, Terada E, Sunairi M, Wakita H, Shinmachi F, Noguchi A, Nakajima M, Yazaki J (1997) Genetic improvement of heavy metal tolerance in plants by transfer of the yeast metallothionein gene (CUP1). Plant Soil 196:277–281

Hasset JJ, Miller JE, Koeppe DDE (1976) Interaction of Pb and Cd on maize root growth and uptake of Pb and Cd by roots. Environ Pollut 11:297–302

Hattori J, Labba H, Miki BL (1994) Construction and expression of a metallothionein β glucoronidase gene fusion. Genome 37:508–512

Haug A, Caldwell CR (1985) Aluminium toxicity in plants: the role of the root plasma membrane and calmodulin. In: John JB, Berlin E, Jackson PC (eds) Frontiers of membrane research in agriculture. Totowa, pp 359–381

Haung CY, Bazzaz FA, Vanderhoef LN (1974) The inhibition of soybean metabolism by cadmium and lead. Plant Physiol 54:122–124

Hegedus A, Erdei S, Horvath G (2001) Comparative studies of H_2O_2 detoxifying enzymes in green and greening barley seedlings under cadmium stress. Plant Sci 160:1085–1093

Heiss S, Schäfer HJ, Kerwer AH, Rausch T (1999) Cloning S-assimilation genes of *Brassica juncea* L.: cadmium differentially affects the expression of a putative low affinity sulphate transporter and isoforms of ATP sulphurylase and APS reductase. Plant Mol Biol 39:847–857

Heiss S, Wachter A, Bogs J, Cobbett C, Rausch T (2003) Phytochelatin synthase (PCS) protein is induced in *Brassica juncea* leaves after prolonged Cd exposure. J Exp Bot 54(389):1833–1839

Hendry GAF, Baker AJM, Ewart CF (1992) Cadmium tolerance and toxicity: oxygen radical processes and molecular damage in cadmium tolerant and cadmium sensitive clones of *Holcus lanatus*. Acta Bot Neerl 41:271–281

Hernandez JA, Olmas E, Copras FJ, Sevilla F, del Rio LA (1995) Salt induced oxidative stress in chloroplast of pea plants. Plant Sci 105:151–167

Hirt H, Casari G, Barta A (1989) Cadmium enhanced gene expression in suspension culture cells of tobacco. Planta 179:414–420

Holtman WL, Heistek JC, Mattem KA, Bakhuisen R, Douma AC (1994) β-Oxidation of fatty acids is linked to the glyoxylate cycle in the aleurone but not in the embryo of germinating barley. Plant Sci 99:43–53

Horvath G, Droppa M, Oraveez A, Raskin VI, Marder JB (1996) Formation of the photosynthetic apparatus during greening and cadmium-poisoning barley leaves. Planta 199:238–243

Howe G, Merchant S (1992) Heavy metal activated synthesis of peptides in *Chlamydomonas reinhartii*. Plant Physiol 98:127–136

Howlett NG, Avery SV (1997) Relationship between cadmium sensitivity and degree of plasma membrane fatty acid unsaturation in *Saccharomyces cerevisiae*. Appl Microbiol Biotechnol 48(4):539–546

Islam E, Yang X, Li T, Liu D, Jin X, Meng F (2007) Effect of Pb toxicity on root morphology, physiology and ultrastructure in the two ecotypes of *Elsholtzia argyi*. J Hazard Mater 147(3):806–816

Islam E, Liu D, Li T, Yang X, Jin X, Mahmood Q, Tian S, Li J (2008) Effect of Pb toxicity on leaf growth, physiology and ultrastructure in the two ecotypes of *Elsholtzia argyi*. J Hazard Mater 154(1-3):914–926

Jiang W, Liu D (2010) Pb-induced cellular defense system in the root meristematic cells of *Allium sativum* L. BMC Plant Biol 10:40

John R, Ahmad P, Gadgil K, Sharma S (2009) Heavy metal toxicity: effect on plant growth, biochemical parameters and metal accumulation by *Brassica juncea* L. Int J Plant Prod 3(3):65–76

Jozefizak M, Remans T, Vangronsveld J, Cuypers A (2012) Glutathione is a key player in metal induced oxidative stress defenses. Int J Mol Sci 13:3145–3175

Kagi JHR (1991) Overview of metallothioneins. Method Enzymol 205:613–626

Kahle H (1993) Response of roots of trees to heavy metals. Environ Expt Bot 33:99–119

Kakar SUR, Wahid A, Tareen RB, Kakar SA, Tariq M, Kayani SA (2010) Impact of municipal wastewater of Quetta city on biomass, physiology and yield of canola (*B. napus* L.). Pak J Bot 42(1):317–328

Kang W, Shamsi IH, Zhang GP (2007) Synergistic interaction of NaCl and Cd on growth and photosynthetic parameters in soybean genotypes differing in salinity tolerance. J Zhejiang Univ Sci B8 (4):266–271

Kang MS, Kim SM, Park SW, Lee JJ, Yoo KH (2007) Assessment of reclaimed wastewater irrigation impact on water quality, soil and rice cultivation in paddy fields. J Environ Sci Health 42(4):439–445

Karami A, Zulkifili HJS (2010) Phytoremediation of heavy metals with several efficiency enhancer methods. Afr J Biotechnol 9(25):3689–3698

Kärenlampi S, Schat H, Vangronsveld J, Verkleij JAC, Lelia D, Mergeay M, Tervahauta AI (2000) Genetic engineering in the improvement of plants for phytoremediation of metals from polluted soils. Environ Pollut 107:225–231

Katare DP, Nabi G, Azooz MM, Aeri V, Ahmad P (2012) Biochemical modifications and enhancement of psoralen content in salt-stressed seedlings of *Psoralea corylifolia* Linn. J Funct Environ Bot 2(1):65–74

Kavi Kishor PB, Sangam S, Amrutha RN, Sri Laxmi P, Naidu KR, Rao KRSS, Rao S, Reddy KJ, Theriappan P, Sreenivasulu N (2005) Regulation of proline biosynthesis degradation, uptake and transport in higher plants: its implications in plant growth and abiotic stress tolerance. Curr Sci 88:424–438

Kawashima I, Inokuchi Y, China M, Kimure M, Shimizu N (1991) Isolation of a gene for a metallothionein like protein from soybean. Plant Cell Physiol 32(6):913–916

Kawashima I, Kennedy TO, China M, Lane BG (1992) Wheat Ec metallothionein genes: like mammalian Zn^{2+} are conspicuously experessed during embryogenesis. Eur J Biochem 209:971–976

Khan S, Khan NN (1983) Influence of lead and cadmium on growth and nutrition concentration of tomato (*Lycopersicon esculentum*) and egg plant (*Solanum melongena*). Plant Soil 74:387–394

Khan MA, Shaukat SS, Khan MA (2009) Growth, yield and nutrient content of sunflower (*Helianthus annus* L.) using treated wastewater from waste stabilization ponds. Pak J Bot 41(3):1391–1399

Kishor PBK, Hong Z, Miao GH, Hu CAA, Verma DPS, Hong ZL, Miao GH (1995) Overexpression of DELTA-1 pyroline 5′-carboxylate synthetase increases proline production and confers osmo-tolerance to transgenic plant. Plant Physiol 108:1387–1394

Klein BP, Grossman S, King D, Cohen BS, Pinsky A (1984) Pigment bleaching, carbonyl production and antioxidant effects during anaerobic lipoxygenase reaction. Biochem Biophy Acta 793:72–79

Krupa Z, Baszynski T (1989) Acyl lipid composition of thylakoid membranes of cadmium treated tomato plants. Acta Physiol Plant 11(2):111–116

Krupa Z, Seidlecka A, Maksymiec W, Baszynski T (1993a) In vivo response of photosynthetic apparatus of *Phaseolus vulgaris* to nickle toxicity. J Plant Physiol 142:664–668

Krzesłowska M, Lenartowska M, Samardakiewicz S, Bilski H, Wo´zny A (2010) Lead deposited in the cell wall of *Funaria hygrometrica* protonemata is not stable-A remobilization can occur. Environ Pollut 158(1):325–338
Kubo A, Saji H, Tannaka K, Kondo N (1995) Expression of Arabidopsis cytosolic ascorbic peroxidase gene in response to ozone or SO_2. Plant Mol Biol 29:479–486
Kumar D, Yusuf MA, Singh P, Sardar M, Sarin NB (2013) Modulation of antioxidant machinery in α-tocopherol-enriched transgenic *Brassica juncea* plants tolerant to abiotic stress conditions. Protoplasma 250(5):1079–1089
Lane B, Kajoika R, Kennedy T (1987) The wheat germ Ec protein is a zinc containing metallothioneine. Biochem Cell Biol 65:1001–1005
Larsson EH, Bornman JF, Aspo H (1998) Influence of UV-B radiation and Cd^{2+} on chlorophyll fluorescence, growth and nutrient content in *Brassica napus*. J Exp Bot 49:1031–1039
Lee S, Leustek T (1998) APS kinase from *Arabidopsis thaliana*: genomic organisation, expression and kinetic analysis of the recombinant enzyme. Biochem Biophys Res Commun 247:171–175
Lefebvre DD, Miki BL, Laliberk JF (1987) Mammalian metallothionein functions in plants. Biotechnology 5:1053–1058
Leustek T, Satio K (1999) Sulphate transport and assimilation in plants. Plant Physiol 120:637–643
Leustek T, Martin MN, Bick JA, Davies JP (2000) Pathways and regulations of sulphur metabolism revealed through molecular and genetic studies. Annu Rev Plant Physiol Plant Mol Biol 51:141–165
Li H, Luo H (2012) Antioxidant enzyme activity and gene expression in response to lead stress in perennial ryegrass. J Amer Soc Hort Sci 137(2):80–85
Liu D, Jiang W, Gao X (2004) Effects of cadmium on root growth, cell division and nucleoli in root tips of garlic. Physiol Plant 47:79–83
Liu D, Li T, Jin X, Yang X, Islam E, Mahmood Q (2008) Lead induced changes in the growth and antioxidant metabolism of the lead accumulating and non-accumulating ecotypes of *Sedum alfredii*. J Integr Plant Biol 50(2):129–140
Lozano Rodriguez E, Hernandez LE, Boney P, Carpena Ruiz RO (1997) Distribution of cadmium in shoot and root tissues of maize and pea plants: physiological disturbances. J Exp Bot 48 (306):123–128
Luna CM, Gonzalez CA, Trippi VS (1994) Oxidative damage caused by an excess of Cu in oat leaves. Plant Cell Physiol 35:11–15
Maiti IB, Wagner GJ, Yeargan R, Hunt AG (1989) Inheritance and expression of the mouse metallothionein gene in tobacco. Plant Physiol 91:1020–1024
Malan Hl FJM (1998) Effect of the metal pollutants cadmium and nickel on soybean seed development. Seed Sci Res 8:445–453
Malan C, Greyling MM, Gressel J (1990) Correlation between Cu-Zn, SOD and glutathione reductase and environmental and xenobiotic stress tolerance in maize inbreds. Plant Sci 69:157–166
Masuda T, Ohta H, Shioi Y, Takamiya KI (1996) Light regulation of 5′ALA synthesis system in *Cucumis sativus*. Light stimulates activity of glutamyl tRNA reductase during greening. Plant Physiol Biochem 34:11–16
McCord JM, Fridovich I (1969) Superoxide dismutase an enzyme function for erythrocuprein (Hemocuprein). J Biol Chem 244:6049–6055
McEven B, Seyyedi M, Younis S, Sundquist C (1996) Formation of short wavelength chlorophyll (ide) after brief irradiation is corrected with the occurrence of protochlorophyll (ide) in dark grown epi-and hypocotyls of bean (*Phaseolus vulgaris*). Physiol Plant 96:51–58
Mehta SK, Gaur JP (1999) Heavy metal induced proline accumulation and its role in ameliorating metal toxicity in *Chlorella vulgaris*. New Phytol 143:253–259
Mejare M, Bulow L (2001) Metal binding properties and peptides in bioremediation and phytoremediation of heavy metals. Trend Biotechnol 19(2):67–73
Mendoza-Cózatl DG, Butko E, Springer F, Torpey JW, Komives EA, Kehr J, Schroeder JI (2008) Identification of high levels of phytochelatins, glutathione and cadmium in the phloem sap of *Brassica napus*. A role for thiol-peptides in the long-distance transport of cadmium and the effect of cadmium on iron translocation. Plant Cell 54(2):249–259

Mishra S, Srivastava S, Tripathi R, Kumar R, Seth C, Gupta D (2006) Lead detoxification by coontail (*Ceratophyllum demersum* L.) involves induction of phytochelatins and antioxidant system in response to its accumulation. Chemosphere 65(6):1027–1039
Misra S, Gedamu L (1989) Heavy metal tolerant transgenic *Brassica napus* L. and *Nicotiania tobaccum* L. plants. Theor Appl Genet 78:161–168
Mobin M, Khan NA (2007) Photosynthetic activity, pigment composition and antioxidative response of two mustard (*Brassica juncea*) cultivars differing in photosynthetic capacity subjected to cadmium stress. J Plant Physiol 5(3):601–610
Mohamed AA, Castagna A, Ranieri A, Sanità di Toppi L (2012) Cadmium tolerance in *Brassica juncea* roots and shoots is affected by antioxidant status and phytochelatin biosynthesis. Plant Physiol Biochem 57:15–22
Mohan BS, Hosetti BB (1997) Potential phytotoxicity of lead and cadmium to *Lemna minor* L. growth in sewage stabilization ponds. Environ Pollut 98:233–236
Morsy AA, Hamid K, Salama A, Kamal HA, Mansour MMF (2012) Effect of heavy metal on plasma membrane lipids and antioxidant enzymes of *Zygophyllum* species. Eurasia J Biosci 6:1–10
Muneer S, Qadri TN, Mahmooduzaffar, Siddiqi TO (2011) Cytogenetic and biochemical investigations to study the response of *Vigna radiata* to cadmium stress. Afr J Plant Sci 5 (3): 183–192
Munné-Bosch S, Alegre L (2002) The function of tocopherols and tocotrienols in plants. Crit Rev Plant Sci 21:31–57
Murphy A, Zho JM, Goldsbrough PB, Taiz L (1997) Purification and immunological identification of metallothioneins 1 and 2 from *A.thailiana*. Plant Physiol 113:1293–1301
Murphy DJ (1994) Designer oil crops-breeding, processing and biotechnology. Verlagsgesells Chaft mbH, Weinheim
Nagajyoti PC, Dinakar N, Prasad TN, Suresh C, Damodharam T (2008) Heavy metal toxicity: Industrial effluent effect on groundnut (*Arachis hypogea* L.) seedlings. J Appl Sci Res 4(1):110–121
Nanba O, Satoh K (1987) Isolation of a photosystem II reaction centre consisting of D1 and D2 polypeptide and cytochrome b 559. Proc Natl Acad Sci U S A 84:109–112
Nandakumar PBA, Shivanna KR, Prakash S (1988) Wide hybridization in Brassica: crossability barriers and studies on the hybrid and synthetic amphidiploid of *B. fruticulosa* × *B. campestris*. Sex Plant Reprod 1:234–239
Nandi DL, Shemim D (1968) δ Aminolevulinic acid dehydratase of *Rhodopsedomonas phaeroides*. J Biol Chem 243:1236–1242
Nicholson FA, Smith SR, Alloway BJ, Carlton-Smith C, Chambers BJ (2003) An inventory of heavy metals inputs to agricultural soils in England and Wales. Sci Total Environ 311:205–219
Nijs D, Kelley PM (1991) Vitamins C and E donate single hydrogen atoms in vivo. FEBS Lett 284:147–151
Nikolopoulos D, Manetas Y (1991) Compatible solutes and in vitro stability of sabola sada enzymes: proline incompatibility. Phytochemistry 30:411–413
Nishio JN (2000) Why are higher plants green? Evolution of the higher plant photosynthetic pigment complement. Plant Cell Environ 23:539–548
Noctor G, Foyer CH (1998) Ascorbate and glutathione: keeping active oxygen under control. Annu Rev Plant Physiol Plant Mol Biol 49:249–79
Okamoto OK, Pinto E, Latorie LR, Becharie EJH, Colepicola P (2001) Antioxidant modulation in response to metal induced oxidative stress in algal chloroplasts. Arch Environ Contam Toxicol 40:18–24
Oquist G, Hunner NPA (1993) The temperature dependence of the redox-state of QA and the susceptibility of photosynthesis to photoinhibition. Plant Physiol Biochem 31:683–691
Ouarti O, Bussama N, Zarrouk M, Cherif A, Ghorbal MH (1997) Cadmium and copper induced changes in tomato membrane lipids. Phytochemistry 45:1343–1350
Ouzounidou G, Moustakai M, Eleftheriou EP (1997) Physiological and ultra structural effect of cadmium on wheat (*Triticum aestivum* L.) leaves. Arch Environ Contam Toxicol 32:154–160
Ouzounidou G, Ilias I, Transpoulou H, Karataglis S (1998) Amelioration of copper toxicity by iron on spinach physiology. J Plant Nutr 21:209–221

Padh H (1990) Cellular functions of ascorbic acid. Biochem Cell Biol 68:1166–1173

Pal R, Rai JP (2010) Phytochelatins: peptides involved in heavy metal detoxification. Appl Biochem Biotech 160(3):945–963

Paleg LG, Douglas TJ, Van Daal A, Keech DB (1981) Proline, betaine and other organic solutes protect enzymes against heat inactivation. Aust J Plant Physiol 8:107–114

Paleg LG, Stewart GR, Bradbeer JW (1984) Proline and glycine betaine influence solvation. Plant Physiol 75:975–978

Panda SK, Choudhary S (2005) Chromium stress in plants. Braz J Plant Physiol 17:19–102

Parmar P, Dave B, Sudhir A, Panchal K, Subramanian RB (2013) Physiological, biochemical and molecular response of plants against heavy metal stress. Int J Curr Res 5(1):80–89

Patra J, Panda MM (1998) A comparison of biochemical responses to oxidative and metal stress in seedlings of barley *Hordeum vulgare* L. Environ Pollut 196:99–105

Pehlivan E, Özkan AM, Dinç S, Parlayici S (2009) Adsorption of Cu_{2+} and Pb_{2+} ion on dolomite powder. J Hazard Mater 167(1-3):1044–1049

Peralta JR, Gardea Torresdey JL, Tiemann KJ, Gomes E, Asteaga S, Rascon E, Parsonns JG (2000) Proceedings of the 2000 conference on Hazardous Waste Research, pp 135–140

Peško M, Kráľovál K, Masarovičová E (2011) Phytotoxic effect of some metal ions on selected rapeseed cultivars registered in Slovakia. Proceedings of ECOpole 5(1)

Pilon-Smits EAH, Le Duc DL (2009) Phytoremediation of selenium using transgenic plants. Curr Opin Biotechnol 20:207–212

Pilon-Smits EAH, Hwang S, Lytle CM, Zhu Y, Tai JC, Bravo RC, Chen Y, Leustek T, Terry N (1999) Overexpression of ATP sulfurylase in Indian mustard leads to increased selenate uptake, reduction and tolerance. Plant Physiol 119:123–132

Pilon-Smits EAH, Zhu Y, Sears T, Terry N (2000) Overexpression of glutathione reductase in *Brassica juncea*: Effects on cadmium accumulation and tolerance. Physiol Plant 110:455–460.

Pilon-Smits EAH, Valdez Barillas JR, Van Hoewyk D, Lin ZQ (2010) Phytoremediation of Selenium. In: Plaza G (ed) Trends in bioremediation and phytoremediation. Research Signpost, Kerala, pp 355–372

Piotrowska A, Bajguz A, Godlewska-Zylkiewicz B, Czerpak R, Kaminska M (2009) Jasmonic acid as modulator of lead toxicity in aquatic plant *Wolffia arrhiza* (Lemnaceae). Environ Exp Bot 66(3):507–513

Piqueras A, Olmas E, Martinez -SolanoJR, Hellin E (1999) Cadmium induced oxidative burst in tobacco BY cells, time course, subcellular location and antioxidant response. Free Radic Res 31(Suppl):33–38

Pleines S, Marquard R, Friedt W (1987) Recurent selection for modified polyenoic fatty acid composition in rapeseed (*Brassica napus* L). 7th Rapessed Congress Poznan (Poland), pp 140–145

Possingharri JV (1956) The effect of mineral nutrition on the content of free amino acid and amides in tomato plants-A comparison of the effect of deficiencies of copper, zinc, mangnese, iron, and molybdenum. Aust J Biol Sci 9:539–551

Pourrut B, Perchet G, Silvestre J, Cecchi M, Guiresse M, Pinelli E (2008) Potential role of NADPH-oxidase in early steps of lead-induced oxidative burst in *Vicia faba* roots. J Plant Physiol 165(6):571–579

Prasad MNV (1995) Cd toxicity and tolerance in vascular plants. Environ Exp Bot 35 (4):525–545

Prasad DDK, Prasad ARK (1987a) Altered δ-aminolevulinic acid (ALA) metabolism by lead and mercury in germinating seedlings of bajra (*Pennisetum typhoideum*) seedlings. J Plant Physiol 127:241–249

Prasad DDK, Prasad ARK (1987b) Effect of lead and mercury on chlorophyll synthesis in mungbean seedlings. Phytochem 26:881–883

Prasad KVSK, Saradhi PP, Sharmila P (1999) Concerted action of antioxidant enzymes and curtailed growth under zinc toxicity in *Brassica juncea*. Environ Exp Bot 42:1–10

Prins CN, Hantzis LJ, Quinn CF, Pilon-Smits EAH (2011) Effects of selenium accumulation on reproductive functions in *Brassica juncea* and *Stanleya pinnata*. J Exp Bot 62:5233–5239

Qadir S (2003) Biochemical response of *Brassica juncea* (Czern and Coss) genotypes to cadmium stress. PhD Thesis, Jamia Hamdard, New Delhi, India

Qadir S, Qureshi MI, Javed S, Abdin MZ (2004) Genotypic variation in phytoremediation potential of *Brassica juncea* cultivars exposed to Cd stress. Plant Sci 167:1171–1181

Qufei LA, Fashui H (2009) Effects of Pb2+ on the structure and function of photosystem II of *Spirodela polyrrhiza*. Biol Trace Elem Res 129(1):251–260

Qureshi M, Abdin MZ, Qadir S, Iqbal M (2007) Lead-induced oxidative stress and metabolic alterations in *Cassia angustifolia* Vahl. Biol Plant 51(1):121–128

Ralph PJ, Burchertt MD (1998) Photosynthetic response of Halophila ovalis to heavy metal stress. Environ Pollut 103:91–101

Rascio N, Dalla Vecchia D, Ferretti M, Merlo L, Ghisi R (1993) Some effects of cadmium on maize plants. Arch Environ Contam Toxicol 25:244–249

Rasool R, Hameed A, Azooz MM, Rehman M, Siddiqi TO, Ahmad P (2013) Salt stress: causes, types and responses of plants. Ecophysiology and responses of plants under salt stress. Springer, New York, pp 1–24

Rauser WE (1990) Phytochelatins. Annu Rev Biochem 59:61–86

Rauser WE, Meuwly P (1995) Retention of cadmium in roots of maize seedlings: role of complexation by phytochelatins and related thiol peptides. Plant Physiol 109:195–202

Raziuddin, Farhatullah, Hassan G, Akmal M, Shah SS, Mohammad F, Shafi M, Bakht J, Zhou W (2011) Effects of cadmium and salinity on growth and photosynthesis parameters of *Brassica* species. Pak J Bot 43(1):333–340

Reddy AM, Kumar SG, Jyothsnakumari G, Thimmanaik S, Sudhakar C (2005) Lead induced changes in antioxidant metabolism of horsegram (*Macrotyloma uniflorum* (Lam.) Verdc.) and bengal gram (*Cicer arietinum* L.). Chemosphere 60(1):97–104

Rivai IF, Koyama H, Suzuki S (1990) Cadmium content in rice and rice field soils in China, Indonesia and Japan with special reference to soil type and daily intake from rice. Jap J Health Human Ecol 56:168–177

Robinson NT, Tommey AM, Luske C, Jackson PJ (1993) Plant metallothioneins. Biochem J 295:1–10

Root RA, Miller RJ, Koeppe DE (1975) Uptake of cadmium-its toxicity and effect on the iron to zinc ratio in hydroponically growth corn. J Plant Quality 4:473–476

Rose RC, Bode AM (1993) Biology of free radical scavengers: an evaluation of ascorbate. FASEB J 7:1135–1142

Ruegsegger A, Brunold C (1992) Effect of cadmium on g EC synthesis in maize seedlings. Plant Physiol 99:428–433

Ruegsegger A, Schmutz D, Brunold C (1990) Regulation of glutathione synthesis by cadmium in *Pisum sativum* L. Plant Physiol 93:1579–1584

Salt DE, Smith RD, Raskin I (1998) Phytoremediation. Annu Rev Plant Physiol Plant Mol Biol 49:643–668

Sandalio LM, Dalura HC, Gomez M, Romero-Puertas MC, del Rio LA (2001) Cadmium induced changes in the growth and oxidative metabolism of pea plant. J Exp Bot 52 (364):2114–2126

Saradhi A, Saradhi PP (1991) Proline accumulation under heavy metal stress. J Plant Physiol 138:554–558

Saradhi PP, Alia, Vani B (1993) Inhibition of mitochondrial electron transport is the prime cause behind proline accumulation during mineral deficiency in *Oryza sativa*. Plant Soil 155/156:465–468

Sasa T, Sugahara K (1976) Photoconversion of protochlorophyll to chlorophylla in a mutant of *Chlorella regularis*. Plant Cell Physiol 17:273–279

Scandalias JG (1993) Oxygen stress and superoxide dismutase. Plant Physiol 101:7–12

Schäfer HJ, Haag Kerwer A, Rausch T (1998) cDNA cloning and expression analysis of genes encoding GSH synthesis in roots of heavy metal accumulator *Brassica juncea* L: evidence or Cd induction of a putative mitochondrial γ GCS isoform. Plant Mol Biol 37:87–97

Schat H, Sharma SS, Vooijs R (1997) Heavy metal induced accumulation of free proline in metal tolerant and non-tolerant ecotypes of *Siline vulgaris*. Physiol Plant 101:477–482

Schlegel H, Godbold DI, Hutlerman A (1987) Whole plant aspects of heavy metal induced changes in CO_2 uptake and water relations of spruce (*Picea abies*) seedlings. Physiol Plant 69:265–270

Schutzendubel A, Schwanz P, Teiichmamm T, Gross K, Langerfeld-Heyser R, Goldbold DL, Polle A (2001) Cadmium induced changes in antioxidative systems H_2O_2 content and differentiation in scots pine roots. Plant Physiol 127:887–898

Sekara A, Poniedzia M, Ciura J, Jêdrszczy E (2005) Cadmium and lead accumulation and distribution in the organs of nine crops: implications for phytoremediation. Polish J Environ Stud 14:509–516

Sengar RS, Gautam M, Sengar RS, Sengar RS, Garg SK, Sengar K, Chaudhary R (2009) Lead stress effects on physiobiochemical activities of higher plants. Rev Environ Contam Toxicol 196:1–21

Sepehr FM, Ghorbanli M (2006) Physiological responses of *Zea mays* seedlings to interactions between cadmium and salinity. J Integr Plant Biol 48(7):807–813

Seth CS, Misra V, Chauhan HS (2012) Accumulation, detoxification and genotoxicity of heavy metals in Indian mustard (*Brassica Juncea* L.). Int J Phytorem 14(1):1–13

Setia RC, Bala R, Setia N, Malik CP (1993) Photosynthetic characteristics of heavy metal treated wheat (*T. aestivum* L.) plants. J Pant Sci Res 9:47–49

Seyyedi M, Timko MP, Sundguist C (1999) Protochlorophyllide, NADPH- Protochlorophyllide oxidoreductase and chlorophyll formation in the lip1 mutant of pea. Physiol Plant 106:344–354

Sezgin N, Ozcan HK, Demir G, Nemlioglu S, Bayat C (2003) Determination of heavy metal concentrations in street dusts in Istanbul E-5 highway. Environ Int 29:979–985

Shafi M, Bakht J, Hassan MJ, Raziuddin M, Zhang G (2009) Effect of cadmium and salinity stresses on growth and antioxidant enzyme activities of wheat (*Triticum aestivum* L.). Bull Environ Contam Toxicol 82(6):772–776

Shafi M, Zhang GP, Bakht J, Khan MA, Islam E, Dawood MK, Raziuddin (2010) Effect of cadmium and salinity stresses on root morphology of wheat. Pak J Bot 42(4):2747–2754

Shafiq M, Iqbal MZ (2005) The toxicity effects of heavy metals on germination and seedling growth of *Cassia siamea* Lamark. J New Seeds 7:95–105

Shah K, Kumar RG, Verma S, Dubey RS (2001) Effect of cadmium on lipid peroxidation, superoxide anion generation and activities of antioxidant enzymes in growing rice seedlings. Plant Sci 161:1135–1144

Shahid M, Pinelli E, Pourrut B, Silvestre J, Dumat C (2011) Lead-induced genotoxicity to *Vicia faba* L. roots in relation with metal cell uptake and initial speciation. Ecotoxicol Environ Saf 74(1):78–84

Shanmugaraj BM, Chandra HM, Srinivasan B, Ramalingam S (2013) Cadmium induced physiobiochemical and molecular response in *Brassica juncea*. Int J Phytoremed 15(3):206–218

Sharma P, Dubey RS (2005a) Lead toxicity in plants. Braz J Plant Physiol 17(1):35–52

Sharma P, Dubey RS (2005b) Modulation of nitrate reductase activity in rice seedlings under aluminium toxicity and water stress: role of osmolytes as enzyme protectant. J Plant Physiol 162:854–864

Sharma A, Sainger M, Dwivedi S, Srivastava S, Tripathi RD, Singh RP (2010) Genotypic variation in *Brassica juncea* (L.) Czern. cultivars in growth, nitrate assimilation, antioxidant responses and phytoremediation potential during cadmium stress. J Environ Biol 31(5):773–780

Shaw BP (1995) Effects of mercury and cadmium on the activities of antioxidative enzymes in the seedlings of *Phaseolus aureus*. Biol Plant 3:587–596

Shekhawat K, Rathore SS, Premi OP, Kandpal BK, Chauhan JS (2012) Advances in Agronomic Management of Indian Mustard (*Brassica juncea* (L.) Czernj. Cosson): An Overview. Int J Agron 2012, Article ID 408284

Shiyab S, Chen J, Han FX, Monts DL, Matta FB, Gu M, Su Y (2009) Phytotoxicity of mercury in Indian mustard (*Brassica juncea* L.). Ecotoxicol Environ Saf 72(2):619–625

Singh RP, Dabas S, Chaudhary A, Maheshwari R (1998) Effect of lead on nitrate reductase activity and alleviation of lead toxicity by inorganic salts and 6-benzylaminopurine. Biol Plant 40:399–404

Singh R, Tripathi RD, Dwivedi S, Kumar A, Trivedi PK, Chakrabarty D (2010) Lead bioaccumulation potential of an aquatic macrophyte *Najas indica* are related to antioxidant system. Bioresour Technol 101:3025–3032

Sinha P, Dube BK, Srivastava P, Chatterjee C (2006) Alteration in uptake and translocation of essential nutrients in cabbage by excess lead. Chemosphere 65:651–656

Sinha S, Sinam G, Mishra RK, Mallick S (2010) Metal accumulation, growth, antioxidants and oil yield of *Brassica juncea* L. exposed to different metals. Ecotoxicol Environ Saf 73(6):1352–1361

Skorzynska-Polit E, Baszynski T (1997) Differences in sensitivity in cadmium stressed runner bean plants in relation to their age. Plant Sci 128:11–21

Smirnoff N, Cumbes QJ (1989) Hydroxyl radical scavenging activity of compatible solutes. Phytochemistry 28:1057–1060

Smith GC, Brenann E (1984) Response of silver maple seedlings to an acute dose of root applied cadmium. Forest Sci 30:582–586

Somashekaraiah SV, Padmaja K, Prasad ARK (1992) Phytotoxicity of cadmium ions on germinating seedlings of mungbean (*Phaseolus vulgaris*): involvement of lipid peroxidation in chlorophyll degradation. Physiol Plant 85:85–89

Stroinski A, Kozlowska M (1997) Cadmium induced oxidative stress in potato tuber. Acta Soc Bot Pol 66(2):189–195

Suter M, Von Ballmoos P, Kupriva S, Opden Camp R, Schaller J, Kuhlemeier C, Schurmann P, Brunold C (2000) Adenosine 5'phosphosulphosulphate-sulphotransferase and adenosine 5'phosphosulphate reductase are identical enzymes. J Biol Chem 275:930–936

Suzuki N, Kasumi K, Sana H (2001) Screening of cadmium responsive genes in Arabidopsis thaliana. Plant Cell Environ 24:1177–1188

Sylwia W, Anna R, Ewa B, Stephan C, Maria AD (2010) The role of subcellular distribution of cadmium and phytochelatins in the generation of distinct phenotypes of AtPCS1- and CePCS3-expressing tobacco. J Plant Physiol 167(12):981–988

Szabados L, Savoure S (2010) Proline: a multifunctional amino acid. Trend Plant Sci 15(2):89–97.

Szollosi R, Varga IS, Erdei L, Mihalik E (2009) Cadmium-induced oxidative stress and antioxidative mechanisms in germinating Indian mustard (*Brassica juncea* L.) seeds. Ecotoxicol Environ Saf 72(5):1337–1342

Tamoutsidis EI, Papadopoulos S, Zotis J, Mavropoulos T (2002) Wet sewage sludge effect on soil properties and element content of leaf and root vegetables. J Plant Nutr 25:1941–1955

Tangahu BV, Abdullah SRS, Basri H, Idris M, Anuar N, Mukhlisin M (2011) A review on heavy metals (As, Pb, and Hg) uptake by plants through phytoremediation. Int J Chem Eng. doi:org/10.1155/2011/939161

Theriappan P, Gupta AK, Dasarathan P (2011) Accumulation of proline under salinity and heavy metal stress in cauliflower seedlings. J Appl Sci Environ Manage 15 (2):251–255

Thomas JC, Bohnert HJ (1993) Salt stress perception and plant growth regulators in the halophyte, *M. crystallinum*. Plant Physiol 103:1299–1304

Touiserkani T, Haddad R (2012) Cd-induced stress and antioxidant response in different *B. napus* cultivars. J Agri Sci Technol 14:929–937

Tremolieres A, Dubacq JP, Drapier D (1982) Unsaturated fatty acids in maturing seeds of sunflower and rape: regulation by temperature and light intensity. Phytochem 21:41–45

Tukendorf A, Baszynski T (1991) The in vivo effect of cadmium on photochemical activities in chloroplasts of runner bean plants. Acta Physiol Plant 13(1):51–57

Tyler G, Pahlsson AM, Bengtsson G, Baath E, Tranvik L (1989) Heavy metal ecology and terrestrial plants, micro organisms and invertebrates: a review. Water Air Soil Pollut 47:189–215

Tzivelka G, Kaldis A, Hegedus A, Kissimon J, Prombona A, Horvath J, Akoyounoglou A (1999) The effect of Cd on chlorophyll and light harvesting complex II biosynthesis in greening plants. Z Naturforsch 54c:740–745

UN (1935) Genomic analysis in Brassica with special reference to the experimental formation of *Brassica napus* and peculiar mode of fertilization. Japan J Bot 7:389–452

USDA (2010) The PLANTS database. National plant data center, Natural resources conservation service, United States department of agriculture. Baton Rouge, LA

Vaglio A, Landriscina C (1999) Changes in liver enzyme activity in the teleost *Sparus aurata* in response to cadmium intoxication. Ecotoxicol Environ Safety 43(1):111–116

Vogeli-Lange R, Wagner GJ (1990) Subcellular localization of cadmium and cadmium-binding peptides in tobacco leaves. Plant Physiol 92:1086–1093

Van Assche F, Clijsters H (1990) Effects of metals on enzyme activity in plants. Plant Cell Environ 13:195–206

Vassilev V, Tsonev T, Yordanov I (1998) Physiological responses of barley plants (*Hordeum vulgare*) to cadmium contamination in soil during ontogenesis. Environ Pollut 103:287–293

Vatehova Z, Kollarova K, Zelko I, Richterova-Kucerova D, Bujdos M, Liskova D (2012) Interaction of silicon and cadmium in *Brassica juncea* and *Brassica napus*. Biologia 67(3):498–504

Verma S, Dubey RS (2003) Lead toxicity induces lipid peroxidation and alters the activities of antioxidant enzymes in growing rice plants. Plant Sci 164:645–655

Vitória AP, Lea PJ, Azevedo RA (2001) Antioxidant enzymes responses to cadmium in radish tissues. Phytochem 57:701–710

Voetberg GS, Sharp RE (1991) Growth of the maize primary root at low water potential III. Role of increased proline deposition in osmotic adjustment. Plant Physiol 96:1125–1130

Warwick SI (1993) Guide to wild germplasm of Brassica and allied crops Part IV: wild species in the tribe Brassicaceae (cruciferae) as sources of agronomic trait. Technical Bulletin 17E-1993. Centre for Land and Biological Resources Research, Agriculture Canada, Ottawa, Ontaria

Wensheng S, Chonytu L, Zhiquan Z (1997) Analysis of major constraints on plant colonization at Fankou Pb/Zn mine tailings. Chinese J Appl Ecol 8:314–318

Wierzbicka M, Antosiewiez D (1993) How lead can easily enter the food chain. A study of plant roots. Sci Total Environ Suppl Pt (1):423–429

Wingate VPM, Lawton MA, Lamb CJ (1988) Glutathione causes a massive and selective induction of plant defence genes. Plant Physiol 87:206–210

Winglse G, Karpinski S (1996) Differential redox regulation by glutathione and glutathione reductase and Cu-Zn SOD gene expression in *Pinus sylvestris* L. needles. Planta 198:151–157

Wuana RA, Okieimen FE (2010) Phytoremediation potential of maize (*Zea mays* L.): a review. Afr J Gen Agric 6(4):275–287

Xiong Z, Zhao F, Li M (2006) Lead toxicity in *Brassica pekinensis* Rupr: effect on nitrate assimilation and growth. Environ Toxicol 21(2):147–153

Yadav S (2010) Heavy metals toxicity in plants: an overview on the role of glutathione and phytochelatins in heavy metal stress tolerance of plants. S Afr J Bot 76(2):167–179

Yilmaz K, Akinci IE, Akinci S (2009) Effect of lead accumulation on growth and mineral composition of egg plant seedlings (*Solanum melongena*). New Zealand J Crop Horticul Sci 37:189–199

Young AJ, Britton G (1990) Carotenoids and antioxidative stress. Curr Res Photosyn 4:587–590

Yousuf MA, Kumar D, Rajawanshi R, Strasser RJ, Michael MT, Govindgee, Sarin NB (2010) Overexpression of γ-tocopherol methyl transferase gene in transgenic *Brassica juncea* plants alleviates abiotic stress: physiological and chlorophyll a fluorescence measurements. Biochim Biophy Acta (BBA)-Bioenergetica 1797(8):1428–1438

Zengin FK, Munzuroglu O (2005) Effects of some heavy metals on content of chlorophyll, proline and some antioxidant chemicals in bean (*Phaseolus vulgaris* L.) seedlings. Acta Biol Cracov Ser Bot 47/2:157–164

Zhang F, Li X, Wang C, Shen Z (2000) Effect of Cd on autoxidation of free proline in seedling of mung bean. J Plant Nutr 23 (3):357–368

Zhu LY, Pilon-Smits EAH, Jouanin L, Terry N (1999) Overexpression of glutathione synthetase in Indian mustard enhances cadmium accumulation and tolerance. Plant Physiol 119:73–79

Chapter 2
Recent Advances in Rapid and Sensitive Screening For Abiotic Stress Tolerance

Nitin Mantri, Vikas Patade and Edwin Pang

Abstract Traditionally, screening for abiotic stress tolerance at field level was based on necrosis scores and shoot biomass reduction on stress exposure, relative to unstressed controls. However, such a measure of tolerance screening is laborious, destructive, and time consuming, and results are subjected to environmental variation. Recently, noninvasive, high-throughput screening techniques have been developed for screening abiotic stress tolerance in crops. In this direction, some physiological, biochemical, and/or molecular indicators/markers have been identified for rapid and sensitive indirect screening of germplasm. Physiological markers like membrane damage based on electrolyte leakage, stomatal conductance, chlorophyll content and so on are currently available. In addition, quick and sensitive screening in crop plants is possible with biochemical markers like status of reactive oxygen species and oxidative damage to biological macromolecules like lipids, proteins, and nucleic acids. Identification of molecular markers associated with the tolerance response has also made rapid and sensitive indirect selection possible in a few crop species. Thus, development of such methods is valuable in breeding for abiotic stress tolerance in plants.

Keywords Abiotic stresses · Screening · Indirect selection · Physio-biochemical and molecular markers

N. Mantri (✉) · E. Pang
Health Innovations Research Institute, School of Applied Sciences, RMIT University, Melbourne, VIC, 3000, Australia
e-mail: nitin.mantri@rmit.edu.au

E. Pang
e-mail: eddie.pang@rmit.edu.au

V. Patade
Defence Institute of Bio-Energy Research, Defence Research and Development Organisation, Nainital, Uttarakhand, 263139, India
e-mail: vikaspatade1981@gmail.com

P. Ahmad et al. (eds.), *Improvement of Crops in the Era of Climatic Changes,* Volume 2,
DOI 10.1007/978-1-4614-8824-8_2,

1 Introduction

Plant growth, productivity, and distribution are greatly affected by environmental stresses such as high and low temperature, drought, and high salinity. In response to abiotic stresses, plants undergo a variety of changes at the molecular level (gene expression) leading to physiological adaptation (Zheng et al. 2010; Patade et al. 2011a, b; Mantri et al. 2012). Salinity and drought are the major abiotic stresses which severely affect yield and quality in many regions of the world endangering the food security. The situation has become more serious with concerns of global climate change. Therefore, studies on abiotic stress tolerance have become one of the main areas of research worldwide. In this direction, efforts are being made to breed tolerant varieties using conventional breeding and contemporary biotechnological tools. Recent advances in this area include unraveling the physiological, biochemical, and molecular mechanism of abiotic stress tolerance and corresponding development of tolerant cultivars through transgenic technology or molecular breeding (Ashraf 2010; Patade et al. 2011a, b, c; Niu et al. 2012). Desired tolerant genotypes/varieties so developed need to be screened at laboratory as well as field levels for functional validation. Numerous physio-biochemical indicators for tolerance screening have been suggested (Chen et al. 2007; Smethurst et al. 2009). In addition, indirect selection using molecular markers linked to desired loci is being deployed for accelerating the production of stress-tolerant varieties (Ribaut and Ragot 2007; Wei et al. 2009; Ashraf and Foolad 2013). Here, advancement in various abiotic stress tolerance-screening techniques is discussed in light of recent publications in this area.

2 Field Screening For Abiotic Stress Tolerance Based On Conventional Techniques

Earlier, screening the breeding population/developed varieties for abiotic stress tolerance at field level was based on visual symptoms and/or shoot/root biomass reduction on exposure to the stress(es). Further, to quantitate the tolerance level, necrosis scores on stress exposure relative to unstressed controls were proposed for tolerance screening (Coram et al. 2007; Mantri et al. 2010a). Moses et al. (2008) screened 600 accessions of chickpea (*Cicer arietinum* L.) for salt tolerance under greenhouse conditions based on necrosis scores and shoot biomass reduction compared to unstressed controls at harvest stage. The results indicated wide variation in salinity tolerance determined by both measures. In addition, increase in grain yield on exposure to stress has been commonly used to screen for tolerance in the field. However, these measures of screening for tolerance are laborious, destructive, and time consuming, and the results are subject to environmental variation. Therefore, nondestructive biomass measurement techniques based on satellite remote sensing have been recently developed (Heiskanen 2006; Masuka et al. 2012).

3 Indirect Screening for Abiotic Stress Tolerance in Laboratory and Field

Indirect screening of breeding population/varieties for stress tolerance may be performed at the laboratory level (in vitro or in greenhouse) or in open field conditions. Various physio-biochemical indicators for tolerance screening have been suggested including photosynthesis rate based on CO_2 assimilation, stomatal conductance, chlorophyll fluorescence, and various electrophysiological characteristics (Munns and James 2003; Chen et al. 2007; Smethurst et al. 2009). Plant phenotyping greatly helps the genetic analysis of abiotic stress tolerance to further elucidate the stress tolerance mechanisms. However, conventional methods of plant phenotyping are laborious and destructive as compared to the recently developed high-throughput, nondestructive imaging technologies (reviewed by Roy et al. 2011; Yang et al. 2013). The recent phenotyping techniques, being nondestructive, enable acquiring quantitative data on plant growth, health, and water use under abiotic stress by taking multiple images of the same plant at different time points and at different wavelengths (Morison et al. 2008; Jones et al. 2009). Therefore, these technologies are being routinely applied to quantify traits related to salt and drought tolerance in a number of crop plants (Berger et al. 2010; Rajendran et al. 2009; White et al. 2012).

3.1 In Vitro Tolerance Screening

The major advantages of in vitro screening are: controlled environment, large population that can be handled in a lesser space within a short span of time, and the plant material being kept disease free (Patade et al. 2008; Patade and Suprasanna 2008). Patade and Suprasanna (2009) in vitro screened radiation-induced sugarcane cv. Co 86032 mutants for salt tolerance by exposing embryogenic calli to different salt (NaCl) concentrations. The screening was performed based on relative growth rate, cell viability, and membrane damage. The results suggested in vitro mutagenesis–selection as a powerful tool for efficient screening for salt tolerance in sugarcane to enable commercial cultivation in saline areas. Recently, Sorkheh et al. (2011) screened wild almond species based on root growth characteristics using sorbitol and polyethylene glycol (PEG) as an osmoticum and concluded in vitro screening as an effective system for screening drought tolerance.

Screening for aluminum (Al) tolerance using nutrient solution culture is the most common method as it allows nondestructive measurement of tolerance, and provides easy access to root systems and tight control over nutrient availability and pH (reviewed by Wang et al. 2006; Arenhart et al. 2013). In nutrient solution culture, root length measurement and root staining are the major criteria for evaluation of Al tolerance.

3.2 Physiological and Biochemical Markers For Tolerance Screening

Chlorophyll fluorescence and thermal imaging are well-established, powerful, non-destructive, and rapid techniques for detecting and diagnosing plant stresses in the field by providing information on both stomatal and photosynthesis-related parameters, the key factors that determine plant yield (West et al. 2005). Studies comparing the chlorophyll fluorescence with the conventional techniques indicated that in vivo chlorophyll fluorescence can be a useful tool for screening biotic and abiotic stress tolerance in various crops (Matous et al. 2006; Chaerle et al. 2007a; Mishra et al. 2011). Chaerle et al. (2007b) reviewed the relative advantages and disadvantages of thermal and chlorophyll fluorescence imaging for the study of spatial and temporal heterogeneity of leaf transpiration and photosynthetic performance. The combined thermal and chlorophyll fluorescence imaging can highlight presymptomatic responses before appearing in visual spectrum images and thus may increase the power of disease diagnosis and the potential for screening of stress-tolerant genotypes. Each thermal and spectral sensor detects a different basic physiological response; therefore, combining information generated from a broad range of sensors may enhance sensitivity of diagnosing and quantifying different stresses. Jones and Schofield (2008) reviewed the potential applications of multi-sensor imaging in diagnosis and quantification of both abiotic and biotic stresses in plants. The multi-sensor imaging for stress diagnosis and monitoring may be a simple combination of thermal and reflectance sensors, or visible reflectance and fluorescence sensors, through to combined fluorescence, reflectance, and thermal imaging sensors. Jiang et al. (2006) screened barley genotypes for salinity tolerance by measuring net photosynthesis, stomatal conductance, gas exchange, chlorophyll fluorescence parameters, dry matter, and carbon isotope discrimination in saline conditions relative to control plants. Among the various attributes measured, stomatal conductance was the best to screen in barley genotypes for absolute performance on exposure to salinity stress. Salinity susceptibility indices (SSI) used to estimate the relative salinity tolerance also varied considerably between the parameters and could not provide useful information on performance under saline conditions. However, according to Chen et al. (2007), only the chlorophyll fluorescence method has often been the most attractive tool for rapid and sensitive screening with fully automated fluorometers. Positive correlation of grain carbon isotope discrimination under post-anthesis drought stress with economic yield has been established in wheat (*Triticum aestivum* L.); therefore, these indices may be used as indirect selection criteria for wheat grown under stress environments (Monneveux et al. 2005, 2006; Zhu et al. 2009). However, a study conducted to investigate the relationships between seed cotton yield and carbon isotope discrimination concluded that the leaf physiological traits could not be reliably used for yield selection in cotton (*Gossypium hirsutum* L.) due to site-specific effects on the yield–physiological trait relationship (Tsiatas et al. 2008).

El-Shabrawi et al. (2010), based on biochemical analysis on redox homeostasis and antioxidant defense in salt-tolerant and salt-sensitive rice cultivars, suggested that the status of reactive oxygen species and ascorbate and glutathione homeostasis can serve as quick and sensitive biomarkers for screening salt tolerance in crop plants. A recent study using salt-tolerant and salt-sensitive genotypes (Gomathi and Rakkiyapan 2011) identified reliable indices, viz. higher membrane stability, and maintenance of high chlorophyll fluorescence ratio (fv/fm) and lower lipid peroxidation for salt tolerance screening in sugarcane at various stages of crop growth. Biochemical analyses in response to root zone salinity at various growth stages in salt-tolerant and salt-sensitive genotypes indicated that lower lipid peroxidation and higher phenolic contents were associated with tolerance response in hexaploid bread wheat (Ashraf et al. 2010); hence, they may be used for tolerance screening. Thakur (2004) screened fruit crops for drought tolerance based on indices, namely xylem water potential, relative water content, chlorophyll stability index, drought injury index, and rapid test for drought tolerance. The results indicated that fruit crops may be rapidly screened for drought tolerance based on these simple, cost-effective, and reliable indices at all phenological phases. Siddiqi et al. (2009) screened ten accessions of safflower (*Carthamus tinctorius* L.) for salt tolerance based on biomass (shoot and root dry mass) and other physio-biochemical parameters, viz., photosynthesis, transpiration, stomatal conductance, and chlorophyll *a* and *b* at the vegetative stage. Positive association with biomass was observed only for net photosynthetic rate among the various parameters examined. Hence, it may be used as an effectual indicator of salinity tolerance in safflower.

Mishra et al. (2011) studied cold tolerance based on electrolyte leakage and chlorophyll fluorescence in *Arabidopsis* accessions. The results indicated easy applicability of the fluorescence technique over the conventional electrolyte leakage methods. Further, it can be employed to detect cold tolerance at mild subzero temperatures by including the resolving power of several fluorescence features, thus avoiding plant freezing to the largely damaging temperatures of around −15 °C for screening. Root length measurement and root staining are the major criteria for evaluation of Al tolerance. Stodart et al. (2007) screened 250 accessions of bread wheat (*Triticum aestivum* L.) for Al tolerance based on hematoxylin staining of root tips and root regrowth measurement. The accessions classified as tolerant based on the root tip staining test also exhibited increased root length on exposure to Al. Thus, the results indicated hematoxylin staining of root tips as a simple technique to screen large number of accessions for Al tolerance. Measurement of change in stomatal conductance is a reliable and useful screening technique for abiotic stress tolerance. Results of the experiment in durum wheat indicated that stomatal conductance could be a means of screening for osmotic stress tolerance in cereals (Rahnama et al. 2010). Results of the high-temperature screening of common bean genotypes in greenhouses and fields indicated that a superior heat-tolerant genotype may be identified based on geometric mean and stress tolerance index (Porch 2006).

3.3 Molecular Marker-Assisted Indirect Selection For Stress Tolerance

Molecular marker-assisted selection (MAS) is desirable, if visual selection is difficult and cost/time ineffective. MAS is a strategy for accelerating the crop breeding for biotic and abiotic stress tolerance (Ribaut and Ragot 2007; Wei et al. 2009; Mantri et al. 2010b). Identification of molecular markers linked to the desired traits has made it possible to examine their usefulness in crop improvement (Ashraf 2010; Delannay et al. 2012). Indirect selection using molecular markers linked to desired loci is highly regarded as an efficient selection tool. In the recent past, efforts were made to develop molecular markers such as restriction fragment length polymorphism (RFLPs), random amplified polymorphic DNA (RAPDs), amplified fragment length polymorphism (AFLPs), and simple sequence repeats (SSRs) for efficient MAS in breeding programs. Lang et al. (2008) investigated the genetic basis for salinity tolerance using SSR markers in rice and tagged the locus imparting salt tolerance (RM223) for marker-assisted salt tolerance screening. Huseynova and Rustamova (2010) screened drought-tolerant, semi-tolerant, and sensitive wheat genotypes using RAPD primers associated with drought tolerance. The results indicated amplification of a specific product of 920 bp only in the drought-tolerant and semi-tolerant (absent in sensitive) genotypes. This molecular marker may be useful in screening for drought tolerance in wheat. Wang et al. (2006) reviewed the identification of molecular markers linked with the Al tolerance gene(s) in barley (*Hordeum vulgare* L.). Identification of tightly linked RFLP (Tang et al. 2000), SSR (Raman et al. 2003, Wang et al. 2007), and AFLP (Raman et al. 2002) markers to a major Al tolerance locus has enabled fast-tracking of the tolerance alleles in different breeding programs (Wang et al. 2006). However, these markers are usually not developed from the target genes. On the contrary, functional markers (FMs) are usually designed based on polymorphism for transcribed regions of the functional target genes. Therefore, these markers completely correlate with gene function and may facilitate accurate selection of target genes (Andersen and Lübberstedt 2003; Wei et al. 2009).

Indirect selection based on molecular markers has been mostly limited to improving traits with marker-assisted backcrossing (MABC) of major genes (Natarajkumar et al. 2010). Improvement of abiotic stress tolerance and other quantitatively inherited traits involves introgression of many genes; therefore, it may not be logically feasible for most breeding programs (Wang et al. 2007; Xu and Crouch 2008). Further, the requirement of prior mapping of significant marker–trait associations across breeding pools, in different environments, or after several cycles of selection is another drawback of MAS strategies. A recent marker-assisted recurrent selection (MARS) strategy, which involves multiple cycles of the indirect selection, is suggested for obtaining the desired frequency of target quantitative trait locus (QTL) alleles (Bernardo 2008). Further, genome-wide selection (GS) is another recent strategy that uses the combined effect of genome-wide markers on a trait, to pyramid favorable alleles for minor-effect QTLs (Bernardo 2009; Heffner et al. 2009;

Bernardo 2010). The major advantage of GS is that it does not require any prior knowledge about the QTL controlling the target traits. GS predicts the breeding values of lines in a population by analyzing their phenotypes and high-density marker scores. Unlike MABC or MARS, in GS, the marker effects across the entire genome that explain the entire phenotypic variation are calculated. The genome-wide marker data on the progeny lines are used to calculate genomic breeding values as the sum of the effects of all QTLs across the genome, thereby potentially exploiting all the genetic variance for a trait (Heffner et al. 2009). Thus, once a marker–trait association is developed, performance of progeny from selected cross may be predicted even before phenotyping, enabling indirect selection of the desired traits. Recently, single nucleotide polymorphisms (SNPs), a new-generation marker, are rapidly taking over the conventional molecular markers owing to their abundance, stability, and cost-effectiveness (McNally et al. 2009). In addition, these markers are amenable to automation and efficient in screening large population (Tung et al. 2010). Efforts are being made to discover SNPs, to develop functional SNPs for foreground selection, and to develop high-resolution SNP chips through deep sequencing for association genetics studies (Duran et al. 2009; McNally et al. 2009; McCouch et al. 2010). To this end, the availability of cost-effective, next-generation sequencing platforms and high-throughput marker genotyping may greatly enhance genome-wide selection for crop improvement in the near future (Varshney et al. 2009; Bernardo et al. 2010; Akpınar et al. 2013).

4 Conclusion and Future Perspectives

Multiple physio-biochemical traits contribute to the abiotic stress tolerance in plants. The success of the breeding (conventional and/or molecular) program for abiotic stress tolerance was previously limited by the availability of rapid, nondestructive, sensitive, and efficient screening techniques. Further, the ability to screen large sample population was challenging in abiotic stress tolerance screening using the conventional techniques. Recent advances have led to identification of highly predictive, simple, low-cost techniques for abiotic stress tolerance screening. Further, recently developed molecular markers have improved efficiency of plant breeding through precise and rapid foreground and background selection. Thus, the recent breakthroughs in the development of physio-biochemical–molecular indicators/markers for rapid and sensitive tolerance screening have revolutionized a new variety of development process.

References

Akpınar BA, Lucas SJ, Budak H (2013) Genomics approaches for crop improvement against abiotic stress. Sci World J 2013. Article ID 361921,9

Andersen JR, Lübberstedt T (2003) Functional markers in plants. Trends Plant Sci 8:554–560
Ashraf M (2010) Inducing drought tolerance in plants: recent advances. Biotechnol Adv 28:169–183
Ashraf MA, Ashraf M, Ali Q (2010) Response of two genetically diverse wheat cultivars to salt stress at different growth stages: leaf lipid peroxidation and phenolic contents. Pak J Bot 42(1):559–565
Ashraf M, Foolad MR (2013) Crop breeding for salt tolerance in the era of molecular markers and marker-assisted selection. Plant Breeding 132(1):10–20
Arenhart RA, Lima JC, Pedron M, Carvalho FE, Silveira JA, Rosa SB, Caverzan A, Andrade CM, Schünemann M, Margis R, Margis-Pinheiro M (2013) Involvement of ASR genes in aluminium tolerance mechanisms in rice. Plant Cell Environ 36(1):52–67
Berger B, Parent B, Tester M (2010) High-throughput shoot imaging to study drought responses. J Exp Botany 61:3519–3528
Bernardo R (2010) Genomewide selection with minimal crossing in self-pollinated crops. Crop Sci 50:624–627
Bernardo R (2008) Molecular markers and selection for complex traits in plants: learning from the last 20 years. Crop Sci 48:1649–1664
Bernardo R (2009) Genomewide selection for rapid introgression of exotic germplasm in maize. Crop Sci 49:419–425
Bernardo R (2010) Genomewide selection with minimal crossing in self-pollinated crops. Crop Sci 50:624–627
Chaerle L, Hagenbeek D, De Bruyne E, Van Der Straete D (2007a) Chlorophyll fluorescence imaging for disease-resistance screening of sugar beet. Plant Cell Tiss Organ Cult 91:97–106
Chaerle L, Leinonen I, Jones HG, Van Der Straeten D (2007b) Monitoring and screening plant populations with combined thermal and chlorophyll fluorescence imaging. J Exp Bot 58(4):773–784
Chen ZH, Zhou MX, Newman IA, Mendham NJ, Zhang GP, Shabala S (2007) Potassium and sodium relations in salinised barley tissues as a basis of differential salt tolerance. Funct Plant Biol 34:150–162
Coram TE, Mantri N, Ford R, Pang ECK (2007) Evans Review No. 4: functional genomics in chickpea: an emerging frontier for molecular-assisted breeding. Funct Plant Biol 34:861–873
Delannay X, McLaren G, Ribaut JM (2012) Fostering molecular breeding in developing countries. Mol Breeding 29 (4):857–873
Duran C, Appleby N, Clark T, Wood D, Imelfort M, Batley J, Edwards D (2009) AutoSNPdb: an annotated single nucleotide polymorphism database for crop plants. Nucleic Acids Res 37:D951–953
El-Shabrawi H, Kumar B, Kaul T, Reddy MK, Singla-Pareek SL, Sopory SK (2010) Redox homeostasis, antioxidant defense, and methylglyoxal detoxification as markers for salt tolerance in Pokkali rice. Protoplasma 245(1–4):85–96
Gomathi R, Rakkiyapan P (2011) Comparative lipid peroxidation, leaf membrane thermostability, and antioxidant system in four sugarcane genotypes differing in salt tolerance. Int J Plant Physiol Biochem 3(4):67–74
Heffner EL, Sorrells ME, Jannink JL (2009) Genomic selection for crop improvement. Crop Sci 49:1–12
Heiskanen J (2006) Estimated aboveground tree biomass and leaf area index in a mountain birch forest using ASTER satellite data. Int J Remote Sens 27:1135–1158
Huseynova IM, Rustamova SM (2010) Screening for drought stress tolerance in wheat genotypes using molecular markers. Proc ANAS (Biol Sci) 65(5–6):132–139
Jiang Q, Roche D, Monaco TA, Durham S (2006) Gas exchange, chlorophyll fluorescence parameters and carbon isotope discrimination of 14 barley genetic lines in response to salinity. Field Crops Res 96:269–278
Jones HG, Schofield P (2008) Thermal and other remote sensing of plant stress. Gen Appl Plant Physiol 34(1–2):19–32 (Spl Issue)

Jones HG, Serraj R, Loveys BR, Xiong L, Wheaton A, Price AH (2009) Thermal infrared imaging of crop canopies for the remote diagnosis and quantification of plant responses to water stress in the field. Funct Plant Bio 36:978–989

Lang NT, Buu BC, Ismail A (2008) Molecular mapping and marker assisted selection for salt tolerance in rice (*Oryza sativa* L.). Omonrice 16:50–56

Mantri N, Ford R, Coram TE, Pang ECK (2010a) Evidence of unique and shared responses to major biotic and abiotic stresses in chickpea. Environ Exp Bot 69(3):286–292

Mantri N, Pang ECK, Ford R (2010b) Molecular biology for stress management. In: Yadav S, McNeil D, Redden R, Patil S (eds) Climate change and management of cool season grain legume crops. Springer, Science + Business Media NY, USA, pp 377–408

Mantri N, Patade V, Penna S, Ford R, Pang ECK (2012) Abiotic stress responses in plants—present and future. In: Ahmad P, Prasad MNV (eds) Abiotic stress responses in plants: metabolism to productivity. Springer, Science + Business Media NY, USA, pp 1–19

Matous K, Benediktyová Z, Berger S, Roitsch T, Nedbal L (2006) Case study of combinatorial imaging: what protocol and what chlorophyll fluorescence image to use when visualizing infection of Arabidopsis thaliana by Pseudomonas syringae? Photosynth Res 90(3):243–253

Masuka B, Araus JL, Das B, Sonder K, Cairns JE (2012) Phenotyping for abiotic stress tolerance in maize. J Integr Plant Biol 54(4):238–249

McCouch SR, Zhao K, Wright M, Tung CW, Ebana K, Thomson M, Reynolds A, Wang D, DeClerck G, Ali ML, McClung A, Eizenga G, Bustamante C (2010) Development of genome-wide SNP assays for rice. Breed Sci 60:524–535

McNally KL, Childs KL, Bohnert R, Davidson RM, Zhao K, Ulat VJ, Zeller G, Clark RM, Hoen DR, Bureau TE, Stokowski R, Ballinger DG, Frazer KA, Cox DR, Padhukasahasram B, Bustamante CD, Weigel D, Mackill DJ, Bruskiewich RM, Rätsch G, Buell CR, Leung H, Leach JE (2009) Genomewide SNP variation reveals relationships among landraces and modern varieties of rice. Proc Natl Acad Sci U S A 106(30):12273–12278

Mishra A, Mishra KB, Höermiller II, Heyer AG, Nedbal L (2011) Chlorophyll fluorescence emission as a reporter on cold tolerance in *Arabidopsis thaliana* accessions. Plant Signal Behav 6(2):301–310

Monneveux P, Reynolds MP, Trethowan R, González-Santoyo H, Peña RJ, Zapata F (2005) Relationship between grain yield and carbon isotope discrimination in bread wheat under four water regimes. European J Agron 22(2):231–242

Monneveux P, Rekika D, Acevedo E, Merah O (2006) Effect of drought on leaf gas exchange, carbon isotope discrimination, transpiration efficiency and productivity in field grown durum wheat genotypes. Plant Sci 170(4):867–872

Morison JIL, Baker NR, Mullineaux PM, Davies WJ (2008) Improving water use in crop production. Philos Trans R Soc B Bio Sci 363:639–658

Moses FA, McNeil MD, Redden B, Kollmorgen JF, Pittock C (2008) Sampling strategies and screening of chickpea (*Cicer arietinum* L.) germplasm for salt tolerance. Genet Resour Crop Evol 55:53–63

Munns R, James RA (2003) Screening methods for salinity tolerance: a case study with tetraploid wheat. Plant Soil 253:201–218

Natarajkumar P, Sujatha K, Laha GS, Viraktamath BC, Reddy CS, Mishra B, Balachandran SM, Ram T, Srinivasarao K, Hari Y, Sundaram RM (2010) Identification of a dominant bacterial blight resistance gene from Oryza nivara and its molecular mapping. Rice Genet Newsl 25:22

Niu CF, Wei W, Zhou QY, Tian AG, Hao YJ, Zhang WK, Ma B, Lin Q, Zhang ZB, Zhang JS, Chen SY (2012) Wheat WRKY genes TaWRKY2 and TaWRKY19 regulate abiotic stress tolerance in transgenic Arabidopsis plants. Plant Cell Environ 35(6):1156–1170

Patade VY, Suprasanna P (2008) Radiation induced *in vitro* mutagenesis for sugarcane improvement. Sugar Tech 10(1):14–19

Patade VY, Suprasanna P (2009) An *in vitro* radiation induced mutagenesis-selection system for salinity tolerance in sugarcane. Sugar Tech 11(3):246–251

Patade VY, Suprasanna P, Bapat VA (2008) Gamma irradiation of embryogenic callus cultures and *in vitro* selection for salt tolerance in sugarcane (*Saccharum offcinarum* L.). Agril Sci China 7(9):101–105

Patade VY, Bhargava S, Suprasanna P (2011a) Transcript expression profiling of stress responsive genes in response to short-term salt or PEG stress in sugarcane leaves. Mol Biol Rep doi:10.1007/s11033-011-1100-z

Patade VY, Bhargava S, Suprasanna P (2011b) Effects of NaCl and iso-osmotic PEG stress on growth, osmolytes accumulation and antioxidant defense in cultured sugarcane cells. Plant Cell Tiss Organ Cult doi:10.1007/s11240-011-0041-5

Patade VY, Bhargava S, Suprasanna P (2011c) Salt and drought tolerance of sugarcane under iso-osmotic salt and water stress: growth, osmolytes accumulation and antioxidant defense. J Plant Interact doi:10.1080/17429145.2011.557513

Porch TG (2006) Application of stress indices for heat tolerance screening of common bean. J Agron Crop Sci 192:390–394

Raman H, Moroni JH, Saito K, Read BJ, Scott BJ (2002) Identification of AFLP and microsatellite markers linked with an aluminium tolerance gene in barley (*Hordeum vulgare* L.). Theor Appl Genet 105(2–3):458–464

Raman H, Karakousis A, Moroni JS, Raman R, Read B, Garvin DF, Kochian LV, Sorrells ME (2003) Development and allele diversity of microsatellite markers linked to the aluminium tolerance gene *Alp* in barley. Aust J Agric Res 54(12):1315–1321

Rahnama A, James R, Poustini K, Munns R (2010) Stomatal conductance as a screen for osmotic stress tolerance in durum wheat growing in saline soil. Funct Plant Bio 37(3):255–263

Rajendran K, Tester M, Roy SJ (2009) Quantifying the three main components of salinity tolerance in cereals. Plant Cell Environ 32:237–249

Ribaut JM, Ragot M (2007) Marker-assisted selection to improve drought adaptation in maize: the backcross approach, perspectives, limitations, and alternatives. J Exp Bot 58(2):351–360

Roy SJ, Tucker EJ, Tester M (2011) Genetic analysis of abiotic stress tolerance in crops. Current Opinion in Plant Biology 14(3):232–239

Siddiqi EH, Ashraf M, Hussain M, Jamil A (2009) Assessment of inter-cultivar variation for salt-tolerance in safflower (*Carthamus tinctorius* L.) using gas exchange characteristics as selection criteria. Pak J Bot 41(5):2251–2259

Smethurst CF, Gill WM, Shabala S (2009) Using excised leaves to screen lucerne for salt tolerance: Physiological and cytological evidence. Plant Signal Behav 4(1):39–41

Sorkheh K, Shiran B, Khodambshi M, Rouhi V, Ercisli S (2011) *In vitro* assay of native Iranian almond species (*Prunus* L. spp.) for drought tolerance. Plant Cell Tiss Organ Cult 105:395–404

Stodart BJ, Raman H, Coombes N, Mackay M (2007) Evaluating landraces of bread wheat *Triticum aestivum* L. for tolerance to aluminium under low pH conditions. Genetic Resor Crop Evol 54(4):759–766

Tang Y, Sorrells ME, Kochian LV, Gravan DF (2000) Identification of RFLP markers linked to the barley aluminium tolerance gene Alp. Crop Sci 40:778–782

Thakur A (2004) Use of easy and less expensive methodology to rapidly screen fruit crops for drought tolerance. Acta Hort 662:231–235

Tsiatas JT, Tokatlidis IS, Tsikrikoni C, Lithourgidis AS (2008) Leaf carbon isotope discrimination, ash content and *K* relationships with seedcotton yield and lint quality in lines of *Gossypium hirsutum* L. Field Crops Res 107(1):70–77

Tung CW, Zhao K, Wright MH, Ali ML, Jung J, Kimball J, Tyagi W, Thomson MJ, McNally K, Leung H (2010) Development of a research platform for dissecting phenotype–genotype associations in rice (*Oryza* spp.). Rice 3(4):205–217

Varshney RK, Nayak SN, May GD, Jackson SA (2009) Next-generation sequencing technologies and their implications for crop genetics and breeding. Trends Biotechnol 27(9):522–530

Wang JP, Raman H, Zhang GP, Mendham N, Zhou MX (2006) Aluminium tolerance in barley (*Hordeum vulgare* L.): physiological mechanisms, genetics and screening methods. J Zhejiang Univ Sci B 7(10):769–787

Wang JK, Chapman SC, Bonnett DG, Rebetzke GJ, Crouch J (2007) Application of population genetic theory and simulation models to efficiently pyramid multiple genes via marker-assisted selection. Crop Sci 47:582–590

Wang J, Raman H, Zhou M, Ryan PR, Delhaize E, Hebb DM, Coombes N, and Mendham N (2007) High-resolution mapping of the Alp locus and identification of a candidate gene Hv-MATE controlling aluminium tolerance in barley (Hordeum vulgare L.). Theoretical and Applied Genetics 115:265–276

White JW, Andrade-Sanchez P, Gore MA, Bronsona KF, Coffelt TA, Conley MM, Feldmann KA, French AN, Heun JT (2012) Field-based phenomics for plant genetics research. Field Crops Res 133:101–112

Wei B, Jing R, Wang Ch, Chen J, Mao X, Chang X, Jia J (2009) Dreb1 genes in wheat (*Triticum aestivum* L.): development of functional markers and gene mapping based on SNPs. Mol Breed 23:13–22

West JD, Peak D, Peterson JQ, Mott KA (2005) Dynamics of stomatal patches for a single surface of *Xanthium strumarium* L. leaves observed with fluorescence and thermal images. Plant Cell Environ 28:633–641

Xu Y, Crouch JH (2008) Marker-assisted selection in plant breeding: from publications to practice. Crop Sci 48:391–407

Yang W, Duan L, Chen G, Xiong L, Liu Q (2013) Plant phenomics and high-throughput phenotyping: accelerating rice functional genomics using multidisciplinary technologies. Curr Opin Plant Biol 16(2):180–187

Zheng J, Fu J, Gou M, Huai J, Liu Y, Jian M, Huang Q, Guo X, Dong Z, Wang H, Wang G (2010) Genome-wide transcriptome analysis of two maize inbred lines under drought stress. Plant Mol Biol 72:407–423

Zhu L, Liang ZS, Xu X, Li SH, Monneveux P (2009) Evidences for the association between carbon isotope discrimination and grain yield—Ash content and stem carbohydrate in spring wheat grown in Ningxia (Northwest China). Plant Sci 176(6):758–767

Chapter 3
Transcriptomics of Heat Stress in Plants

Boghireddy Sailaja, Satendra K. Mangrauthia, Neelamraju Sarla and Sitapati Rao Voleti

Abstract High-temperature stress is a major abiotic stress that affects various biological processes of plants such as biochemical and physiological response, growth, development, and yield. High-temperature stress has critical effects at cellular and molecular levels also. The increased concentration of regulatory proteins such as heat shock transcription factors (Hsfs) is a major molecular response that occurs during heat stress. These regulatory proteins in turn regulate the expression of heat shock protein (HSP) genes that act as critical players during stress to maintain cell homeostasis. Besides HSPs, the other metabolic and regulatory genes, signaling compounds, compatible osmolytes, and antioxidants too play an important role during heat stress in plants. Apart from the protein-coding genes, recent studies have shown that noncoding microRNAs (miRNAs) also play a key role during heat stress by modulating the gene expression at the transcription and post-transcriptional level. The transcriptome approaches are important to understand the molecular and cellular changes occurring in response to heat stress. The approaches rely mostly by adopting the traditional methods like Northern blot/RNA blot and reverse transcription PCR (RT-PCR), where the expression of the genes can be studied in different tissues and cells, whereas the extent of their expression can be achieved by quantitative PCR or real time PCR. Further, the genome-wide expression profiling tools such as microarray analysis, next-generation sequencing, and RNA sequencing offer a great potential in this direction. This chapter primarily provides the current understanding on the role of regulatory genes (transcription factors), HSP genes, metabolic genes, signaling compounds, osmolytes, reactive oxygen species,

S. K. Mangrauthia (✉) · B. Sailaja · N. Sarla
Crop Improvement Section, Directorate of Rice Research, Rajendranagar,
500030 Hyderabad, India
e-mail: skmdrr@gmail.com

S. R. Voleti
Plant Physiology Section, Directorate of Rice Research, Rajendranagar,
500030 Hyderabad, India
e-mail: voletisr58@rediffmail.com

B. Sailaja (✉)
e-mail: sailajaprasadd@gmail.com

N. Sarla
e-mail: sarla_neelamraju@yahoo.com

P. Ahmad et al. (eds.), *Improvement of Crops in the Era of Climatic Changes,* Volume 2,
DOI 10.1007/978-1-4614-8824-8_3,

and miRNAs as well as other small RNAs of plants under high temperature. In addition, it gives a brief account of various transcriptome approaches to study the expression profiling of genes during heat stress.

Keywords High temperature · HSPs · Strategies · Transcription factors · miRNAs

1 Introduction

Elevation of global mean temperature beyond the optimum by various means is termed as high-temperature stress or heat stress. In a changing climate scenario, heat stress is considered as a serious threat for food crop productivity (Hall 2001) and is a major challenge in attaining food security. The extent of rise in temperature in specific climatic zones depends on the period of high temperature occurring during the day and the night. Apart from the day temperature, increase in night temperature also shows a serious effect on crop productivity. A considerable rise in gaseous emissions is thought to be one of the reasons for increasing concentrations of greenhouse gases like carbon dioxide, methane, nitrous oxides, chlorofluorocarbons, etc. It is predicted that rise in these greenhouse gases will ultimately increase the global ambient temperature (Wahid et al. 2007). A report of the Intergovernmental Panel on Climatic Change (IPCC) suggested global mean temperature rises between about 0.15 and 0.3 °C per decade for 1990–2005. It says that "continued greenhouse gas emissions at or above current rates would cause further warming and induce many changes in the global climate system during the 21st century that would very likely be larger than those observed during the 20th century."

Heat stress can cause severe damage to almost all the stages of crop growth and may lead to death of the plants if it causes an irreversible damage in cellular homeostasis, destruction of metabolic pathways, and degradation of structural and functional proteins/metabolites and membranes which ultimately lead to cell death (Vinocur and Altman 2005; Bohnert et al. 2006). As plants are sessile in nature, they have adapted different mechanisms to tolerate stress caused by high temperature. Avoidance mechanism is one of the important mechanisms adapted by the plants to cope with the stress which includes change in leaf orientation, transpirational cooling, alteration of membrane lipid profile, and excessive rooting (Lehman and Engelke 1993; Bonos and Murphy 1999). The other mechanism called tolerance mechanism includes activation of free radical scavengers, osmoprotectants, ion transporters, increased concentration of heat shock transcription factors (Hsfs), heat shock proteins (HSPs), late embryogenesis abundant (LEA) proteins, switching on different signal transduction cascades, production of metabolites, etc. (Wahid et al. 2007). In several crop plants, it is reported that under high temperature, plants attain the reproductive stage faster with compromise in total yield which is described as an escape mechanism (Adams et al. 2001).

Morphologically, high temperature causes significant reduction in relative growth and dry weight. Reduced size of internodes, more tillering, early senescence, and reduced biomass in sugarcane plants was observed during high temperature (Ebrahim et al. 1998). It may lead to production of polymorphic leaves to reduce water loss by transpiration (Sayed 1996). Anatomical changes such as cell size reduction, closing of stomata, and condensed water loss are reported during heat stress (Banon et al. 2004). Crop plants affected by heat during anthesis and grain filling stage show yield reduction. In wheat, heat stress increases the period of grain filling with kernel growth reduction that leads to the reduction in density and weight of kernel by 7% (Guilioni et al. 1997) and ultimately a reduction in grain number (Ferris et al. 1998). Similarly, in maize, decreased accumulation of proteins, starch, and lipid content of the kernel was observed (Wilhelm et al. 1999). In rice, heat stress affects anthesis causing irreversible effects on reproductive tissues and leads to spikelet sterility (Prasad et al. 2006; Oh-e et al. 2007; Jagadish et al. 2008). Yield loss due to heat stress was reported in groundnut (Vara Prasad et al. 1999) and Phaseolus (Rainey and Griffiths 2005).

Heat stress affects the physiological and biochemical processes of plants. Plant hormones such as cytokinins and ethylene play regulatory roles in heat stress tolerance (Veselov et al. 1998; Musatenko et al. 2003; Xu and Huang 2009). Stress shows significant impact on protein biochemistry such as proteins synthesis, folding, posttranslational modifications, targeting, and degradation depending on the level and duration of high temperature. Further, heat injury causes loss of membrane integrity and inactivates chloroplast and mitochondrial enzymes (Howarth 2005). Increased respiration during high temperature causes synthesis of reactive oxygen species (ROS) in plants (McDonald and Vanlerberghe 2005). Heat stress damages the mesophyll cells and increases the plasma membrane permeability (Zhang et al. 2005). At the subcellular level, major changes can be observed in chloroplasts such as reduced photosynthesis by alteration in the thylakoids structural organization (Karim et al. 1997). It also affects microtubule organization and spindle formation in mitotic cells (Smertenko et al. 1997).

Biological membrane integrity and function are sensitive to high temperature which changes the tertiary and quaternary structures of membrane proteins. Regular function of biological membranes under stress is crucial for important physiological processes such as photosynthesis and respiration (Blum 1988). Rapid movement of molecules across membranes during heat stress disrupts the chemical bonds present in cellular membranes. Fluidity of the lipid bilayer in membranes also occurs by protein denaturation and increased concentration in unsaturated fatty acids (Savchenko et al. 2002). These changes increase the permeability of membranes resulting in electrolyte leakage. Thus, cell membrane thermostability (MTS) is a measure of increased electrolyte leakage and has been used in a wide variety of species like wheat (Blum et al. 2001), rice (Mohammed and Tarpley 2009), cotton (Ashraf et al. 1994), soybean (Martineau et al. 1979), sorghum (Marcum 1998), barley (Wahid and Shabbir 2005), etc. to understand the heat tolerance mechanism. The positive correlation between MTS and yield was observed in crops like wheat (Reynolds et al. 1994) and sorghum (Sullivan and Ross 1979). Early effects of high-temperature stress on the plasmalemma leads to fluidity of the membrane lipid bilayer which in turn increases Ca^{2+} influx and reorganization of cytoskeleton. This helps in transducing the activation of various signal molecules (Sung et al. 2003).

The heat stress-induced morphological, cellular, physiological, and biochemical changes described above are ultimately governed by the expression of a set of genes or transcripts. Change in osmotic level, ion concentrations, and membrane fluidity during heat stress activates several genes encoding transcription factors and signaling molecules to activate different pathways involved in the production of various compounds to maintain cellular homeostasis. The study of transcriptome during heat stress provides information about the regulatory roles of different genes involved in stress tolerance and susceptibility. Transcriptomics can be defined as a study of gene expression through mRNA profiling. Increase in the availability of genome and transcriptome sequence data for model crop species has helped in understanding the molecular events leading to various pathways of stress response. Complete genome sequences are available for several plants species. *Arabidopsis thaliana* was the first plant genome sequenced (Kaul et al. 2000) followed by rice (Barry 2001; Yu et al. 2002; International rice genome 2005) and sorghum (Bedell et al. 2005; Paterson et al. 2009). Whole genome sequencing has been done in other important food crops like pigeonpea and soybean. The advancement of genomic tools, particularly next-generation sequencing (NGS) technologies, has helped in understanding the genome structure and information of many plant species. Besides the genomic structure and genes organization, understanding the expression pattern of the genes or transcriptomics is very crucial to fully appreciate the usefulness of genome sequence information. Unlike genomics, transcriptomics is highly dynamic in nature as the expression of genes is unique to tissue, stage of plant growth, environmental stimuli, etc. Further, the complexity of transcriptomics is increased due to different levels of regulations at the posttranscriptional level, particularly alternate splicing, alternate polyadenylation, genes/transcripts fusion, RNA editing, and posttranscriptional gene silencing (PTGS).

Traditional methods like northern blot analysis and RT-PCR have helped to understand the gene expression of plants to a certain extent. The efforts on this direction were further augmented with the advent of semiquantitative and qPCR / real-time PCR. Gene expression can be quantified in terms of relative and absolute numbers by using real-time qPCR. However, these methods can help in understanding the expression of only few known genes at a time. Genome-wide transcription profiling is an important and powerful tool to understand the regulation of genes at the molecular level during stress which can be achieved by using medium- and high-throughput methods. Medium-throughput methods include complementary DNA (cDNA) clones and expressed sequence tags (ESTs) for transcriptomic study, whereas microarray and NGS are high-throughput methods. EST and cDNA sequences provide a direct evidence for all the generated transcripts and they are the most important resources for transcriptome study (Nagaraj et al. 2007). Microarray technology provided a boost to plant biology in understanding genome-wide gene expression in several plant species. With the help of this tool, many stress-associated genes could be discovered in the last decade. Microarray analyses have been demonstrated in investigations of transcriptional networks occurring in a variety of developmental processes (Lee et al. 2002; Anisimov 2008). It provides a pattern in gene expression as well as metabolic network models (Xiang et al. 2011). However, microarray was difficult to use in crops where genome sequence data were not available. Further, it could identify the expression of only known genes, hence several novel and rare genes playing important roles in plant metabolism were unattended while studying transcriptome using microarray.

In addition, microarray could not specify the posttranscriptional changes in genes which are very important to understand transcriptomics. One step further in this direction, RNA-Seq technologies provide genome-wide information of transcriptomics where limitations of microarray could be sorted out. These tools facilitate the investigations of structural and functional complexity of the transcriptome. The expression level of almost all transcripts in a sample is quantified by measuring the number of individual mRNA molecules transcribed from each gene. RNA-Seq is a more reliable technique for comparative gene expression profiling studies as it gives an overall view of transcripts expression (Wang et al. 2009; Garber et al. 2011). This technology has been utilized in transcriptome studies of various crops such as maize, rice, soybean, etc. (Eveland et al. 2010; Zhang et al. 2010; Severin et al. 2010). This book chapter provides an overview of important heat-stress-regulated genes/transcripts encoding proteins such as chaperones, Hsfs, antioxidants, signaling molecules, osmolytes, etc. Also, the genes influencing the photosynthesis process during high temperature are discussed. In addition to protein-coding mRNAs, this chapter describes the regulatory role of small RNAs (sRNAs) during heat stress. Further, different transcriptomic approaches to study the expression of transcriptomics during stress are also given.

2 Gene Expression During Heat Stress

High-temperature stress induces the alteration of transcriptome in different tissues and stages of plant growth. Heat-responsive genes have been studied in various crops which constitute a significant proportion of the genome. With the progress in genomic technologies, a remarkable progress has been made towards deciphering the transcriptional networks during heat stress in plants. Heat stress response (HSR) in plants and other organisms is generally characterized by the induced expression of a set of proteins known as molecular chaperones or HSPs. Expression of the HSPs is in turn governed by another important heat-stress-induced gene family encoding Hsfs. Plant genome possesses a highly complex multigene family encoding HSPs and Hsfs. In addition to HSPs and Hsfs, transcripts encoding enzymes involved in ROS scavenging, osmolyte accumulation, synthesis of signaling molecules, and photosynthesis machinery are greatly influenced by heat stress. Recently, 25, 29, 26, 9, and 10 genes have been identified in the rice genome encoding Hsfs, small HSP (sHSP), HSP70, HSP90, and HSP100 families, respectively (Hu et al. 2009).

2.1 Heat Shock Proteins

Synthesis and accumulation of HSPs are ascertained during high-temperature stress in plants. Biosynthesis of HSPs during temperature stress is a common response in all living organisms starting from bacteria to plants and human beings (Vierling 1991; Gupta et al. 2010). Acquired thermo-tolerance due to the accumulation of HSPs has been studied (Bowen et al. 2002). HSPs are induced in plants at all the stages of development during heat stress (Vierling 1991). These HSPs are categorized into gene families based on their molecular weight, sequence homologies, and functions: HSP60, HSP100, HSP90, HSP70, and small HSP family (Gupta et al. 2010). These HSPs act like molecular chaperones by inhibiting irreversible aggregation of other proteins as well as participating in refolding of proteins during high-temperature stress to maintain cellular homeostasis (Tripp et al. 2009). Further, HSPs help in preventing the denaturation of other proteins caused by heat stress. Also, HSPs help in shuttling and transporting of other proteins inside the cell. Plants differ significantly in the expression of different kinds and number of HSPs (Hamilton et al. 1996). Not only individually but also in combinations with other proteins the HSPs play a crucial role during heat stress. For example, in many plant species, HSP100 family proteins acquired thermo-tolerance due to the induction of HSP70 as well as HSP101 during heat stress (Schoffl et al. 1999).

2.1.1 HSP60

HSP60 are also called as chaperonins which are evolutionarily homologous to the GroEL protein of *Escherichia coli*. These proteins are present in prokaryotes as well as eukaryotes. In eukaryotes, HSP60 proteins are found in cytosol, mitochondria, and chloroplasts. Chaperonins present in bacteria, mitochondria, and chloroplasts are categorized under group I, such as GroE chaperonins and chCpn60, whereas in Archaea and the cytosol of eukaryotes, they are categorized into group II, such as CCT (chaperonins containing t-complex polypeptide 1; TCP1) chaperonins (Ranson et al. 1998). Generally, these chaperonins assist plastid proteins like Rubisco (Wang et al. 2004) and play an important role in folding and aggregation of other proteins. HSP60 helps in the translocation of proteins to different organelles such as chloroplasts and mitochondria (Lubben et al. 1989). In a mutant of *Arabidopsis,* chloroplast chaperonin Cpn60α showed defects in development of chloroplast (Apuya et al. 2001).

2.1.2 HSP70

Role of HSP70 in different plant species has been studied. These proteins are expressed constitutively as well as in response to environmental stimuli. Constitutively expressed HSP70 helps in the folding of newly synthesized proteins and the transport of precursor proteins. Environmentally induced HSP70 chaperones along with other co-chaperones (e.g., GrpE and DnaJ/HSP40) act as a protein complex involved in protein folding processes and in preventing aggregation of nonnative proteins under abiotic stress conditions in almost all cellular compartments (Hart 1996; Wang et al. 2004). Furthermore, it has a wide variety of functions in important phases of protein metabolism such as protein synthesis, transport, degradation, folding, and activation of denatured proteins (Zhang et al. 2005). They facilitate the lysosome- or proteasome-mediated degradation of unstable proteins (Wang et al. 2004). HSP70 together with sHSPs act as molecular chaperones to protect the plant cells from deleterious effects caused by heat stress (Rouch et al. 2004). HSP70 participates in adenosine triphosphate (ATP)-dependent protein assembly to prevent protein denaturation during high-temperature stress (Iba 2002). The cells where HSP70 synthesis was blocked were more susceptible to heat injury (Burke 2001). A recent study in *A. thaliana* showed that chloroplast HSP70 is involved in heat tolerance (Su and Li 2008). Schroda et al. (1999) showed that chloroplast HSP70B participates in photo-protection by repairing photosystem II (PSII) proteins during photo-inhibition. HSP70 also plays a key role in the expression of other stress-associated genes and in the modulation of signal transducers. The genes encoding HSP70 are constitutively expressed in plants cells; however, they get upregulated during stress.

2.1.3 HSP90

HSP90 class proteins are one of the most abundant proteins present in a cell which acts on proteins associated with signal transduction such as kinases and steroid hormone receptors. These proteins work in an ATP-dependent fashion and mainly regulate the cellular signals of glucocorticoid receptor activity (Pratt et al. 2004). Besides their major role as molecular chaperones to assist protein folding, these proteins are also associated with signal transduction cascades, cell-cycle regulation, degradation, and trafficking of proteins (Pratt and Toft 2003). In plants, HSP90 is present in the cytosol, ER, and plastids. In *A. thaliana,* it helps in stress adaptation. It forms a multiprotein complex with HSP70 and other co-chaperones to perform its function. Yamada et al. (2007) suggested that under normal conditions, cytoplasmic HSP90 inhibits the activity of Hsfs, while during heat stress, the temporary suspension of HSP90 activity leads to the activation of Hsfs.

2.1.4 HSP100

HSP100, also called Clp proteins, are ATP-dependent chaperones. They solubilize aggregated proteins and help in protein degradations (Boston et al. 1996). In order to maintain cellular homeostasis, degradation of nonfunctional and harmful polypeptides synthesized due to misfolding, denaturation, or aggregation is important (Wang et al. 2004). These proteins in association with HSP70 (ATP-dependent chaperone system) perform the function of protein disaggregation and refolding. HSP100 plays an important role in plants during severe heat stress (Hong and Vierling 2000). Induced expression of HSP100 proteins have been reported in several plant species, such as *Arabidopsis* (Schirmer et al. 1994), rice (Pareek et al. 1995), maize (Nieto-Sotelo et al. 2002), soybean (Lee et al. 1994), and lima bean (Keeler et al. 2000). Similar to other HSP proteins, HSP100 family chaperones are constitutively expressed in general; however, their expression gets upregulated during stress. In rice, the immunological homologue of yeast HSP104 gets accumulated during heat shock (Singla and Grover 1993). The disappearance of protein granules in yeast was associated with the production of rice HSP100. Moreover, dissolution of electron-dense granules by HSP101 was shown after heat stress, implicating its role during recovery of cell stress (Agarwal et al. 2003). Thermo-tolerance activity of HSP100 family proteins was proved through genetic approaches also (Lee et al. 1994; Schirmer et al. 1994). Increased expression of mitochondrial HSP68 in maize, soybean, tomato, and barley was observed during heat stress (Neumann et al. 1993).

2.1.5 Small HSPs

Low molecular weight (LMW) HSPs of about 12–40 kDa are designated as sHSPs; Vierling 1991; Sun et al. 2002). In comparison to other HSPs, these proteins are more diverse in terms of sequence similarity, localization, and functions. sHSPs cannot refold nonnative proteins by themselves but they can provide stability and prevent nonnative protein aggregation through binding to nonnative proteins, thereby helping ATP-dependent chaperones for subsequent refolding (Wang et al. 2004). During normal conditions, sHSPs are usually not detectable in plant tissues, but during stress conditions, they are induced to impart acquired stress tolerance (Scharf et al. 2001; Zhang et al. 2008a). Based on the abundance and diversification of sHSPs, probably plants have the differential ability to adapt during heat stress conditions. Plant sHSPs are encoded by six nuclear multigene families which are present in different cellular compartments. Class I and class II gene products are present in cytosol, whereas others are present in the chloroplast, endoplasmic reticulum, mitochondria, and membranes (Waters et al. 1996). In *A. thaliana* and *Lycopersicon esculentum,* sHSPs were divided into three subclasses such as subclass CI, CII, and CIII (Scharf et al. 2001; Siddique et al. 2003). A recent study on *A. thaliana* reported other groups in the cytoplasm that were categorized into CIV, CV, CVI, and CVII. Each subclass has its own distinct characteristic specified role (Siddique et al. 2008). Higher plants usually possess up to 20 sHSPs and each different species may contain up to 40 sHSPs (Vierling 1991). It was reported that during heat stress conditions, the expression of class I sHSPs in soybean can increase up to 1% (Hsieh et al. 1992). In the absence of stress, the expression of some sHSPs in plants is confined to certain developmental stages (Sun et al. 2002).

3 Heat Shock Transcription Factors

In plants, the expression of HSPs is a common phenomenon observed during heat stress. The expression of HSPs is in turn regulated by Hsfs and the process is termed as HSR. Generally, eukaryotes possess one to four HSF family members, whereas plants possess a large number of Hsfs. For instance, *Arabidopsis* has 21 HSF members, while rice genome possesses 25 HSF members (Nover et al. 2001). Hsfs control the expression of HSPs in plants by binding specifically to the heat shock element (HSE), a highly conserved region having palindromic motifs of nGAAn (Miller and Mittler 2006). Plant Hsfs are characterized by possessing a DNA-binding domain at the N-terminal (helix-turn-helix type) followed by two hydrophobic heptad repeats (HR-A and HR-B) known as the oligomerization domain, a nuclear localization signal (NLS) for nuclear uptake of the protein, and a nuclear export signal (NES) for exportation. Apart from this, a C-terminal activation domain (CTD) rich in hydrophobic, aromatic, and acidic amino acid residues commonly known as AHA motifs, is essential for the activation of HSF (Nover et al. 2001). Based on the protein structures, three classes of Hsfs have been identified in plants, namely HsfA, HsfB, and HsfC (Nover et al. 2001). HsfA and HsfC have a long HR-A/B region. HsfA possesses the insertion of 21 amino acids, whereas HsfC possess the insertion of 7 amino acids between the hydrophobic regions HR-A and HR-B. HsfB and HsfC do not possess AHA motifs at their C-terminal ends (Nover et al. 2001; Kotak et al. 2004).

Hsfs are transcriptional regulators, having a well-defined role in heat stress signaling and regulation of several downstream target genes. Expression of genes encoding Hsfs is induced during elevated temperature (Liu et al. 2005, 2009). Upregulation of Hsfs in *Arabidopsis* was reported in response to high-temperature stress (Swindell et al. 2007). Overexpression of HSF genes in transgenic plants resulted in the upregulation of heat-stress-associated genes as well as enhanced thermo-tolerance, whereas the downregulation of HSF genes results in the reduction of thermo-tolerance (Schramm et al. 2008). Accumulation of HSPs during the overexpression of HsfA1 suggested a unique function of this HSF as the master regulator for induced thermo-tolerance in tomato (Mishra et al. 2002). The expression of OsHsfA2a gene was greatly stimulated by high-temperature stress in root and shoot tissues of rice (Chauhan et al. 2011a). In *Arabidopsis,* HsfA1 was shown to be the primary regulator of heat stress, while HsfA2 was essential for prolonged heat stress (Charng et al. 2007). Ogawa et al. (2007) reported that the expression of HsfA2 showed higher expression among all 21 *Arabidopsis* Hsfs in response to heat stress. In tomato, constitutive expression of LeHsfA1 showed improved tolerance to high-temperature stress, whereas silencing of LeHsfA resulted in decreased thermo-tolerance (Mishra et al. 2002). Similarly in *Arabidopsis,* overexpression of AtHsfA2 gene in transgenic plants increases tolerance to the environmental stresses and altered expression of some heat-responsive genes. In addition, acquired thermo-tolerance by using the knockout mutant of AtHsfA2 (Li et al. 2005; Schramm et al. 2006; Charng et al. 2007) was also demonstrated. In the same way, overexpression of rice Hsf OsHsfA2e in transgenic *Arabidopsis* responded to high temperature and showed acquired thermo-tolerance.

4 Reactive Oxygen Species

ROS are reactive chemicals derived from molecular oxygen either by energy transfer or by electron transfer reactions (Gill and Tuteja 2010). These are produced as normal metabolites during cellular metabolism in chloroplasts, mitochondria, and peroxisomes. ROS play very important role in various aspects of plant metabolism such as growth, development, cell cycle, hormone signaling, stress response, and programmed cell death. The production and removal of ROS is tightly regulated; however, the equilibrium gets disturbed during heat stress and other abiotic as well as biotic stresses. In *Arabidopsis,* at least 152 genes are involved in maintaining the steady-state level of ROS (Mittler 2002). Increased concentration of ROS in plants causes oxidative damage by damaging the cellular and membrane proteins, lipids, carbohydrates, and DNA. ROS include free radicals such as superoxide radical ($O_2^{\bullet -}$), hydroxyl radical ($OH^{\bullet}$), hydroperoxyl radical ($HO_2^{\bullet}$), hydrogen peroxide (H_2O_2), alkoxy radical ($RO^{\bullet\bullet}$), peroxy radical ($ROO^{\bullet}$), singlet oxygen ($^1O_2$), and excited carbonyl ($RO^{\bullet}$). Under steady-state conditions, these free radicals are scavenged by different antioxidative defense mechanisms (Foyer and Noctor 2005). Stress-induced ROS are efficiently scavenged by either enzymatic or nonenzymatic antioxidants defense system. Major enzymes involved in ROS scavenging are superoxide dismutase (SOD), ascorbic peroxidase (APX), catalase (CAT), monodehydroascorbate reductase (MDHAR), dehydroascorbate reductase (DHAR), glutathione peroxidase (GPX), glutathione reductase (GR), glutathione S-transferase (GST), and guaiacol peroxidase (GOPX). Nonenzymatic antioxidant defense system includes ascorbic acid, glutathione, alkaloids, phenolic compounds, α-tocopherols, and nonprotein amino acids (Apel and Hirt 2004; Gill and Tuteja 2010).

4.1 Super Oxide Dismutase

SOD is a metalloenzyme present in subcellular compartments of all aerobic organisms that catalyzes the dismutation of superoxide into O_2 and H_2O_2, which is a vital antioxidant defense mechanism. This enzyme provides the first line of defense mechanism against the toxic effects produced by ROS and imparts crucial role in stress tolerance (Apel and Hirt 2004; Gill and Tuteja 2010). SODs have been classified into three groups based on the metal cofactor: Fe-SOD (ferrous SOD), Mn-SOD (manganese SOD), and Cu/Zn-SOD (copper/zinc SOD). Fe-SOD, generally not detected in plants, is localized in chloroplasts, while Mn-SOD is localized in mitochondria and peroxisomes. Cu/Zn-SOD has been detected in fractions of chloroplasts and cytosol (Del Rio et al. 1996; Gill and Tuteja 2010). Increased activities of SODs have been shown in plant cells under stress conditions, which is a part of the defensive mechanism under oxidative stress (Ushimaru et al. 1995). In the *A. thaliana* genome, three Cu/Zn-SOD genes, namely CSD1, CSD2, and CSD3, three Fe-SOD genes, namely FSD1, FSD2, and FSD3, and one Mn-SOD gene (MSD1) have been identified (Kliebenstein et al. 1999). In rice seedlings, the expression of cytosolic Cu/Zn-SOD was stimulated under heat stress (Shah and Nahakpam 2012).

4.2 Catalase

Decomposition of H_2O_2 into H_2O and O_2 is catalyzed by the heme-containing enzyme called catalase. It has the highest turnover rate by converting ~6 million of H_2O_2 molecules per minute. It plays a very crucial role during oxidative stress by removing H_2O_2 produced by oxidases during β-oxidation of fatty acids, photorespiration, and purine catabolism in peroxisomes (Mittler 2002; Vellosillo et al. 2010). It is a light-sensitive enzyme that has different isoforms. *Arabidopsis* possesses three genes, namely CAT1, CAT2, and CAT3, encoding polypeptides that associate to form at least six isozymes (Frugoli et al. 1996). In *Brassica,* 12 isozymes and in *Helianthus annuus* cotyledons, four isozymes were reported (Azpilicueta et al. 2007). Three isoforms of maize are differentially expressed and independently regulated. CAT1 and CAT2 are present in peroxisomes and cytosol, whereas CAT3 is present in mitochondria (Scandalias et al. 1990). In order to improve salinity tolerance of *Oryza sativa,* the rice cultivar Nipponbare was genetically transformed with a catalase gene of *E. coli*. The transgenic rice plants showing constitutive expression of catalase gene could grow for > 14 days in the presence of 250 mM NaCl (Nagamiya et al. 2007). Expression of wheat catalase gene in transgenic rice showed tolerance to cold stress, which may be attributed to the effective detoxification of H_2O_2 by increased catalase activities (Matsumura et al. 2002).

4.3 Ascorbate Peroxidase

APX performs a similar function as catalase. APX performs its action in chloroplast, glyoxisome, and cytosol of plant cells in scavenging ROS and protecting cells in higher plants during stress. It uses ascorbic acid as a hydrogen donor to break down H_2O_2 to form H_2O and monodehydroascorbate (Asada 2000) and is also involved in the electron transport through the ascorbate–glutathione cycle (Foyer and Noctor 2005). It has five different isoforms showing higher affinity towards its substrate when compared to CAT. In cytosol, it has two different isoforms (cAPX), while in chloroplasts soluble (sAPX) and thylakoid-bound forms (tAPX) are reported (Asada 2000, 2006). Another form in the membrane of glyoxisomes (gmAPX) is also present. It was reported that cytosolic APX1 plays an important role in protecting the plants from heat stress (Koussevitzky et al. 2008). Overexpression of the APX-like 1 gene (CAPOA1) of *Capsicum annuum* in transgenic tobacco plants conferred tolerance to oxidative stress (Sarowar et al. 2005). Hsu and Kao (2007) showed that pretreatment of rice seedlings with H_2O_2 resulted in enhanced APX activity and protected seedlings from cadmium (Cd) stress.

4.4 Monodehydroascorbate Reductase and Dehydroascorbate Reductase

MDHAR is a flavin adenine dinucleotide (FAD)-containing enzyme having high specificity towards its substrate monodehydroascorbate (MDHA). Its isoforms are located in chloroplast, cytosol, mitochondria, and peroxisomes. It is involved in the regeneration of reduced ascorbate. MDHA is a very good electron acceptor (Noctor and Foyer 1998; Asada 2000) that accepts electrons from nicotinamide adenine dinucleotide phosphate (NADPH). Reduction of MDHA to ascorbate is attained by using electrons derived from the photosynthetic electron transport chain. It has been reported that increased concentration of MDHAR activity scavenges harmful ROS (Karuppanapandian et al. 2011) and confers chilling stress tolerance in tomato (Stevens et al. 2008). Dehyroascorbate (DHA) is an oxidized form of DHAR which regenerates ascorbic acid. Overexpression of DHAR conferred abiotic stress tolerance in *Arabidopsis* and tobacco (Ushimaru et al. 2006; Eltayeb et al. 2006).

4.5 Glutathione Reductase

GR, mainly present in chloroplast, is a tripeptide flavoprotein oxidase. It catalyzes the reduction of glutathione (GSH) by disulfide bond formation in glutathione disulfide (GSSG) and is crucial in maintaining the equilibrium of GSH via NADPH-dependent reaction. Enhanced GR activity in plants due to the accumulation of GSH provides stress tolerance (Rao and Reddy 2008). Further, GR plays an important role in the regeneration of GSH that provides resistance against oxidative stress (Ding et al. 2009b). Expression of GR is greatly affected by various stresses including high temperature, chilling, exposure to heavy metals, etc. (Apel and Hirt 2004; Karuppanapandian et al. 2011). Sharma and Dubey (2005) reported increased activity of MDHAR, DHAR, and GR in rice seedlings under drought stress.

4.6 Ascorbic Acid

Ascorbic acid is the most abundant water-soluble antioxidant present in almost all types of plant cells. Biosynthesis of ascorbic acid from L-galactono-γ-lactone dehydrogenase and regeneration from oxidized ascorbate takes place in mitochondria. It detoxifies ROS by donating its electrons to a wide range of reactions. It has the ability to scavenge $O_2^{\bullet -}$, $OH^{\bullet}$, and $^1O_2$ directly, and can reduce the concentration of H_2O_2 through APX. Its concentration is maximum in matured leaves with developed chloroplasts and high chlorophyll content. It gives protection to membranes by regenerating α-tocopherol from tocopheroxyl radical. High content of ascorbic acid showed improved tolerance to oxidative stress in tobacco and *Populus* (Aono et al. 1993; Foyer et al. 1995).

4.7 Glutathione

GSH (γ-glutamyl cysteinyl glycine) is an important metabolite involved in scavenging ROS. It occurs as a reduced form localized in cytosol, endoplasmic reticulum, vacuole, mitochondria, chloroplasts, peroxisomes, and in apoplast (Mittler and Zilinskas 1992; Jimenez et al. 1998) and plays a crucial role in several physiological events such as detoxification of xenobiotics, transport of sulfate, conjugation of metabolites, and signal transduction (Xiang et al. 2001). A central nucleophilic cysteine residue is critical for the high reductive potential of GSH. It potentially scavenges cytotoxic H_2O_2 along with other ROS molecules, such as $^1O_2$, $O_2^{\bullet-}$, and $OH^{\bullet}$ (Noctor and Foyer 1998). Additionally, GSH plays a major role in the antioxidative defense system in combination with ascorbate through the AA–GSH cycle (Foyer and Halliwell 1976). Plants with increased GSH concentrations were found to be more tolerant to oxidative stress (Pietrini et al. 2003).

4.8 Tocopherols and Carotenoids

Tocopherols (TOCs) are lipophilic antioxidants mainly present in the thylakoid membrane of chloroplasts of plants. In higher plants, chloroplast membranes containing tocopherols protect lipids and other components of membrane from physical quenching and ROS, thus guarding the structure and function PSII (Igamberdiev et al. 2004). Among the four isoforms (α, β, γ, and δ), α-TOC is having the highest antioxidative activity.

Carotenoids are also lipophilic in nature and play multifunctional roles during stress in plants. They act as detoxifying agents during various environmental stresses and protect photosystem complexes (Karuppanapandian et al. 2011). Carotenoids like β-carotene and zeaxanthin play important role by dissipating excess excitation energy as heat and by scavenging ROS. β-Carotene inhibits oxidative damage by preventing the formation of $^1O_2$through direct quenching of triplet sensitizer, Chl^{3*} (Collins 2001).

5 Signaling Compounds

High-temperature stress in plants induces various molecules and ions for sensing and signaling. Rise in cytosolic Ca^{2+} is the primary signal during temperature stress (Larkindale and Knight 2002), which leads to several changes in the plant gene expression and metabolism. Induced concentration of Ca^{2+} along with calcium-dependent protein kinases (CDPK) regulates the expression of HSPs (Sangwan and Dhindsa 2002). Further, increase in cytosolic Ca^{2+} content facilitates plant cells to better survive during heat injury by increasing the activity of antioxidants and turgor maintenance of the guard cells (Webb et al. 1996; Gong et al. 1997). It activates other signaling pathways through mitogen-activated protein kinase (MAPK) cascade system (Larkindale and Knight 2002). In plants, MAPK is most the important signal transduction cascade that responds to external signal (Kaur and Gupta 2005). Gong et al. (1997) reported that high-temperature stress induces Ca^{2+} uptakes and activates calmodulin (CaM)-related genes in plants. Other signaling compounds known to be involved in HSR are CaM, inositol-3-phosphate (IP3), abscisic acid (ABA), and ethylene. Specific groups of signaling molecules like salicylic acid, calcium chloride ($CaCl_2$), ABA, H_2O_2, and 1-aminocyclopropane-1-carboxylic acid (ACC) may enhance heat stress tolerance by protecting from oxidative damage (Larkindale and Huang 2004).

6 Osmolyte Accumulation

Genes associated with osmolytes synthesis and accumulations are involved in HSR. The increased accumulation of compatible solutes (osmolytes) helps in osmoprotection through maintaining cellular turgidity. In addition, it facilitates antioxidation and chaperoning through direct stabilization of membranes and/or proteins (Yancey et al. 1982; Hare et al. 1998). Different plants may accumulate different types of osmolytes such as sugars, sugar alcohols, proline, tertiary sulfonium compounds, and ammonium compounds (Sairam and Tyagi 2004). Accumulation of these solutes confers heat stress tolerance to the plants. These osmolytes may function as buffering agents during heat stress conditions and other abiotic stresses (Wahid and Close 2007). Glycine betaine (quaternary amine), a compatible solute, plays a crucial role under different abiotic stresses (Sakamoto and Murata 2002). Increased concentration of glycine betaine was observed in maize (Quan et al. 2004) and sugarcane (Wahid and Close 2007) during high-temperature and water-logging conditions.

Proline is an osmolyte which usually accumulates under environmental stress in plants (Kavi Kishore et al. 2005). Proline is synthesized from L-glutamic acid through Δ^1-pyrroline-5-carboxylate, which is catalyzed by two enzymes Δ^1-pyrroline-5-carboxylate synthetase (P5CS) and reductase (P5CR) in plants (Verbruggen and Hermans 2008). Free proline plays the role of osmoprotectants for scavenging different ROS species (Ashraf and Foolad 2007; Trovato et al. 2008). Transgenic tobacco transformed with proline dehydrogenase gene showed more proline accumulation and enhanced osmotolerance (Teteishi et al. 2005). In rice, transgenic plants possessing P5CS gene showed the accumulation of proline that conferred tolerance during water deficit and high salt conditions (Su and Wu 2004). In addition to proline, accumulation of soluble sugars under high-temperature stress was observed in sugarcane (Wahid and Close 2007). Sugars like trehalose, fructans, or mannitol get accumulated during stressed conditions. Several efforts have been made for the production of these sugars through transgenic approaches (Hare et al. 1998).

7 Photosynthesis and Heat Stress

Photosynthesis is considered as one of the most heat-sensitive phenomena of plants. The effect of heat on photosynthesis may occur through reduction in chlorophyll accumulation, disruption in chloroplast membrane, and proteins/enzymes and stomata functioning. The expression of transcripts/genes associated with the photosynthetic machinery of plants is altered during heat stress.

7.1 Effect on Photosystem II and Oxygen Evolving Complex

Heat stress affects greatly the PSII as well as the electron transport chain (Mathur et al. 2011). Primary effect of heat stress is observed on photosynthetic reactions of thylakoid membrane and stroma of chloroplast (Wise et al. 2004). Heat stress may result in the dissociation of oxygen evolving complex (OEC) that leads to imbalance in the flow of electrons from OEC to PSII (De Ronde et al. 2004). The repair mechanism of PSII also gets damaged due to the production of ROS during heat stress that results in the reduction of carbon fixation (Allakhverdiev et al. 2008). Sites in the PSII reaction center were damaged due to high temperature and excessive light in wheat (Sharkova 2001). In barley, damage of PSII units leads to the loss of activity of OEC and restriction of electron transport which was aborted completely after 4 h (Toth et al. 2005).

7.2 Chlorophyll Fluorescence and Reduction

The relation between heat tolerance and fluorescence patterns has been revealed in several plant species (Moffatt et al. 1990). Maximum quantum efficiency of PSII (Fv/Fm) is used to measure the stress in various abiotic stresses including heat stress (Baker and Rosenqvist 2004). It gives information of the maximum energy input into leaf and functions upstream of other potential stress responses. Decrease in Fv/Fm reflects the reduction of maximum quantum yield in photosynthesis (Ogren 1988). Chlorophyll a and b degradation during high temperature was reported more in matured leaves (Karim et al. 1997, 1999). Moreover, decrease in chlorophyll to carotenoids ratio and increase in chlorophyll a to b ratio was reported in tomato implicating that these changes would make the plants tolerant to high temperatures (Camejo et al. 2005; Wahid and Ghazanfar 2006). Chlorophyll a quantification is one of the important methods to measure plant response to heat stress (Maxwell and Johnson 2000; Baker and Rosenqvist 2004).

7.3 Photosynthetic Enzymes

High-temperature stress causes thermal instability of Rubisco (Feller et al. 1998) and impairs the process of carbon assimilation which in turn affects carbohydrate reserves. This leads to the shortening of grain filling duration by fastening the rate of development, and early induction of flag leaf senescence, thus causing reduction in total yield (Yang et al. 2002). Effect of high temperature is observed more in the photosynthetic activity of C3 plants than in C4 plants. Heat stress also affects the synthesis of other metabolites such as starch and sucrose, which are in turn greatly influenced by reduced activity of enzymes such as sucrose phosphate synthase (SPS; Chaitanya et al. 2001), invertase (Vu et al. 2001), and ADPglucose pyrophosphorylase (AGPase). The instability of AGPase, a key regulator of starch biosynthesis, results in starch accumulation during high-temperature stress (Singletary et al. 1994; Linebarger et al. 2005; Yamakawa et al. 2007). Other enzymes like branching enzymes and starch synthases are negatively regulated by heat (Singletary et al. 1994; Yamakawa et al. 2007).

8 Small NonCoding RNAs

Endogenous noncoding sRNAs of 21–25 nts in size are also involved in the plant stress response through a silencing mechanism. These are broadly classified into four categories: microRNAs (miRNAs), natural antisense transcripts small interfering RNAs (nat-siRNAs), trans-acting siRNAs (ta-siRNAs), and repeat-associated siRNAs (Jamalkandi and Masoudi-Nejad 2009). Mostly, sRNAs are produced from long precursor RNAs of double strands or single strand forming a self-complementary hairpin structures. These precursor RNAs are used as a substrate to produce sRNAs by four different dicer-like proteins (DCL) proteins, DCL1, 2, 3, and 4 (Chapman and Carrington 2007). miRNAs are generated from DCL1 protein; nat-siRNAs are produced by DCL1 and DCL2, while DCL3 is involved in the production of heterochromatic siRNAs. DCL4 is involved in the formation of ta-siRNAs. It has been well established that besides the protein-coding mRNAs, noncoding sRNAs play a key regulatory role in abiotic stress response in plants.

8.1 MicroRNAs

miRNAs are endogenous single-stranded sRNAs of 21–25 nt in size produced from single-stranded primary transcript termed as pri-miRNAs (Tang et al. 2008). These sRNAs regulate gene expression. The miRNAs get into the posttranscriptional gene silencing pathway, leading either to degradation of the target mRNA or to translational repression. miRNAs suppress the expression of target mRNA with the help of a protein complex known as RNA-induced silencing complex (RISC). Since the discovery of the first miRNA lin-4 in *Caenorhabditis elegans,* which regulates the larval timing during development (Lee et al. 2003; Reinhart et al. 2000), a large number of miRNAs have been reported and characterized in both plants and animals. Plant miRNAs were first reported in early 2002 (Llave et al. 2002; Park et al. 2002; Reinhart et al. 2002). Initially, miRNAs were considered to regulate largely transcription factor genes involved in a variety of plant developmental processes (Llave et al. 2002; Reinhart et al. 2002; Rhoades et al. 2002). A number of reports suggested a regulatory role of miRNAs in plant growth and development such as leaf and flower differentiation, flowering time, floral identity, and auxin response (Sunkar et al. 2005; Mallory and Vaucheret 2006). Further research in this area showed that these miRNAs play a vital role in regulating genes associated with abiotic and biotic stresses also (Sunkar and Zhu 2004; Sunkar et al. 2006; Mishra et al. 2009).

Plant miRNAs have been shown to regulate gene expression under heat stress also. Solexa sequencing revealed differential expression of miRNAs in wheat in response to heat (Xin et al. 2010). Out of the 32 identified miRNA families in wheat, 9 were conserved miRNAs and found to be putatively heat responsive. miRNAs miRs156, 159, 160, 166, 168, 169, 393, and 827 were upregulated under heat stress, whereas miR172 showed significant downregulation. In *Brassica rapa,* sRNA library was constructed from the seedlings exposed to high temperature. By deep sequencing, 35 miRNA families were found to be conserved with *A. thaliana*. Within those families, five miRNA families were heat stress responsive. Two miRNAs, miR398a and bra-miR398b, were downregulated during heat stress, whereas the corresponding targets genes—BracCSD1 (copper superoxide dismutase)—were upregulated. Similarly, miR156h and miR156g were upregulated by heat and their putative target BracSPL2 was downregulated (Yu et al. 2011). In *Populus,* solexa sequencing of two sRNA libraries generated from heat stress and control tissue revealed 52 heat-stress-responsive miRNAs from 15 families that included 16 novel miRNAs (Chen et al. 2012). In rice, 62 libraries of sRNAs were constructed and sequenced using Illumina sequencing. These sRNA libraries were constructed from control as well as stress tissues subjected to various stress treatments. Approximately 94 million reads matched with genome resulting in 16 million diverse sRNA sequences. Out of these, 400 were annotated miRNAs, 150 were siRNA like, and 76 new miRNAs were found. In this study, miRNAs involved in regulation in response to water, nutrient, and temperature stress were identified. miR397b.2 expression was upregulated and its target gene, L-ascorbate oxidase, was downregulated during heat stress (Jeong et al. 2011). Another miRNA, mir444, targets the Hsf-type transcription factor (Koskull-Doring et al. 2007). Similarly, miR169 targets nuclear transcription factor Y subunit (a drought-induced protein) and CCAAT-binding transcription factor (Li et al. 2010; Zhou et al. 2010).

8.2 Small Interfering RNAs

siRNAs are sRNAs involved in transcript silencing from which they originate (Bartel 2004). Double-stranded RNAs (dsRNAs) of diverse origins such as viruses, transposons, transgenes, etc. are cleaved into 21–24-nts siRNAs by multiple DCL proteins. These siRNAs are then loaded into RISC complex containing the argonaute protein. This complex binds to the mRNA from which they originate and silences its expression. Further, siRNAs bind to mRNA and convert single-stranded RNA into dsRNA by RNA-dependent RNA polymerase (RDRP), thus amplifying siRNA production. Change in the expression of four siRNAs in wheat seedlings during cold, heat, salt, and drought stresses was reported. Furthermore, siRNA007927_0100_2975.1 was downregulated by all stresses except heat stress (Yao et al. 2010) and the remaining three siRNAs—siRNA002061_0636_3054.1, siRNA 005047_0654_1904.1, and siRNA080621_1340_0098.1—were downregulated during heat stress conditions.

8.3 Trans-acting siRNAs

Ta-siRNAs (21-nt RNAs) are generated by processing of miRNA from TAS gene transcript, with respect to the miRNA cleavage site. In plants, the target mRNA expressed from ta-siRNA loci is cleaved by miR173 and miR390. After cleavage, they are modified into dsRNA by RDRP enzymes and processed into siRNAs which ultimately target the degradation of mRNA different from the transcript of ta-siRNA from which they originated. In *Arabidopsis,* miR173 recognizes TAS1 and TAS2 transcripts, whereas TAS3 and TAS4 are recognized by miR390 and miR828, respectively (Allen et al. 2005). Ta-siRNAs showed enhanced expression in hypoxia-treated tissues of *Arabidopsis* (Moldovan et al. 2009).

9 Approaches/Tools to Study Transcriptomics

Increase in the availability of complete genome sequence information from several plant species such as *Arabidopsis,* rice, *Populus,* chickpea, pigeonpea, etc. has paved the way to perform genome-wide function analysis of plant genes contributing to heat stress tolerance. Transcriptomic study of an organism can be performed by using different throughput methods like low, medium, and high (Fig. 3.1). Low-throughput method comprises single gene expression analysis, medium-throughput methods are ESTs and cDNA clones, and high-throughput methods include microarray and deep sequencing approaches. In order to perform single gene expression analysis, various techniques like northern blot analysis, RT-PCR, and qPCR can be used (Lockhart and Winzeler 2000). Genome-wide transcriptome analysis can be performed using high-throughput technologies such as microarray and NGS platforms. This has an advantage over individual gene analysis tools as interactions

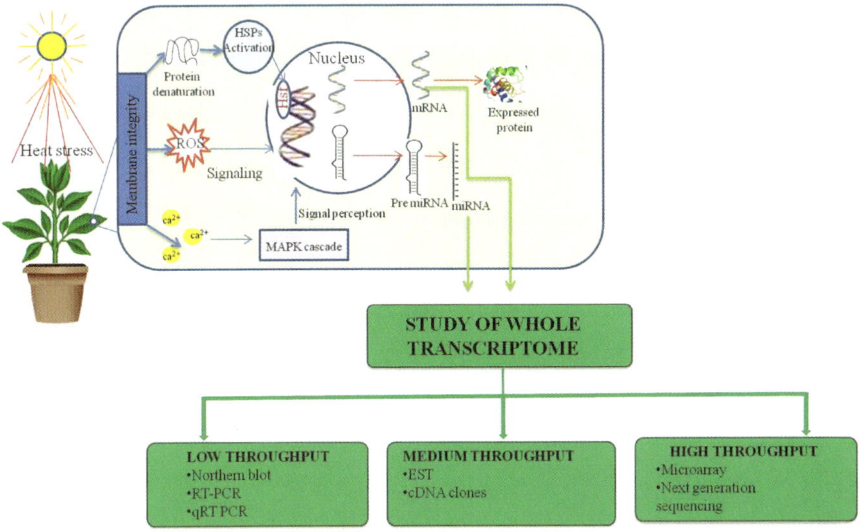

Fig. 3.1 An overview of molecular regulation of heat stress response of plants and approaches of transcriptomics

between different genes, pathway analysis, and action of regulatory elements can easily be understood by using high-throughput technologies. Several studies have used these methods in different plants to study the effect of heat stress on transcriptome (Table 3.1). In addition, recently, these tools have been successfully utilized in sRNA profiling of plants during abiotic stresses (Table 3.2).

9.1 Northern Blot Hybridization

Study of the expression of a particular gene during stressed conditions in plants can be achieved by using northern blotting. It involves the separation of RNA on agarose gel, transfer of RNA from the electrophoresed gel to Hybond-N^+ membrane, and detection by means of a hybridized probe, sequence which is complementary to a part or the entire target gene. This technique is generally used to detect the expression of targeted mRNA as well as sRNAs. It can be used to study the expression of only those genes for which sequence information is available. Northern blot analysis was used to demonstrate that OsBADH1 mRNA expression was upregulated by different abiotic stresses, whereas downregulated by heat stress (Hasthanasombut et al. 2011). The expression of heat-inducible genes such as Hsf (HsfB2a), HSP (HSP101), and cytosolic ascorbate peroxidase (APX1) was reduced in wrky25 mutants. Overexpression of WRKY25 enhanced the expression level of heat-responsive genes HsfA2, HsfB1, HsfB2a, and HSP101 (Li et al. 2009). Also, expression pattern of glutathione reductases was studied in rice, wheat, barley, and maize during photooxidative stress using northern blot analysis (Melchiorre et al. 2009).

Table 3.1 Heat-stress-induced transcriptomics studies of plants using various approaches/tools

S. No	Plant species	Approach used	Differentially regulated genes	Reference
1	Rice	RT-PCR	Spl7 (rice spotted leaf gene)	Yamanouchi et al. 2002
2	Rice	RT-PCR	*OsHsfA4b, OsHsfA5, OsHsfA7, OsHsfA4d, OsHsf A2a, OsHsfA2c, and OsHsfA2d*	Liu et al. 2010
3	Rice	RT-PCR	*OsHSP80.2, OsHSP74.8, OsHSP71.1, OsHSP26.7, OsHSP24.1,OsHSP17.0, OsHSP58.7, OsHSP50.2, and OsHSP23.7*	Zou et al. 2009
4	Rice	RT-PCR	*HSP70* gene	Goswami et al. 2010
5	Rice	Microarray	Hsfs, HSPs, chitinase, cellulase, cell wall invertase 2, beta-expansin, chalcone synthase, and isoflavone reductase family protein genes	Zhang et al. 2008b
6	Rice	Microarray	Hsfs, sHSPs, members of HSP70, HSP90, and HSP100 gene families	Hu et al. 2009
7	Rice	Microarray	*OsHsfA1a, OsHsfA2a, OsHsfA2c, OsHsfA2d, OsHsfA2f, OsHsfA4b, OsHsfB2a, OsHsfB4b,* and *OsHsfC1a* genes	Mittal et al. 2009
8	Rice	Microarray	*HSP17.4-CI, HSP17.9B-CIX, HSP23.2-ER, HSP18.6-CI, HSP24.0-MI, HSP26.2-MI, HSP16.6-CVIII, HSP16.9C-CI,* and *HSP18.0-CII*	Sarkar et al. 2009
9	Rice	Microarray	*OsClpB-cyt, OsClpB-m,* and *OsClpB-c*	Singh et al. 2010
10	Rice	Microarray	*OsHsfA2a, OsHsfA2c, OsHsfA2d, OsHsfB2a, OsHsfB2b, OsHsfB2c,* and *OsHsfC1a*	Chauhan et al. 2011a
11	Rice	Microarray	Hsfs, bZIP TFs, *HSP10s,* and *HSP20s*	Jung et al. 2013
12	Rice	Microarray	*sHSP (HSP17.4), HSP30, HSP70, HSP90, cytochrome P450,* and *CBL-1* gene	Mittal et al. 2012
13	Rice	Microarray	*HSP20, HSP40, HSP70, HSP90, clpB 1, HSP101, OsSTI1, OsSTI2a,* and phosphosulfolactate synthase related	Jung and An 2012
14	Rice	Microarray	*HsfA2a, HsfA2d, HsfA2f, HsfA3, HsfB2a, Hsfb, Hsfc, DREB, ERF,* and members of HSP70, HSP90, and HSP100	Zhang et al. 2012
15	*Arabidopsis*	Microarray	*HsfHsfA2, HsfB1, Hsf-A4a, HsfB2a, HsfB2b,* and *HsfA7a*	Busch et al. 2005
16	*Arabidopsis*	Microarray	*DREB2A, DREB2B, DREB2C,* and *DREB2H*	Lim et al. 2006

Table 3.1 (continued)

S. No	Plant species	Approach used	Differentially regulated genes	Reference
17	*Arabidopsis*	Microarray	*HSP90, HSP70, and HSP101*	Yamada et al. 2007
18	*Arabidopsis*	Microarray	Members of *HSP20, HSP70, HSP90,* and *HSP100*	Swindell et al. 2007
19	*Arabidopsis*	Microarray	*HSP101, HSE7, APX2, HSFA7, NF-XI, Pro oxidase, SGT1a HSP110* (HSP70-15), and choline kinase	Larkindale and Verling 2008
20	*Arabidopsis*	Illumina sequencing	*SR45a*	Gulledge et al. 2012
21	*Arabidopsis*	qRT-PCR	Rubisco's Chaperone activase (RCA), *AtRCAβ2, AtRCAα,* and *AtRCAβ*	Deridder et al. 2012
22	*Arabidopsis*	Illumina	*HSP25.3-P* and *HSP22.0-ER*	Li et al. 2013
23	Wheat	Microarray	*Hsfs, HSPs, DREB2B and DREB6A, ERETC,* and member of *MBF1*	Qin et al. 2008
24	Wheat	Microarray	b-ZIP transcription factors and TaCAM3-1(zinc finger with calmodulin)	Chauhan et al. 2011b
25	Wheat	Microarray	HSPs, transporters, protein modifiers, and signaling molecules	Khurana et al. 2011
26	Maize	qRT-PCR	*ZmHsf-01, ZmHsf-03, ZmHsf-04, ZmHsf-06, ZmHsf-10, ZmHsf-11, ZmHsf-14, ZmHsf-15, ZmHsf-19, ZmHsf-20, ZmHsf-21, ZmHsf-22, ZmHsf-23, ZmHsf-24, and ZmHsf-25*	Lin et al. 2011
27	Maize	EST	*HSP22*	Lund et al. 1998
28	Tomato	Microarray	sHSP genes, members of *HSP70, HSP101,* and *HSP90* families, *HSFA2* and *HSFA3*	Frank et al. 2009
29	Tomato	Microarray	Class I sHSP 17.6, class II sHSP 17.6, class III sHSP, DNA-J and mitochondrial sHSP, protein similar to AthHSP22.3, and cytosolic ascorbate peroxidase	Bita et al. 2011
30	Barley	Microarray	Raffinose synthase 1, UDP-D-glucose 4-epimerase 1, UDP-D-glucose 4-epimerase 3, trehalose-6-phosphate synthase, trehalose-6-phosphate phosphatase, invertase inhibitor, heat shock transcription factor A2d, hexokinase 2, and SNF1-related protein kinases 2.6	Mangelsen et al. 2011
31	*Populus*	RT-PCR	*Hsf-03, Hsf-13, Hsf-15, Hsf-2, Hsf-22,* and *Hsf-23*	Wang et al. 2011

Table 3.2 Expression studies of plant microRNAs during abiotic stresses using different transcriptomic approaches

S. No	Stress	Plant species	Approach	Expressed miRNAs (up-/downregulated)	References
1	Heat	Wheat	Solexa sequencing	miR156, miR159, miR160, miR166, miR168, miR169, miR172, miR827, and miR2005	Xin et al. 2010
2	Heat	*Brassica*	Illumina sequencing	miR156h, miR398a, miR398b, miR399b, and miR827	Yu et al. 2011
3	Heat	*Populus*	Illumina sequencing	miR156, miR166, miR167, miR168, miR396, miR397, miR408, miR1444, miR473, miR530, miR160, miR394, miR395, miR408, miR472, miR482, miR530, and miR1450	Chen et al. 2012
4	Heat	*Arabidopsis*	RNA sequencing	miR156/miR157 and miR172	May et al. 2013
5	Heat	Rice	MPSS	miR397b	Jeong et al. 2011
6	Cold	*Brachypodium*	Deep sequencing	miR169e, miR172b, and miR397	Zhang et al. 2009
7	Cold	*Arabidopsis*	Microarray	miR156, miR159, miR164, miR165, miR169, miR172, miR393, miR394, miR396, miR397, and miR398	Zhou et al. 2008
8	Cold	Rice	Microarray	miR156k, miR166k, miR166m, miR167a/b/c, miR168b, miR169e, miR169f, miR169h, miR171a, miR535, miR319a/b, miR1884b, miR444a.1, miR1850, miR1868, miR1320, miR1435, and miR1876	Lv et al. 2010
9	Drought	Rice	Microarray	miR169g, miR171a, and miR393	Jian et al. 2010; Zhou et al. 2010
10	Drought	*Arabidopsis*	Microarray	miR157, miR167, miR168, miR171, miR408, miR393, and miR396	Liu et al. 2008
11	Drought	*Medicago*	Microarray	miR398a/b and miR408	Trindade et al. 2010
12	Drought	*Populus*	Microarray	miR1446a-e, miR1444a, miR1447, and miR1450	Lu et al. 2008

Table 3.2 (continued)

S. No	Stress	Plant species	Approach	Expressed miRNAs (up-/downregulated)	References
13	Drought	*Prunus persica*	Deep sequencing	miR156, miR157, miR159, miR160, miR165, miR167, miR168, miR169, miR171, miR390, miR393, miR395, miR396, miR397, miR398, and miR408	Eldem et al. 2012
14	Drought	*Populus*	Illumina sequencing	miR159a-c, miR472a, miR472b, miR473a, miR160a-d, miR164a-e, miR394a/b-5p, miR408, and miR1444b-c	Shuai et al. 2013
15	Salinity	Rice	Northern blot	miR169g, miR169n, and miR169o	Zhao et al. 2009
16	Salinity	Maize	Microarray	miR162, miR168, miR395, and miR47	Ding et al. 2009a

9.2 Reverse Transcription PCR

RT-PCR technique is commonly used to detect RNA expression. Here, the amplification using cDNA derived from mRNA template can be highly specific, fast, and sensitive (McDowell et al. 1996; Wang et al. 1999). It is useful to detect gene expression qualitatively through synthesizing cDNA from RNA. To know the presence and expression level of rice spotted leaf gene (spl7), a transgene in rice, RT-PCR was performed which suggested an increase in spl7 mRNA levels with increase in temperature (Yamanouchi et al. 2002). During the study of effect of abscisic acid-insensitive$_3$(ABI$_3$) on HsfA9 and HSP accumulation in *Arabidopsis,* RT-PCR revealed that HsfA9 transcripts are not inducible by heat stress and could not be detected in mutant line, whereas the synthesis of other HSP transcripts such as HSP17.4-CI, HSP17.6A-CI, HSP17.6-CII, HSP17.7-CII, and HSP101 was induced during heat stress (Kotak et al. 2007). In *Arabidopsis,* expression analysis of different members of HSP70 in response to temperature stress was reported using RT-PCR analysis (Sung et al. 2001).

9.3 Real-Time PCR Analysis

Real-time PCR, also called qPCR, is used to detect a gene quantitatively, where amplification and simultaneous quantification of a particular target gene can be achieved. In other words, it enables both detection and quantification. Expression and quantification of both mRNA and sRNAs can be achieved by using this technique. Plenty of reports have utilized real-time expression analysis of genes during heat stress. Overexpression of HsfA1 gene in transgenic *Arabidopsis,* soybean, and tomato showed upregulation of stress-regulated genes and increased thermo-tolerance (Lohmann et al. 2004; Zhu et al. 2006). Expression of OsHsfA2a gene was greatly induced by heat stress. During high temperatures and high light conditions, it has been reported that the expression of HsfA2 was maximum among all the class A Hsfs in *Arabidopsis* (Schramm et al. 2006; Nishizawa et al. 2007). Apart from these transcription factors, expression of other heat-stress-responsive genes has been reported using qPCR. In rice, increased expression of nine OsHSPs during heat shock was demonstrated using real-time PCR (Zou et al. 2009). Further, expression of HSP70 genes was increased by 2- to 20-fold in *Arabidopsis* during heat stress (Sung et al. 2001). Downregulation of miR398a and b and upregulation of their corresponding targets CSD1 and CSD2 was shown in *B. rapa* and *Populus* under heat stress (Yu et al. 2011; Chen et al. 2012). Elevated expression of miRNAs under high-temperature stress was shown in plants. Expression of miR169j was upregulated in Populus, whereas in wheat, the members of miR156 family were upregulated during heat stress. Similar expression pattern of miR156s and its corresponding target SPL2 was recorded in *B. rapa* (Yu et al. 2011).

9.4 cDNA Clones and ESTs

ESTs are generated by high-throughput single-pass sequencing of cDNA clones (Adams et al. 1991). RNA (mRNA) from desired tissue or organism has to be isolated and transcribed to cDNA, using reverse transcriptase enzyme. cDNA thus formed is cloned to make libraries (cDNA libraries) representing the set of transcribed genes of our interest. Further, these cDNA clones are sequenced for obtaining ESTs. In rice, cDNA-AFLP analysis of heat-tolerant and heat-sensitive rice lines revealed 54 differentially expressed transcript-derived fragments (TDFs) playing an important role during heat stress in tolerant cultivar. Functions of 28 of the 54 TDFs were annotated based on their homologous genes. These genes were classified into different groups: metabolism, biosynthesis, transport, transcriptional regulation, oxidation, and signal transduction (Liao et al. 2012). To date, more than 2 million plant-derived ESTs are available at public databases from various species and these data are useful as a rich source for gene discovery and annotation (Rudd 2003). In addition, in some cases, genome sequencing is less attractive for plants such as maize due to genome complexity and abundance of repetitive sequence elements (Bennetzen 2002). In such cases, EST data set sources can be treated as an alternative to the complete genome sequencing (Rudd 2003). Plant ESTs are versatile and have multiple functions. They are very much useful in gene discovery, the expression of a particular gene, and the extent of its expression during particular stress. In addition, plant ESTs are a very useful resource for the identification of gene families which are conserved among different plant species, to detect alternative splicing, genome structure, discovery and characterization of SNPs, and mapping of gene-based site markers (Dong et al. 2005; Nagaraj et al. 2007). An efficient and faster way to identify novel genes induced by various environmental stresses using EST approach has been demonstrated (Markandeya et al. 2005; Gorantla et al. 2007). In maize, mitochondrial small heat shock protein HSP22 was induced during heat stress which was identified and cloned using this approach (Lund et al. 1998). cDNA library was constructed and more than 30,000 ESTs were sequenced in *Populus* subjected to different environmental stresses. This helped in analyzing cDNAs encoding for the ERF/AP2 domain transcription factor (Nanjo et al. 2004). Recent study in *Jatropha curcas,* two HSP genes—HSP-1 and HSP-2—were identified and characterized from developing seeds using the EST approach (Omar et al. 2011). In rice, high-density physical maps were developed based on the availability of EST resources of rice cultivar N22 (Markandeya et al. 2005). Another study in rice identified 589 candidate genes sharing the drought response through EST approach by comparative analysis of plants species. Comparative analysis with five EST libraries revealed that expression of tentative Uni Genes (TUGs) was significantly overexpressed in *Zostera noltii* (seagrass) during recovery from heat shock exposure. These were identified as molecular chaperones (Massa et al. 2011).

9.5 Microarray

Microarray is a hybridization-based technique where nucleic acid is hybridized to a probe. Basically, two microarray-based approaches are available—one is oligonucleotide based (in situ synthesis of oligonucleotides) and the other is cDNA based (depositing of DNA fragments on solid surface). The former is termed as the gene chip method commercially prepared by Affymetrix having 25-bp-long oligonucleotides complementary to the 3′ end of the expressed sequence in the genome (Aharoni and Vorst 2001), whereas the later one consists of PCR-amplified cDNA fragments spotted on the glass surface. After the discovery of microarray, significant progress has been achieved in plant biology to characterize transcript expression profiling during stress. To decipher the changes in plant transcriptome under heat stress, microarray studies were carried out in many of the model crop species such as *Arabidopsis,* rice, wheat, maize, soybean, and tomato. Transcript profiling of *A. thaliana* revealed that a significant number of HsfA1a/1b-regulated genes were expressed during the study of mutants and wild species during heat stress (Busch et al. 2005). Hu et al. (2009) studied the regulation of Hsfs and HSPs during different abiotic stresses including heat. The important role of Hsfs in regulation of downstream targets including HSPs under heat stress was revealed using microarray (Mittal et al. 2009). Genome-wide expression profiling of Hsfs demonstrated an important role of OsHsfA2a in HSR in root, shoot, and panicle tissues of rice (Chauhan et al. 2011). In *Arabidopsis,* differential response of genome to photoperiod and thermal induction was demonstrated using microarray. The study suggested that a slight rise in ambient temperature beyond the optimum laboratory conditions triggers flowering of *Arabidopsis* in the absence of photoperiodic cues which is in turn governed by a set of genes (Balasubhramanian et al. 2006). While exposing barley seedlings to high temperature, many of the genes were found to be upregulated during the seedling stage rather than the panicle stage (Oshino et al. 2011). A similar study in wheat revealed that the altered gene expression was more during short-term stress rather than prolonged heat stress (Qin et al. 2008). Microarrays were also used extensively to study the gene response mechanisms in other abiotic stresses such as drought, cold, and salinity. Expression profiling of sRNAs using microarray has been reported in many crop species like *Arabidopsis* (Zhou et al. 2008), rice (Lv et al. 2010), *Populus* (Lu et al. 2008), *Medicago* (Trindade et al. 2010), maize (Ding et al. 2009), etc. Thus, the application of microarrays in the area of plant transcriptome is very wide. It is a highly useful tool to study the expression, regulation, and role of different genes/pathways involved in stress tolerance mechanisms.

9.6 Next-Generation Sequencing

Sequencing refers to the identification of nucleotides in DNA or RNA and it provides a better option for gene expression studies. Sequencing of whole transcriptome includes isolation of total RNA from tissue of interest and enrichment of mRNA by reducing rRNA abundance followed by construction of cDNA library from total mRNA, sequencing using PCR amplification from either one end (single-end) or both ends (paired-end). The sequenced read length may vary from 30 to 400 nucleotides depending on the sequencing platform used. The sequencing technologies available for transcriptome sequencing include Roche 454 (pyrosequencing-based), Illumina/Hiseq, and Sequencing by Oligo Ligation and Detection (SOLiD) sequencing (Varshney et al. 2009). In addition to above-mentioned NGS technologies, single-molecule sequencing (SMS) can also be used to sequence the nucleic acids. More number of reads helps in the accurate measurement of transcript expression. Usually, shorter read platforms provide deep sequencing and good coverage, which is highly useful in detecting rare transcripts (Jain 2012). NGS is a very useful tool to decipher sRNA profiling also. Furthermore, identification of novel transcripts, transcript isoforms, alternative splice variants, and differentially expressed transcripts in a genome during stress can be identified using this tool. It is very useful in few other applications like single nucleotide polymorphism (SNP) detection, mutational analysis, transcriptional regulation, and identification of RNA editing sites.

Development of the NGS technology has provided a novel method of transcriptome studies under stress conditions. The availability of the complete genome sequence for several plant species facilitates expression profiling of the whole transcriptome using this tool. Whole genome transcript profile of two subspecies of rice was unraveled through sequencing technology, where 15,708 novel transcriptional and 3,464 differentially expressed genes were reported (Lu et al. 2010). Furthermore, whole genome transcript expression (including sRNAs) and integrated epigenome analysis were unraveled in two subspecies of rice and their reciprocal hybrids using Illumina technology. This was aimed to study epigenetic variations in different genetic backgrounds that lead to phenotypic variability (He et al. 2011). NGS has great potential to be utilized in understanding the transcriptome and gene regulation during heat stress in various plant species. Recent study using NGS revealed that stress-regulated miRNAs play a crucial role during the inflorescence stage in rice (Barrera-Figueroa et al. 2012). Using sequencing technology, heat-stress-responsive miRNAs have been studied in different crops like rice (Jeong et al. 2011), *Brassica* (Yu et al. 2011), *Populus* (Chen et al. 2012), and wheat (Xin et al. 2010). Apart from the heat-stress-regulated miRNAs, sRNAs involved in other abiotic stresses were also studied using NGS (Eldem et al. 2012; Shuai et al. 2013). The revolution in sequencing technology enables the researchers to resolve the ambiguity towards the plant genome in understanding genomic, development, physiological, and evolutionary processes.

10 Conclusion and Future Perspectives

Heat stress shows serious impact on plant growth and development. It affects physiological, morphological, biochemical, and cellular mechanisms of plants. High temperature may lead to cell membranes disintegration, disturbing leaf water relations, and retarded photosynthesis due to the increased concentrations of ROS. In response to heat stress, plants show various adaptation strategies through the induction of Hsfs signaling and accumulation of HSPs, increased production of antioxidant compounds, synthesis of various osmolytes, and activation of signal transduction cascades through induction of signaling molecules. However, not all the plants are equally efficient in employing these adaptation strategies. Plant species show a high degree of variation in stress tolerance and they differ significantly in adaptation capability during high-temperature stress. The difference in heat stress tolerance at cellular or biochemical level is ultimately a response of different genetic constitution of different organisms/species. Further, difference in genetic constitution leads to change in the expression of genes and sRNAs involved in synthesis of metabolites associated with heat stress tolerance. Hence, to understand the heat response at molecular level, transcriptome analysis is very much crucial to unravel the complexity of mechanisms involved in heat stress tolerance. Study of the transcriptome may be aimed at specific genes or whole transcript level which can be achieved by various low- and high-throughput approaches. The progress in transcriptome analysis has enabled researchers to identify entire gene sets and pattern of expression, regulation, and function of potential transcripts. Use of NGS tools in transcriptome analysis has facilitated the identification of SNPs, splice junction sites and splice variants, differentially expressed genes (DEGs), and novel genes. The fast progress in transcriptome studies has helped biologists to understand the regulatory mechanisms of HSR. Particularly, transcriptome studies could establish the overlapping response of plants to various abiotic and biotic stresses. The integration of gene expression data with proteomic studies and sRNA profiling will help in understanding the network of biological processes operated during stress. With the progress in transcriptome tools, such as direct RNA sequencing, heat-stress-associated gene expression and regulation mechanisms will be further advanced.

Acknowledgments The authors are highly thankful to the project director (Directorate of Rice Research) for his kind support. Financial support received from NICRA (National Initiative of Climate and Resilient Agriculture) project is highly acknowledged.

References

Adams MD, Kelley JM, Gocayne JD, Dubnick M, Polymeropoulos MH, Xiao H, Merril CR, Wu A, Olde B, Moreno RF, Kerlavage AR, McCombie WR, Venter JC (1991) Complementary DNA sequencing: expressed sequence tags and human genome project. Science 252:159–164

Adams SR, Cockshull KE, Cave CRJ (2001) Effect of temperature on the growth and development of tomato fruits. Ann Bot 88:869–877

Agarwal M, Sahi C, Katiyar-Agarwal S, Agarwal S, Young T, Gallie DR, Sharma VM, Ganesan K, Grover A (2003) Molecular characterization of rice hsp101: complementation of yeast hsp104 mutation by disaggregation of protein granules and differential expression in indica and japonica rice types. Plant Mol Biol 51(4):543–553

Aharoni A, Vorst O (2001) DNA microarrays for functional plant genomics. Plant Mol Biol 48:99–118

Allakhverdiev SI, Kreslavski VD, Klimov VV, Los DA, Carpentier R, Mohanty P (2008) Heat stress: an overview of molecular responses in photosynthesis. Photosyn Res 98(1–3):541–550

Allen Z, Xie AM, Gustafson JC, Carrington JC (2005) MicroRNA-directed phasing during trans-acting siRNA biogenesis in plants. Cell 121:207–221

Anisimov SV (2008) Serial analysis of gene expression (SAGE): 13 years of application in research. Curr Pharm Biotechnol 9:338–350

Aono A, Kubo A, Saji H, Tanaka K, Kondo N (1993) Enhanced tolerance to photooxidative stress of transgenic Nicotiana tabacum with high chloroplastic glutathione reductase activity. Plant Cell Physiol 34:129–135

Apel K, Hirt H (2004) Reactive oxygen species: metabolism, oxidative stress, and signal transduction. Annu Rev Plant Biol 55:373–399

Apuya NR, Yadegari R, Fischer RL, Harada JJ, Zimmerman JL, Goldberg RB (2001) The Arabidopsis embryo mutant schlepperless has a defect in the chaperonin-60α Gene. Plant Physiol 126(2):717–730

Asada K (2000) The water-water cycle as alternative photon and electron sinks. Phil Trans R Soc Lond B Biol Sci 355:1419–1431

Asada K (2006) Production and scavenging of reactive oxygen species in chloroplasts and their functions. Plant Physiol 141:391–396

Ashraf M, Foolad MR (2007) Roles of glycine betaine and proline in improving plant abiotic stress resistance. Environ Exp Bot 59:206–216

Ashraf M, Saeed MM, Qureshi MJ (1994) Tolerance to high temperature in cotton (*Gossypium hirsutum* L.) at initial growth stages. Environ Exp Bot 34:275–283

Azpilicueta CE, Benavides MP, Tomaro ML, Gallego SM (2007) Mechanism of CATA3 induction by cadmium in sunflower leaves. Plant Physiol Biochem 45:589–595

Baker NR, Rosenqvist E (2004) Applications of chlorophyll fluorescence can improve crop production strategies: an examination of future possibilities. J Exp Bot 55(403):1607–1621

Balasubramanian S, Sureshkumar S, Lempe J, Weigel D (2006) Potent induction of *Arabidopsis thaliana* flowering by elevated growth temperature. Plos Genet 2:106

Banon S, Fernandez JA, Franco JA, Torrecillas A, Alarcon JJS, Anchez-Blanco MJ (2004) Effects of water stress and night temperature preconditioning on water relations and morphological and anatomical changes of *Lotus creticus* plants. Sci Hortic 101:333–342

Barrera-Figueroa BE, Gao L, Wu ZG, Zhou XF, Zhu JH, Jin HL, Liu RY, Zhu JK (2012) High throughput sequencing reveals novel and abiotic stress-regulated microRNAs in the inflorescences of rice. BMC Plant Biol 12:132

Barry G (2001) The use of the Monsanto draft rice genome sequence in research. Plant Physiol 125:1164–1165

Bartel DP (2004) MicroRNAs: genomics, biogenesis, mechanism and function. Cell 116:281–297

Bedell JA, Budiman MA, Nunberg A, Citek RW, Robbins D, Jones J, Flick E, Rholfing T, Fries J, Bradford K, McMenamy J, Smith M, Holeman H, Roe BA, Wiley G, Korf IF, Rabinowicz PD, Lakey N, McCombie WR, Jeddeloh JA (2005) Sorghum genome sequencing by methylation filtration. Plos Biol 3:103–115

Bennetzen JL (2002) Mechanisms and rates of genome expansion and contraction in flowering plants. Genetica 115:29–36

Bita CE, Zenoni S, Vriezen WH, Mariani C, Pezzotti M, Gerats T (2011) Temperature stress differentially modulates transcription in meiotic anthers of heat-tolerant and heat-sensitive tomato plants. Bio Med Cen 12:384

Blum A (1988) Plant breeding for stress environments. CRC Press, Boca Raton, p 223

Blum A, Klueva N, Nguyen HT (2001) Wheat cellular thermotolerance is related to yield under heat stress. Euphytica 117:117–123

Bohnert HJ, Gong Q, Li P, Ma S (2006) Unraveling abiotic stress tolerance mechanisms getting genomics going. Curr Opin Plant Biol 9:180–188

Bonos SA, Murphy JA (1999) Growth responses and performance of Kentucky bluegrass under summer stress. Crop Sci 39(3):770–774

Boston RS, Viitanen PV, Vierling E (1996) Molecular chaperones and protein folding in plants. Plant Mol Bio 32:191–222

Bowen J, Michael LY, Plummer KIM, Ferguson IAN (2002) The heat shock response is involved in thermotolerance in suspension-cultured apple fruit cells. J Plant Physiol 159:599–606

Burke JJ (2001) Identification of genetic diversity and mutations in higher plant acquired thermotolerance. Physiol Plant 112:167–170

Busch W, Wunderlich M, Schoffl F (2005) Identification of novel heat shock factor-dependent genes and biochemical pathways in *Arabidopsis thaliana*. Plant J 41:1–14

Camejo D, Rodrıguez P, Morales MA, Dellamico JM, Torrecillas A, Alarcon JJ (2005) High temperature effects on photosynthetic activity of two tomato cultivars with different heat susceptibility. J Plant Physiol 162:281–289

Chaitanya KV, Sundar D, Reddy AR (2001) Mulberry leaf metabolism under high temperature stress. Biol Plant 44:379–384

Chapman EJ, Carrington JC (2007) Specialization and evolution of endogenous small RNA pathways. Nat Rev Genet 8:884–896

Charng YY, Liu HC, Liu NY, Chi WT, Wang CN, Chang SH, Wang TT (2007) A heat-inducible transcription factor, HsfA2, is required for extension of acquired thermotolerance in Arabidopsis. Plant Physiol 143(1):251–262

Chauhan H, Khurana N, Agarwal P, Khurana P (2011a) Heat shock factors in rice (*Oryza sativa* L.) genome-wide expression analysis during reproductive development and abiotic stress. Mol Genet Gen 286:171–187

Chauhan H, Khurana N, Tyagi AK, Khurana JP, Khurana P (2011b) Identification and characterization of high temperature stress responsive genes in bread wheat (*Triticum aestivum* L.) and their regulation at various stages of development. Plant Mol Biol 75:35–51

Chen L, Ren Y, Zhang Y, Xu J, Sun F, Zhang Z, Wang Y (2012) Genome-wide identification and expression analysis of heat-responsive and novel microRNAs in *Populus tomentosa*. Gene 504:160–165

Collins A (2001) Carotenoids and genomic stability. Mutat Res 475:1–28

De Ronde JAD, Cress WA, Kruger GHJ, Strasser RJ, Staden JV (2004) Photosynthetic response of transgenic soybean plants containing an Arabidopsis *P5CR* gene, during heat and drought stress. J Plant Physiol 61:1211–1244

Del Rio LA, Palma JM, Sandalio LM, Corpas FJ, Pastori GM, Bueno P, Lopez-Huertas E (1996) Peroxisomes as a source of superoxide and hydrogen peroxide in stressed plants. Biochem Soc Trans 24:434–438

DeRidder BP, Shybut ME, Dyle MC, Kremling KAG, Shapiro MB (2012) Changes at the 3′-untranslated region stabilize rubisco activase transcript levels during heat stress in Arabidopsis. Planta 236:463–476

Ding D, Zhang L, Wang H, Liu Z, Zhang Z, Zheng Y (2009a) Differential expression of miRNAs in response to salt stress in maize roots. Ann Bot 103(1):29–38

Ding S, Lu Q, Zhang Y, Yang Z, Wen X, Zhang L, Lu C (2009b) Enhanced sensitivity to oxidative stress in transgenic tobacco plants with decreased glutathione reductase activity leads to a decrease in ascorbate pool and ascorbate redox state. Plant Mol Biol 69:577–592

Dong Q, Kroiss L, Oakley FD, Wang BB, Brendel V (2005) Comparative EST analyses in plant systems. Methods Enzymol 395:400–418

Ebrahim MK, Zingsheim O, El-Shourbagy, MN, Moore PH, Komor E (1998) Growth and sugar storage in sugarcane grown at temperature below and above optimum. J Plant Physiol 153:593–602

Eldem V, Çelikkol Akçay U, Ozhuner E, Bakır Y, Uranbey S, Unver T (2012) Genome-wide identification of miRNAs responsive to drought in peach (*Prunus persica*) by high-throughput deep sequencing. PLos One 7(12):e50298. doi:10.1371journal.pone.0050298 (Epub)

Eltayeb AE, Kawano N, Badawi GH, Kaminaka H, Sanekata T, Morishima I, Shibahara T, Inanaga S, Tanaka K (2006) Enhanced tolerance to ozone and drought stresses in transgenic tobacco overexpressing dehydroascorbate reductase in cytosol. Physiol Plant 127:57–65

Eveland AL, Satoh-Nagasawa N, Goldshmidt A, Meyer S, Beatty M, Sakai H, Ware D, Jackson D (2010) Digital gene expression signatures for maize development. Plant Physiol 154:1024–1039

Feller U, Crafts-Brandner SJ, Salvucci ME (1998) Moderately high temperatures inhibit ribulose-1,5-bisphosphate carboxylase/oxygenase (rubisco) activase-mediated activation of rubisco. Plant Physiol 116(2):539–546

Ferris R, Ellis RH, Wheeeler TR, Hadley P (1998) Effect of high temperature stress at anthesis on grain yield and biomass of field grown crops of wheat. Plant Cell Environ 34:67–78

Foyer CH, Halliwell B (1976) The presence of glutathione and glutathione reductase in chloroplasts: a proposed role in ascorbic acid metabolism. Planta 133:21–25

Foyer CH, Noctor G (2005) Redox homeostis and antioxidant signaling: a metabolic interface between stress perception and physiological responses. Plant Cell 17:1866–1875

Foyer CH, Souriau N, Perret S, Lelandais M, Kunert KJ, Pruvost C, Jouanin L (1995) Overexpression of glutathione reductase but not glutathione synthetase leads to increases in antioxidant capacity and resistance to photoinhibition in poplar trees. Plant Physiol 109:1047–1057

Frank G, Pressman E, Ophir R, Althan L, Shaked R, Freedman M, Shen S, Firon N (2009) Transcriptional profiling of maturing tomato (*Solanum lycopersicum* L.) microsporesreveals the involvement of heat shock proteins, ROS scavengers, hormones, and sugars in the heat stress response. J Exp Bot 60:3891–3908

Frugoli JA, Zhong HH, Nuccio ML, McCourt P, McPeek MA, Thomas TL, McClung CR (1996) Catalase is encoded by a multigene family in *Arabidopsis thaliana* (L.). Plant Physiol 112:327–336

Garber M, Grabherr MG, Guttman M, Trapnell C (2011) Computational methods for transcriptome annotation and quantification using RNA-Seq. Nat Methods 8:469–477

Gill SS, Tuteja N (2010) Reactive oxygen species and antioxidant machinery in abiotic stress tolerance in crop plants, India. J Plant Physiol Bioch 48:909–930

Gong M, Chen SN, SongYQ, Li ZG (1997) Effect of calcium and calmodulin on intrinsic heat tolerance in relation to antioxidant systems in maize seedlings. Aust J Plant Physiol 24:371–379

Gorantla M, Babu PR, Lachagari VB, Reddy AM, Wusirika R, Bennetzen JL, Reddy AR (2007) Identification of stress-responsive genes in an indica rice (*Oryza sativa* L.) using ESTs generated from drought-stressed seedlings. J Exp Bot 58(2):253–265

Goswami A, Banerjee R, Raha S (2010) Mechanism of plant adaptation/memory in rice seedling under arsenic and heat stress: expression of heats-shock protein gene HSP70. AoB Plants. doi:10.1093/aobpl/plq023

Guilioni L, Wery J, Tardieu F (1997) Heat stress-induced abortion of buds and flowers in pea: is sensitivity linked to organ age or to relations between reproductive organs? Ann Bot 80:159–168

Gulledge AA, Roberts AD, Vora H, Patel K, Loraine AE (2012) Mining *Arabidopsis thaliana* RNA-seq data with integrated genome browser reveals stress-induced alternative splicing of the putative splicing regulator SR45a. Am J Bot 99:219–231

Gupta SC, Sharma A, Mishra M, Mishra R, Chowdhuri DK (2010) Heat shock proteins in toxicology: how close and how far? Life Sci 86:377–384

Hall AE (2001) Crop responses to environment. CRC Press LLC, Boca Raton

Hamilton EW, Heckathorn SA, Downs CA, Schwarz TE, Coleman JS, Hallberg RL (1996) Heat shock proteins are produced by field-grown naturally occurring plants in the summer in the temperate northeast. US Bulletin of the Ecologic Soc Am.77, Suppl. Part 2:180 (Abstr.)

Hart FU (1996) Molecular chaperones in cellular protein folding. Nature 381:571–580

Hare PD, Cress WA, Staden JV (1998) Dissecting the roles of osmolyte accumulation during stress. Plant Cell Environ 21:535–553

Hasthanasombut S, Paisarnwipatpong N, Triwitayakorn K, Kirdmanee C, Supaibulwatana K (2011) Expression of OsBADH1 gene in indica rice (*Oryza sativa* L.) in correlation with salt, plasmolysis, temperature and light stresses. Plant Omics 4(7):75–81

He G, Zhu X, Elling AA, Chen L, Wang X, Guo L, Liang M, He H, Zhang H, Chen F, Qi Y, Chen R, Deng XW (2011) Global epigenetic and transcriptional trends among two rice subspecies and their reciprocal hybrids. Plant Cell 22:17–33

Hong SW, Vierling E (2000) Mutants of *Arabidopsis thaliana* defective in the acquisition of tolerance to high temperature stress. Proc Natl Acad Sci U S A 97(8):4392–4397

Howarth CJ (2005) Genetic improvements of tolerance to high temperature. In: Ashraf M, Harris PJC (eds) Abiotic stresses: plant resistance through breeding and molecular approaches. Howarth Press, New York, pp 277–300

Hsieh MH, Chen JT, Jinn TL, Chen YM, Lin CY (1992) A class of soybean low molecular weight heat shock proteins, immunological study and quantification. Plant Physiol 99:1279–1284

Hsu YT, Kao CH (2007) Heat shock-mediated H_2O_2 accumulation and protection against Cd toxicity in rice seedlings. Plant Soil 300:137–147

Hu WH, Hu GC, Han B (2009) Genome-wide survey and expression profiling of heat shock proteins and heat shock factors revealed overlapped and stress specific response under abiotic stresses in rice. Plant Sci 176:583–590

Iba K (2002) Acclimative response to temperature stress in higher plants: approaches of gene engineering for temperature tolerance. Annu Rev Plant Biol 53:225–245

Igamberdiev AU, Serege´lyes C, Manac'h N, Hill RD (2004) NADH-dependent metabolism of nitric oxide in alfalfa root cultures expressing barley haemoglobin. Planta 219:95–102

International Rice Genome Sequencing Project (2005) The map based sequence of the rice genome. Nature 436:793–800

Jagadish SVK, Craufurd PQ, Wheeler TR (2008) Phenotyping parents of mapping populations of rice (*Oryza sativa* L.) for heat tolerance during anthesis. Crop Sci 48:1140–1146

Jain M (2012) Nextgeneration sequencing technologies for gene expression profiling in plants. Brief Funct Genom 11:63–70

Jamalkandi SA, Masoudi-Nejad A (2009) Reconstruction of *Arabidopsis thaliana* fully integrated small RNA pathway. Funct Integr Genomic 9:419–432

Jeong DH, Park S, Zhai J, Gurazada SG, De Paoli ED, Meyers BC, Green PJ (2011) Massive analysis of rice small RNAs: mechanistic implications of regulated microRNAs and variants for differential target RNA cleavage. Plant Cell 23:4185–4207

Jian X, Zhang L, Li G, Zhang L, Wang X, Cao X, Fang X, Chen F (2010) Identification of novel stress-regulated microRNAs from *Oryza sativa* L. Genomics 95(1):47–55.
Jimenez A, Hernandez JA, Pastori G, Del Rio LA, Sevilla F (1998) Role of the ascorbate-glutathione cycle of mitochondria and peroxisomes in the senescence of pea leaves. Plant Physiol 118:1327–1335
Jung KH, An G (2012) Application of MapMan and Rice Net drives systematic analyses of the early heat stress transcriptome in rice seedlings. J Plant Biol 55:436–449
Jung KH, Ko HJ, Nguyen MX, Kim SR, Ronald P, An G (2013) Genome-wide identification and analysis of early heat stress responsive genes in rice. J Plant Biol 55:458–468. doi:10.1007/s12374-012-0271-z
Karim MA, Fracheboud Y, Stamp P (1997) Heat tolerance of maize with reference of some physiological characteristics. Ann Bangladesh Agric 7:27–33
Karim MA, Fracheboud Y, Stamp P (1999) Photosynthetic activity of developing leaves of *Zea mays* is less affected by heat stress than that of developed leaves. Physiol Plant 105:685–693
Karuppanapandian T, Moon JC, Kim C, Manoharan K, Kim W (2011) Reactive oxygen species in plants: their generation, signal transduction, and scavenging mechanisms. Aus J Crop Sci 5(6):709–725
Kaul S, Koo HL, Jenkins J, Rizzo M, Rooney T, Tallon LJ, Feldblyum T, Nierman W, Benito MI, Lin XY, Town CD, Venter JC, Fraser CM, Tabata S, Nakamura Y, Kaneko T, Sato S, Asamizu E, Kato T, Kotani H, Sasamoto S, Ecker JR, Theologis A, Federspiel NA, Palm CJ, Osborne BI, Shinn P, Conway AB, Vysotskaia VS, Dewar K, Conn L, Lenz CA, Kim CJ, Hansen NF, Liu SX, Buehler E, Altafi H, Sakano H, Dunn P, Lam B, Pham PK, Chao QM, Nguyen M, Yu GX, Chen HM, Southwick A, Lee JM, Miranda M, Toriumi MJ, Davis RW, Wambutt R, Murphy G, Dusterhoft A, Stiekema W, Pohl T, Entian KD, Terryn N, Volckaert G, Choisne N, Rieger M, Ansorge W, Unseld M, Fartmann B, Valle G, Artiguenave F, Weissenbach J, Quetier F, Wilson RK, de la Bastide M, Sekhon M, Huang E, Spiegel L, Gnoj L, Pepin K, Murray J, Johnson D, Habermann K, Dedhia N, Parnell L, Preston R, Hillier L, Chen E, Marra M, Martienssen R, McCombie WR, Mayer K, White O, Bevan M, Lemcke K, Creasy TH, Bielke C, Haas B, Haase D, Maiti R, Rudd S, Peterson J, Schoof H, Frishman D, Morgenstern B, Zaccaria P, Ermolaeva M, Pertea M, Quackenbush J, Volfovsky N, Wu DY, Lowe TM, Salzberg SL, Mewes HW, Rounsley S, Bush D, Subramaniam S, Levin I, Norris S, Schmidt R, Acarkan A, Bancroft I, Quetier F, Brennicke A, Eisen JA, Bureau T, Legault BA, Le QH, Agrawal N, Yu Z, Martienssen R, Copenhaver GP, Luo S, Pikaard CS, Preuss D, Paulsen IT, Sussman M, Britt AB, Eisen JA, Selinger DA, Pandey R, Mount DW, Chandler VL, Jorgensen RA, Pikaard C, Juergens G, Meyerowitz EM, Ecker JR, Theologis A, Dangl J, Jones JDG, Chen M, Chory J, Somerville C (2000) Analysis of the genome sequence of the flowering plant *Arabidopsis thaliana*. Nature 408:796–815
Kaur N, Gupta AK (2005) Signal transduction pathways under abiotic stresses in plants. Curr Sci 88:1771–1780
Kavi Kishore PB, Sangam S, Amrutha RN, Laxmi PS, Naidu, KR, Rao KRSS, Rao S, Reddy KJ, Theriappan P, Sreenivasulu N (2005) Regulation of proline biosynthesis, degradation, uptake and transport in higher plants: its implications in plant growth and abiotic stress tolerance. Curr Sci 88:424–438
Keeler SJ, Boettger CM, Haynes JG, Kuches KA, Johnson MM, Thureen DL, Keeler CL Jr, Kitto SL (2000) Acquired thermotolerance and expression of the HSP100/ClpB genes of lima bean. Plant Physiol 123:1121–1132
Khurana P, Chauhan H, Khurana N (2011) Characterization and expression of high temperature stress responsive genes in bread wheat. J Genet Plant Breed 47:S94–S97
Kliebenstein DJ, Dietrich RA, Martin AC, Last RL, Dangl JL (1999) Regulates salicylic acid induction of copper zinc superoxide dismutase in Arabidopsis thaliana. Mol Plant Microbe Interact 12:1022–1026
Koskull-Doring PV, Scharf KD, Nover L (2007) The diversity of plant heat stress transcription factors. Trends Plant Sci 12:452–457

Kotak S, Port M, Ganguli A, Bicker F, von Koskull-Döring P (2004) Characterization of C-terminal domains of Arabidopsis heat stress transcription factors (Hsf) and identification of a new signature combination of plant class A Hsfs with AHA and NES motifs essential for activator function and intracellular localization. Plant J 39:98–112

Kotak S, Vierling E, Baumlein H, von Koskull-Doring P (2007) A novel transcriptional cascade regulating expression of heat stress proteins during seed development of Arabidopsis. Plant Cell 19:182–195

Koussevitzky S, Suzuki N, Huntington S, Armijo L, Sha W, Cortes D, Shulaev V, Mittler R (2008) Ascorbate peroxidase 1 plays a key role in the response of *Arabidopsis thaliana* to stress combination. J Biol Chem 283:34197–34203

Larkindale J, Knight MR (2002) Protection against heat stress-induced oxidative damage in Arabidopsis involves calcium, abscisic acid, ethylene and salicylic acid. Plant Physiol 128:682–695

Larkindale J, Huang B (2004) Thermotolerance and antioxidant systems in *Agrostis stolonifera*: involvement of salicylic acid, abscisic acid, calcium, hydrogen peroxide, and ethylene. J Plant Physiol 161:405–413

Larkindale J, Vierling E (2008) Core genome responses involved in acclimation to high temperature. Plant Physiol 146:748–761

Lee RC, Feinbaum RL, Ambros V (1993) The C. elegans heterochronic gene lin-4 encodes small RNAs with antisense complementarity to lin-14. Cell 75:843–854

Lee YJ, Nagao RT, Key JL (1994) A soyabean 101-kD heat stress protein complements yeast HSP104 deletion mutant in acquiring thermotolerance. Plant Cell 6:1889–1897

Lee JM, Williams ME, Tingey SV, Rafalski JA (2002) DNA array profiling of gene expression changes during maize embryo development. Funct Integr Genomics 2:13–27

Lehman G, Engelke MC (1993) Heat resistance and rooting potential of Kentucky bluegrass cultivars. Int Turfgrass Soc Res J 7:775–779

Li C, Chen Q, Gao X, Qi B, Chen N, Xu S, Chen J, Wang X (2005) AtHsfA2 modulates expression of stress responsive genes and enhances tolerance to heat and oxidative stress in Arabidopsis. Sci China C Life Sci 48:540–550

Li SJ, Fu QT, Huang WD, Yu DQ (2009) Functional analysis of an *Arabidopsis* transcription factor WRKY25 in heat stress. Plant Cell Rep 28:683–693

Li YF, Zheng Y, Addo-Quaye C, Zhang L, Saini A, Jagadeeswaran G, Michael J, Zhang WA, Sunkar R (2010) Transcriptome-wide identification of microrna targets in rice. Plant J 62:742–759

Li Z, Zhang L, Wang A, Xu X, Li J (2013) Ectopic overexpression of slhsfa3, a heat stress transcription factor from tomato, confers increased thermotolerance and salt hypersensitivity in germination in transgenic Arabidopsis. PLos One 8(1):e54880. doi:10.1371/journal.pone.0054880

Liao JL, Zhang HY, Liu JB, Zhong PA, Huang YJ (2012) Identification of candidate genes related to rice grain weight under high-temperature stress. Plant Sci 196:32–43

Lim CJ, Yang KA, Hong JK, Choi JS, Yun DJ, Hong JC, Chung WS, Lee SY, Cho MJ, Lim CO (2006) Gene expression profiles during heat acclimation in *Arabidopsis thaliana* suspension-culture cells. J Plant Res 119:373–383

Lin YX, Jiang HY, Chu ZX, Tang XL, Zhu SW, Cheng BJ (2011) Genome-wide identification, classification and analysis of heat shock transcription factor family in maize. BMC Genomics 12:76. doi:10.1186/1471-2164-12-7

Linebarger CRL, Boehlein SK, Sewell AK, Shaw J, Hannah LC (2005) Heat stability of maize endosperm ADPglucose pyrophosphorylase is enhanced by insertion of a cysteine in the N terminus of the small subunit. Plant Physiol 139:1625–1634

Liu JG, Yao QH, Zhang Z, Peng RH, Xiong AS, Xu F, Zhu H (2005) Isolation and characterization of a cDNA encoding two novel heat-shock factor OsHSF6 and OsHSF12 in *Oryza sativa* L. J Biochem Mol Biol 38(5):602–608

Liu HH, Tian X, Li YJ, Wu CA, Zheng CC (2008) Microarray-based analysis of stress-regulated microRNAs in *Arabidopsis thaliana*. RNA 14(5):836–843

Liu JG, Qin QL, Zhang Z, Peng RH, Xiong AS, Chen JM, Yao QH (2009) OsHSF7 gene in rice, *Oryza sativa* L, encodes a transcription factor that functions as a high temperature receptive and responsive factor. BMB Rep 42(1):16–21

Liu AL, Zou J, Zhang XW, Zhou XY, Wang WF, Chen L, Xiong X, Chen X (2010) Expression profiles of class a rice heat shock transcription factor genes under abiotic stresses. J Plant Biol 53:142–149

Llave C, Kasschau KD, Rector MA, Carrington JC (2002) Endogenous and silencing-associated small RNAs in plants. Plant Cell 14:1605–1619

Lockhart DJ, Winzeler EA (2000) Genomics, gene expression and DNA arrays. Nature 405:827–836

Lohmann C, Eggers-Schumacher G, Wunderlich M, Schoffl F (2004) Two different heat shock transcription factors regulate immediate early expression of stress genes in *Arabidopsis*. Mol Genet Genomics 271:11–21

Lu S, Sun YH, Chiang VL (2008) Stress-responsive microRNAs in Populus. Plant J 55(1):131–151

Lu T, Lu G, Fan D, Zhu C, Li W, Zhao Q, Feng Q, Zhao Y, Guo Y, Li W, Huang X, Han B (2010) Function annotation of the rice transcriptome at single-nucleotide resolution by RNA-seq. Genome Res 20:1238–1249

Lubben TH, Donaldson GK, Viitanen PV, Gatenby AA (1989) Several proteins imported into chloroplasts form stable complexes with the GroEL-related chloroplast molecular chaperone. Plant Cell 1:1223–1230

Lund AA, Blum PH, Bhattramakki D, Elthon TE (1998) Heat-stress response of maize mitochondria. Plant Physiol 116:1097–1110

Lv DK, Bai X, Li Y, Ding XD, Ge Y, Cai H, Ji W, Wu N, Zhu YM (2010) Profiling of cold-stress-responsive miRNAs in rice by microarrays. Gene 459(1–2):39–47

Mallory AC, Vaucheret H, (2006) Functions of microRNAs and related small RNAs in plants. Nat Genet 38:S31–S36

Mangelsen E, Kilian J, Harter K, Jansson C, Wanke D, Sundberg E (2011) Transcriptome analysis of high-temperature stress in developing barley caryopses: early stress responses and effects on storage compound biosynthesis. Mol Plant 4:97–115

Marcum KB (1998) Cell membrane thermostability and whole plant heat tolerance of Kentucky bluegrass. Crop Sci 38:1214–1218

Markandeya G, Babu PR, Lachagari VBR, Feltus FA, Paterson AH, Reddy AR (2005) Functional genomics of drought-stress response in rice: transcript mapping of annotated unigenes of an indica rice (*Oryza sativa* L. cv. Nagina 22). Curr Sci 89:496–514

Martineau JR, Specht JE, Williams JH, Sullivan CY (1979) Temperature tolerance in soybeans I. Evaluation of technique for assessing cellular membrane thermostability. Crop Sci 19:75–78

Massa SI, Pearson GA, Aires T, Kube M, Olsen JL, Reinhardt R, Serrão EA, Arnaud-Haond S (2011) Expressed sequence tags from heat-shocked seagrass Zostera noltii (Hornemann) from its southern distribution range. Mar Genomics 4(3):181–188

Mathur S, Jajoo A, Mehta P, Bharti S (2011) Analysis of elevated temperatureinduced inhibition of photosystem II using chlorophyll a fluorescence induction kinetics in wheat leaves (*Triticum aestivum*). Plnt. Plant Biol 13(1):1–6

Matsumura T, Tabayashi N, Kamagata Y, Souma C, Saruyama H (2002) Wheat catalase expressed in transgenic rice can improve tolerance against low temperature stress. Physiol Plant 116:317–327

Maxwell K, Johnson GN (2000) Chlorophyll fluorescence—a practical guide. J Exp Bot 51(345):659–668

May P, Liao W, Wu Y, Shuai B, McCombie WR, Zhang MQ, Liu QA (2013) The effects of carbon dioxide and temperature on microRNA expression in Arabidopsis development. Nat Commun 4:2145. doi:10.1038/ncomms3145

McDonald AE, Vanlerberghe GC (2005) Alternative oxidase and plastoquinol terminal oxidase in marine prokaryotes of the Sargasso Sea. Gene 349:15–24

McDowell JM, An YQ, Huang S, McKinney EC, Meagher RB (1996) The Arabidopsis ACT7 actin gene is expressed in rapidly developing tissues and responds to several external stimuli. Plant Physiol 111:699–711

Melchiorre M, Robert G, Trippi V, Racca R, Lascano HR (2009) Superoxide dismutase and glutathione reductase overexpression in wheat protoplast: photooxidative stress tolerance and changes in cellular redox state. Plant Grow Regul 57:57–68
Miller G, Mittler R (2006) Could heat shock transcription factors function as hydrogen peroxide sensors in plants? Ann Bot (Lond) 98(2):279–288
Mishra SK, Tripp J, Winkelhaus S, Tschiersch B, Theres K, Nover L, Scharf KD (2002) In the complex family of heat stress transcription factors, HsfA1 has a unique role as master regulator of thermotolerance in tomato. Genes Dev 16(12):1555–1567
Mishra NS, Kumar V, Sopory SK, Mukherjee SK (2009) Cloning and validation of novel miRNA from basmati rice indicates cross talk between abiotic and biotic stresses. Mol Genet Genomics 282:463–474
Mittal D, Chakrabarti S, Sarkar A, Singh A, Grover A (2009) Heat shock factor gene family in rice: genomic organization and transcript expression profiling in response to high temperature, low temperature and oxidative stresses. Plant Physiol Biochem 47:785–795
Mittal D, Madhyastha DA, Grover A (2012) Genome-wide transcriptional profiles during temperature and oxidative stress reveal coordinated expression patterns and overlapping regulons in rice. PLos One 7(7):e40899. doi:10.1371/journal.pone.0040899
Mittler R, Zilinskas BA (1992) Molecular cloning and characterization of a gene encoding pea cytosolic ascorbate peroxidase. J Biol Chem 267:21802–21807
Mittler R (2002) Oxidative stress, antioxidants and stress tolerance. Trends Plant Sci 7:405–410
Moffatt JM, Sears RG, Paulsen GM (1990) Wheat high-temperature tolerance during reproductive growth 1. Evaluation by chlorophyll fluorescence. Crop Sci 30:881–885
Mohammed AR, Tarpley L (2009) Impact of high nighttime temperature on respiration, membrane stability, antioxidant capacity, and yield of rice plants. Crop Sci 49:313–322
Moldovan D, Spriggs A, Yang J, Pogson BJ, Dennis ES, Wilson IW (2009) Hypoxia responsive microRNAs and trans-acting small interfering RNAs in Arabidopsis. Exp Bot 61:165–177
Musatenko I, Vedenicheva NP, Vasyuk VA, Generalova VN, Martyn GI, Sytnik KM (2003) Phytohormones in seedlings of maize hybrids differing in their tolerance to high temperature. Russ J Plant Physiol 50(4):444–448
Nagamiya K, Motohashi T, Nakao K, Prodhan SH, Hattori E, Hirose S, Ozawa K, Ohkawa Y, Takabe T, Takabe T, Komamine A (2007) Salt tolerance in transgenic rice expressing an *Escherichia coli* catalase gene, katE. Plant Biotechnol Rep 1:49–55
Nagaraj SH, Gasser RB, Ranganathan S (2007) A hitchhiker's guide to expressed sequence tag (EST) analysis. Brief Bioinform 8:6–21
Nanjo T, Futamura N, Nishiguchi M, Igasaki T, Shinozaki K, Shinohara K (2004) Characterization of full-length enriched expressed sequence tags of stress-treated poplar leaves. Plant Cell Physiol 45(12):1738–1748
Neumann DM, Emmermann M, Thierfelder JM, Zur Nieden U, Clericus M, Braun HP, Nover L, Schmitz UK (1993) HSP68-a DNAK-like heat-stress protein of plant mitochondria. Planta 190:32–43
Nieto-Sotelo J, Martinez LM, Ponce G, Cassab GI, Alagon A, Meeley RB, Ribaut JM, Yang R (2002) Maize HSP101 plays important rolesin both induced and basal thermotolerance and primary root growth. Plant Cell 14:1621–1633
Nishizawa A, Yabuta Y, Yoshida E, Maruta T, Yoshimura K, Shigeoka S (2007) Arabidopsis heat shock transcription factor A2 as a key regulator in response to several types of environmental stress. Plant J 48:535–547
Noctor CH, Foyer A (1998) Re-evaluation of the ATP: NADPH budget during C3 photosynthesis. A contribution from nitrate assimilation and its associated respiratory activity? J Exp Bot 49:1895–1908
Nover L, Bharti K, Doring P, Mishra SK, Ganguli A, Scharf KD (2001) Arabidopsis and the heat stress transcription factor world: how many heat stress transcription factors do we need? Cell Stress Chaperones 6(3):177–189

Ogawa D, Yamaguchi K, Nishiuchi T (2007) High-level overexpression of the Arabidopsis HsfA2 gene confers not only increased themotolerance but also salt/osmotic stress tolerance and enhanced callus growth. J Exp Bot 58(12):3373–3383
Ogren E (1988) Photoinhibition of photosynthesis in willow leaves under field conditions. Planta 175(2):229–236
Oh-e I, Saitoh K, Kuroda T (2007) Effects of high temperature on growth, yield and dry-matter production of rice grown in the paddy field. Plant Prod Sci 10:412–422
Omar SA, Fu QT, Chen MS, Wang GJ, Song SQ, Elsheery NI, Xu ZF (2011) Identification and expression analysis of two small heat shock protein cDNAs from developing seeds of biodiesel feedstock plant Jatropha curcas. Plant Sci 181(6):632–637
Oshino T, Miura S, Kikuchi S, Hamada K, Yano K, Watanabe M, Higashitani A (2011) Auxin depletion in barley plants under high-temperature conditions represses DNA proliferation in organelles and nuclei via transcriptional alterations. Plant Cell Environ 34:284–290
Pareek A, Singla SL, Grover A (1995) Immunological evidence for accumulation of two high-molecular-weight (104 and 90 kDa) HSPs in response to different stresses in rice and in response to high temperature stress in diverse plant genera. Plant Mol Biol 29:293–301
Park W, Li J, Song R, Messing J, Chen X (2002) CARPEL FACTORY, a Dicer homolog, and HEN1, a novel protein, act in micro-RNA metabolism in *Arabidopsis thaliana*. Curr Biol 12:1484–1495
Paterson AH, Bowers JE, Bruggmann R, Dubchak I, Grimwood J, Gundlach H, Haberer G, Hellsten U, Mitros T, Poliakov A, Schmutz J, Spannagl M, Tang H, Wang X, Wicker T, Bharti AK, Chapman J, Feltus FA, Gowik U, Grigoriev IV, Lyons E, Maher CA, Martis M, Narechania A, Otillar RP, Penning BW, Salamov AA, Wang Y, Zhang L, Carpita NC, Freeling M, Gingle AR, Hash CT, Keller B, Klein P, Kresovich S, McCann MC, Ming R, Peterson DG, Mehboob-ur-Rahman, Ware D, Westhoff P, Mayer KF, Messing J, Rokhsar DS (2009) The Sorghum bicolor genome and the diversification of grasses. Nature 457:551–556
Pietrini F, Iannelli MA, Pasqualini S, Massacci A (2003) Interaction of cadmium with glutathione and photosynthesis in developing leaves and chloroplasts of *Phragmites australis* (Cav.) Trin. ex Steudel. Plant Physiol 133:829–837
Prasad PVV, Boote KJ, Allen LH, Sheehy JE, Thomas JMG (2006) Species, ecotype and cultivar differences in spikelet fertility and harvest index of rice in response to high temperature stress. Field Crops Res 95:398–411
Pratt WB, Toft DO (2003) Regulation of signaling protein function and trafficking by the hsp90/hsp70-based chaperone machinery. Exp Biol Med 228:111–133
Pratt WB, Galigniana MD, Harrell JM, Deranco DB (2004) Role of hsp90 and the hsp90-binding immunophilins in signaling protein movement. Cell Signal 16:857–872
Qin D, Wu H, Peng H, Yao Y, Ni Z, Li Z, Zhou C, Sun Q (2008) Heat stress-responsive transcriptome analysis in heat susceptible and tolerant wheat (*Triticum aestivum* L.) by using Wheat Genome Array. BMC Genomics 9:432
Quan R, Shang M, Zhang H, Zhao Y, Zhang J (2004) Engineering of enhanced glycine betaine synthesis improves drought tolerance in maize. Plant Biotechnol J 2:477–486
Rainey K, Griffiths P (2005) Evaluation of *Phaseolus acutifolius* A. Gray plant introductions under high temperatures in a controlled environment. Genet Resour Crop Evol 52:117–120
Ranson NA, White HE, Saibil HR (1998) Chaperonins. Biochem J 333:233–242
Rao ASVC, Reddy AR (2008) Glutathione reductase: a putative redox regulatory system in plant cells. In: Khan NA, Singh S, Umar S (eds) Sulfur assimilation and abiotic stresses in plants. Springer, The Netherlands, pp 111–147
Reinhart BJ, Slack FJ, Basson M, Pasquinelli AE, Bettinger JC, Rougvie AE, Horvitz HR, Ruvkun G (2000) The 21- nucleotide *let-7* RNA regulates developmental timing in *Caenorhabditiselegans*. Nature 403:901–906
Reinhart BJ, Weinstein EG, Rhoades MW, Bartel B, Bartel DP (2002) MicroRNAs in plants. Genes Dev 16:1616–1626

Reynolds MP, Balota M, Delgado MIB, Amani I, Fisher RA (1994) Physiological and morphological traits associated with spring wheat yield under hot irrigated conditions. Aust J Plant Physiol 21:717–730
Rhoades MW, Reinhart BJ, Lim LP, Burge CB, Bartel B, Bartel DP (2002) Prediction of plant microRNAs targets. Cell 110:513–520
Rouch JM, Bingham SE, Sommerfeld MR (2004) Protein expression during heat stress in thermo-intolerance and thermotolerance diatoms. J Exp Mar Biol Ecol 306:231–243
Rudd S (2003) Expressed sequence tags: alternative or complement to whole genome sequences? Trends Plant Sci 8:321–329
Sairam RK, Tyagi A (2004) Physiology and molecular biology of salinity stress tolerance in plants. Curr Sci 86:407–421
Sakamoto A, Murata N (2002) The role of glycine betaine in the protection of plants from stress: clues from transgenic plants. Plant Cell Environ 25:163–171
Sangwan V, Dhindsa RS (2002) In vivo and in vitro activation of temperatureresponsive plant map kinases. FEBS Lett 531:561–564
Sarkar NK, Kim YK, Grover A (2009) Rice sHsp genes: genomic organization and expression profiling under stress and development. BMC Genomics 10:393
Sarowar S, Kim EN, Kim YJ, Ok SH, Kim KD, Hwang BK, Shin JS (2005) Overexpression of a pepper ascorbate peroxidase-like 1 gene in tobacco plants enhances tolerance to oxidative stress and pathogens. Plant Sci 169:55–63
Savchenko GE, Klyuchareva EA, Abrabchik LM, Serdyuchenko EV (2002) Effect of periodic heat shock on the membrane system of etioplasts. Russ J Plant Physiol 49:349–359
Sayed OH (1996) Adaptational responses of *Zygophyllum qatarense* Hadidi to stress conditions in a desert environment. J Arid Environ 32:445–452
Scandalias JG (1990) Response of plant antioxidant defense genes to environmental stress. Adv Genet 28:1–41
Scharf KD, Siddique M, Vierling E (2001) The expanding family of *Arabidopsis thaliana* small heat stress proteins and a new family of proteins containing alpha-crystalline domains (Acd proteins). Cell Stress Chaperon 6:225–237
Schirmer EC, Lindquist S, Vierling E (1994) An *Arabidopsis* heat stress protein complements a thermotolerance defect in yeast. Plant Cell 6:1899–1909
Schoffl F, Prandl R, Reindl A (1999) Molecular responses to heat stress. In: Shinozaki K, Yamaguchi-Shinozaki K (eds) Molecular responses to cold, drought, heat and salt stress in higher plants. Landes, Austin, pp 81–98
Schramm F, Ganguli A, Kiehlmann E, Englich G, Walch D, Von Koskull-Do ring P (2006) The heat stress transcription factor HsfA2 serves as a regulatory amplifier of a subset of genes in the heat stress response in Arabidopsis. Plant Mol Biol 60:759–772
Schramm F, Larkindale J, Kiehlmann E, Ganguli A, Englich G, Vierling E, von Koskull-Doring P (2008) A cascade of transcription factor DREB2A and heat stress transcription factor HsfA3 regulates the heat stress response of *Arabidopsis*. Plant J 53:264–274
Schroda M, Vallon V, Wollman F, Beck CF (1999) A chloroplast-targeted heat shock protein 70 (HSP70) contributes to the photoprotection and repair of photosystem II during and after photoinhibition. Plant Cell 11:11165–11178
Severin AJ, Woody JL, Bolon YT, Joseph B, Diers BW, Farmer AD, Muehlbauer GJ, Nelson RT, Grant D, Specht JE, Graham MA, Cannon SB, May GD, Vance CP, Shoemaker RC (2010) RNA-Seq atlas of *Glycine max*: a guide to the soybean transcriptome. BMC Plant Biol 10:160
Shah K, Nahakpam S (2012) Heat exposure alters the expression of SOD, POD, APX and CAT isozymes and mitigates low cadmium toxicity in seedlings of sensitive and tolerant rice cultivars. Plant Physiol Biochem 57:106–113
Sharkova VE (2001) The effect of heat shock on the capacity of wheat plants to restore their photosynthetic electron transport after photoinhibition or repeated heating. Russ J Plant Physiol 48:793–797

Sharma P, Dubey RS (2005) Modulation of nitrate reductase activity in rice seedlings under aluminium toxicity and water stress: role of osmolytes as enzyme protectan. J Plant Physiol 162:854–864
Shuai P, Liang D, Zhang Z, Yin W, Xia X (2013) Identification of drought-responsive and novel; *Populus trichocarpa*; microRNAs by high-throughput sequencing and their targets using degradome analysis. BMC Genomics 14:233. doi:10.1186/1471-2164-14-233
Siddique M, Port M, Tripp J, Weber C, Zielinski D, Calligaris R, Winkelhaus S, Scharf KD (2003) Tomato heat stress protein Hsp16.1-CIII represents a member of a new class of nucleocytoplasmic small heat stress proteins in plants. Cell Stress Chaperon 8:381–394
Siddique M, GernhardS, Von Koskull-Doring P, Vierling E, Scharf KD (2008) The plant sHSP superfamily: five new members in *Arabidopsis thaliana* with unexpected properties. Cell Stress Chaperon 13:183–197
Singh A, Singh U, Mittal D, Grover A (2010) Genome-wide analysis of rice ClpB/HSP100, ClpC and ClpD genes. BMC Genomics 11:95
Singla SL, Grover A (1993) Antibodies raised against yeast HSP104 cross-react with a heat- and abscisic acid-regulated polypeptide in rice. Plant Mol Biol 22:1177–1180
Singletary G, Roshie B, Keeling P (1994) Heat stress during grain filling in maize: effects on carbohydrate storage and metabolism. Aust J Plant Physiol 21:829–841
Smertenko A, Draber P, Viklicky V, Opatrny Z (1997) Heat stress affects the organization of microtubules and cell division in *Nicotiana tabacum* cells. Plant Cell Environ 20:1534–1542
Stevens R, Page D, Gouble B, Garchery C, Zamir D, Causse M (2008) Tomato fruit ascorbic acid content is linked with monodehydroascorbate reductase activity and tolerance to chilling stress. Plant Cell Environ 31:1086–1096
Su J, Wu R (2004) Stress inducible synthesis of proline in transgenic rice confers faster growth under stress conditions than with constitutive synthesis. Plant Sci 166:941–948
Su PH, Li HM (2008) Arabidopsis stromal 70-kD heat shock proteins are essential for plant development and important for thermotolerance of germinating seeds. Plant Physiol 146:1231–1241
Sullivan CY, Ross WM (1979) Selecting for drought and heat resistance in grain sorghum. In: Mussell H, Staple R (eds) Stress physiology in crop plants. Wiley, New York, pp 263–281
Sun W, Motangu MV, Verbruggen N (2002) Small heat shock proteins and stress tolerance in plants. Biochim Biophys Acta 1577:1–9
Sung DY, Vierling E, Guy CL (2001) Comprehensive expression profile analysis of the Arabidopsis hsp70 gene family. Plan Physiol 126(2):789–800
Sung DY, Kaplan F, Lee KJ, Guy CL (2003) Acquired tolerance to temperature extremes. Trends Plant Sci 8:179–187
Sunkar R, Zhu JK (2004) Novel and stress regulated microRNAs and other small RNAs from Arabidopsis. Plant Cell 16:2001–2019
Sunkar R, Girke T, Jain PK, Zhu JK (2005) Cloning and characterization of microRNAs from rice plant. Cell 17(5):1397–1411
Sunkar R, Kapoor A, Zhu JK (2006) Posttranscriptional induction of two Cu/Zn superoxide dismutase genes in Arabidopsis is mediated by downregulation of miR398 and important for oxidative stress tolerance. Plant Cell 18:2051–2065
Swindell WR, Huebner M, Weber AP (2007) Transcriptional profiling of Arabidopsis heat shock proteins and transcription factors reveals extensive overlap between heat and non-heat stress response pathways. BMC Genomics 8:125
Tang G, Tang X, Mendu V, Jia X, Chen QJ, He L (2008) The art of microRNA: various strategies leading to gene silencing via an ancient pathway. Biochim Biophys Acta 1779:655–662
Tateishi Y, Nakagawa T, Esaka M (2005) Osmotolerance and growth stimulation of transgenic tobacco cells accumulating free proline by silencing proline dehydrogenase expression with double-stranded RNA interference technique. Physiol Plant 125:224–234
Toth SZ, Schansker G, Kissimon J, Kovacs L, Garab G, Strasser RJ (2005) Biophysical studies of photosystem II-related recovery processes after a heat pulse in barley seedlings (*Hordeum vulgare* L.). J Plant Physiol 162:181–194

Trindade I, Capitão C, Dalmay T, Fevereiro MP, Santos DM (2010) miR398 and miR408 are up-regulated in response to water deficit in *Medicago truncatula*. Planta 231(3):705–716.
Tripp J, Mishra SK, Scharf KD (2009) Functional dissection of the cytosolic chaperone network in tomato mesophyll protoplasts. Plant Cell Environ 32:123–133
Trovato M, Mattioli R, Costantino P (2008) Multiple roles of proline in plant stress tolerance and development. Rendiconti Lincei 19:325–346
Ushimaru T, Ogawa K, Ishida N, Shibasaka M, Kanematsu S, Asada K, Tsuji H (1995) Changes in organelle superoxide dismutase isoenzymes during air adaption of submerged rice seedlings: differential behaviour of isoenzymes in plastids and mitochondria. Planta 196:606–613
Ushimaru T, Nakagawa T, Fujioka Y, Daicho K, Naito M, Yamauchi Y, Nonaka H, Amako K, Yamawaki K, Murata N (2006) Transgenic Arabidopsis plants expressing the rice dehydroascorbate reductase gene are resistant to salt stress. J Plant Physiol 163:1179–1184
Vara Prasad PV, Craufurd PQ, Summerfield RJ (1999) Fruit number in relation to pollen production and viability in groundnut exposed to short episodes of heat stress. Ann Bot 84:381–386
Varshney RK, Nayak SN, May GD, Jackson SA (2009) Next-generation sequencing technologies and their implications for crop genetics and breeding. Trends Biotechnol 27 (9):522–530
Vellosillo T, Vicente J, Kulasekaran S, Ha mbergM, Castresana C (2010) Emerging complexity in reactive oxygen species production and signaling during the response of plants to pathogens. Plant Physiol 154:444–448
Verbruggen N, Hermans C (2008) Proline accumulation in plants: a review. Amino Acids 35:753–759
Veselov P, Lobov VP, Olyunina LN (1998) Phytohormones during heat shock and recovery. Russ J Plant Physiol 45(5):611–616
Vierling E (1991) The role of heat shock proteins in plants. Annu Rev Plant Phys 42:579–620
Vinocur B, Altman A (2005) Recent advances in engineering plant tolerance to abiotic stress: achievements and limitations. Curr Opin Biotechnol 16:123–132
Vu JCV, Gesch RW, Pennanen AH, Allen LHJ, Boote KJ, Bowes G (2001) Soybean photosynthesis, rubisco and carbohydrate enzymes function at supra-optimal temperatures in elevated CO2. J Plant Physiol 158:295–307
Wahid A, Shabbir A (2005) Induction of heat stress tolerance in barley seedlings by pre-sowing seed treatment with glycinebetaine. Plant Growth Regul 46:133–141
Wahid A, Ghazanfar A (2006) Possible involvement of some secondary metabolites in salt tolerance of sugarcane. J Plant Physiol 163:723–730
Wahid A, Close TJ (2007) Expression of dehydrins under heat stress and their relationship with water relations of sugarcane leaves. Biol Plant 51:104–109
Wahid A, Gelani S, Ashraf M, Foolad MR (2007) Heat tolerance in plants: an overview. Environ Exper Bot 61 (3):199–223
Wang SM, Khandekar JD, Kaul KL, Winchester DJ, Morimoto RI (1999) A method for the quantitative analysis of human heat shock gene expression using multiplex RT-PCR assay. Cell Stress Chaperones 4:153–161
Wang W, Vinocur B, Shoseyov O, Altman A (2004) Role of plant heat-shock proteins and molecular chaperones in the abiotic stress response. Trends Plant Sci 9:244–252
Wang Z, Gerstein M, Snyder M (2009) RNA-Seq: a revolutionary tool for transcriptomics. Nat Rev Genet 10:57–63
Wang F, Dong Q, Jiang H, Zhu S, Chen B, Xiang Y (2011) Genome-wide analysis of the heat shock transcription factors in *Populus trichocarpa* and *Medicago truncatula*. Mol Biol Rep 39(2):1877–1886. doi:10.1007/s11033-011-0933-9
Waters E, Lee G, Vierling E (1996) Evolution, structure and function of the small heat shock proteins in plants. J Exp Bot 47:325–338
Webb AAR, Mcainsh MR, Taylor JE, Hetherington AM (1996) Calcium ions as intercellular second messengers in higher plants. Adv Bot Res 22:45–96
Wilhelm EP, Mullen RE, Keeling PL, Singletary GW (1999) Heat stress during grain filling in maize: effects of kernel growth and metabolism. Crop Sci 39:1733–1741

Wise RR, Olson AJ, Schrader SM, Sharkey TD (2004) Electron transport is the functional limitation of photosynthesis in field-grown Pima cotton plants at high temperature. Plant Cell Environ 27:717–724

Xiang BL, Werner EM, Christensen DJ, Oliver DJ (2001) The biological functions of glutathione revisited in Arabidopsis transgenic plants with altered glutathione levels. Plant Physiol 126:564–574

Xiang D, Venglat P, Tibiche C, Yang H, Risseeuw E, Cao Y, Babic V, Cloutier M, Keller W, Wang E, Selvaraj G, Datla R (2011) Genome-wide analysis reveals gene expression and metabolic network dynamics during embryo development in Arabidopsis. Plant Physiol 156:346–356

Xin M, Wang Y, Yao Y, Xie C, Peng H, Ni1 Z, Sun Q (2010) Diverse set of microRNAs are responsive to powdery mildew infection and heat stress in wheat (*Triticum aestivum* L.). BMC Plant Biol 10:123

Xu Y, Huang B (2009) Effects of foliar-applied ethylene inhibitor and synthetic cytokinin on creeping bentgrass to enhance heat tolerance. Crop Sci 49(5)1876–1884

Yamada K, Fukao Y, Hayashi M, Fukazawa M, Suzuki I, Nishimura M (2007) Cytosolic HSP90 regulates the heat shock response that is responsible for heat acclimation in *Arabidopsis thaliana*. J Biol Chem 282:37794–37804

Yamakawa H, Hirose T, Kuroda M, Yamaguchi T (2007) Comprehensive expression profiling of rice grain filling related genes under high temperature using DNA microarray. Plant Physiol 144:258–277

Yamanouchi U, Yano M, Lin H, Ashikari M, Yamada K (2002) A rice spotted leaf gene, Spl7, encodes a heat stress transcription factor protein. Proc Natl Acad Sci U S A 99:7530–7535

Yancey PH, Clark ME, Hand SC, Bowlis RD, Somero GN (1982) Living with water stress: evolution of osmolyte system. Science 217:1214–1222

Yang J, Sears RG, Gill BS, Paulsen GM (2002) Growth and senescence characteristics associated with tolerance of wheat-alien amphiploids to high temperature under controlled conditions. Euphytica 126(2):185–193

Yao Y, Ni Z, Peng H, Sun F, Xin M, Sunkar R, Zhu JK, Sun Q (2010) Non-coding small RNAs responsive to abiotic stress in wheat (*Triticum aestivum* L.). Funct Integr Genomics 10:187–190

Yu J, Hu S, Wang J, Wong GK, Li S, Liu B, Deng Y, Dai L, Zhou Y, Zhang X, Cao M, Liu J, Sun J, Tang J, Chen Y, Huang X, Lin W, Ye C, Tong W, Cong L, Geng J, Han Y, Li L, Li W, Hu G, Huang X, Li W, Li J, Liu Z, Li L, Liu J, Qi Q, Liu J, Li L, Li T, Wang X, Lu H, Wu T, Zhu M, Ni P, Han H, Dong W, Ren X, Feng X, Cui P, Li X, Wang H, Xu X, Zhai W, Xu Z, Zhang J, He S, Zhang J, Xu J, Zhang K, Zheng X, Dong J, Zeng W, Tao L, Ye J, Tan J, Ren X, Chen X, He J, Liu D, Tian W, Tian C, Xia H, Bao Q, Li G, Gao H, Cao T, Wang J, Zhao W, Li P, Chen W, Wang X, Zhang Y, Hu J, Wang J, Liu S, Yang J, Zhang G, Xiong Y, Li Z, Mao L, Zhou C, Zhu Z, Chen R, Hao B, Zheng W, Chen S, Guo W, Li G, Liu S, Tao M, Wang J, Zhu L, Yuan L, Yang H (2002) A draft sequence of the rice genome (*Oryza sativa* L.ssp. indica). Science 296:79–92

Yu X, Wang H, Lu Y, Ruiter M, Cariaso M, Prins M, Tunenn AV, He Y (2011) Identification of conserved and novel microRNAs that are responsive to heat stress in *Brassica rapa*. J Exp Bot. 63(2):1025–1038

Zhang JH, Huang WD, Liu YP, Pan QH (2005) Effects of temperature acclimation pretreatment on the ultrastructure of mesophyll cells in young grape plants (*Vitis vinifera* L. cv. Jingxiu) under cross-temperature stresses. J Integr Plant Biol 47:959–970

Zhang JH, Wang LJ, Pan QH, Wang YZ, Zhan JC, Huang WD (2008a) Accumulation and subcellular localization of heat shock proteins in young grape leaves during cross-adaptation to temperature stresses. Sci Horti 117:231–240

Zhang X, Liu A, Zou J, Zhou X, Wang W, Chen L, Xiong X, Chen X (2008b) Expression profile of young rice panicle under heat stress using DNA microarray. Chin Sci Tech Papers online. A200812–3129

Zhang J, Xu Y, Huan Q, Chong K (2009) Deep sequencing of Brachypodium small RNAs at the global genome level identifies microRNAs involved in cold stress response. BMC Genomics 10:449

Zhang G, Guo G, Hu X, Zhang Y, Li Q, Li R, Zhuang R, Lu Z, He Z, Fang X, Chen L, Tian W, Tao Y, Kristiansen K, Zhang X, Li S, Yang H, Wang J, Wang J (2010) Deep RNA sequencing at single base-pair resolution reveals high complexity of the rice transcriptome. Genome Res 20:646–654

Zhang X, Li J, Liu A, Zou J, Zhou X, Xiang J, Rerksiri W, Peng Y, Xiong X, Chen X (2012) Expression profile in rice panicle: insights into heat response mechanism at reproductive stage. PLos One 7(11):e49652. doi:10.1371/journal.pone.0049652

Zhao B, Ge L, Liang R, Li W, Ruan K, Lin H, Jin Y (2009) Members of miR-169 family are induced by high salinity and transiently inhibit the NF-YA transcription factor. BMC Mol Biol 10:29

Zhou X, Wang G, Sutoh K, Zhu JK, Zhang W (2008). Identification of cold inducible microRNAs in plants by transcriptome analysis. Biochem Biophys Acta 1779(11):780–788

Zhou L, Liu Y, Liu Z, Kong D, Duan M, Luo L (2010) Genome-wide identification and analysis of drought-responsive microRNAs in Oryza sativa. J Exp Bot 61(15):4157–4168

Zhu B, Ye C, Lü H, Chen X, Chai G, Chen J, Wang C (2006) Identification and characterization of a novel heat shock transcription factor gene, *GmHsfA1*, in soybeans (*Glycine max)*. J Plant Res 119:247–256

Zou J, Liu A, Chen X, Zhou X, Gao G, Wang W, Zhang X (2009) Expression analysis of nine rice heat shock protein genes under abiotic stresses and ABA treatment. J plant Physiol 166(8):851–861

Chapter 4
Biotic Stress and Crop Improvement

Irada Huseynova, Nargiz Sultanova, Alamdar Mammadov, Saftar Suleymanov and Jalal A. Aliyev

Abstract Biotic stress is one of the major environmental factors affecting plants. Viruses, fungi, bacteria, weeds, insects, and other pests and pathogens represent a major constraint to agricultural productivity and a serious threat to vegetable and grain crops. Plant protection against pathogens and pests is a commercially important issue and one of the main directions of researches. Almost half of new infectious diseases identified in plants during the past 10 years have a viral nature. The number and distribution areas of some plant viruses in Europe have been increasing rapidly during the past 35 years that caused big problems from an economic point of view. Viral diseases have also become a real threat for different cultivars of vegetables, grains, and other agricultural crops in our country. It causes extensive leaf yellowing, stem and leaf deformation, reducing the fruit quality, substantial yield loss, and shortening the lifecycle of crops. The probable cause of decay of virus-infected plants is not only the virus activity itself but also reduced tolerance to unfavorable environmental conditions. The identification of the viruses affecting plants and the study of the plant responses are very important for the better understanding of the plant–virus interactions and for developing the tendency to reduce the plant virus-associated risks in Azerbaijan. Therefore, the main goal of the present study is to identify the most widespread plant viruses in Azerbaijan using different molecular techniques and to evaluate some characteristics of plant response to viral stress.

Keywords Biotic stress · Vegetable crops · Virus infections · ELISA assay · RCA · PCR · Protein and chlorophyll content · Thylakoid membrane polypeptides · Antioxidant enzymes · Photochemical activity · Photosystem II · Photosystem I · Fluorescence · ROS

I. Huseynova (✉) · N. Sultanova · A. Mammadov · S. Suleymanov · J. Aliyev
Institute of Botany, Azerbaijan National Academy of Sciences,
40 Badamdar Highway, AZ1073 Baku, Azerbaijan
e-mail: huseynova-i@botany-az.org

J. Aliyev
Department of Plant Physiology and Biotechnology, Research Institute of Crop Husbandry,
Ministry of Agriculture of the Republic of Azerbaijan,
AZ1098 Baku, Azerbaijan

P. Ahmad et al. (eds.), *Improvement of Crops in the Era of Climatic Changes,* Volume 2,
DOI 10.1007/978-1-4614-8824-8_4,

Abbreviations

APX	Ascorbate peroxidase
CAT	Catalase
Chl	Chlorophyll
CP	Coat protein
ELISA	Enzyme-linked immunosorbent assay
F	Fluorescence
GR	Glutathione reductase
LHC	Light-harvesting complex
MV	Methyl viologen
ORF	Open reading frame
PCR	Polymerase chain reaction
PS	Photosystem
RCA	Rolling circle amplification
ROS	Reactive oxygen species
SOD	Superoxide dismutase
ssRNA	Single-stranded RNA

1 Introduction

1.1 Crop Production Affected by Biotic Stresses

Plants as well as other living organisms are subjected to adverse biotic and abiotic environmental factors that negatively affect their vital activity and viability. Biotic stress factors comprise fungi, bacteria, viruses, nematodes, insects, phytoplasmas, etc. The main cause for serious interest in the study of plant pathogens is their wide dissemination and high pestiferousness. Many plant diseases are chronic in nature and dramatically reduce the overall plant productivity. In addition to the sharp decline in productivity, plants become more susceptible to the other unfavorable environmental factors. The number of plant viruses is constantly growing. The climate change affects plant diseases together with other components of global change, i.e., anthropogenic processes such as air, water and soil pollution, long-distance introduction of exotic species, urbanization (Gurr et al. 2011; Bradley et al. 2012; Matyssek et al. 2012; Régnière 2012; Siebold and von Tiedemann 2012) and drier conditions may also have direct effects on pathogens (Marco Pautasso et al. 2012; Fischer et al. 2011; Parks and Bernier 2010; Tomback and Achuff 2010; Carnicer et al. 2011; McDowell et al. 2011). The global warming that causes the expansion of the area of virus-carrying insects along with globalization of agricultural system and the international exchange of seeds and seedling materials create the conditions for the introduction of viral infections into the new areas

(Navas-Castillo et al. 2011). Integrated plant protection against pest organisms is the main condition for obtaining a stable product and high quality (Watt et al. 2011). Therefore, it is necessary to eliminate the harm caused by plant infections or completely clean the seeding materials from viruses using radical methods. It was found that the degree of plant resistance to various natural phyto-diseases and abiotic stress factors provided by a number of physiological and biochemical parameters is responsible for changes in plant metabolism and viability in stressful situations (Siebold and von Tiedemann 2012). Based on these results, it becomes possible to assess the genetic potential of different genotypes of plants against diseases. These indicators are used as a data source for biotechnological and genetic engineering techniques to create transgenic plants resistant to various infections. Determination of plant genotypes resistant to biotic stress factors is the most pressing problem of modern times, and these genotypes stored in a gene bank are the source of genes of resistance to diseases and can be used as parental forms in the subsequent breeding.

1.2 Impact of Plant Diseases on Food Security

Biotic stress factors include all influences from the living environment, and occur as a result of damage done to plants by other living organisms, such as bacteria, viruses, fungi, parasites, beneficial and harmful insects, weeds, and cultivated or native plants (Atkinson and Urwin 2012). To date, the technology has largely been used to counter weeds and pests, despite abundant evidence that plant pathogens can evolve rapidly (Fitter 2012). Crop plants can be attacked at any phase of their life cycle: during seedling establishment, plant maturation, or grain or fruit setting. Plant diseases affect the existence, adequate growth, and productivity of all kinds of plants, thereby affecting one or more of the basic prerequisites for a healthy and safe life for humans and resulting in reduced crop yield (Britton et al. 2010).

The environmental protection and solving the environmental problems are one of the important factors of sustainable development. From this perspective, the sustainable development in our republic requires protection of the environment and consistent use of natural resources. Thus, a large part of the twenty-first century's agenda is related to biodiversity, environmental protection, and rational use of natural resources. Effective solution of such strategic problems as providing the population with high-quality ecologically pure foods and other plant products depends on the creation of agricultural plants that are new and more productive and resistant to abiotic and biotic stress factors. At the present time, we can say that there is no plant virus database characterizing the viruses infecting agricultural crops, particularly vegetable plants in Azerbaijan. One of the most real problems we are facing today is to obtain varieties resistant to biotic stress factors using various methods of biotechnology, plant breeding, genetic engineering, genetics, and molecular biology. The identification of the main pathogens that infect plants in Azerbaijan and their inclusion in the databases are the first significant steps for the subsequent researches. It is known that selection of healthy plants, health control, seed certification, and continuous monitoring against viral infections should be carefully developed as methods of mass diagnosis in order to restrict the area of virus spreading.

1.3 Plant Viruses: Major Threat to Food Safety

Global climate change and the unprecedented rate of infectious disease emergence represent formidable ecological problems of our time (Rohr et al. 2011). Agriculture is the traditional production sector in Azerbaijan, which is among the countries where greater portion of the population depends on the scale farming for their income and subsistence. Crops are frequently affected by a wide array of diseases showing varying degrees and kinds of symptoms. Most of the causal agents of the diseases are biotic, not ruling out the involvement of abiotic factors too. Among the biotic factors, virus diseases constitute a main part of the diseases observed in all plant types, with variable symptoms including the leaf curling and distortion, green or yellow foliar mosaic, stunting of plants and reduced yields (Yamaji et al. 2013). Viruses are a major cause of productivity loss all over the world, including our country, and distribution areas and types of viral diseases differ from year to year (Luna et al. 2012). Jones (2009) has identified up to nine different scenarios for emergence in which introduced plants are exposed for the first time to indigenous viruses and vectors associated with the native flora. These scenarios represent situations in which the donor and recipient hosts, the vector and the virus, may interplay, and involve jumps from the native flora to the introduced crop and vice versa. We can say that in our country, most food, forage, technical, and ornamental plant crops which are vegetatively reproduced are chronically infected with viruses and other pathogens; thus, there is a large amount of product losses, and overall performance of agriculture in this regard is strongly reduced. The selection of healthy plants and their reproduction is one of the main purposes of the economy. However, due to the highly infectious background, viral contamination of healthy plants could not be excluded. Development of methods of mass diagnosis for genetic engineering and breeding processes has acquired a great importance because of the wide spread of viral diseases of plants and harm to the economy of our country.

1.4 Viruses Infecting Vegetable Crops in Azerbaijan

1.4.1 Cucumber Mosaic Virus

One of the most common viruses found in many vegetable crops is *cucumber mosaic virus* (CMV). This disease is transmitted by several aphid species in a nonpersistent manner. There are several strains of CMV that can infect more than 1,287 different plant species belonging to 100 families including snap beans, peppers, tomato, lettuce, tobacco, and sugar beets. It is geographically widespread and has been reported in Europe, Asia, Australia, and North America (Mochizuki et al. 2012). Symptoms depend upon the plant species infected, but vary from yellow-green mottling of leaves and fruits to stunting of plants (Lin et al. 2003). It is transmitted by numerous species of aphids, through the sap, the seeds, and mechanically (Dijkstra and Khan 2006). Morphologically, CMV has rather characteristic about 30-nm polyhedral particles with a hollow center. The genome consists of three (+) sense single-stranded RNAs (ssRNA), packaged in separate particles. CMV particles contain about 18% RNA. The RNA consists of four RNAs. Only the largest RNA3 is required for infectivity. A great number of different CMV strains, serogroups, subgroups, and biological variations have been described (Aramburu et al. 2007; Pita and Roossinck 2013; Sheikha et al. 2013).

1.4.2 Tomato Mosaic Virus

Tobamoviruses, including *tobacco* and *tomato mosaic virus* (TMV and ToMV, respectively), belong to the α-like supergroup of viruses. They consist of a characteristic proteinaceous rod, made up of 2,140 coat protein (CP) copies, which envelop the positive-stranded linear RNA genome. After infection of the plant cell, the RNA genome is uncoated and the viral gene products, the RNA-dependent RNA polymerase (RdRP), the movement protein (MP), and the CP are produced. Infection of neighboring cells commences with the movement of RNA–MP complexes through plasmodesmata with MP-induced altered size exclusion limits. Long-distance transport of the virus proceeds through the vascular tissue and depends on both MP and CP. Tobamoviruses, especially TMV, are the classical model system for the study of virus infection in plants. The foliage of affected tomato plants shows mottling, with alternate yellowish and dark green areas, the latter often appearing thicker and raised giving it a blister-like appearance. The leaves tend to be fern-like in appearance with pointed tips and younger leaves may be twisted. The fruit may be distorted, yellow blotches and necrotic spots may occur on both ripe and green fruit and there may be internal browning of the fruit wall. In young plants, the infection reduces the set of fruit and may cause distortions and blemishes. Besides solanaceous plants, ToMV affects a wide range of other crop and ornamental plants, including pepper, petunia, snapdragon, delphinium, marigold, and many other plants to a lesser extent. The virus may be introduced into an infected seed. Only a small number of seedlings need to be infected for the virus to spread rapidly.

1.4.3 Tomato Chlorosis Virus

Tomato chlorosis virus (ToCV) is a causal agent of a tomato disease called “yellow leaf disorder” reported for the first time in Florida (USA), also occurring in the Mediterranean area (Portugal, Italy, Greece, Spain, Cyprus, Lebanon, and Israel), Réunion Island, and Taiwan (Lozano et al. 2006). ToCV is a typical member of the *Closteroviridae,* with long flexuous virions varying from 800 to 850 nm in length. All members of the *Closteroviridae* have large genomes of positive ssRNA, but those belonging to the genus *Crinivirus* are divided into two RNAs: RNA1 is organized into four open reading frames (ORFs) and encodes proteins involved in replication; RNA2 is composed of nine ORFs and encodes among others a HSP70 homolog and two proteins involved in the encapsidation of viral RNA and referred to as the major CP and the minor coat protein (CPm). The CPm of ToCV is probably involved in determining the transmissibility of the virus by its vectors (Wintermantel et al. 2005). Vectors include the greenhouse whitefly (*Trialeurodes vaporariorum* Westwood), the banded winged whitefly (*T. abutilonea* Haldeman), and sweet potato whitefly (*Bemisia tabaci* Gennadius) biotypes A and B. ToCV-infected tomato plants show intervening leaf yellowing and older leaves become thicker and crispy, flowers abort, and the number and size of fruits decrease (Wintermantel et al. 2005; Lozano et al. 2006).

1.4.4 Melon Necrotic Spot Virus

Melon necrotic spot virus (MNSV) is a *Carmovirus* within the family *Tombusviridae,* which is present in cucurbit crops worldwide. MNSV can be transmitted mechanically, by the zoospores of the fungus *Olpidium bornovanus* (syn. *O. radicale*; Chytridiomycota: Olpidiaceae), and through the seed. The symptoms induced by MNSV in melon include local necrotic spots or large necrotic lesions on leaves and necrosis on stems and petioles. The MNSV genome is a (+) sense ssRNA molecule of 4.3 kb with at least five ORFs (Sela et al. 2013). The MNSV host range is restricted to members of the Cucurbitaceae family, where severity of the symptoms depends on the species and cultivars and the growing conditions. The hosts include melon (*Cucumis melo*), cucumber (*Cucumis sativus*), and watermelon (*Citrullus lanatus*). Disease is prevalent in Europe (France, Greece, Crete, Italy, Sardinia, Netherlands, Norway, Spain, Canary islands, mainland Spain, Sweden, UK, England, and Wales), Asia (China, Jiangsu, Iran, Israel, Japan, Hokkaido, Honshu, Kyushu, Shikoku, Korea Republic, Syria, and Turkey), Africa (Tunisia), North America (Canada, Mexico, USA), Central America and the Caribbean (Guatemala, Honduras, and Panama), and South America (Uruguay). The spread of MNSV in cultivated plants can be controlled through the use of virus-free seeds, control of the fungal vector, and the use of resistant cultivars.

1.4.5 Tomato Yellow Leaf Curl Virus

Tomato yellow leaf curl virus (TYLCV) is an Old World geminivirus, first described in the Middle East in the 1960s. Like all geminiviruses, TYLCV has a characteristic particle of twinned morphology which is isometric and approximately 20–30 nm in diameter. The virus consists of 22 capsomeres each containing five units of a

260-amino-acid CP. TYLCV has a single, 2,787-nucleotide (total molecular weight (m.w.) 980 kDa) covalently closed-genomic circular single-stranded DNA (ssDNA; Chen et al. 2011; Ji et al. 2013). This DNA has been shown to be bipartite (DNA A and B; Czosnek et al. 2013). The DNA of several strains has been sequenced (Eybishtz et al. 2009; Duan et al. 2012). The main host of TYLCV is tomato (*Lycopersicon esculentum*), *Datura stramonium,* and *Nicotiana* spp. or can be infected artificially. Ornamentals such as *Eustoma grandiflorum* can be severely damaged by natural infection (Chen et al. 2011). The plants which can be infected by the virus include 15 species in five different families. TYLCV is transmitted by *B. tabaci* in a persistent manner. Biotype B is usually the form of *B. tabaci* involved (Czosnek et al. 2007; Czosnek and Ghanim 2012; Czosnek et al. 2013) which transmits with high frequency. Acquisition and inoculation feeding periods range from 20 to 60 min and from 10 to 30 min, respectively, depending on the isolates (Díaz-Pendón et al. 2010; Luna et al. 2012). The latent period inside the insect is 20–24 h. TYLCV can persist in the vector for 10–12 days and only rarely for up to 20 days. No transovarial transmission of the virus has been found. Most whiteflies lose their ability to transmit within 10–12 days after an acquisition period of 24–28 h. Tomato plants infected at an early stage are severely stunted; their terminal and axillary shoots are erect, and their leaflets are reduced in size and abnormally shaped. Leaves that develop soon after infection are cupped downward, whereas leaves developing later are prominently chlorotic and deformed, with leaf margins rolled upward and curling between the veins. The effect on fruits depends on the age of the plant when infected. If infected early, plants lose vigor and stop producing marketable fruits (Sade et al. 2012). When infections occur at a later stage of development, additional fruits fail to set, but fruits already present ripen in a nearly normal manner. No flower symptoms are observed, but dropping of flowers is common.

1.4.6 Faba Bean Necrotic Yellow Virus

Circumstantial evidence suggests that the genome of *faba bean necrotic yellows virus* (FBNYV; genus *Nanovirus* has a multipartite genome) consists of eight distinct circular ssDNA components that are similar in size, each of about 1 kb and encoding only one protein and individually encapsidated in small isometric particles about 18 nm in diameter (Vetten et al. 2012; Watanabe et al. 2013a; Watanabe et al. 2013b). Early infected plants remain stunted, showing leaf yellowing followed by necrosis and plant death. FBNYV is spread by aphids, most efficiently (in laboratory tests) by *Acyrthosiphonpisum* (Harris) and *Aphis craccivora* (Koch). The virus is contained in the saliva of the aphid and transmitted while feeding takes place on virus-infected summer legumes, and then they introduce the virus into other plants in the autumn. The virus has a wide host range—58 host legume species have been identified (Vetten et al. 2012). It is considered the most economically important disease-inducing agent to the faba bean crop, especially in Middle Egypt (Kumar et al. 2011). It is also economically damaging in the Jordan Valley and in coastal areas of Syria and Turkey where winters are sufficiently mild to sustain the aphid population (Soliman 2013). FBNYV causes severe yield losses and crop failure in food and fodder legumes in Western Asia, North Africa, and Europe and has high variability and rapid evolution (Grigoras et al. 2010a; Grigoras et al. 2010b; Grigoras et al. 2010c; Grigoras et al. 2011).

1.5 Prospects for Reducing Yield Losses Caused by Plant Viral Diseases

Diseases of plants, like diseases of humans and other animals, are complex phenomena. A simple definition of disease is the abnormal functioning of an organism. One important characteristic of plant diseases is that they are injurious, causing harm to plants in some way. Plant diseases pose an enormous problem for global food security and are challenging for those who are interested in maintaining and producing healthy plants. *Azerbaijan entered the twenty-first century* with a very bad phytosanitary state of agricultural production, the omnipresence of the pests, pathogens, and weeds on agricultural land almost reaching the level of emergency. Potential yield losses from pests, diseases, and weeds exceeded several million tons of crop production in grain and protective measures have been undertaken to prevent a 25–30% yield loss. Under these circumstances, protection of plants is the most important for stabilization of agricultural production at the beginning of the twenty-first century. In the last decade, the concept of integrated plant protection, focused on maximizing the natural potential of agricultural lands using nonchemical and limited chemical methods with minimal environmental hazards, has been developing in *Azerbaijan.* A new methodological level of plant protection, using the methods of biotechnology and genetic engineering, provides research and development of the new biological plant protection products (parasites and pests of plants, microbiological preparations based on microorganisms—entomopathogens and antagonists—pathogens, and microbial metabolites; Wang et al. 2012; Eyre et al. 2013). Economic and ecological comparative evaluation of biological and chemical methods determines the advantages for growing vegetables, fruits, and berry crops. Progress in the field of biotechnology and genetic engineering has opened new opportunities by breeding the hybrids resistant to major plant pests and diseases (Kaitao Lai et al. 2012). The main task that is implemented in the course of research on the creation of transgenic plants is the protection of plants from herbicides, pests, and diseases (Britton et al. 2010; Golding et al. 2010; Webber 2010). This direction of transgenetic researches is considered worldwide as the first wave of modern biotechnology revolution, which targets a twofold to threefold increase in agricultural productivity by 2025.

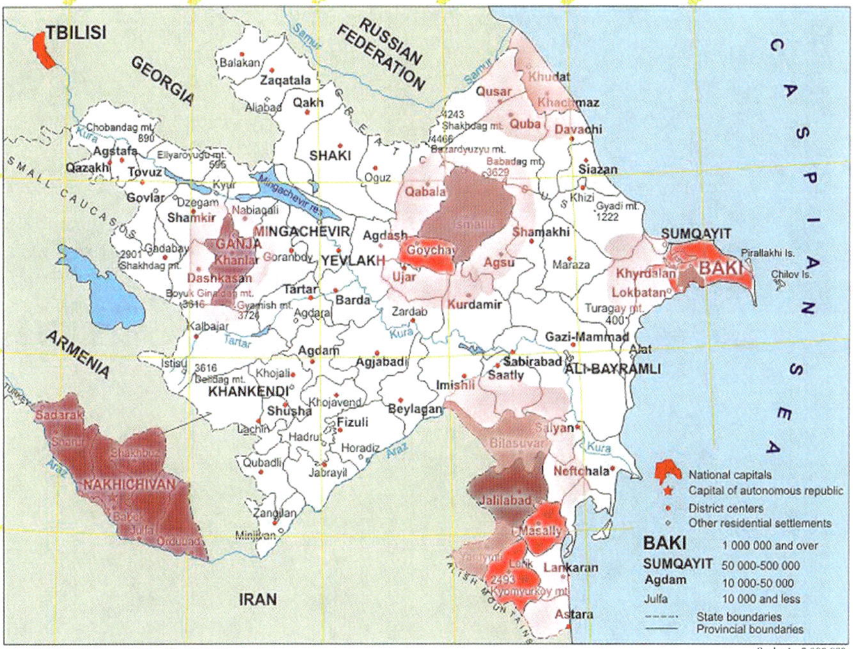

Scale: 1 : 2 000 000

Fig. 4.1 Map of Azerbaijan showing the location of regions where field-grown vegetable crops were surveyed and collected for virus diseases during 2009 and 2013 growing seasons

2 Materials and Methods

2.1 Plant Material: Field Visits and Sample Collections

To determine the presence or absence of virus infection, the vegetable crops (*Solanum lycopersicum* L., *Cucurbita melo* L., *Cucumis sativus* L., *Piper longum* L., *Solánum melongéna* L., *Vicia faba* L.) showing virus-like symptoms were collected from the fields in the main crop production provinces (Fig. 4.1) and the experimental field of the Azerbaijan Research Institute of Vegetable-growing.

Uninfected samples were collected under the same field conditions. The first signs of infection included the deformation of young leaves; the progression of the infection had symptoms of chlorosis, leaf rolling, yellowing, mosaic, stunting, wilting and shortening of the internodes, phloem discoloration, necrosis, and stunted growth (Fig. 4.2a–h).

Each field was evaluated using a standard format by recording location conditions, development stage, virus disease symptoms, and presence or absence of the insect populations. Virus disease incidence in each field was determined on the basis of visual symptoms and by counting the percentage of infected plants at different randomly selected locations in the field. Leaf samples showing virus-like symptoms

Fig. 4.2 Symptoms on vegetables naturally infected with viruses: **a**, **b**, **c**, **d** *Solanum lycopersicum* L. (**a** leaf deformation and discoloration; **b** shortening the life-span, extensive leaf yellowing, and stunting; **c** leaf curling; **d** destroyed and reduced fruit quality); **e** *Vicia faba* L.; **f** *Lens culinaris* L.; **g** *Cucumis sativus* L.; **h** *Piper longum* L.

were taken from vegetable plants exhibiting symptoms, placed in polythene bags and stored at −20 °C or in frozen liquid nitrogen.

2.2 *Virus Identification*

2.2.1 Detection of Viruses Using the Enzyme-Linked Immunosorbent Assay

Virus detection was performed with double and triple antibody sandwich enzyme-linked immunosorbent assay (DAS-ELISA and TAS-ELISA) according to Clark and Adams (1977) and the instructions of the virus antiserum manufacturers (diluted 1:500 and 1:1,000) for the following viruses: TYLCV, ToMV, ToCV, MNSV, CMV, *bean common mosaic virus* (BCMV), and *bean yellow mosaic virus* (BYMV). All antibodies were kindly provided by Dr. S. Winter (DSMZ, Braunschweig, Germany). The ELISA result was measured by recording its absorbance value using an ELISA plate reader (Stat Fax) at A405 nm. Leaf samples with typical symptoms of virus infections were ground (1-g leaf per 5-mL buffer) in extraction buffer, added to microplate wells after coating with virus-specific polyclonal antiserum diluted in carbonate buffer (pH 9.6), and incubated at 4 °C overnight. Plates were washed three times with PBS/Tween-20 buffer and coated with alkaline phosphatase-conjugated antibody diluted in extraction buffer and incubated for 4 h at 37 °C. After washing, p-nitrophenyl phosphate (Sigma) in diethanolamine substrate buffer was added to the wells and incubated at room temperature for 30–180 min. Absorbance values were read at 405 nm using a microplate reader (Stat Fax). Samples with absorbance values higher than the mean value for noninfected control plants plus two or three standard deviations were considered positive.

2.2.2 DNA Extraction

Total genomic DNA was extracted from leaves according to Edwards et al. (1991) with some modifications. About 20 mg of plant tissue (2–3-cm-long piece of fresh leaf material) was placed in an Eppendorf tube and beaked up with pellet pestle in liquid nitrogen. Crushed leaf tissues were ground in a preheated 400 μL of extraction buffer (200-mM Tris–HCl, pH 7.5, 250-mM NaCl, 25-mM ethylenediaminetetraacetic acid, EDTA, 0.5 % sodium dodecyl sulfate, SDS) with a little sterile sand. Homogenization was continued with vigorous shaking on vortex. Then 400 μL of phenol–chloroform (99.8 %) was added to each tube in the ratio 1:1 and the contents of the tube were gently mixed. After the vortex, the tubes were centrifuged in a tabletop centrifuge (Eppendorf 5415, UK) at 13,000 g for 10 min at room temperature. After centrifugation, the supernatant was carefully collected (taking care not to capture the particles of precipitate) and transferred to a clean Eppendorf tube with volume of 1.5 mL. Then, 600 μL of cold isopropanol was added to each tube, mixed well, and left at room temperature for 3–5 min. The tubes were centrifuged at room temperature at 14,000 g for 10 min to precipitate DNA. The DNA was washed several times with 70 % ethanol. Then, it was dried at 56 °C for 5 min and dissolved in 200-µL RTE buffer (10-mM Tris–HCl, pH 7.0, 1-mM EDTA, 10-µg RNase/mL) or TE buffer (10-mM Tris–HCl, pH 8, 1-mM EDTA). To dissolve DNA in the buffer, samples were left overnight in a refrigerator at 4 °C and afterwards stored at −20 °C.

2.2.3 DNA Quantification

The DNA concentration was determined by optical density (OD) at 260 nm with a spectrophotometer ULTROSPEC 3300 PRO (Amersham, USA). The quantification of genomic DNA was determined by the A260/A280 absorbance ratio. The required concentration of DNA (5-µL DNA/2-µL dye) was checked by running the DNA samples on 0.8 % agarose gel stained with 10-mg/mL ethidium bromide in 1 × TBE (Trisbase, Boricacid, EDTA) buffer. The gel was visualized and photographed under UV light using a digital imaging system (Gel Documentation System *UVITEK*).

2.2.4 Rolling Circle Amplification

Circular virus DNA was amplified using the protocol and set of reagents Templi PhiTM (GE Health Care, USA) as follows: 1 µL of DNA (10–20 ng of total nucleic acids) sample was added to 5 µL of test buffer (sample buffer), then the tube was heated at 95 °C for 5 min in tabletop incubator for DNA denaturation. The tubes were cooled down to room temperature in ice and centrifuged at 14,000 g for 5 s to precipitate the content of the bottom of the tubes. 5 µL of the enzyme mix was added to each tube. To prepare the enzyme mix, 0.2 µL enzyme solution containing Phi29 DNA polymerase and random hexamers in 50 % glycerol was added to 5 ml of the reaction buffer. The reaction mixture was run for 18 h with incubation at 30 °C. To inactivate the enzyme and stop the reaction, RCA products were heated for 10 min at 65 °C in tabletop incubator. DNA amplification products were analyzed by electrophoresis on 1.5 % agarose gel. For the determination of the DNA fragments, the gel was stained with ethidium bromide and photographed under UV light.

2.2.5 Polymerase Chain Reaction

Polymerase chain reaction (PCR) was carried out in a total reaction volume of 25 µL using optimized concentration of DNA, virus specific reverse and forward primers (0.2 µM each; Table 4.1), 1.5-mM $MgCl_2$, 0.2-mM dNTPs, 1-U Taq polymerase (Promega, GoTaq A). The thermocycling conditions for different primer sets are given in Table 4.2.

PCR were performed in "Applied Biosystems 2720 Thermal Cycler." The expected bands were analyzed on a 1–1.5 % agarose gel in TAE buffer, includ-

Table 4.1 Primers used for viral DNA amplification

Primer designation	sequence 5′→ 3′	Tm [°C]
MA13 F	AATGCAATCTTCGTCACC	60
MA26 R	CGCCCGTCTCGAAGGTTCG	
MA17 F	GAAAACATTTGTGAATCC	60
MA27 R	TGGAAATGATTATATCGCCTGGTCGC	
V61 F	ATACTTGGACACCTAATGG	55
C473 R	AGTCACGGGCCCTTACAA	
V781 F	CTCACAGAGTGGGTAAGAGG	45
C1256 R	TTAATTTGATATTGAATCATAGAAATAG	
OTYA3 F	GGGTCGACGTCATCAATGACG	55
OTYA6 R	CTACATGAGAATGGGGAACC	
Nano F103	ATTGTATTTGCTAATTTTA	44
Nano R101	TTCCCTTCTCCACCTTGT	
C5-F	TACAGCTGTCTTTGCTTCCT	49
C5-R	CGCGGAGTAATTAAATCAAAT	

ing ethidium bromide and viewed under UV light. Primers were developed from TYLCV and FBNYV sequences provided by B. Gronenborn (personal communication).

2.3 Isolation of Thylakoid Membranes

Some biophysical and biochemical changes caused by virus infection were also investigated in this study. For this purpose, chloroplasts were isolated by homogenization of the leaves with a Waring Laboratory blender for 1–2 min in a cooled buffer containing 0.4-M sucrose, 20-mM Tris, 10-mM NaCl, 1-mM EDTA-Na, 1-mM sodium ascorbate, and 0.1 % PEG, pH 7.8, following the procedure described in (Aliyev et al. 1992). The homogenate was filtered twice through four layers of cheesecloth. The filtrate was first centrifuged at 2,000 g for 5 min, then the precipitate was discarded and supernatant was centrifuged at 1,000 g for 10 min in a centrifuge K-70 (VEB MLW Medizintechnik Leipzig, Germany). The resulting precipitated chloroplasts were diluted in a small volume of hypotonic buffer containing 10-mM $MgCl_2 \times 6\ H_2O$ and 50-mM Tris–HCl, pH 7.2. All steps of this work were carried out at 4 °C. Chloroplasts were used immediately or frozen in liquid nitrogen.

2.4 Electrophoresis Analysis

Analysis of the polypeptide composition and determination of the molecular weight of protein subunits were performed by analytical SDS linear gradient gel electrophoresis (10–25 %; ratio of acrylamide to methylene bisacrylamide was 30: 0.8) by the method of Laemmli (1970). Samples for electrophoresis were prepared as

Table 4.2 The thermocycling conditions for different primer sets

Primer designation	The thermocycling conditions	Number of cycles	PCR fragment size
Nano F103/Nano R101	95 °C, 1 min	1	770–775
	95 °C, 45 s	30	
	54 °C, 30 s		
	72 °C, 30 s		
	72 °C, 10 min	1	
	4 °C,∞		
C-5 F/C-5 R	95 °C, 1 min	1	666
	95 °C,1 min	35	
	41 °C, 1 min		
	72 °C, 1 min		
	72 °C, 10 min	1	
	4 °C, ∞		
MA13F/MA26R	95 °C, 5 min	1	1292
	94 °C, 30 s	30	
	41 °C, 40 s		
	72 °C, 1 min		
	72 °C, 7 min	1	
	4 °C, ∞		
OTYA3F/OTYA6R	95 °C, 1 min	1	649
	94 °C, 30 s	30	
	55 °C, 30 s		
	72 °C, 30 s		
	72 °C, 10 min	1	
	4 °C, ∞		
V61/C473	95 °C, 3 min	1	400
	55 °C, 2 min		
	72 °C, 2 min		
	95 °C, 1 min	30	
	60 °C, 1 min		
	72 °C, 1 min		
	72 °C, 10 min	1	
	4 °C, ∞		

follows: They were treated with 2 % Ds-Na (20 mg detergent per 1 mg Chl) in the presence of 1 % 2-mercaptoethanol at 24 °C for 30 min or in the same medium for 90 s at 100 °C (boiling water bath). Each well was loaded with 4–10 mg of chlorophyll (Chl) or 20–50 μL of protein. The gels were stained for 30 min with 0.04 % (w/v) Coomassie brilliant blue G-250 (France) prepared in 3.5 % perchloric acid.

2.5 Chlorophyll Determination

The ratio between Chl a and Chl b concentration was measured in 80 % and 100 % acetone extract (McKinney 1941). Samples were used immediately or rapidly *frozen in liquid nitrogen* and *stored* at −80 °C until required.

2.6 Measurement of Photochemical Activity

Photochemical activity of chloroplasts was measured by the polarographic method based on the evolution or consumption of oxygen according to Aliyev et al. (1992). Photosystem II (PS II) activity was measured in a reaction medium containing 330-mM sorbitol, 40-mM Hepes-NaOH, pH 7.6, 10-mM NaCl, 5-mM $MgCl_2$, and 0.5-mM $K_3Fe(CN)_6$ as a final electron acceptor. PS I activity was determined in a total reaction volume of 2 ml containing 80-mM sucrose, 30-mM Tris–HCl, pH 8.0, 10-mM NaCl, 10-mM $MgCl_2$, 1-mM sodium ascorbate, 2-µM 3-(3,4-dichlorophenyl)-1,1-dimethylurea (for blocking electron transport from PS II), 0.3-mM 2,6-dichlorophenol-indophenol as the electron donor, and 50 µM of methyl viologen as an electron acceptor. Photochemical activity was measured in µmol O_2 consumption/mg Chl·h (PS I activity) and µmol O_2 evolution/mg Chl·h (PS II activity).

2.7 Assay of Fluorescence Parameters of PS II

Measurements of photo-induced changes of the fluorescence yield from the initial (F_o) level to the maximum (F_{max}) was performed in a laboratory-built setup at room temperature as described earlier (Klimov et al. 1982). Potential quantum yield of PS II was calculated according to the formula: $F_p=F_v/F_m=(F_m-F_o)/F_m$, where F_v, variable fluorescence; F_m, maximum fluorescence; F_o, initial fluorescence; F_p, a potential quantum yield of PS II.

2.8 Protein Content Determination

Protein content was determined according to Sedmak and Grossberg (1977) with slight modification using the dye Coomassie brilliant blue G250 (France) and glycerin (1:1). 0.12% dye solution was prepared in 0.6-N HCl and filtered for separating the undissolved particles. To a measuring cuvette, 750-ml glycerol and the same volume of dye solution were added. Determination of the proteins in the tested samples was carried out on a calibration curve using 100% bovine serum albumin as a standard. The absorbance of samples was measured at 610 nm on spectrophotometers ULTROSPEC 3300 PRO (Amersham, USA).

2.9 Histochemical Staining of Reactive Oxygen Species

Histochemical staining for reactive oxygen species (ROS) accumulation was conducted as described by Fryer et al. (2003), Mahalingam et al. (2005), and Kariola et al. (2006) with some modifications. For superoxide anion radical determination, detached leaves of the virus-infected plants were put in petri plates with 6-mM nitroblue tetrazolium (NBT) solution containing 50-mM sodium phosphate (pH 7.5) and 10-mM sodium azide for 12 h in the dark. To detect hydrogen peroxide, a solution of 5-mM DABS and 10-mM MES, pH 3.8 was used. ROS reaction was stopped by soaking the leaves with lacto-glycerol-ethanol (1:1:4 by vol) and boiling in water for 5 min. The cleared leaves were preserved in 50 % ethanol and photographed.

2.10 Enzyme Extraction and Activity Determination

Healthy and virus-infected leaves were collected and enzyme activities were assayed. Leaves were washed with distilled water and surface moisture was wiped out. For enzyme extraction, the leaves were crushed and homogenized with prechilled pestle in ice-cold mortar using a potassium–phosphate buffer. The homogenates were filtered through four layers of cheesecloth and then centrifuged in Beckman centrifuge at 4 °C. The supernatant was collected and transferred to 50-ml tubes and referred for the assays of enzymatic activities.

2.10.1 Catalase

Catalase (CAT, EC 1.11.1.6) activity was measured according to Kumar and Knowles (1993) with minor modifications. For determination of CAT activity, 1 g of leaf tissue was homogenized in 50-mM potassium–phosphate buffer (KH_2PO_4), pH 7.0. The homogenate was filtered and centrifuged at 8,000 g for 10 min. The activity of CAT was determined by spectrophotometer ULTROSPEC 3300 PRO (Amersham, USA). The reaction was initiated by the addition of 15-mM H_2O_2 and enzyme extract. The decrease of H_2O_2 was measured at 240 nm for 1 min and calculated based on its molar extinction coefficient ($\varepsilon = 36\ mM^{-1}\ cm^{-1}$) and the results expressed as CAT units mg^{-1} of protein (U = 1 mM of H_2O_2 reduction $min^{-1}\ mg^{-1}$ protein).

2.10.2 Ascorbate Peroxidase

Ascorbate peroxidase (APX, EC 1.11.1.11) activity was determined according to Nakano and Asada (1981) with slight modifications. For this purpose, 1 g of the leaf tissue was homogenized in 0.05-M potassium–phosphate buffer (KH_2PO_4), pH 7.6, with 0.3 g of polyvinylpyrrolidone, then filtered and centrifuged at 8,000 g for 15 min. APX activity was measured spectrophotometrically. The reaction mixture contained 0.1-mM EDTA, 0.05-mM ascorbic acid, 0.1-mM hydrogen peroxide, 50-mM phosphate buffer, pH 7.6, and 0.3 mL enzyme extract in a final volume of 3 mL. Enzyme extract was added directly before the measurement. The activity was measured as decrease in absorbance at 290 nm for 30 s and the quantification of ascorbate oxidized was evaluated based on its molar extinction coefficient (ε=2.8 mM^{-1} cm^{-1}) and the results expressed as APX units mg^{-1} of protein (U=1 mM of ascorbate oxidized per min at 25 °C).

2.10.3 Glutathione Reductase

Glutathione reductase (GR, EC 1.6.4.2) activity was determined according to Yannarelli et al. (2007) with slight modifications. For this purpose, 0.8 g of leaf tissue was homogenized in 0.1-M potassium–phosphate buffer (KH_2PO_4), pH 7.8, then filtered and centrifuged at 10,000 g for 20 min. GR activity was measured spectrophotometrically. The reaction mixture contained 0.1-M potassium–phosphate buffer (KH_2PO_4), pH 7.8, 0.001-M EDTA, 0.2-mM nicotinamide adenine dinucleotide phosphate (NADPH), and 0.5 mM of oxidized glutathione in a final volume of 3 mL. The GR activity was measured as decrease in absorbance due to the oxidation of NADPH in the presence of oxidized glutathione at 340 nm for 10 min and calculated based on its molar extinction coefficient (ε=6.2 mM^{-1} cm^{-1}) and the results expressed as GR units mg^{-1} of protein.

2.10.4 Superoxide Dismutase

Superoxide dismutase (SOD, EC 1.15.1.1) activity was determined using the SOD Assay Kit-WST (Sigma-Aldrich, USA) following the manufacturer's instructions. SOD activity was measured as decrease in absorbance at 450 nm for 1 min and one enzyme unit of SOD activity was defined as the amount of enzyme required to cause 50 % inhibition of the rate of NBT reduction.

2.10.5 Qualitative Changes in the Activity of APX and CAT

Qualitative changes in the activity of APX were investigated by polyacrylamide gel electrophoresis (PAGE) according to Laemmli (1970). An equal amount of enzyme extract, bromophenol blue, and glycerol were mixed to a final concentration of 12.5 % (v/v). Electrophoresis was performed on 10 % PAAG. Electrophoresis was performed at 4 °C for 3 h at a steady current of 30 mA, using the set unit SE 250 (Amersham Biosciences, USA). The gel was incubated in a solution containing 50-mM potassium–phosphate buffer (pH 7.0) and 2-mM sodium ascorbate for 30 min. Then gel was incubated again in a solution containing 50-mM potassium–phosphate buffer (pH 7.0), 4-mM sodium ascorbate, and 2-mM H_2O_2 for 20 min. The gel was stained in a solution containing 50-mM potassium–phosphate buffer (pH 7.8), 28-mM TEMED, and 2.45-mM NBT for 15 min with agitation on a shaker.

Qualitative changes in the activity of CAT were examined by PAGE method described earlier. To visualize the isoenzyme lines of CAT, the gel was incubated in a solution containing 3.27-mM H_2O_2 for 25 min. Then the gel was washed twice with distilled water and stained in freshly prepared solution containing 1 % (w/v) $K_3[Fe(CN)_6]$ and 0.1 % (w/v) $FeCl_3$ (Anderson et al. 1995).

3 Effects of Virus Infections on Photosynthetic Activity and Antioxidant System of Vegetable Plants Grown in Azerbaijan

To determine the virus concentration and disease incidence, a total of 126 specimens were collected from different commercial fields and those belonging to governmental institutions. The results obtained from ELISA showed that more than 7.2 % of tomato specimens and 2.5 % of long pepper were positive for ToMV, 2.2 % of tomato specimens were positive for ToCV, 3.6 % of melon specimens were positive for MNSV, 2.8 % of cucumber, and 1.5 % of eggplant specimens were positive for CMV. In order to detect the presence of ssDNA plant viruses, DNA of two chickpeas (*Cicer arietinum*) and two lentils (*Lens culinaris*) with relevant positive results for nanoviruses were amplified by RCA, restricted by endonucleases *Aat*II or *Hind*III, and analyzed on 1.5 % agarose gel. The DNA of lentils was restricted into a major fragment of 1 kb by *Hind*III and AatII digests the DNA of chickpeas and also a minor product of 2 kb was produced (Fig. 4.3).

Experience with nanovirus DNA segments shows that at least some genomic DNAs of a nanovirus have an *Aat*II site. Hence, the predominant 1-kb fragment produced by *Aat*II was a strong indication of the presence of a nanovirus in the two specimens of lentil. The template DNA of lentil specimens was also amplified using specific primer pairs NanoF103/NanoR101 and C5F/C5R for nanoviruses. Lentil specimens as well as the reference nanovirus isolates (SCSV-Australia, MDV-Japan, and FBNYV-Syria) generated amplicons of the expected size 770 bp using the NanoF103/NanoR101 primers and 660 bp using the specific C5F/C5R primers (Fig. 4.4). A PCR test also confirms the presence of nanoviruses in these samples.

We describe an initial assessment of plant viruses with an ssDNA genome in Azerbaijan. Survey of cultivated and wild crops has uncovered the presence of three different nanoviruses, two distinct FBNYV and one *faba bean necrotic stunt virus* (FBNSV).

Tomato (*Lycopersicon esculentum* Mill.) is a popular vegetable crop grown from subsistence to commercial scale in Azerbaijan. There are several factors limiting tomato productivity among which geminiviruses have been unified as very important biotic constraints for tomato cultivation in many tropical and subtropical countries of the world. Geminiviruses are a group of plant pathogens that have circular ssDNA genomes encapsidated within twinned particles. Whitefly-transmitted geminiviruses with mono- or bipartite ssDNA comprise the genus of *Begomovirus* of the family *Geminiviridae*. TYLCV is a monopartite begomovirus which exhibits a range of symptoms and causes serious damage to tomato plants with a greater yield loss and quality reduction. Recently, severe damage occurred in tomatoes associated with heavy infestations of *B. tabaci*. Symptoms consisted of plant stunting,

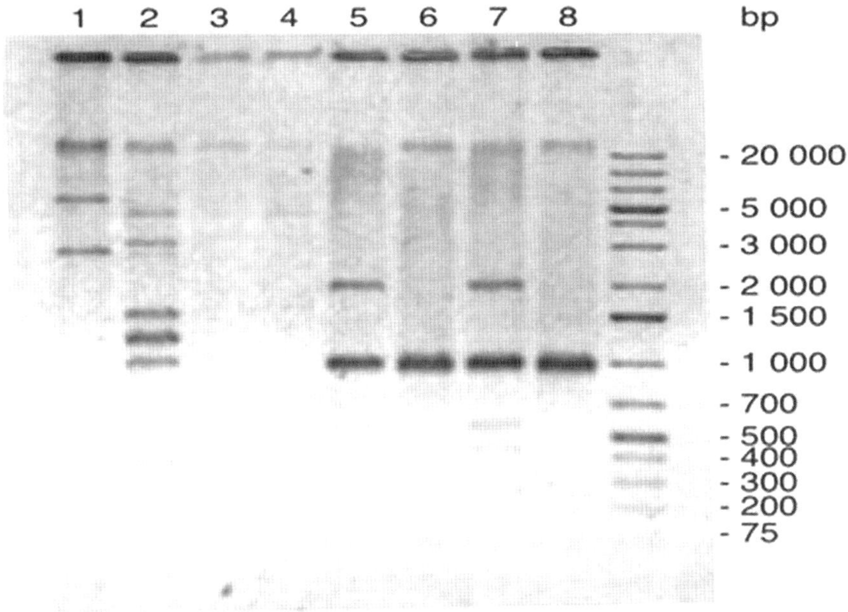

Fig. 4.3 Restriction analysis of RCA-amplified DNAs of chickpea and lentil collected from Lerik. Lanes *1–4*, DNA of two chickpea; lanes *5–8*, DNA of two lentil restricted by endonuclease *Aat*II (lanes *1, 3, 5, 7*); and *Hind*III (lanes *2, 4, 6, 8*). Lane *M*, 1-kb-plus DNA size marker

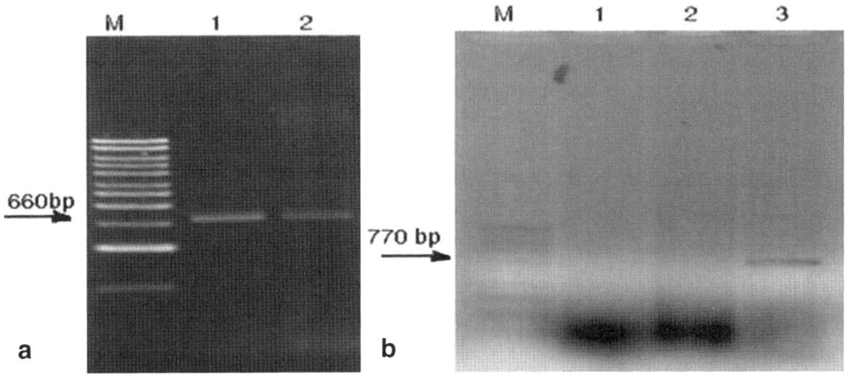

Fig. 4.4 Products of PCR amplification of FBNYV from two infected lentil plants. Reaction products were run on gel and stained with ethidium bromide. Panel A, reaction using the C5F/C5R primer pair; lanes *1–2*, infected lentil plants; *M*, molecular weight marker (1 kb ladder). Panel B, reaction using the NanoF103/NanoR101 primer pair; lanes *1–2*, noninfected lentil plants; lane *3*, infected lentil plant; *M*, molecular weight marker (1-kb ladder)

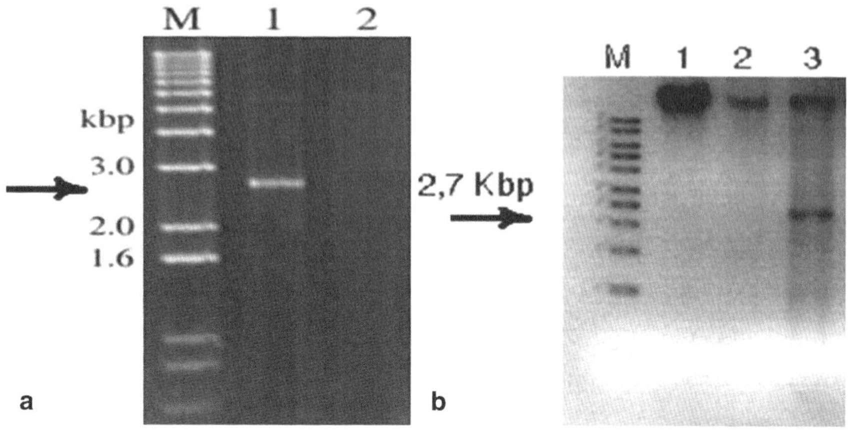

Fig. 4.5 Restriction analysis of RCA-amplified DNAs of tomato plants collected from Masalli. Reaction products were run on gel and stained with ethidium bromide. **a** DNA restricted by endonuclease *xBa*I; lane *1*, infected tomato plant; lane *2*, noninfected tomato plant; Lane *M*, molecular weight marker (1-kb ladder). **b** DNA restricted by endonuclease *Hind*III; lane *1*, RCA product of infected tomato; lane *2*, genomic DNA of infected tomato; lane *3*, DNA restricted by endonuclease *Hind*III; lane M, molecular weight marker (1 kb ladder)

curling, and deformation of leaves that exhibited chlorosis and yellowing. Fruit breaking was also observed in diseased plants. Begomoviruses possess small circular DNAs that are easily multiplied in vitro by rolling circle amplification (RCA) using bacteriophage Phi 29 polymerase. In order to detect the presence of begomoviruses, genomic DNAs of tomato specimens were amplified by RCA. Symptoms recalled begomovirus infections, and a preliminary ELISA analysis suggested the presence of begomovirus genomes in symptomatic tomato plants and RCA analysis confirms it. Restriction fragment length polymorphism (RFLP) is used in combination with RCA to identify differences between viruses based on the presence or absence of restriction enzyme-recognition sites. After RCA amplification, the amplicons were digested with restriction enzymes *Xba*I and *Hind*III and 2.7-kb fragments homologous to the TYLCV were revealed. RFLP is a method that can be used to differentiate isolates of viruses without the expenses of cloning and sequencing. The restriction fragments, together with 1-kb DNA size marker, were separated by 2% agarose gel electrophoresis in 1 × TAE buffer stained with ethidium bromide and visualized under UV (Fig. 4.5).

The template DNAs of tomato samples were also amplified by using specific primer pairs for TYLCV and yielded PCR products of the expected size (Fig. 4.6).

The presence of latent or hidden forms of infection, during which the plants do not express symptoms, contributes to quick spread of the disease and causes serious economic losses. The effective disease control using molecular diagnostic methods allow detecting the presence of the pathogen regardless of the form of the disease. Today, one of the key methods for the identification of pathogens and diagnosis of infectious diseases in plants and animals is a PCR. In order to identify virus diseases easily, we tested infected plants by the RCA method compared with PCR.

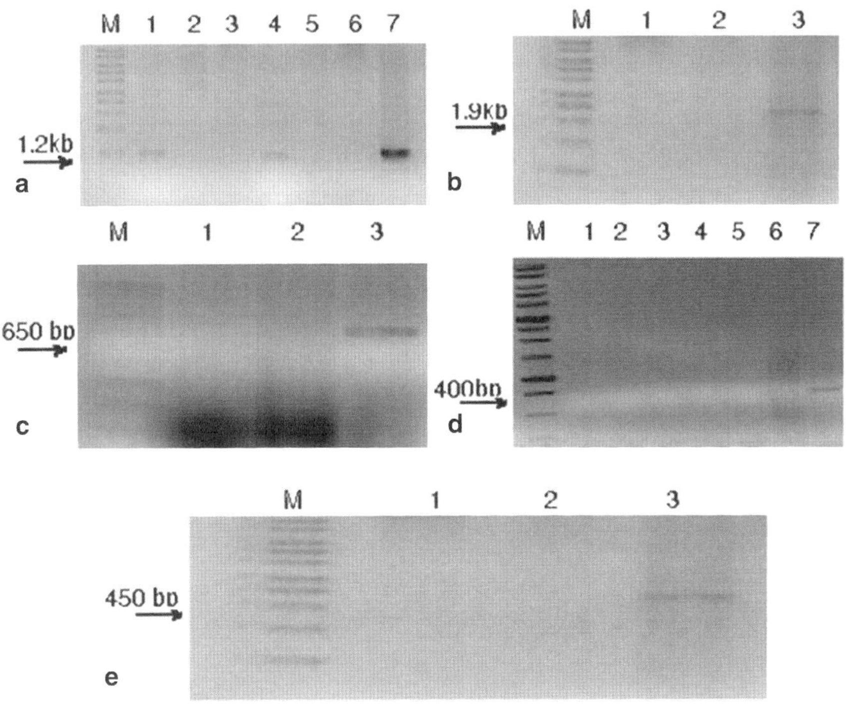

Fig. 4.6 TYLCV detection by PCR using **a** MA13F/MA26R; **b** MA17F/MA27 R; **c** OTYA3F/OTYA6R; **d** V61F/C473 R; **e** V781/C1256R primers. M, molecular weight marker (100-bp and 1-kb ladder). 1–7 tomato samples from Masalli (2010)

Obtained results showed that PCR was a more sensitive method for detection of viruses with a single DNA than RCA method. Furthermore, PCR method requires less expenditure for detection of viruses with circular DNA as compared with RCA method. However, that does not mean that our results contradict the work of Haible et al. (2006) that informs that RCA method is perfect for the identification of ssDNA viruses such as geminiviruses and nanoviruses. RCA method has some advantages compared with PCR: All circular DNAs can be amplified without any sequence knowledge; in addition, RCA has high efficiency and sensitivity (e.g., uses 10 ng of total nucleic acid and generates up to 5 μg of product), high fidelity of amplification with high reproducibility for various samples, and few false-negative or false-positive results. RCA does not require any thermocycler; thus, the reaction takes place at room temperature (Sahu et al. 2013).

Significant differences in the potential quantum yield of PS II photochemical reactions and the activity of PS II based on oxygen evolution were found in isolated chloroplasts from leaves of uninfected and infected plants. The photochemical efficiency of PS II (ratio Fv/Fm) in the infected plants decreased significantly to a minimum and constituted 0.48 in chickpea, 0.52 in faba bean, and 0.61 in lentil, while in the control plants, the efficiency of PS II was equal to 0.74, 0.72, and 0.76, respectively. The lowest F_v/F_m ratio observed in the infected leaves was mainly due

Table 4.3 Photochemical activity of photosystem II and photosystem I in chloroplasts from infected plants (μmol O_2·mg^{-1} chlorophyll·h^{-1})

Plant samples	Photosystem II $H_2O \rightarrow K_3Fe(CN)_6$	In %	Photosystem I DCIP·H→MV	In %
Solanum lycopersicum L. (healthy)	85±6	100	242±12	100
Solanum lycopersicum L. (infected)	35±4	41	189±9	78
Vicia faba L. (healthy)	57±4	100	250±12	100
Vicia faba L. (infected)	25±3	43	225±9	90
Cicer arietinum L. (healthy)	65±4	100	158±6	100
Cicer arietinum L. (infected)	30±3	46	114±6	72
Lens culinaris L. (healthy)	68±6	100	168±5	100
Lens culinaris L. (infected)	33±3	48	148±7	88
Pisum sativum L. (healthy)	72±6	100	190±8	100
Pisum sativum L. (infected)	37±4	52	179±4	94

to the decrease in F_v without increasing the F_o level. This is characteristic for inhibition on the donor side of PS II (Natha et al. 2013; Onda 2013). Activity of PS II was also affected by pathogenesis, constituting 46 % (chickpea), 43 % (faba bean), and 48 % (lentil) of the control value that indicates the processes of photoinhibition in these samples (Table 4.3).

Obtained results also showed that photochemical activity of PS I decreased under viral disease in all tested samples. It may be focused on the primary photosynthetic reactions and inhibition of the electron transport of PSII, because the virus CP seems to accumulate in the membranes of chloroplasts and thylakoids of the infected plants. The present study revealed enormous changes in biochemical components in plants due to the infection with different virus diseases. In the pathogenesis, a gradual reduction in green pigments like chlorophyll (a, b, and total) was observed in all specimens (Fig. 4.7 I).

Similar results were observed in virus-infected apple leaves (Bertamini et al. 2003). The disease development in these plants also altered the ratio between chlorophyll a and b, which might affect the photosynthetic efficiency. The protein concentration was low in infected plants of all species compared to controls (Fig. 4.7 II). Low protein content indicates that the disease might have caused the denaturation or breakdown of proteins, as well as polypeptide chains and protein-bound amino acids, resulting in enhanced free amino acid content of the host tissues.

The most likely explanation for the low PS II activity due to inactivation of the electron transport is that the related proteins are exposed at the thylakoid surface (Muthuchelian et al. 2005). Since the changes in photosynthetic electron transport, activities could be caused primarily by the changes or reorganization of thylakoid components, the thylakoid proteins of control and virus-infected leaves were analyzed using SDS–PAGE in order to investigate the alterations in macromolecular thylakoid protein complexes possibly associated with the observed symptoms and the chloroplast malformations (Fig. 4.8).

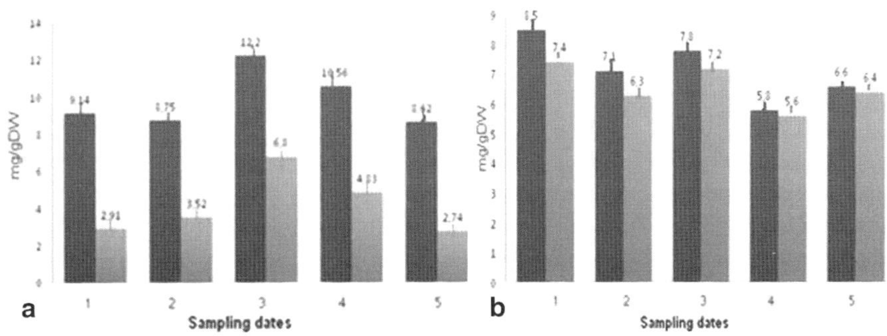

Fig. 4.7 Changes of chlorophyll (*Chl*) concentrations (**a**) and soluble proteins (**b**) in control and virus infected leaves. *1 Solanum lycopersicum* L.; *2 Vicia faba* L.; *3 Cicer arietinum* L.; *4 Lens culinaris* L.; *5 Pisum sativum* L. Mean ± S.E ($n=3$)

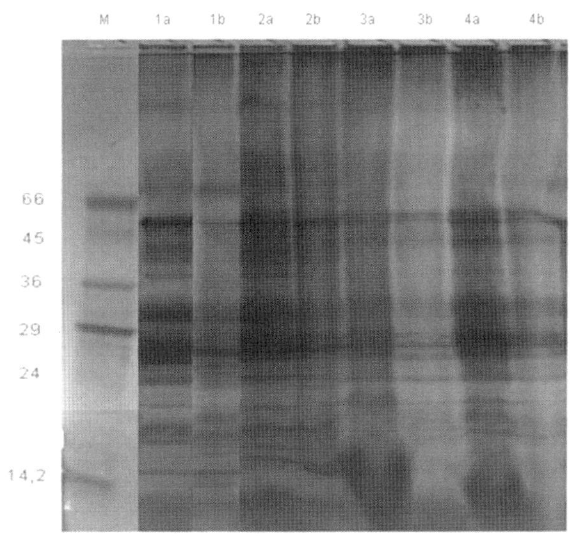

Fig. 4.8 Coomassie blue-stained polypeptide profiles of thylakoid membranes isolated from control and infected leaves. *1 Piper longum* L.; *2 Lycopersicon esculentum* Mill.; *3 Cucumis sativus* L.; *4 Solanum melongena* L., *a* healthy; *b* infected plants; *M* standard proteins (kDa): bovine serum albumin (*66*), albumin egg (*45*), glyceraldehydes-3-phosphate dehydrogenase (*36*), carbonic anhydrase (*29*), trypsinogen (*24*), and α-lactalbumin (*14,2*). Samples were applied to each slot corresponding to 50 μg of protein

Polypeptide analysis of thylakoid membranes revealed about 25 bands which have small changes in the thylakoid protein profile of the infected plants in position of 60–15 kDa. A comparison of thylakoid polypeptides of virus-infected leaves with those of the respective controls indicates a decrease in the amount of 47-, 33-, 28–24-, 17-, and 15-kDa polypeptides. Such changes were associated with a major decrease in the level of 33- and 17-kDa polypeptides. The extrinsic proteins of 33 and 17 kDa associated with the luminal surface of the thylakoid membranes are required for optimal functioning of the oxygen evolving machinery. Virus infection induced not only the loss of extrinsic proteins but also a marked loss of 47- and 43-kDa polypeptides in thylakoid membranes which may be due to greater disruption of the PS II complex.

Light-harvesting complexes play important roles in light absorption, thylakoid stacking, and energy distribution and any damage to these complexes will have multiple effects on the photosynthetic system (Muthuchelian et al. 2005). In our experiment, a significant loss of LHCP II (29–24 kDa) polypeptides was observed. This could be a reason for the observed marked loss of PS II activity in infected leaves.

The content of soluble proteins was reduced markedly in infected leaves. The loss of leaf soluble protein in infected leaves would be partially accounted for damaged chloroplasts or could be the result of inhibition of protein synthesis. Thus, our results suggest that the decrease in photosynthetic pigments, protein contents, photosynthetic electron transport activities (PS II and Fv/Fm ratio), and thylakoid membrane proteins in the infected leaves was due to virus infection, which induced rapid senescence. In addition, it has been found that the photosynthetic apparatus was inhibited on the donor side of PS II; thus, the result could be used in the development of virus resistant varieties.

When plants are exposed to microbial pathogens, they produce ROS that induce programmed cell death in the plant cells surrounding the infection site to effectively wall off the pathogen and terminate the disease process. The possible role of ROS scavenging in plants for protection against oxidative damage induced by viral infection was investigated by detecting the presence of superoxide anion and hydrogen peroxide in places of infection using NBT and 3,3′–diaminobenzidine tetrahydrochloride staining methods (Sultanova and Huseynova 2012). Accumulation of insoluble blue-colored formazan complex (reduced NBT) is an indicator of generation of ROS, particularly superoxide anion. DAB polymerizes to produce a brown precipitate on contact with hydrogen peroxide in the presence of peroxidase, and thus provides a useful marker of peroxide accumulation. In this study, a slightly lower level of DAB staining was observed in healthy leaves compared with infected leaves. Discoloration of the leaves was quantified using a digital imaging system. At the same time, some antioxidant enzymes in symptomatic plants were also studied. It was found that the activities of APX and CAT significantly increased in all symptomatic specimens compared to noninfected plants. Dynamics of GR and SOD (Cu/Zn-SOD) activities differed from those of CAT and APX, their activities slightly increased in stressed samples (Fig. 4.9).

Taking into account the specific distribution and roles of APX and CAT isoenzymes and their potential role in ROS production in each organelle of higher plants, it seems that the isoenzymes are expressed by distinct regulatory mechanisms. Hence, electrophoretic mobility profiling of APX and CAT isoenzymes was also studied. The activities of APX and CAT isoenzymes were predominant under viral pathogenesis, and gel electrophoresis indicated that slow-moving high molecular weight isoenzymes appeared and increased in all infected plants. Pathogenesis under virus infection resulted in increase of APX isoenzyme number up to 5 in leaves of *L. esculentum,* including formation of two new isoforms, and gain in enzyme electrophoretic spectra in infected leaves of *C. sativus* comparing to the control variant (Fig. 4.10 I).

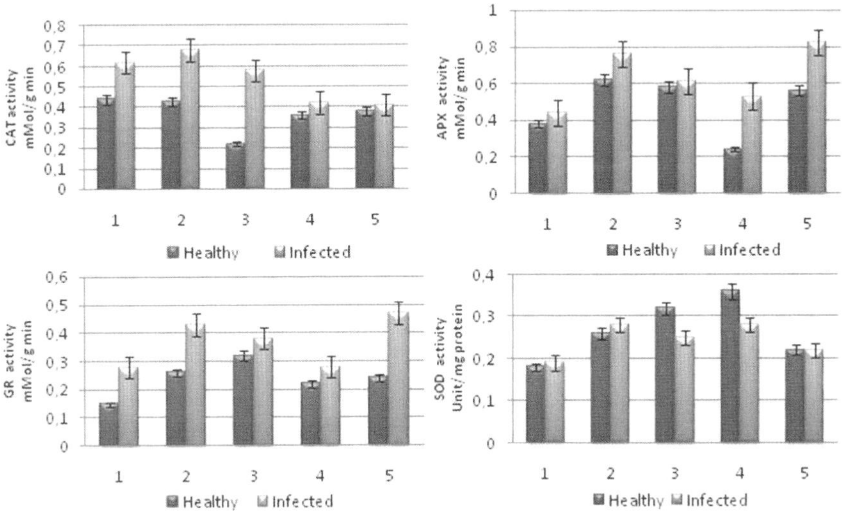

Fig. 4.9 Activities of catalase, ascorbate peroxidase, glutathione reductase, and superoxide dismutase in viral-infected plant leaves. *1 Solanum lycopersicum* L.; *2 Vicia faba* L.; *3 Cicer arietinum* L.; *4 Lens culinaris* L.; *5 Pisum sativum* L. Mean ± S.E (n=3)

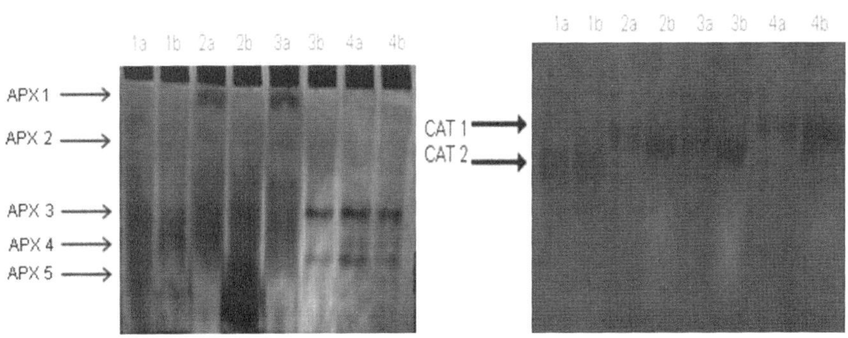

Fig. 4.10 Isoforms of ascorbate peroxidase (I) and catalase (II) activity determined in sodium phosphate buffer (pH 7.6) by 10% native PAGE. *1 Lycopersicon esculentum* Mill.; *2 Solanum melongena* L.; *3 Piper longum* L.; *4 Cucumis sativus* L., **a** healthy; **b** infected plants

Similar results were obtained in the study of *Pisum sativum* infected by *plum pox virus* and during analysis of the antioxidant response of *Nicotiana benthamiana* to infection with two strains of *pepper mild mottle virus* (Díaz-Vivancos et al. 2008; Hakmaoui et al. 2012; Sgherri et al. 2013). The electrophoretic profiles of catalase from infected vegetables identified two isoenzymes (CAT1, CAT2). As shown in Fig. 4.10 II, pathogenesis leads to a decrease in CAT2 and an increase in CAT1 profiles. The data indicated that a high activation of APX and CAT isoenzymes during viral stress may be a key event in hypersensitive response program. Thus, increase of APX and CAT activity and de novo synthesis of their multiple molecular forms under different stresses show their undoubted participation in antioxidant defense system.

However, a final conclusion on the incidence of these viruses can only be made after detailed screening of a large number of fields. The eradication of infected mother plants is a significant matter for the certification system of nursery material for growing crop plants and abovementioned data should be a promising beginning for developing tendency to reduce the virus-associated risks in Azerbaijan.

4 Conclusions and Future Perspective

To evaluate the risk of virus spreading, it is necessary to determine whether the insects/plants are viruliferous or not. Effective control is necessary for the identification of different viral diseases, infection source, and spreading mechanisms. Effective solution of such strategic problems as providing the population with high quality, ecologically pure foods, and other plant products and food safety of the country depends on the creation of new more productive, virus infection-resistant plants. It must be considered that plants are subjected to a combination of adverse conditions which requires understanding of how plants respond to different stress factors. This preliminary consideration is essential for understanding the performance of plants under stress conditions and also for improving stress tolerance and crop production.

The integration of the molecular and biochemical approaches will likely enable scientists to reconstruct the whole cascade of cellular events that lead to rapid responses and adaptation to the various biotic stress factors. Targeted approach that combines molecular, physiological, and metabolic aspects of plant stress tolerance is needed to improve the knowledge on the effects of gene expression and complete study of plants under stress factors. A better understanding of major physiological processes in response to various biotic stresses can lead to the selection of the appropriate promoter or a transcription factor which will be used for different crop improvement strategies.

Acknowledgments This work was financially supported by of the Science Development Foundation under the President of the Republic of Azerbaijan (grant EIF-2011-1(3)-82/48/3).

References

Aliyev JA, Suleymanov SY, Guseinova IM, Asadov A, Ismailov M (1992) Effect of specific translation inhibitors on polypeptide composition and spectral characteristics of wheat thylakoid membrane. Biochemistry 57:679–686

Anderson M, Prasad T, Stewart C (1995) Changes in isozyme profiles of catalase, peroxidase, and glutathione reductase during acclimation to chilling in mesocotyls of maize seedlings. Plant Physiol 109:1247–1257

Aramburu J, Galipienso L, Lopez C (2007) Reappearance of *Cucumber mosaic virus* isolates belonging to subgroup IB in tomato plants in North-eastern Spain. J Phytopathol 155:513–518

Atkinson NJ, Urwin PE (2012) The interaction of plant biotic and abiotic stresses: from genes to the field. J Exp Bot 63:3523–3543

Bertamini M, Grando MS, Muthuchelian K, Nedunchezhian N (2003) Effect of phytoplasmal infection on PS II efficiency and thylakoid membrane protein changes in field grown apple (*Malus pumila*) leaves. Physiol Mol Plant Path 61:349–356

Bradley BA, Blumenthal DM, Early RI, Grosholz ED, Lawler JJ, Miller LP, Sorte CJB, D'Antonio CM, Diez JM, Dukes JS, Ibanez I, Olden JD (2012) Global change, global trade, and the next wave of plant invasions. Front Ecol Environ 10:20–28

Britton KO, White P, Kramer A, Hudler GA (2010) New approach to stopping the spread of invasive insects and pathogens: Early detection and rapid response via global network of sentinel plantings. N. Z. J For Sci 40:109–114

Carnicer J, Coll M, Ninyerola M, Pons X, Sánchez G, Peñuelas J (2011) Widespread crown condition decline, food web disruption, and amplified tree mortality with increased climate change-type drought. Proc Natl Acad Sci U S A 108:1474–1478

Chen H, Fu R, Li X (2011) Tomato yellow leaf virus (TYLCV): The structure, ecotypes and the resistance germplasm resources in tomato. Afr J Biotechnol 10:13361–13367

Clark M, Adams A (1977) Characteristics of the microplate method of enzyme linked immunosorbent assay for the detection of plant viruses. J Gen Virol 34:475–483

Czosnek H (ed) (2007) Tomato yellow leaf curl virus disease: management, molecular biology, breeding for resistance. Springer, Dordrecht, pp 157–170

Czosnek H, Ghanim M (2012) Back to basics: are begomoviruses whitefly pathogens? J Integr Agric 11:225–234 (Agricultural Sciences in China)

Czosnek H, Eybishtz A, Sade D, Gorovits R, Sobol I, Bejarano E, Rosas-Díaz T, Lozano-Durán R (2013) Discovering host genes involved in the infection by the tomato yellow leaf curl virus complex and in the establishment of resistance to the virus using tobacco rattle virus-based post transcriptional gene silencing. Viruses 5:998–1022

Díaz-Pendón JA, Cañizares MC, Moriones E, Bejarano ER, Czosnek H, Navas-Castillo J (2010) Tomato yellow leaf curl viruses: Ménage à trois between the virus complex, the plant, and the whitefly vector. Mol Plant Pathol 11:441–450

Díaz-Vivancos P, Clemente-Moreno MJ, Rubio M, Olmos E, García JA, Martínez-Gómez P, Hernández JA (2008) Alteration in the chloroplast metabolism leads to ROS accumulation in pea plants in response to plum pox virus. J Exp Bot 59:2147–2160

Dijkstra J, Khan JA (2006) Description of positive-sense, single stranded RNA viruses. In: Khan JA, Dijkstra J (eds) Handbook of plant virology. CRC, New York, pp 253–388

Duan CG, Wang CH, Guo HS (2012) Application of RNA silencing to plant disease resistance. Silence 3:3–5

Edwards K, Johnstone C, Thompson CA (1991) Simple and rapid method for the preparation of plant genomic DNA for PCR analysis. Nucleic Acids Res 19:13–49

Eybishtz A, Peretz Y, Sade D, Akad F, Czosnek H (2009) Silencing of a single gene in tomato plants resistant to tomato yellow leaf curl virus renders them susceptible to the virus. Plant Mol Biol 71:157–171

Eyre C, Kozanitas M, Garbelotto M (2013) Population dynamics of aerial and terrestrial populations of *Phytophthora ramorum* in California forest under different climatic conditions. Phytopathology 103:1141–1152

Fischer AR, Tobi H, Ronteltap A (2011) When natural met social: a review of collaboration between the natural and social sciences. Interdiscip Sci Rev 36:341–358

Fitter A (2012) Why plant science matters. New Phytol 193:1–2

Fryer MJ, Ball L, Oxborough K, Karpinski S, Mullineaux PM, Baker NR (2003) Control of ascorbate peroxidase 2 expression by hydrogen peroxide and leaf water status during excess light stress reveals a functional organization of *Arabidopsis* leaves. Plant J 33:691–705

Golding J, Güsewell S, Kreft H, Kuzevanov VY, Lehvävirta S, Parmentier I, Pautasso M (2010) Species-richness patterns of the living collections of the world's botanic gardens: a matter of socio-economic? Ann Bot 105:689–696
Grigoras I, Gronenborn B, Vetten HJ (2010a) First report of a nanovirus disease of pea in Germany. Plant Dis 94:642
Grigoras I, Timchenko T, Grande-Pérez A, Katul L, Vetten HJ, Gronenborn B (2010b) High variability and rapid evolution of a nanovirus. J Virol 84:9105–9117
Grigoras I, Vetten HJ, Sultanova N, Winter S, Huseynova I, Mamedov A, Kheyr-Pour A, Gronenborn B (2010c) Single-stranded DNA plant viruses in Azerbaijan the state of the art. Proc Azerbaijan Natl Acad Sci (Biological Sciences) 65:152–157
Grigoras I, Timchenko T, Gronenborn B, Vetten HJ (2011) European nanoviruses: Identification of three new species and new DNA components. Phytopathology 101(6):64–64
Gurr S, Samalova M, Fisher M (2011) The rise and rise of emerging infectious fungi challenges food security and ecosystem health. Fungal Biol Rev 25:181–188
Haible D, Kober S, Jeske H (2006) Rolling circle amplification revolutionizes diagnosis and genomics of geminiviruses. J Virol Methods 135:9–16
Hakmaoui A, Pérez-Bueno ML, García-Fontana B, Camejo D, Jiménez A, Sevilla F, Barón M (2012) Analysis of the antioxidant response of *Nicotiana benthamiana* to infection with two strains of Pepper mild mottle virus. J Exp Bot 63:5487–5496
Ji YH, Zhang H, Zhang K, Li G, Lian S, Cheng ZB, Zhou YJ (2013) First Report of Tomato yellow leaf curl virus in *Acalypha australis* in China. Plant Dis 97(3):430.3–430.3
Jones RAC (2009) Plant virus emergence and evolution: Origins, new encounter scenarios, factors driving emergence, effects of changing world conditions, and prospects for control. Virus Res 141:113–130
Kariola T, Brader G, Helenius E, Heino JLP, Palva ET (2006) Early responsive to dehydration 15, a negative regulator of ABA-responses in *Arabidopsis*. Plant Physiol 142:1559–1573
Klimov VV, Allakhverdiev SI, Shuvalov VA, Krasnovsky AA (1982) Effect of extraction and re-addition of manganese on light reactions of Photosystem II preparations. FEBS Lett 148:307–312
Kumar G, Knowles N (1993) Changes in lipid peroxidation and lipolytic and free-radical scavenging enzyme during aging and sprouting of potato (*Solanum tuberosum* L.) seed-tubers. Plant Physiol 102:115–124
Kumar PL, Hanna R, Alabi OJ, Soko MM, Oben TT, Vangu GHP, Naidu RA (2011) Banana bunchy top virus in sub-Saharan Africa: Investigations on virus distribution and diversity. Virus Res 59:171–182
Laemmli UK (1970) Cleavage of structural proteins during the assembly of the head of bacteriophage. Nature 227:680–685
Lai K, Lorenc, MT, Edwards D (2012) Genomic databases for crop improvement. J Agron 2:62–73
Lin HX, Rubio L, Smythe A, Jiminez M, Falk BW (2003) Genetic diversity and biological variation among California isolates of *Cucumber mosaic virus*. J Gen Virol 84:249–258
Lozano G, Moriones E, Navas-Castillo J (2006) Complete nucleotide sequence of the RNA2 of crinivirus *tomato chlororis virus*. Arch Virol 151:581–587
Luna AP, Morilla G, Voinnet O, Bejarano ER (2012) Functional analysis of gene silencing suppressors from *Tomato yellow leaf curl* disease viruses. Mol Plant Microbe Interact 25:1294–1306
Mahalingam R, Shah N, Scrymgeour A, Fedoroff N (2005) Temporal evolution of the *Arabidopsis* oxidative stress response. Plant Mol Biol 57:709–730
Matyssek R, Wieser G, Calfapietra C, de Vries W, Dizengremel P, Ernst D, Jolivet Y, Mikkelsen TN, Mohren GMJ, Thiec DL, Tuovinen JP, Weatherall A, Paoletti E (2012) Forests under climate change and air pollution: gaps in understanding and future directions for research. Environ Pollut 160:57–65
McDowell NG, Beerling DJ, Breshears DD, Fisher RA, Raffa KF, Stit M (2011) The interdependence of mechanisms underlying climate-driven vegetation mortality. Trends Ecol Evol 26:523–532
Mc-Kinney G (1941) Absorption of light by chlorophyll solutions. J Biol Chem 140:315–322

Mochizuki T, Ohki ST (2012) *Cucumber mosaic virus*: Viral genes as virulence determinants. Mol Plant Pathol 13:217–225
Muthuchelian K, La Porta N, Bertamini M, Nedunchezhian N (2005) Cypress canker induced inhibition of photosynthesis in field grown cypress (*Cupressus sempervirens* L.) needles. Physiol Mol Plant Pathol 67:33–39
Nakano Y, Asada K (1981) Hydrogen peroxide is scavenged by ascorbate-specific peroxidase in spinach chloroplasts. Plant Cell Physiol 22:867–880
Natha K, Jajooc A, Sharma R, Timilsinab R, Parkd YS, Aroe EM, Namb H, Leea CH (2013) Towards a critical understanding of the photosystem II repair mechanism and its regulation during stress conditions. FEBS Lett 587(21):3372–3381
Navas-Castillo J, Fiallo-Olive E, Sanchez-Campos S (2011) Emerging virus diseases transmitted by whiteflies. Annu Rev Phytopathol 49:219–248
Onda Y (2013) Oxidative protein-folding systems in plant cells. Int J Cell Biol 2013:44–59
Parks CG, Bernier P (2010) Adaptation of forests and forest management to changing climate with emphasis on forest health: a review of science, policies and practices. For Ecol Manage 259:657–659
Pautasso M, Döring TF, Garbelotto M, Pellis L, Jeger MJ (2012) Impacts of climate change on plant diseases-opinions and trends. Eur J Plant Pathol 133:295–313
Pita JS, Roossinck MJ (2013) Fixation of emerging interviral recombinants in *Cucumber Mosaic Virus* populations. J Virol 87(2):1264–1269
Régnière J (2012) Invasive species, climate change and forest health. In: Schlichter T, Montes L (eds) Forests in development: a vital balance. Springer, Berlin, pp 27–37
Rohr JR, Dobson AP, Johnson PTJ, Kilpatrick AM, Paull SH, Raffel TR et al (2011) Frontiers in climate change-disease research. Trends Ecol Evol 26:270–277
Sade D, Eybishtz A, Gorovits R, Sobol I, Czosnek H (2012) A developmentally regulated lipocalin-like gene is overexpressed in *Tomato yellow leaf curl virus*-resistant tomato plants upon virus inoculation and its silencing abolishes resistance. Plant Mol Biol 80:273–287
Sahu AK, Nehra C, Marwal A, Gaur RK (2013) First report of a betasatellite associated with Radish leaf curl virus infecting new host *Cyamopsis tetragonoloba* (Guar bean) in India. Int J Adv Res 1(8):740–743
Sedmak JJ, Grossberg SE (1977) A rapid, sensitive, and versatile assay for protein using Coomassie brilliant blue G 250. Anal Biochem 79:544–552
Sela N, Lachman O, Reingold V, Dombrovsky A (2013) A new cryptic virus belonging to the family Partitiviridae was found in watermelon co-infected with Melon necrotic spot virus. Virus Genes 47(2):382–384
Sgherri C, Ranieri A, Quartacci MF (2013) Antioxidative responses in *Vitis vinifera* infected by grapevine fan leaf virus. J Plant Physiol 170(2):121–128
Sheikha M, Naqvi QA, Mahmooda I (2013) Molecular detection of an isolate of Cucumber mosaic virus infecting *Oxalis corymbosa* in Aligarh, India. Arch Phytopathol Plant Prot 46(15):1816–1820
Siebold M, von Tiedemann A (2012) Potential effects of global warming on oilseed rape pathogens in Northern Germany. Fungal Ecol 5:62–72
Soliman MM (2013) Avoidance of black bean aphids, the vector of necrotic yellow virus (FBNYV) to faba bean plants—Middle Egypt: an overview. Glob J Plant Ecophysiol 3(2):74–82
Sultanova NF, Huseynova IM (2012) Determination of superoxide anion radical and antioxidant enzymes activities of in viral infected vegetable plants. Proc Azerbaijan Natl Acad Sci (Biological Sciences) 67(1):13–19
Tomback DF, Achuff P (2010) Blister rust and western forest biodiversity: ecology, values and outlook for white pines. Forest Pathol 40:186–225
Vetten HJ, Bressan A, Gronenborn B (2012) Aphid transmission of viruses of the family Nanoviridae. In: Brown J (ed) Vector-mediated transmission of plant pathogens. American Phytopathological Society Press, St. Paul
Wang M, Masuta CH, Smith NA, Hanako S (2012) RNA silencing and plant viral diseases. Mol Plant Microbe Interact 25(10):1275–1285

Watanabe S, Bressan A (2013a) Tropism compartmentalisation and retention of *Banana bunchy top virus* (Nanoviridae) in the aphid vector *Pentalonianigronervosa*. J Gen Virol 94:209–219

Watanabe S, Greenwell M, Bressan A (2013b) Localization, concentration, and transmission efficiency of *Banana bunchy top virus* in four asexual lineages of *Pentalonia aphids*. Viruses 5(2):758–776

Watt MS, Ganley RJ, Kriticos DJ, Manning LK (2011) Dothistroma needle blight and pitch canker: the current and future potential distribution of two important diseases of *Pinus* species. Can J For Res 41:412–424

Webber J (2010) Pest risk analysis and invasion pathways for plant pathogens. N. Z. J For Sci 40:45–56

Wintermantel WM, Wisler GC, Anchieta AG, Liu HY, Karsen AV, Tzanetakis IE (2005) The complete nucleotide sequence and genome organization of tomato chlorosis virus. Arch Virol 150:2287–2298

Yamaji Y, Maejima K, Komatsu K, Shiraishi T, Okano Y, Himeno M, Sugawara K, Neriya Y, Minato N, Miura C, Hashimoto M, Namba S (2012) Lectin-mediated resistance impairs plant virus infection at the cellular level. Plant Cell 24:778–793

Yannarelli GG, Fernandez-Alvarez AJ (2007) Glutathione reductase activity and isoforms in leaves and roots of wheat plants subjected to cadmium stress. Phytochemistry 68:505–512

Chapter 5
Salt Stress and Sugar Beet Improvement: Challenges and Opportunities

Abazar Rajabi, Samar Khayamim, Zahra Abbasi and Eric Ober

Abstract Drought and salinity are the major abiotic stresses limiting the production of crops including sugar beet in water-limited environments of the world. With annual precipitation of 240 mm, Iran is classified as a dry region. Furthermore, more than 21 % of the country's area is occupied by saline soil. Therefore, it is necessary to adopt strategies that will maximize yields and economic returns from stressful environments while minimizing environmental impacts. Among these are improved cultural practices; breeding of new varieties, which involves screening and selection of the existing germplasm; utilization of novel genes through transgenic modification; application of exogenous osmoprotectants, etc. Although conventional selection and breeding programs are making achievement in enhancing the abiotic stress tolerance of crops, breeding for stress tolerance should be given high research priority to accelerate these efforts. However, the extent and rate of progress gained through conventional breeding programs is limited due to the interplay of mechanisms of abiotic stress tolerance that are controlled by the expression of many genes. Furthermore, current techniques employed for selecting tolerant plants are often time-consuming and expensive. Using advanced molecular techniques, some researchers are showing promising results in understanding

A. Rajabi (✉)
Department of Plant Breeding, Sugar Beet Seed Institute, 31585-4114, Karaj, Iran
e-mail: rajabi@sbsi.ir

S. Khayamim
Department of Agronomy, Sugar Beet Seed Institute, 31585-4114, Karaj, Iran
e-mail: khayamim@sbsi.ir

Z. Abbasi
Department of Sugar Beet Research, Isfahan Agriculture and
Natural Resources Research Center, 81785-19, Isfahan, Iran
e-mail: abasi@sbsi.ir

E. Ober
The John Bingham Laboratory, National Institute of Agricultural Botany (NIAB)
Huntingdon Road, Cambridge, UK, CB3 0LE
e-mail: eric.ober@niab.com

P. Ahmad et al. (eds.), *Improvement of Crops in the Era of Climatic Changes*, Volume 2,
DOI 10.1007/978-1-4614-8824-8_5,

the molecular basis of tolerance to abiotic stress and increasing stress tolerance in model species and some crops. These findings emphasize that future research should focus on physiological, molecular, and metabolic dimensions of tolerance to stress to facilitate the development of crops with an inherent capacity to withstand abiotic stresses. Additionally, testing the performance of elite genetic materials under realistic field conditions cannot be overlooked. In this chapter, the challenges and opportunities for improvement of salt tolerance in sugar beet (*Beta vulgaris* L.) are discussed. Many of the principles, however, apply to most crop species.

Keywords Sugar beet · Salinity tolerance · Abiotic stress · Morphological · Physiological · Screening · Breeding

1 Introduction

Generally, the human being lives on a salty planet, with the concentration of sodium chloride (NaCl) in most of its water being approximately 30 mgl^{-1}. Due to this salt solution, the cultivated lands have been negatively affected (Flowers 2004). Salinity is one of the most prevalent abiotic stresses, which negatively affects agricultural productivity and restricts the land use. It is reported that 6 % of the total (Talei et al. 2013) and 20 % of the irrigated lands of the world (Negrão et al. 2011) have been adversely affected by soil salinity. This is more severe as one-third of the world's food production comes from the irrigated lands (Linh et al. 2012). In Asia, salinity has influenced 21.5 million hectares of lands due to which up to 50 % of productive lands will be lost by 2050 (Nazar et al. 2011).

The degree to which the plants are affected by salinity depends on the salinity level. At extremely high salt concentrations, plants are killed due to the joint effects of reduced water potential and ion toxicity; leaf burning and severe growth inhibition are caused by high salt concentrations. At lower salinity levels, reduced growth rate is often the sole sign of salt effect, with slightly darkened and/or thickened leaves in some species (Shannon 1985). Plant physiological responses to salinity occur at various levels. At the cellular level, ion distribution, osmolyte accumulation, enzyme reactions, osmotic adjustment, and genetic control are considered; at the tissue level, leaf thickness, salt exclusion, and stomatal conductance; and at the whole plant level, germination ability, vigor, and growth are taken into account (Seaman 2007). Properties such as compartmentalization of organic solutes and ions needed for salt tolerance occur at various plant levels. Thus, salt tolerance involves processes at all levels of plant function, and not just those of isolated cells or tissues (Pitman 1984).

Some general morphological symptoms of salt stress on crop plants are whitening and burning of leaf tip (salinity impact), browning and death of leaf (sodicity impact), leaf rolling, stunted plant growth, reduced root growth, irregular growth, low yield, and low harvest index (Singh 2006). Among the physiological and biochemical symptoms of salinity are: high Na^+ transfer to shoot and its accumulation in older leaves, higher Cl^- and lower K^+ uptake, reduced shoot and root weights, decreased uptake of P and Zn, increased compatible solutes, and increased levels of polyamines (Singh 2006).

Approximately 35% of the world's sugar is supplied by an important crop, the sugar beet (Liu et al. 2008). The performance of the crop is often not fully expressed due to the occurrence of abiotic stresses such as drought and salinity. Sugar beet is considered as a salt-tolerant crop (Shannon 1985; Jamil et al. 2006). However, its yield is reduced by saline soil and saline water: 50% of the yield is decreased at electrical conductivity of soil saturation extract (Ecs) of 13.7 dSm^{-1} (Morillo-Velarde and Ober 2006). Sugar beet shows the highest sensitivity to salinity at germination and early seedling growth stages (Ghoulam and Fares 2001; Jamil et al. 2006). At this stage, the soil saturation extract should be lower than 3 dSm^{-1} (cited in Morillo-Velarde and Ober 2006). The irrigation water quality is also an important issue in many arid and semi-arid regions such as Iran where sugar beet production is wholly reliant on irrigation water. A high atmospheric evaporative demand in these areas results in the accumulation of salts at soil surface and hence reduces germination. This necessitates the leaching of accumulated salts from soil. As sugar beet is able to absorb some of the salts in irrigation water, beet production could help to maintain the quality of soils and sustainability of production systems (Morillo-Velarde and Ober 2006).

Ghoulam et al. (2002) investigated the impact of increasing levels of salinity on agro-physiological characters of sugar beet cultivars and found that such growth characters as leaf area and leaf and root weights were decreased by high NaCl concentrations but the leaf number was not much affected. The highest NaCl concentration (200 mM) revealed varietal differences for almost all of the considered characters. In salt-stressed sugar beet, leaf thickness increased due to declined cellular expansion, increased leaf dry matter, and reduced specific leaf area (Robinson et al. 1983).

2 Management of the Salt-Affected Areas

In general, salt-affected areas could be managed through the following approaches (Singh et al. 2010): (1) modification of environment: Change the environment for the normal growth of plants. This could be done by methods such as leaching of salts from the root zone and proper drainage practices. Such cultural practices as alteration of sowing pattern, proper nutrition, and crop protection against diseases and pests can help to increase crop productivity in stressful environments. Results of the sugar beet transplanting method in salt-affected areas indicated that this method has a higher effect on yield than on quality (Rajabi 2010). Also, early sowing and two-row planting are recommended for salt-affected areas due to the facilitated irrigation and lack of salt accumulation at the seedling emergence zone (Rajabi 2010).

(2) Modification of crop: Select or develop crop varieties that can withstand the salt stress. This could be an environmentally sustainable approach due to the fact that water resources are reserved and destruction of soil structure and salinization are reduced. (3) Hybrid approach which is the combination of the first and second approaches. The advantages of the hybrid approach are that it is more viable, highly productive, and has low resource cost.

3 Challenges in Mitigating Salt Stress

Genetic improvements in salt tolerance have been hindered due to: (1) nonuniformity of chemical and physical soil composition in the salt-affected field, (2) various physiological mechanisms determining the tolerance level, (3) dependence of plant phenotypic responses to growth stage, (4) insufficient or poorly characterized genetic variation, (5) multigenic nature of inheritance and complicated genotype × environment interactions, (6) absence of suitable screening methods and selection tools for assessment of germplasm and segregating populations, and (7) necessity for quick and dependable molecular markers to cull the inferior genotypes (Subbarao and Johansen 1999; Arzani 2008). Salt tolerance is a quantitative trait defined in many ways depending on changes with plant growth and development; thus, it is difficult to assess the salt tolerance of genotypes with different growth or development rate. Also, concentrations of major and minor nutrients in the root zone (Shannon 1996) and field management (Meiri and Plaut 1985) affect the level of salt tolerance. Some other factors which limit the success in salt-affected areas are: (1) Salt stress seldom happens in isolation but occurs together with other stresses such as drought and heat; (2) gaps in understanding mechanisms of tolerance; (3) low priority in breeding programs; and (4) the involvement of few researchers, and new scientists entering the field (Singh 2006).

4 Mechanisms of Salt Tolerance

Responses to salinity are not simple and breeding for salt tolerance requires a combination of various physiological mechanisms. Plants have developed different mechanisms for salt tolerance and adjustment to ionic and osmotic stress induced by high levels of salt stress. Among which most important is osmotic adjustment, which involves the accumulation of solutes such as K^+, sugars, glycine betaine, and proline (Hasegawa et al. 2000; Munns 2005). Effective selection of salt-tolerant plants is needed to understand the impact of salinity on morphological and physiological processes and ways how plants can tolerate salt stress conditions (De Vos et al. 2010).

Three interrelated aspects of plant activity are important for achieving salt tolerance: Damage must be averted, homeostatic conditions must be reestablished, and growth must recommence (Seaman 2007). Plants are affected by high salinity in two major ways: High salt concentrations in soil perturb the root capacity for water extraction, and high salt concentrations inside the plant itself can be noxious, due to which many physiological and biochemical processes are inhibited (Munns 2002). Low osmotic potential of soil solution (water stress), nutritional imbalance, specific ion effects (salt stress), or a combination of these factors are related to the adverse effects of salinity on plant growth (Oliveira et al. 2013).

Therefore, there are two main mechanisms for salt tolerance: those minimizing the entry of salt into the plant and those minimizing the concentration of salt in the cytoplasm. Both types of mechanisms are observed in halophytes; these plants "exclude" salt properly, but efficiently sort out the salt in vacuoles. Thus, they can grow for long periods in salt-affected soils. The salt is also excluded well by some glycophytes, but they are not as efficient as halophytes in compartmentalizing the residual salt. Most glycophytes are poorly able to exclude salt, so it accumulates to toxic levels in the transpiring leaves (Munns 2002). Physiological adaptations to salinity and other stresses such as cold and drought stress show some similarities. Plant adaptation to salinity could be enhanced by improvement of drought tolerance. However, for a plant to be specifically adapted to saline soil, it should regulate the uptake and compartmentalize NaCl when its accumulation reaches to toxic levels (Munns 2002).

5 Genetics of Salt Tolerance

Salt tolerance is a multigenic trait (Munns 2011). In some species, these genes are dispersed across the genome. Breeding for improved salt tolerance is complicated due to epistasis, genotype × environment interaction, and large environmental effects. Utilization of genetic mechanism is the most important approach for dealing with salinity problem (Joshi 2011). Lyon (1941) was probably the first who studied the inheritance of salt tolerance. He crossed tomato (*Lycopersicon esculentum*) with *L. pimpinellifolium* and found that the hybrid was more affected by salt than either parent. Akbar et al. (1986) and Akbar and Yabuno (1977) found that at least three genes control the expression of sterility, an important factor in rice yield under saline conditions. Salt tolerance in pigeonpea (*Cajanus cajan*; Subbarao et al. 1990) and sorghum (Azhar and McNeilly 1988) has also shown to be governed by dominance gene effect. A 6 × 6 diallel cross-analysis in rice revealed that narrow-sense heritability values for salinity tolerance score, K, and Na/K ratio were 0.65, 0.50, and 0.41, respectively, with both the additive and dominance gene effects being significant for all the three characters (Singh 2006). These examples demonstrate that salt tolerance is a genetically complex trait (Shannon 1985) which shows heterosis, dominance, and additive effects.

6 Selection Criteria for Salt Tolerance

6.1 Morphological Characters

Plant height, leaf area, leaf injury, relative growth rate, relative growth reduction, and yield are among the agro-morphological characteristics used for salt tolerance (Ashraf and Harris 2004). Also, root and shoot length, germination percentage, percentage of dead and live leaves, and seedling survival have been used successfully as simple selection criteria in different plants (Sadat Noori and McNeilly 2000). Seed germination, radical and hypocotyl lengths, and leaf succulence are different morphological parameters used as selection criteria under salinity stress for several crops, including sugar beet.

6.1.1 Seed Germination

Seed germination is the most important characteristic of seed quality, and significant correlations between laboratory tests and plant establishment in the field (Sadeghian and Yavari 2004) increase the value of this trait. Seed germination is affected by environment and seed processing (Apostolides and Goulas 1998). In sugar beet seed, the percentage of germination and establishment is affected more by genetic effects than germination rate, so they could be used for genotype selection (Sadeghian and Khodaii 1998). However, germination rate was found to be the best selection index for salt tolerance in sugar beet (Mohammadian 1995).

6.1.2 Percentage of Abnormal Seedlings

Percentage of abnormal seedlings could be a good characteristic for screening sugar beet genotypes to salinity stress at early growth stages because a negative and significant correlation was observed between abnormal seedlings in the laboratory and sand germination in greenhouse and yield in the field (Khayamim et al. 2014). Selection of genotypes based on dead and viable plants and seedling survival under salinity stress would be a reliable method in addition to seed germination and root length (Sadat Noori and McNeilly 2000). The determination of abnormal embryos is difficult and requires technical skill and precision.

6.1.3 Root and Shoot Length

Measurement of sugar beet root length in the laboratory and under high osmotic pressure is suggested as a fast and effective method for evaluating the tolerance of genotypes to drought and salt stresses (Habibi 1993). Generally, salt-tolerant genotypes have greater root and hypocotyl lengths than susceptible types. Saline soils decrease water absorbance and consequently cell development. This may explain how effects of salt stress were found to affect root length to a greater extent than hypocotyl length (Jamil et al. 2006; Jafarzadeh and Aliasgharzadeh 2007; Khayamim et al. 2014). Nevertheless, root length may be a reliable index for early season screening of sugar beet genotypes for salt tolerance. Shoot growth shows more sensitivity to salt-induced osmotic stress than root growth. Salinity-induced reduction in shoot growth is usually expressed as reduced leaf area and stunted shoots. Both cell division and cell elongation govern the final leaf size. Salt stress does not affect leaf initiation in sugar beet, but leaf expansion showed salt sensitivity (Carillo et al. 2011).

6.1.4 Succulence

Succulence is a trait which is generally observed in salt-tolerant species growing in salt stress conditions (Hajiboland et al. 2009). Sugar beet leaves become thicker during the growth because of decreased cell expansion and greater dry matter increment of salinized plants. More water per unit leaf area is stored in succulent leaves. Indeed, some halophytic species have evolved special water-storage cells in leaves (Voznesenskaya et al. 2010).

6.1.5 Yield

Yield of field crops is limited by soil salinity. Low salt concentrations (less than 150 mM) have a low or no impact on yield. Consequently, there has been some success with irrigation of sugar beet with brackish water (Azzazy 2004), although care is required with this practice to prevent salt buildup in soil. With the increase of salt concentrations, yields decrease to zero, since the majority of glycophytes are not able to grow at high salinity levels (Carillo et al. 2011). The proportional decrease of root and white sugar yields of sugar beet under environmental stresses such as salinity and drought can be used as selection criteria for tolerant genotypes (Ranji and Parvizi 1996; Abdel and Zanovny 2004; Ebrahimian et al. 2008; Abbasi and Rezaei 2014). Although yield is the most direct criterion for assessing crop response to abiotic stresses such as salinity, the inheritance of yield is complex and considerably influenced by the environment (Ashraf 2004).

6.2 Physiological and Biochemical Characters

In sugar beet, osmotic adjustment is achieved by accumulation of ions and organic compounds (Geissler et al. 2009). Nontoxic solutes (osmoprotectants or compatible solutes) are synthesized and accumulated in some plants to withstand osmotic stresses. Osmoprotectants can accumulate to considerable levels without disturbing metabolism, some of which can also prevent enzymes and membranes from being damaged by high salt concentrations (Yancey 1994). Some other osmoprotectants (especially polyols) protect plants against reactive oxygen species (ROS; Smirnoff 1998). These compounds are located in the cytoplasm where it is important not to disrupt biochemical reactions while maintaining osmotic balance, so water potential will be maintained and cellular structure, protein, and enzyme conformations will be conserved, probably through phenomena related to macromolecular crowding and preservation of water structure at protein interfaces (Fayer 2012). Thus, they are known as "compatible" compounds because of these properties (Kafi et al. 2010). The main biochemical parameters of stress tolerance in plants contain trehalose, proline, glycine betaine, ROS scavengers, and polyols (Ashraf and Harris 2004; Sayfzadeh et al. 2011).

6.2.1 Proline and Glycine Betaine

Osmotic adjustment is mediated and the subcellular structures of stressed plants are protected by glycine betaine and proline (Ahmad and Sharma 2008). The accumulation of these two osmolytes has been reported to be positively correlated with stress tolerance (Yamada et al. 2009). Proline is an important osmoprotectant which in addition to adaptation of plants to stress may play a role in many biological processes such as osmotic adjustment, cell protection and preservation of the structural strength of the cell (membrane and protein), antioxidant action, energy transport, carbon and nitrogen storage, and several other roles which are needed for stability of the cell (Kuznetsov and Shevyakova 1999).

Although some plants like *Paspalum vaginatum* accumulate proline, no apparent differences among salt-tolerant and -intolerant genotypes were found in these plants (Lee et al. 2008). This suggests that proline may not be the best physiological or biochemical marker for salt screening. However, it was shown in maize roots growing at low water potentials that deposition of proline increased in elongating cells, and therefore accumulation was not a consequence of growth inhibition (Ober and Sharp 1994). Proline accumulation in most tolerant plants is greater than sensitive ones under environmental stresses (Ashraf and Foolad 2007; Ahmad et al. 2012a, b, 2013).

Osmolyte accumulations differ among plant species, plant organs, stress levels, and rate of stress development in plants. Long-term water cessation induces accumulation of different osmolytes and osmoprotectors in sugar beet organs. Glycine betaine and sucrose play a major role in the youngest leaves, whereas sucrose, proline, and betaine have a dominant role in older leaves but in storage roots the main role is played by soluble sugars (Choluj et al. 2008). It was reported that proline accumulation in sugar beet under salinity stress in greenhouse experiments was greater in the first 5 h of stress, which then decreased after 24 h of stress (Ranji et al. 1996). Water and salt stresses provoke the accumulation of many amino acids in sugar beet, but the increase in proline was greater than the others. However, it was noticed that sugar beet was tolerant to low levels of salt (about 150 mM NaCl) and more salt could increase proline accumulation (Gzik 1996). In contrast, it was reported that the contents of proline were too low to have a significant role in osmotic adjustment and therefore could not be used to reliably identify salt-tolerant sugar beet varieties.

Glycine betaine accumulation played the main role in osmotic adjustment of sugar beet under osmotic stress (Ghoulam et al. 2002) but it could mask the contribution of other nitrogenous components. There are reports that show proline concentration increased in stressed sugar beet and the increase was more in tolerant genotypes than that in sensitive types (Pakniyat and Armion 2007; Monreal et al. 2007). In sugar beet roots, proline level was found to be positively correlated with the glucose level (Monreal et al. 2007). Proline is also a good indicator of abiotic stress, because its level decreases when the stress is relieved (Sharma et al. 2011). The greater proline level of spring-sown beets in summer than in the subsequent autumn suggests that the proline accumulated in summer could have been used to supply nitrogen and energy requirements following stress relief later in the season (Monreal et al. 2007). The growth stage of plants also has an important role in osmolyte accumulation, such that the proline content in leaves of sugar beet increased about 20 % at leaf development stage but was reduced by about 10 % at harvest time (Monreal et al. 2007). Although the changes were not significant at harvest time, it can be speculated that at the late stages of plant growth, all materials and supplies were spent for sugar production (Khayamim 2010).

6.2.2 Antioxidants

When the plants are subjected to adverse environmental conditions, production of ROS increases (Ahmad et al. 2008, 2010a, b, 2011a, b; Ahmad and Umar 2011; Rasool et al. 2013). ROS are scavenged by both enzymatic and nonenzymatic mechanisms. The concentrations of superoxide radical and hydrogen peroxide are regulated by enzymatic mechanisms. The main enzymes include glutathione reductase (GR), ascorbate peroxidase (APX), superoxide dismutase (SOD), and glutathione-synthesizing enzymes (Mittova et al. 2003; Ahmad et al. 2008, 2010a, b, 2011a, b; Ahmad and Umar 2011; Rasool et al. 2013). Membrane is damaged in plants due to the stress-induced ROS. Stress-induced damage at the cellular level is manifested as free radical-induced peroxidation of lipid membranes. Peroxidation of membrane lipids results in increased level of malondialdehyde (MDA) which is often used as sign of oxidative damage (Nagesh Babu and Devaraj 2008). So, production of less MDA in tolerant plants under stress condition should be indicative of less damage to membranes (Ahmad et al. 2012a, b; Rasool et al. 2013).

6.2.3 Carbohydrates

The other class of osmoprotectants which is important in osmotic adjustment is carbohydrates. Simple and complex sugars accumulate under salt stress (Parida et al. 2004) and play a major role in osmotic adjustment, osmoprotection, carbon storage, and radical scavenging. Carbohydrates such as trehalose protect membranes and proteins in stressed cells and decrease aggregation of denatured proteins (Singer and Lindquist 1998; Fayer 2012). Salt stress decreases soluble carbohydrates in most plants, including sugar beet (Niazi et al. 2004; Geissler et al. 2009). Plants need energy under salinity stress conditions for the production of high-energy materials such as sugar and proteins (Niazi et al. 2004). Decreased concentrations of soluble carbohydrates in some plants under salt stress are often associated with decreased protein, but in some halophytes stress increases total proteins and decreases soluble carbohydrates (Geissler et al. 2009). Stress can affect the C to N balance and partitioning of energy between protein and carbohydrate synthesis (Coello et al. 2011). In salt-stressed sugar beet, soluble carbohydrate reduction was 15, 28, 8, and 2 % at four-leaf, establishment, leaf development, and harvesting growth stages of sugar beet, respectively (Khayamim 2010). It was concluded that there was a greater requirement for carbohydrates for osmotic adjustment and salt tolerance of sugar beet at the establishment stage.

6.2.4 Ionic Adjustment

Salinity results in both ionic and osmotic stress on plants. Initially, plants exposed to salinity stress (first minutes to hours of initial tension) responded to osmotic stress as reduced leaf expansion (Munns 2002). These effects can continue during the exposure; consequently, cell expansion and cell division are inhibited and stomata are closed (Munns 2002). When the plants are exposed to salinity for a long time, they experience ionic stress, due to which the adult leaves senesce prematurely, and thus the photosynthetic area is reduced (Carillo et al. 2011).

Ionic stress is caused by Na^+ accumulation in leaf tissues, which leads to growth inhibition and eventually necrosis. Leaves show more susceptibility than roots to Na^+ probably because Na^+ and Cl^- accumulation is higher in shoots than in roots (Tester and Davenport 2003). There are reports that Na^+ and Cl^- are accumulated in sugar beet leaves by salt stress (Ghoulam et al. 2002; Pakniyat and Armion 2007). Maintaining better Na^+ and K^+ balance is more apparent in tolerant than susceptible plants. This balance could be achieved by distribution of ions into vacuoles and production of organic osmotica (Jampeetong and Brix 2008). In *Sorghum bicolor*, Ca^{2+} can ameliorate some of the effects of Na^+ by helping to preserve K^+ to Na^+ balance (Colmer et al. 1996) and overcoming the inhibition of cell expansion (Bernstein et al. 1993). Salinity inhibits Ca^+ transport; hence, supplemental Ca^{2+} can help maintain processes such as the apoplastic ROS levels required for growth (Shoresh et al. 2011). Sugar beet genotypes that showed greater sugar concentrations and better yields under saline conditions were associated with greater accumulation of K^+ and α-amino N compounds in roots (Tsialtas and Maslaris 2009).

Osmotic adjustment appears to play a role in salinity tolerance in sugar beet (Katerji et al. 1997). Although drought has a lower effect on ionic stress, water shortage resulted in a significantly higher ratio of univalent to divalent cations in young leaves and petioles of sugar beet and it was slightly decreased in taproots (Ashraf 2004). Only with a greater understanding of the regulation of cellular ion concentrations would ion concentration measurements function as selection criteria for salt tolerance.

6.2.5 Photosynthesis and Respiration

Photosynthesis and respiration are processes that can be directly affected by environmental stresses such as drought and salinity, which can have an impact on growth and crop productivity. The ability to maintain photosynthetic rates in saline conditions may be an important indicator of stress tolerance (Ashraf et al. 2007). The degree to which photosynthesis is reduced depends on the concentration of salts, and differences in intrinsic susceptibility to salt stress (Robinson et al. 1983). Products of photosynthesis are required for the synthesis and accumulation of storage and structural materials as well as solutes for osmotic adjustment. In some saline conditions (200 mM NaCl), photosynthesis was unaffected while growth increased in fodder and wild beet genotypes (Niazi et al. 2000). In other experiments, photosynthesis was decreased to a greater extent than growth (Niazi et al. 2004). Salinity also inhibits photosynthesis via stomatal closure (Robinson et al. 1983; Orcutt and Nilsen 2000) and non-stomatal factors (Gordon and Duniway 1982; Harley et al. 1992; Delfine et al. 1999). The decreases in transpiration and daily water stress indices have been suggested as methods to classify varieties according to salt tolerance (Katerji et al. 2003). Regulation of photosynthetic enzymes such as ribulose bisphosphate carboxylase (Plaut and Heuer 1985; Orcutt and Nilsen 2000) also may be an important indicator. Light reactions are susceptible to salt stress, and photosystem II (PSII) may be damaged in plant species susceptible to salinity (Orcutt and Nilsen 2000). In a report comparing two sugar beet genotypes, photosynthetic rate increased during 4 days of exposure to 180 mM NaCl, but then decreased in one genotype while the other remained unaffected (Heuer and Plaut 1989). The stress effects depend on the age and position of leaves within the canopy (Ober et al. 2005), so these variables must be taken into account when comparing different genotypes or experiments.

During the past several decades, extensive research has been conducted to examine the impact of salt stress on photosynthesis. However, many of them have only correlative nature, and further research is still needed to elucidate the specific changes occurring during photosynthesis under saline conditions. These processes are tightly interlinked, so that a minor change in one process may induce a series of events that finally inhibit the overall rate of photosynthesis (Ashraf 2004).

6.2.6 Chlorophyll and Chlorophyll Fluorescence

There are many reports concerning the effects of environmental stress using chlorophyll fluorescence techniques. The ratio of variable to maximum fluorescence yield (F_v/F_m), known as maximum quantum yield of PSII, can reflect salinity effects on photochemistry, although it may not be the most sensitive indicator of stress effects (Kovar et al. 2001; James et al. 2002). Measurement of PSII efficiency in sugar beet during early growing season under drought stress (Mohammadian et al. 2003) and low levels of salinity up to 9 dSm^{-1} (Park et al. 2006) represented a suitable physiological parameter to characterize sugar beet genotypes under stress conditions. However, there are different reports in which PSII efficiency did not have a significant effect on stress tolerance, for example, on maize under drought stress (Ashraf et al. 2007) and low levels of salinity on sugar beet (Netondo et al. 2004; Hajiboland et al. 2009).

7 Genetic Resources for Salt Tolerance

In semiarid regions such as Iran, particularly during the last decade, selection for salt tolerance has been one of the important objectives of sugar beet breeding programs (Ahmadi et al. 2011). Sugar beet is considered a moderately salt-tolerant crop (Jamil et al. 2006); however, development of varieties with improved yield and quality in salt-stressed environments is deemed appropriate. Genetic variation for salt tolerance is an essential prerequisite for the development of more stress-tolerant varieties. There is high genotypic variation in sugar beet for drought (Ober et al. 2004; Rajabi et al. 2008; Rajabi et al. 2009; Rajabi and Ober 2012; Rajabi et al. 2013a, b) and salt tolerance (Marschner et al. 1981a; Uno et al. 1996; Ghoulam et al. 2002; Rajabi 2010; Ober and Rajabi 2010) which could be exploited by plant breeders to develop stress-tolerant varieties for drought and saline environments. The extent and distribution of genetic variation for salt tolerance is difficult to assess by conventional field methods. Rather, molecular markers, especially simple sequence repeat (SSR, or microsatellite) markers, are suited to describe genetic diversity at the DNA level (Kalia et al. 2011). By using 18 microsattelite (SSR) markers, Abbasi et al. (2014) analyzed genetic variation within a set of 168 sugar beet breeding lines and found a high SSR marker polymorphism within this set of material. In the long run, this work can contribute to the establishment of marker-assisted selection (MAS) for salt tolerance in sugar beet. Marschner et al. (1981b) found considerable genotypic differences for sensitivity to sodium within a limited set of sugar beet genotypes. Mostafavi (2012) found significant genetic variation among sugar beet cultivars for germination and seedling growth traits under saline conditions, while Khavari-Nejad et al. (2008) identified a beet genotype that showed greater tolerance to 50 mM NaCl than three other local varieties. Zein et al. (2002) also found varietal differences in field experiments under a range of salinity stress levels.

Since improvement of most crop species took place outside their native habitats or under favorable agricultural conditions, their gene pools of breeding germplasm have become limited (Tal 1985). Compared with other open-pollinated crop

Fig. 5.1 *Beta vulgaris* L. ssp.*maritima* growing on a shingle beach estuary near Maldon, Essex, UK

species, sugar beet has a relatively narrower genetic base (Panella and Lewellen 2007). This situation is exacerbated by the fact that the hybrid commercial varieties of sugar beet should integrate both the monogerm and O-type (the maintainer of the cytoplasmically male sterile line) characters (Panella and Lewellen 2007). This may make it necessary to use the wild relatives, especially *Beta vulgaris* ssp. *maritima* (the wild progenitor of all domesticated beets), to enrich the gene pool of sugar beet with novel sources of variation. Many populations of *maritima* beet thrive near the sea in saline conditions (Fig. 5.1). However, the potential breeding value of wild relatives for improving stress tolerance is low because it is difficult to transfer and introgress desirable traits from the wild relatives into high-yielding germplasm (Biancardi et al. 2010). Among the unwanted traits introgressed from wild beet along with the desired genes were annual growth habit (sugar beet is a biennial crop), red color in the root, branched roots, elongated or multiple crowns, and low sugar content and sugar extractability (reviewed in Lewellen 1992; Panella and Lewellen 2005). For more distant *Beta* species as sources, there is the additional difficulty of transferring genes to domesticated cultivars due to reproductive barriers (Turan et al. 2012). However, rhizomania and beet cyst nematode resistance genes from wild beet have already been commercialized (Panella and Lewellen 2007).

8 Screening Methodologies

Artificial selection and conventional breeding approaches in the past few years have resulted in significant improvement in salinity tolerance in a number of field crops. However, selection has been mainly based on in agronomic traits such as yield and survival in saline conditions (Subbarao and Johansen 1999; Ashraf 2004; Noble and Rogers 2004). It is difficult to use yield-based response measurements in field studies as a measure of tolerance. This is due to (1) the complexity of interactions with biotic and abiotic stress factors, (2) variability of NaCl in the soil profile, and (3) differential responses according to both growth stage and genotype (Munns et al. 2006; Leonforte et al. 2013). Rather, physiological traits associated with yield have been considered as criteria for indirect selection to improve salt tolerance. More reliable information can be obtained from physiological criteria than from agronomic traits. However, these attributes have not been extensively used as selection tools to improve salt tolerance in breeding programs (Ashraf 2004). This may be due to the fact that the following questions still remain unanswered: the extent of genetic diversity, the importance of the targeted mechanism, the ease of selection of the physiological mechanism, and the breeding strategy (Noble and Rogers 2004).

In order for a secondary trait to be an efficient tool in screening genotypes for abiotic stress tolerance, it should meet four requirements: (1) a high genetic correlation with yield and salinity tolerance; (2) a high heritability; (3) practical on a large, breeding scale; (4) cost-effective (in breeding programs with limited resources and competing breeding targets; Ober et al. 2005). For example, carbon isotope discrimination ratio (Δ) was shown to be an effective selection tool for improving water use efficiency in wheat (Farquhar and Richard 1984), and with potential in sugar beet (Rajabi et al. 2008, 2009). It has also been of potential value as a physiological indicator for salt tolerance in rice (Shaheen and Hood-Nowotny 2008). The advantages of this criterion in plant breeding programs are: (1) it is a time-integrated measure of plant response to stress; (2) it has a low genotype $\times$ environment interaction and high heritability values, and (3) sample preparation is easy and isotope analysis is automated (Farquhar and Richard 1984; Condon and Richards 1992; Rajabi et al. 2008). It has been shown that the leaf Δ value of the wheat grown in the greenhouse at 16 dSm^{-1} is positively correlated with field yield of cultivars grown under saline conditions. This suggests that screening for salinity tolerance could be conducted by the simple, early leaf sampling for Δ analysis (Shaheen and Hood-Nowotny 2008). However, selection for Δ in breeding programs requires advanced conventional methodologies, varieties and elite lines, and adequate resources for efficient use of the technology (Hall et al. 1994).

Other physiological traits such as osmotic adjustment could be used to discriminate superior genotypes, but their reliability as selection criterion for abiotic stress tolerance deserves further investigation (Ober et al. 2005). Ahmad et al. (2012a) suggested that photosynthetic and transpiration rates, stomatal conductance, chlorophyll content, and relative water content could be used to select for salt tolerance in mustard. Leaf thickness is a morphological adaptation against stress so that thicker leaves reserve more water per unit area (Hajiboland et al. 2009). This trait, however, is manifested at later growth stages (leaf development and technological maturity) of sugar beet (Khayamim 2010) which makes it impossible to quickly screen the salt-tolerant genotypes.

9 Screening Environments

Controlled environments such as greenhouse or growth chamber are often used by researchers for the initial screening of germplasm lines and to reduce a large number of breeding lines to a manageable level for more precise evaluations at later stages in the field (Subbarao and Johansen 1999). Also, early-generation selection of breeding materials for salt tolerance is conducted in controlled environments which helps minimize the environmental variance and maximize the genetic variance.

It is a laborious task to select for salt tolerance in the field (Arzani 2008). Also, salinity levels are highly variable in the field, so it is very difficult to screen germplasm resources under field conditions (Shaheen and Hood-Nowotny 2008). This makes selection for genetic improvement difficult, as environmental variance exceeds the genetic component (Subbarao and Johansen 1999). Alternatively, testing could be conducted under relatively controlled conditions in the field, as done by the US Salinity Laboratory at Riverside, California (reviewed in Subbarao and Johansen 1999). Relatively uniform salinity levels can be created by using nonsaline soil and by irrigating with saline water. Similar facilities are used at Roudasht Salinity Research Station of Isfahan Agriculture Research Center in Iran (Ebrahimian et al. 2008). Sugar beet lines are evaluated at different salinity levels provided at the station. A nonsaline treatment is usually used in order to compare genotypes on a relative yield basis (Subbarao and Johansen 1999). Since the salinity is highly inconsistent in soil due to the dynamic status of soluble salts, one should go for more replications over the years and locations (judiciously partitioning the resources) for more precise estimates of genotypic differences (Singh 2006). Precision increases when replications, locations, and years increase, with the latter having the highest influence.

10 Proper Growth Stages for Screening

Sugar beet is sensitive to salinity at germination. Different thresholds of salt damage to the sugar beet crop vary according to environmental conditions (Abrol et al. 1988). Khayamim (2010) found that the best salinity level to differentiate sugar beet genotypes in greenhouse conditions was 16 dSm^{-1}. Further experiments showed that sugar beet can germinate at greater levels of soil solution electrical conductivity in the laboratory using only NaCl ($EC=20–24$ dSm^{-1}) compared with germination rates obtained using saline soil in the greenhouse or field (EC=16 dSm^{-1}; Khayamim et al. 2014). These findings showed that perhaps the side effect of other soil ions in addition to NaCl may inhibit germination (Duan et al. 2004). Except for one cultivar, PP22, the adverse effect of salinity of the irrigation water on sugar beet seed germination and seedling root length was higher for NaCl alone than for the salt mixture, which refers to lower salt stress in the field conditions with natural salt composition (Jafarzadeh and Aliasgharzadeh 2007). Leaf relative water content in sugar beet was affected by salinity at the seedling to four-leaf growth stages, which implies an osmotic stress, whereas during germination the stress may be chiefly ionic because relative water content was not affected by salinity (Khayamim 2010).

Sugar beet seeds germinated under salinity stress in greenhouse experiments, but seedlings were damaged by increasing salt stress. It was reported that salinity (EC=16 dSm^{-1}) decreased seed germination by 50% in field experiments, and establishment was decreased up to 75% (Fotuhi et al. 2006). In greenhouse experiments, salinity (EC=16 dSm^{-1}) decreased germination up to 35% and increased seedling mortality to 80% (Khayamim et al. 2014). These saline conditions delayed seedling growth and development, which has a negative impact on seedling establishment (Durrant et al. 1974). Under greenhouse conditions (22°C/16 h light period), it was concluded that seeds with greater germination in soil produced healthier and stronger seedlings (Sadeghian and Khodaii 1998). Acceptable yields under saline conditions depend not only on good germination of seeds and seedling development but also on the further survival of seedlings for better establishment (Sadat Noori and McNeilly 2000; Khayamim et al. 2014). The greatest impact of salinity was observed at the establishment stage according to photosynthetic characteristics, osmolyte, and inorganic ion accumulation measured in greenhouse tests. Therefore, salt-tolerant genotypes could be identified by screening for seedling growth during the establishment stage (Khayamim 2010).

11 Breeding Strategies

Attempts to improve salinity tolerance have involved conventional breeding approaches: pooling physiological traits, use of the variation present in the cultivated germplasm, interspecific hybridization, use of recurrent selection to generate variation within existing crops, mutagenesis or in vitro selection, use of halophytes as alternative crops, use of MAS, and breeding for yield rather than tolerance (Flowers and Yeo 1995; Flowers 2004; Joshi 2011). These are all potential solutions to the problem. However, breeding for salinity tolerance will likely be increasingly accompanied by transgenic strategies, the success of which needs extensive "metabolic engineering," which requires the transfer of many genes (Bohnert and Jensen 1996).

Breeding for salt tolerance could be carried out at three stages (Singh 2006): (1) identification of the genotypes based on the intrinsic physiological mechanisms determining salt tolerance (Na^+ exclusion, K^+ uptake, tissue tolerance, high initial vigor, etc.); (2) inter-mating of genotypes showing contrasting expression of salinity tolerance mechanisms for quantitative trait loci(QTL) discovery; (3) identification of the recombinants for pyramiding of the mechanisms (Singh 2006).

11.1 Conventional Breeding

Plant breeding has a long history in producing high-yielding and stress-tolerant crops. In order to develop salt-tolerant lines, the genetic diversity in crops has been exploited by plant breeders at intraspecific, interspecific, and intergeneric levels. Given the difficulties involved in breeding salt-tolerant varieties, selection for yield rather than tolerance has been suggested, based on the idea that selection for yield will result in yield improvement in both normal and stress environments (reviewed in Joshi 2011). The best strategy for moderately saline soils is to breed for yield; however, for highly saline conditions, breeding for yield and salinity tolerance together is important (Shannon 1985; Isla et al. 2003). Salt-tolerant genotypes could be developed by a crossing scheme which results in maximum recombination followed by single-seed descent and concurrent selection for tolerance and agronomic traits (Flowers and Yeo 1995). The capacity for enhancing salt tolerance could be estimated using statistical techniques that identify the components of variance attributable to genotype and environment.

One of the most important objectives of sugar beet research in arid and semi-arid areas such as Iran is to develop drought- and salt-tolerant cultivars that can produce reasonable yields in fields. As almost all the current sugar beet varieties are hybrids, tolerance in both the male and female parents should be taken into account. The ultimate goal is to develop the improved varieties with high yield and good technical quality for processing into white sugar. However, due to a long procedure for development of tolerant female parents, the main emphasis has been made on development of male parents (Rajabi 2010).

11.2 Nonconventional Approaches

11.2.1 Molecular Breeding

Molecular-based breeding could serve as an example for improvement in salinity tolerance (McGrath et al. 2000; De los Reyes and McGrath 2003; McGrath et al. 2008). These studies aimed at improving sugar beet establishment by selection for germination under oxidative stress conditions. The results of De los Reyes and McGrath (2003) showed that one germin-like protein gene (*BvGer165*) was differentially regulated, and was induced only in the good emerger. Selection for stress-induced germin expression provides the first straight objective to facilitate breeding of sugar beet for improved field emergence.

Proteomics is a means progressively used to examine stressed plants. The implementation of proteomics in plant breeding is generally started by the discovery of stress-reactive proteins through comparison between stressed and nonstressed plants. Detection of these expressed candidate proteins may then disclose that their functions are associated with stress tolerance (Salekdeh and Komatsu 2007). Analysis of proteome and transcriptome in a sugar beet monosomic addition line revealed some salt-responsive candidate genes and proteins with different functions which open the ways for further investigation of salt tolerance mechanism in this crop (Yang et al. 2012). A proteomics study on drought-stressed sugar beet genotypes highlighted some candidate proteins associated with drought tolerance (Hajheidari et al. 2005). Thirteen protein spots responded to salt stress (16 dSm^{-1}) at both establishment and maturity growth stages (Khayamim 2010). However, a study of seedlings grown on 125 mM NaCl did not show any strong candidate proteins associated with salinity tolerance (Wakeel et al. 2011).

11.2.2 Mapping Salinity Tolerance Genes

QTLs are segments of the genome linked with a particular trait. Salt tolerance-related QTLs play an important role in understanding the stress response and developing stress-tolerant crops (Turan et al. 2012). QTLs associated with salt tolerance have been recognized in different field crops (Flowers 2004; Ammar et al. 2004; Ren et al. 2005; Cakmac 2005; Thomson et al. 2007; Pandit et al. 2010) but there is no published report for sugar beet.

QTL mapping is an approach to develop molecular markers related to genes or QTLs which control the traits. By using molecular marker technology, the chromosomal position, number, and individual and interactive effects of QTLs controlling a trait can be determined. When identified, QTLs of interest can be transferred into desirable genetic materials by MAS. Salt tolerance of crops could be substantially improved by pooling desirable traits through MAS. In addition, such an approach may improve the efficiency of pyramiding of tolerance components from diverse genetic backgrounds. A QTL mapping study was carried out to find the number, genomic position, and individual effects of QTL influencing salt tolerance at different growth stages of tomato (Foolad 2004).

By using such molecular markers as randomly amplified polymorphic DNA (RAPD); restriction fragment length polymorphisms (RFLP); SSR; amplified fragment length polymorphisms (AFLP); single nucleotide polymorphisms (SNP), as well as morphological and isozyme markers, many genetic maps have been produced for sugar beet (reviewed in Biancardi et al. 2010). Most of the maps are first-generation maps which are used to detect the genetic basis of complex traits. Second-generation maps are to be constructed based on a common platform (McGrath 2011). The first publication of the sugar beet genome sequence was imminent at the time of this writing.

Linkage mapping studies were conducted in sugar beet to elucidate the genetic basis of agronomic traits (e.g., root and sugar yields, sugar content, and impurity levels; Weber et al. 1999; Weber et al. 2000; Schneider et al. 2002; Schneider et al.

2007). QTLs affecting root yield and sugar content were mapped by using association mapping approach (Stich et al. 2008a, b) and a genome-wide approach revealed epistatic interactions (Würschum et al. 2011). Quantitative trait locus analysis in sugar beet has identified genomic positions linked to resistance to cercospora leaf spot (Setiawan et al. 2000), rhizomania (Gidner et al. 2005; Lein et al. 2007), powdery mildew (Grimmer et al. 2007), and rhizoctonia crown and root rot (Lein et al. 2008). Although a number of genetic linkage maps of sugar beet have been developed since the 1990s, applying this approach to abiotic stresses has not yet been attempted.

As root function is an important component of salinity tolerance, studies on root morphology and architecture could identify main root characters as criteria for selecting high-yielding varieties under edaphic stresses (Gahoonia and Nielsen 2004). Limited information is available on genes governing sugar beet root characters (Stevanato et al. 2010). A mapping study conducted on a sugar beet segregating population derived from a cross between two lines characterized for root elongation rate (RER) at the seedling stage and root yield in the field made it possible to identify two AFLP markers linked to the RER (Stevanato et al. 2010). This molecular analysis and QTL mapping study opens the way for MAS as revealed for the "RER" under stress conditions.

Since the genetics of most quantitative traits including salt tolerance is complex, with multiple genes and loci scattered at various places throughout the genome of sugar beet, the combination of genes and alleles and their interactions may be more important than the expression of the genes themselves (Setiawan et al. 2000; Schneider et al. 2002; Lein et al. 2007; Stevanato et al. 2010). This is where molecular markers can be used to gain clarity into the quantitative genetic contributions (e.g., QTLs) and to provide some context in which they can be deployed effectively, prior to applying markers for assisted selection (McGrath 2010).

11.2.3 Genetic Engineering and its Application in Creating Salt Tolerance

Breeding salt-tolerant cultivars is one of the objectives of most research programs. It is believed that salinity tolerance will be realized only when several traits are pooled in a specific genotype, whereas a significant increase in salinity tolerance could not be expected from a single trait alone (Yeo et al. 1988). Salt tolerance is a polygenic trait affected by several salt stress-responsive genes (Zhu 2000). Recently, salt tolerance was significantly increased by the overexpression of vacuolar Na^+/H^+ antiporter gene in engineered sugar beet plants (Liu et al. 2008). A lower amount of Na^+ in the leaves and higher K^+ amount in the roots was observed in the AtNHX3 transgenic lines than non-transgenic sugar beet plants under saline conditions (300 or 500 mM NaCl). Salt-induced ion-specific stresses result from the altered K^+/Na^+ ratios. Therefore, Na^+ accumulation in the vacuole is essential for plant salt tolerance. This process depends on the activity of *NHX* genes (vacuolar Na^+/H^+ antiporters; Xia et al. 2002). Creation of salt-tolerant transgenic plants via the overexpression of *NHX* genes in different species has revealed the main role of *NHX* genes (reviewed in Liu et al. 2008). The salt accumulation in leaves and the increased root yield and sugar content under salinity condition display a large potential use of the *AtNHX3* gene in enhancing the yield and quality of field crops.

An efficient transformation system plays a vital role for large-scale production of transgenic sugar beet plants (Mutasa-Göttgens et al. 2009). Improvements in transformation protocols for beet and the development of genetic engineering methodologies has made it possible to introduce many genes into beet, particularly for herbicide and disease resistance (Yang et al. 2005). Improvements in transgene expression (e.g., a GUS gene) using *Agrobacterium*-mediated transformation (Lindsey and Gallois 1990; Krens et al. 1996) have been realized more recently (Hisano et al. 2004; Pavli and Skaracis 2010). The high-yielding varieties can be grown on saline soil and will be aided for further improvements in the efficiency through the plant transformation systems. Manipulation of gene expression conferring salinity tolerance by transgenic methods has made it possible to develop crops with improved salt tolerance in various species (Serrano et al. 1999; Zhang and Blumwald 2001; Hasthanasombut et al. 2010; Zhang et al. 2011). These findings show that salt-tolerant crops could be produced with a combination of breeding and transgenic approaches with far fewer target genes than had been expected.

11.3 Varietal Improvement

Due to the increase in salt-affected areas and difficulties in rehabilitating many soils, and reduced opportunities to increase yields through increasing the acreage of the sugar beet crop, it is necessary to develop and use salt-tolerant varieties. Therefore, one of the important objectives of sugar beet research in arid and semiarid areas such as Iran is to develop drought- or salt-tolerant cultivars that can produce a reasonable yield in fields. Fortunately, there are opportunities for varietal improvements in sugar beet since significant genotypic diversity for tolerance to abiotic stresses such as heat, cold, drought, and salinity exist within sugar beet germplasm (Ober and Rajabi 2010).

In one of the salinity projects, hybrids derived from crosses between single cross and pollinator parents displayed higher root and white sugar yields than the check variety (Rajabi 2010; Fig. 5.2). Interestingly, the drought-tolerant populations were found to be salt-tolerant as well. This may be due to similarities in the physiological adaptations to salinity and other stresses such as drought stress, so improvements in drought tolerance may also improve yields in saline soil (Munns 2002). In another salinity project, two tetraploid populations were identified as salt-tolerant pollinators, which were used to develop triploid hybrids. Among these, three hybrids were superior to the salt-tolerant check varieties and were introduced as promising hybrids for further evaluation (Rajabi 2010; Abbasi and Rezaei 2014).

Fig. 5.2 Sugar beet monogerm, diploid hybrids (three rows in the *right*) displayed higher root and white sugar yields than the check variety (three rows in the *left*) in the salt-stressed field condition in Isfahan, Iran. (Rajabi 2010)

12 Conclusion and Future Perspectives

Abiotic stresses occur in nearly all arable areas, although with inconsistent intensity and duration. Despite abiotic stresses being so deleterious for crop production, tolerance mechanisms are not well identified and attempts to improve the tolerance of crops are still far from satisfactory. Among the reasons for this slow progress is the lack of clear-cut traits conferring tolerance, the multigenic inheritance of salt tolerance, and the complication caused by the concurrent incidence of more than one stress when genetic materials are phenotyped. However, various researchers have been successful in correlating some morphophysiological characters with stress tolerance, leading to a greater understanding of the genetic mechanisms involved, and further advances in functional genomics should accelerate the pace of discovery. Over the last century, several salt-tolerant varieties for different crops have been developed by conventional plant breeding. But this approach has been limited due to reproductive barriers with related wild species and the narrow genetic base of most field crops. Furthermore, the genetic variation for salt tolerance present in certain wild relatives of crops, or indeed other organisms, can be utilized through genetic engineering for the development of salinity-tolerant varieties. There appear to be many genes of unknown function that can impart tolerance to multiple abiotic stresses. Despite progress in engineering salt tolerance in the laboratory (where these efforts must necessarily begin), there are few reports of successful field trials (Jewell et al. 2010); but experiments in this environment are eventually required.

The salinity tolerance of today's varieties is not yet optimized; therefore, this will continue to be an objective for plant breeders for current and future markets (Bosemark 2006). Looking ahead, it is possible that as pressure increases to produce food for a growing world population, agriculture will be pushed further onto marginal and often saline land. Combined with increasingly less available fresh water to keep salinity levels low in many arid and semiarid areas, breeders will have to face the challenge of improving salt tolerance in plants through different biological means in near future.

References

Abbasi Z, Rezaei M (2014) Development of sugar beet salt tolerant triploid hybrids. Sugar Tech. DOI:10.1007/s12355-014-0309-2.

Abbasi Z, Arzani A, Majidi MM (2014) Evaluation of genetic diversity of sugar beet (*Beta vulgaris* L.) crossing parents using agro - morphological traits and molecular markers. J Agric Sci Technol (Accepted).

Abdel MSE, Zanovny I (2004) Response of sugar beet (*Beta vulgaris*) to potassium application and irrigation with saline water. Ass Univ Bull Environ Res 7(1):123–136

Abrol IP, Yadav JSP, Massoud FI (1988) Crops in saline soils in: salt-affected soils and their management. FAO soils bulletin 39. http://www.fao.org/docrep/x5871e/x5871e00.htm#Contents Accessed 15 August 2013

Ahmad P, Sharma S (2008) Salt stress and phyto-biochemical responses of plants—a review. Plant Soil Environ 54:89–99

Ahmad P, Umar S (2011) Antioxidants: oxidative stress management in plants. Studium, New Delhi, p 369

Ahmad P, Sarwat M, Sharma S (2008) Reactive oxygen species, antioxidants and signaling in plants. J Plant Biol 51(3):167–173

Ahmad P, Jaleel CA, Salem MA, Nabi G, Sharma S (2010a) Roles of enzymatic and non-enzymatic antioxidants in plants during abiotic stress. Crit Rev Biotechnol 30(3):161–175

Ahmad P, Umar S, Sharma S (2010b). Mechanism of free radical scavenging and role of phytohormones during abiotic stress in plants. In: Ashraf M, Ozturk M, Ahmad MSA (eds) Plant adaptation and phytoremediation. Springer, Dordrecht, pp 99–108

Ahmad P, Nabi G, Ashraf M (2011a) Cadmium-induced oxidative damage in mustard [*Brassica juncea* (L.) Czern. & Coss.] plants can be alleviated by salicylic acid. South Afr J Bot 77:36–44

Ahmad P, Nabi G, Jeleel CA, Umar S (2011b) Free radical production, oxidative damage and antioxidant defense mechanisms in plants under abiotic stress. In: Ahmad P, Umar S (eds) Oxidative stress: role of antioxidats in plants. Studium, New Delhi, pp 19–53

Ahmad P, Hakeem K, Kumar A, Ashraf M, Akram NA (2012a) Salt-induced changes in photosynthetic activity and oxidative defense system of three cultivars of mustard (*Brassica juncea* L.). Afr J Biotechnol 11(11):2694–2703

Ahmad P, Ozturk M, Gucel S (2012b) Oxidative damage and antioxidants induced by heavy metal stress in two cultivars of mustard plants. Fresen Environ Bull 21(10):2953–2961

Ahmad P, Ashraf M, Azooz MM, Rasool S, Akram NA (2013) Potassium starvation-induced oxidative stress and antioxidant defense responses in *Brassica juncea*. J Plant Interact (In Press, doi:10.1080/17429145.2012.747629)

Ahmadi M, Majidi Heravan E, Sadeghian SY, Mesbah M, Darvish F (2011) Drought tolerance variability in S1 pollinator lines developed from a sugar beet open population. Euphytica 178:339–349

Akbar M, Yabuno T (1977) Breeding saline-resistant varieties of rice. IV. Inheritance of delayed type panicle sterility induced by salinity. Jap J Breed 27:237–240

Akbar M, Khush GS, Hille Ris Lambers D (1986) Genetics of salt tolerance in rice. Proceedings of the international rice genetics symposium IRRI 399–409

Ammar MHM, Pandit A, Singh RK, Sameena S, Chauhan MS, Singh AK, Sharma PC, Gaikwad K, Sharma TR, Mohapatra T, Singh NK (2004) Mapping of QTLs controlling Na^+, K^+ and Cl^- ion concentrations in salt tolerant indica rice variety CSR27. J Plant Biochem Biotechnol 18:139–150

Apostolides G, Goulas C (1998) Seed crop environment and processing effects on sugar beet (*Beta vulgaris* L.) certified by hybrid variety seed quality. Seed Sci Technol 26:223–235

Arzani A (2008) Improving salinity tolerance in crop plants: a biotechnological view. Vitro Cell Dev Biol Plant 44:373–383

Ashraf M (2004) Some important physiological selection criteria for salt tolerance in plants. Flora 199:361–376

Ashraf M, Harris PJC (2004) Potential biochemical indicators of salinity tolerance in plants. Plant Sci 166:3–16

Ashraf M, Foolad MR (2007) Roles of glycine betaine and proline in improving plant abiotic stress resistance. Environ Exp Bot 59:206–216

Ashraf M, Nawazish SH, Athar HUR (2007) Are chlorophyll fluorescence and photosynthetic capacity potential physiological determinants of drought tolerance in Maize (*Zea mays* L.). Pak J Bot 39(4):1123–1131

Azhar FM, McNeilly T (1988) The genetic basis for salt tolerance in *Sorghum bicolor* (L.) Moench seedlings. Plant Breed 101:114–121

Azzazy NB (2004) Yield and quality of some sugar beet varieties as affected by water quality and nitrogen fertilization. Egypt J Agric Res 82:1733–1745

Bernstein N, Läuchli A, Silk WK (1993) Kinematics and dynamics of sorghum (*Sorghum bicolor* L.) leaf development at various Na/Ca salinities. Plant Physiol 103:1107–1114

Biancardi E, McGrath JM, Panella LW, Lewellen RT, Stevanato P (2010) Sugar Beet. In: Bradshaw JE (ed) Root and tuber crops. Springer, New York, pp 173–219

Bohnert HJ, Jensen RG (1996) Metabolic engineering for increased salt tolerance-the next step. Aust J Plant Physiol 23:661–666

Bosemark NO (2006) Genetics and breeding. In: Draycott S (ed) Sugar beet. Blackwell, Oxford, pp 50–88

Cakmac I (2005) The role of potassium in alleviating detrimental effects of abiotic stresses in plants. J Plant Nutr Soil Sci 168:521–530

Carillo P, Annunziata MG, Pontecorvo G, Fuggi A, Woodrow P (2011) Salinity stress and salt tolerance. In: Shanker A, Venkateswarlu B (eds) Abiotic stress in plants: mechanisms and adaptations. InTech, Rijeka, pp 21–38

Choluj D, Karwowska R, Ciszewska A, Jasinska M (2008) Influence of long-term drought stress on osmolyte accumulation in sugar beet (*Beta vulgaris* L.) plants. Acta Physiol Plant 30:679–687

Coello P, Hey SJ, Halford NG (2011) The sucrose non-fermenting-1-related (SnRK) family of protein kinases: potential for manipulation to improve stress tolerance and increase yield. J Exp Bot 62:883–893

Colmer TD, Fan TW-M, Higashi RM, Läuchli A (1996) Interactive effects of Ca^{2+} and NaCl salinity on the ionic relations and proline accumulation in the primary root tip of *Sorghum bicolor*. Physiol Plantarum 97:421–424

Condon AG, Richards RA (1992) Broad-sense heritability and genotype × environment interaction for carbon isotope discrimination in field-grown wheat. Aust J Agric Res 43:921–934

Delfine S, Alvino A, ConcettaVilani M, Loreto F (1999) Restriction to carbon dioxide conductance and photosynthesis in spinach leaves recovering from salt stress. Plant Physiol 119:1101–1106

De los Reyes BG, McGrath JM (2003) Cultivar-specific seedling vigor and expression of a putative oxalate oxidase germin-like protein in sugar beet (*Beta vulgaris* L.). Theor Appl Genet 107:54–61

De Vos AC, Broekman R, Groot MP, Rozema J (2010) Ecophysiological response of *Crambe maritima* to airborne and soil-borne salinity. Ann Bot 105(6):925–937

Duan D, Liu X, Ajmal Khan M, Gul B (2004) Effect of salt and water stress on the germination of *Chenopodium glaucum* L. seed. Pak J Bot 36(4):793–800

Durrant MJ, Draycott AP, Payne PA (1974) Some effect of sodium chloride on germination and seedling growth of sugar beet. Ann Bot 38:1045–1051

Ebrahimian HR, Ranji ZA, Rezaee MA, Abbasi Z (2008) Screening sugar beet genotypes under salinity stress in the greenhouse and field conditions. J Sugar Beet 24(1):1–21 (In Persian)

Farquhar GD, Richard RA (1984) Isotopic composition of plant carbon correlates with water use efficiency of wheat genotypes. Aust J Plant Physiol 11:539–552

Fayer MD (2012) Dynamics of water interacting with interfaces, molecules and ions. Acc Chem Res 45:3–14

Flowers TJ (2004) Improving crop salt tolerance. J Exp Bot 55(396):307–319

Flowers TJ, Yeo AR (1995) Breeding for salinity resistance in crop plants-where next? Aust J Plant Physiol 22:875–884

Foolad MR (2004) Recent advances in genetics of salt tolerance in tomato. Plant Cell Tissue Org Culture 76:101–119
Fotuhi K, Mesbah M, Sadeghian SY, Ranji ZA, Orazizadeh MR (2006) Assessment salt tolerance in sugar beet lines. J Sugar Beet 22(2):1–18 (In Persian)
Gahoonia TS, Nielsen NE (2004) Root traits as tools for creating phosphorus efficient crop varieties: new challenges for rhizosphere research at the entrance of the 21st Century. Plant Soil 260:47–57
Geissler N, Hussin S, Koyro HW (2009) Interactive effect of NaCl salinity and elevated atmospheric CO_2 concentration on growth, photosynthesis, water relations and chemical composition of the potential cash crop halophyte *Aster tripolium* L. Environ Exp Bot 65:220–231
Ghoulam C, Fares K (2001) Effect of salinity on seed germination and early seedling growth of sugar beet (*Beta vulgaris* L.). Seed Sci Technol 29:357–364
Ghoulam C, Ahmed F, Fares K (2002) Effects of salt stress on growth, inorganic ions and proline accumulation in relation to osmotic adjustment in five sugar beet cultivars. Environ Exp Bot 47(1):1–93
Gidner S, Lennefors BL, Nilsson NO, Bensenfelt J, Johansson E, Gyllenspetz U, Kraft T (2005) QTL mapping of BNYVV resistance from the WB41 source in sugar beet. Genome 48:279–285
Gordon TR, Duniway JM (1982) Effects of Powdery Mildew infection on the efficiency of CO_2 fixation and light utilization by sugar beet leaves. Plant Physiol 69:139–142
Grimmer MK, Bean KMR, Asher MJC (2007) Mapping of five resistance genes to sugar-beet powdery mildew using AFLP and anchored SNP markers. Theor Appl Genet 115:67–75
Gzik A (1996) Accumulation of proline and pattern of α amino acids in sugar beet plants in response to osmotic water and salt stress. Environ Exp Bot 36(1):29–38
Habibi D (1993) Selection of sugar beet tolerant progenies to drought and salt stress in germination phase. MSc Thesis. Islamic Azad University, Science and Research Branch (In Persian)
Hajheidari M, Abdollahian-Noghabi M, Askari H, Heidari M, Sadeghian SY, Ober ES, Salekdeh GH (2005) Proteome analysis of sugar beet leaves under drought stress. Proteomics 5(4):950–960
Hajiboland R, Joudmand A, Fotouhi K (2009) Mild salinity improves sugar beet (*Beta vulgaris* L.) quality. Acta Agric Scand Sect B Soil Plant Sci 59:295–305
Hall A, Rivhards RA, Condon AG, Wright GC, Farquhar GD (1994) Carbon isotope discrimination and plant breeding. In: Janick J (ed) Plant breeding reviews, vol. 12. Wiley, New York, pp 81–113
Harley PC, Loreto F, Dimacro G, Sharkey TD (1992) Theoretical consideration when estimating the mesophyll conductance to CO_2 flux by analysis of the responses of photosynthesis to CO_2. Plant Physiol 98:1429–1436
Hasegawa PM, Bressan RA, Zhu JK, Bohnert HJ 2000. Plant cellular and molecular responses to high salinity. Annu Rev Plant Physiol Plant Mol Biol 51(1):463–499
Hasthanasombut S, Ntui V, Supaibulwatana K, Mii M, Nakamura I (2010) Expression of Indica rice OsBADH1 gene under salinity stress in transgenic tobacco. Plant Biotechnol Rep 4:75–83
Heuer B, Plaut Z (1989) Photosynthesis and osmotic adjustment of two sugar beet cultivars grown under saline condition. J Exp Bot 40:437–440
Hisano H, Kimoto Y, Hayakawa H, Takeichi J, Domae T, Hashimoto R, Abe J, Asano S, Kanazawa A, Shimamoto Y (2004) High frequency Agrobacterium-mediated transformation and plant regeneration via direct shoot formation from leaf explants in *Beta vulgaris* and *Beta maritima*. Plant Cell Rep 22:910–918
Isla R, Aragues R, Royo A (2003) Spatial variability of salt-affected soils in the middle Ebro valley (Spain) and implications in plant breeding for increased productivity. Euphytica 134:325–334
Jafarzadeh AA, Aliasgharzadeh N (2007) Salinity and salt composition effects on seed germination and root length of four sugar beet cultivars. Biol Bratislava 62(5):562–564
James RA, Rivelli AR, Munns R, von Caemmerer S (2002) Factors affecting CO_2 assimilation, leaf injury and growth in salt-stressed durum wheat. Funct Plant Biol 29:1393–1403
Jamil M, Lee DB, Jung KY, Ashraf M, Lee SC, Rha ES (2006) Effects of salt (NaCl) stress on germination and early seedling growth of four vegetables species. J Cent Eur Agric 7(2):273–282

Jampeetong A, Brix H (2008) Effects of NaCl salinity on growth, morphology, and photosynthesis and proline accumulation of *Salvinia natans*. Aquatic Bot. doi:10.1016/j.aquabot.2009.05.003

Jewell M, Campbell B, Godwin I (2010) Transgenic plants for abiotic stress resistance. In: Kole C, Michler C, Abbott A, Hall T (eds) Transgenic crop plants. Springer, Berlin, pp 67–132

Joshi AK (2011) Genetic factors affecting abiotic stress tolerance in crops. In: Pessarakli M (ed). Handbook of plant and crop stress. CRC, Boca Raton, pp 803–850

Kafi M, Asadi H, Ganjeali A (2010) Possible utilization of high-salinity waters and application of low amounts of water for production of the halophyte *Kochia scoparia* as an alternative fodder in saline agroecosystems. Agric Water Manag 97:139–147

Kalia RK, Rai M, Kalia K, Singh S, Dhawan RAK (2011) Microsatellite markers: an overview of the recent progress in plants. Euphytica 177:309–334

Katerji N, Vanhoorn JW, Hamdy A, Matrorilli M, Karzel EM (1997) Osmotic adjustment of sugar beets in response to soil salinity and its influence on stomatal conductance, growth and yield. Agric Water Manag 34:57–69

Katerji N, Van Hoorn TW, Hamdy A, Mastroilli M (2003) Salinity effect on crop development and yield, Analysis of salt tolerance according to several classification methods. Agric Water Manag 62:37–66

Khavari-Nejad RA, Najafi F, Khavari-Nejad S (2008) Growth and some physiological parameters of four sugar beet (*Beta vulgaris* L.) cultivars as affected by salinity. Pak J Biol Sci 11 (10):1390–1393

Khayamim S (2010) Study on some physiological parameters of sugar beet lines in salt stress and nonstress conditions. PhD thesis University of Tehran Iran (In Persian)

Khayamim S, Tavakkol Afshar R, Sadeghian SY, Poustini K, Rouzbeh F, Abbasi Z (2014) Seed germination, plant establishment and yield in sugar beet genotypes under salinity stress. J Agric Sci Technol.16(4):779–790

Kovar M, Brestic M, Olsovska K (2001) Chlorophyll a fluorescence as a bioindicator of the plant environmental stress. Acta fytothechnica Et zootechnicavol 4. Special number. Proceedings of the international scientific conference on the occasion of the 55th anniversary of Slovak Agricultural University in Nitra

Krens FA, Trifonova A, Keizer LCP, Hall RD (1996) The effect of exogenously-applied phytohormones on gene transfer efficiency in sugar beet (*Beta vulgaris* L.). Plant Sci 116:97–106

Kuznetsov VIV, Shevyakova NI (1999) Proline under stress: biological role, metabolism and regulation. Russ J Plant Physiol 46:274–278

Lee G, Carrow RN, Duncan RR, Eiteman MA, Rieger MW (2008) Synthesis of organic osmolytes and salt tolerance mechanisms in *Paspalum vaginatum*. Environ Exp Bot 63:19–27

Lein JC, Asbach K, Tian Y, Schulte D, Li C, Koch G, Jung C, Cai DG (2007) Resistance gene analogues are clustered on chromosome 3 of sugar beet and cosegregate with QTL for rhizomania resistance. Genome 50:61–71

Lein JC, Sagstetter CM, Schulte D, Thurau T, Varrelmann M, Saal B, Koch G, Borchardt DC, Jung C (2008) Mapping of rhizoctonia root rot resistance genes in sugar beet using pathogen response-related sequences as molecular markers. Plant Breed 127:602–611

Leonforte A, Forster JW, Redden RJ, Nicolas ME, Salisbury PA (2013) Sources of high tolerance to salinity in pea (*Pisum sativum* L.). Euphytica 189:203–216

Lewellen RT (1992) Use of plant introductions to improve populations and hybrids of sugarbeet. Use of plant introductions in cultivar development Part 2 Crop Science Society of America Madison WI (USA), pp 117–135

Lindsey K, Gallois P (1990) Transformation of sugar beet (*Beta vulgaris* L.) by *Agrobacterium tumefaciens*. J Exp Bot 41:529–536

Linh LH, Linh TH, Xuan TD, Ham LH, Ismail AM, Khanh TD (2012) Molecular breeding to improve salt tolerance of rice (*Oryza sativa* L.) in the Red River Delta of Vietnam. Int J Plant Genomics. doi:10.1155/2012/949038

Liu H, Wang Q, Yu M, Zhang Y, Wu Y, Zhang H (2008) Transgenic salt-tolerant sugar beet (*Beta vulgaris* L.) constitutively expressing an *Arabidopsis thaliana* vacuolar Na^+/H^+ antiporter gene, AtNHX3, accumulates more soluble sugar but less salt in storage roots. Plant Cell Environ 31:1325–1334

Lyon C (1941) Responses of two species of tomatoes and the F1 generation to sodium sulphate in the nutrient medium. Bot Gaz 103:107–122

Marschner H, Kylin A, Kuiper PJC (1981a) Differences in salt tolerance of three sugar beet genotypes. Physiol Plant 51(2):234–238

Marschner H, Kuiper PJC, Kylin A (1981b) Genotypic differences in the response of sugar beet plants to replacement of potassium by sodium. Physiol Plant 51(2):239–244

McGrath JM (2010) Assisted breeding in sugar beets. Sugar Tech. 12:187–193.

McGrath JM, Derrico CA, Morales M, Copeland LO, Christenson DR (2000) Germination of sugar beet (*Beta vulgaris* L.) seed submerged in hydrogen peroxide and water as a means to discriminate cultivar and seedlot vigor. Seed Sci Technol 28:607–620

McGrath JM, Elawady A, El-Khishin D, Naegele RP, Carr KM, de los Reyes B (2008) Sugar beet germination: phenotypic selection and molecular profiling to identify genes involved in abiotic stress response. Proceedings of the IVth International Sympoium on Seed Transplant and Stand Establishment of Hort Crops Acta Hort 782 ISHS

Meiri A, Plaut Z (1985) Crop production and management under saline conditions. Plant Soil 89:253–271

Mittova V, Theodoulou FL, Kiddle G, Gomez L, Volokita M, Tal M, Foyer CH, Guy M (2003) Coordinate induction of glutathione biosynthesis and glutathione-metabolizing enzymes is correlated with salt tolerance in tomato. FEBS Lett 554:417–421

Mohammadian R (1995) Effects of seed exhaustion on germination and establishment of seven sugar beet genotypes under salinity stress. Ms Thesis Tabriz University Iran (In Persain)

Mohammadian R, Rahimian H, Moghaddam M, Sadeghian SY (2003) The effect of early season drought on chlorophyll a fluorescence in sugar beet (*Beta vulgaris* L.). Pak J Biol Sci 6(20):1763–1769

Monreal JA, Jim´enez ET, Remesal E, Morillo-Velarde R, Garc´ıa-Maurino S, Echevarrıa C (2007) Proline content of sugar beet storage roots: response to water deficit and nitrogen fertilization at field conditions. Environ Exp Bot 60:257–267

Morillo-Velarde R, Ober ES (2006) Water use and irrigation. In: Draycott P (ed) Sugar beet. Blackwell Oxford, pp 50–255

Mostafavi Kh (2012) Effect of salt stress on germination and early seedling growth stage of sugar beet cultivars. Am-Eur J Sust Agric 6(2):120–125

Munns R (2002) Comparative physiology of salt and water stress. Plant Cell Environ 25:239–250

Munns R (2005) Genes and salt tolerance: bringing them together. New Phytol 167(3):645–663

Munns R (2011) Plant adaptations to salt and water stress: differences and commonalities. In: Ismail T (ed) Advances in botanical research. Academic, New York, pp 1–32

Munns R, James R, Lauchli A (2006) Approaches to increasing the salt tolerance of wheat and other cereals. J Exp Bot 57(5):1025–1043

Mutasa-Göttgens E, Qi A, Mathews A, Thomas S, Phillips A, Hedden P (2009) Modification of gibberellin signalling (metabolism & signal transduction) in sugar beet: analysis of potential targets for crop improvement. Transgenic Res 18:301–308

Nagesh Babu R, Devaraj VR (2008) High temperature and salt stress response in French bean (*Phaseolus vulgaris*). Aust J Crop Sci 2:40–48

Nazar R, Iqbal N, Masood A, Syeed S, Khan NA (2011) Understanding the significance of sulfur in improving salinity tolerance in plants. Environ Exp Bot 70(2-3):80–87

Negrão S, Courtois B, Ahmadi N, Abreu I, Saibo N, Oliveira MM (2011) Recent updates on salinity stress in rice: from physiological to molecular responses. Crit Rev Plant Sci 30(4):329–377

Netondo GW, Onyango JC, Beck E (2004) Sorghum and salinity: II. Gas exchange and chlorophyll fluorescence of sorghum under salt stress. Crop Sci 44:806–811

Niazi BH, Broekman RA, Salim M (2000) Dynamics of growth and water relations of fodder beet and sea beet in response to salinity. J Agron Crop Sci 184:101–109

Niazi BH, Athar M, Rozema J (2004) Salt tolerance in the fodder beet and sea beet: analysis of biochemical relations. Bulg J Plant Physiol 30(1–2):78–88

Noble CL, Rogers ME (2004) Arguments for the use of physiological criteria for improving the salt tolerance in crops. Plant Soil 146(1-2): 99–107

Ober ES, Rajabi A (2010) Abiotic stress in sugar beet. Sugar Tech 12 (3-4): 294–298
Ober ES, Sharp RE (1994) Proline accumulation in maize (*Zea mays* L.) primary roots at low water potentials. I. Requirement for increased levels of abscisic acid. Plant Physiol 105:981–987
Ober ES, Clark CJA, Le Bloa M, Royal A, Jaggard KW, Pidgeon JD (2004) Assessing the genetic resources to improve drought tolerance in sugar beet: agronomic traits of diverse genotypes under drought and irrigated conditions. Field Crop Res 90:213–234
Ober ES, Le Bloa M, Clark CJA, Royal A, Jaggard KW, Pidgeon JD (2005) Evaluation of physiological traits as indirect selection criteria for drought tolerance in sugar beet. Field Crops Res 91:231–249
Oliveira AB, Alencar NLM, Gomes-Filho E (2013) Comparison between the water and salt stress effects on plant growth and development. In: Akinic S (ed) Response of organisms to water stress. InTech, Rijeka, pp 67–94
Orcutt DM, Nilsen ET (2000) Physiology of plants under stress: soil and biotic factors. Wiley, New York, pp 683
Pakniyat H, Armion M (2007) Sodium and proline accumulation as osmoregulators in tolerance of sugar beet genotypes to salinity. Pak J Biol Sci 10:4081–4086
Pandit A, Rai V, Bal S, Sinha S, Kumar V, Chauhan M, Gautam RK, Singh R, Sharma PC, Singh AK, Gaikwad K, Sharma TR, Mohapatra T, Singh NK (2010) Combining QTL mapping and transcriptome profiling of bulked RILs for identification of functional polymorphism for salt tolerance genes in rice (*Oryza sativa* L.). Mol Genet Genom 284:121–136
Panella L, Lewellen RT (2005) Plant introduction and genetic diversity. In: Biancardi E, Campbell LG, Skaracis GN, De Biaggi M (eds) Genetics and breeding of sugar beet. Science Publishers, Enfield, pp 34–38
Panella L, Lewellen RT (2007) Broadening the genetic base of sugar beet: introgression from wild relatives. Euphytica 154:383–400
Parida AK, Das AB, Mohanty P (2004) Defense potentials to NaCl in a mangrove, *Bruguiera parviflora*: differential changes of isoforms of some antioxidative enzymes. J Plant Physiol 161:531–542
Park SJ, Lee JY, Lee SE, Yoo SY, Shim MY (2006) Detection of salt tolerance using chlorophyll fluorescence photometer. 18th World Congress of Soil Science, pp 104–106
Pavli OI, Skaracis GN (2010). Fast and efficient genetic transformation of sugar beet by Agrobacterium rhizogenes. Nature Protocol Exchange Year published. doi:10.1038/nprot.2010.98
Pitman MG (1984) Transport across the root and shoot/root interactions. In: Staples RC, Toenniessen GH (eds) Salinity tolerance in plants-strategies for crop improvement. Wiley, New York, pp 93–124
Plaut Z, Heuer B (1985) Adjustment, growth, photosynthesis and transpiration of sugar beet plants exposed to saline conditions. Field Crops Res 10:1–13
Rajabi A (2010) Abiotic stresses (drought and salinity) of sugar beet in Iran: physiology, breeding, agronomy and biotechnology. Research final report Sugar Beet Seed Institute Karaj Iran No: 90.123. p 75 (In Persian)
Rajabi A, Ober ES (2012) Breeding for improved drought tolerance. In: Ashraf M, Ozturk M, Aqil Ahmad MS, Aksoy A (eds) Crop production for agricultural improvement. Springer, London, p 796
Rajabi A, Griffiths H, Ober ES, Kromdijk W, Pidgeon JD (2008) Genetic characteristics of water-use related traits in sugar beet. Euphytica 160:175–187
Rajabi A, Ober ES, Griffiths H (2009) Genotypic variation for water use efficiency, carbon isotope discrimination, and potential surrogate measures in sugar beet. Field Crops Res 112:172–181
Rajabi A, Ober ES, Norouzi P, Razavi Kh, Sadeghzade Hemayati S, Sadeghian SY (2013a) Molecular aspects of crop response to abiotic stress with an emphasis on drought and salinity. In: Guar RK, Sharma P (eds) Molecular approaches in plant abiotic stress. CRC Press, Boca Raton
Rajabi A, Vahidi H, Haj S Hadi MR, Fathollah Taleghani D (2013b) Study on drought tolerance and interrelationships among some agronomic and morphophysiological traits in sugar beet lines. Int J Agric Crop Sci 5-7:761–768

Ranji ZA, Parvizi M (1996) Selection of sugar beet progeny hybrid to salinity stress with comparison of potential production and sensitivity index in saline and normal soil. Sugar beet J 12(1–2):19–29 (In Persian)

Ranji ZA, Parvizi M, Yavari N (1996) Evaluation of sugar beet progeny hybrid response for proline synthesis in salt stress. Research final report Sugar Beet Seed Research Institute Karaj Iran No: 75.310. p 23 (In Persian)

Rasool S, Ahmad A, Siddiqi TO, Ahmad P (2013) Changes in growth, lipid peroxidation and some key antioxidant enzymes in chickpea genotypes under salt stress. Acta Physiol Plant 35(4):1039–1050

Ren ZH, Gao JP, Li LG, Cai XL, Huang W, Chao DY, Zhu MZ, Wang ZY, Luan S, Lin HX (2005) A rice quantitative trait locus for salt tolerance encodes a sodium transporter. Nat Genet 37:1141–1146

Robinson SP, John W, Downton S, Millhouse JA (1983) Photosynthesis and ion content of leaves and isolated chloroplasts of salt stressed spinach. Plant Physiol 73:238–242

Sadat Noori SA, McNeilly T (2000) Assessment of variability in salt tolerance based on seedling growth in *Triticum durum* Desf. Genet Res Crop E 47:285–291

Sadeghian SY, Khodaii A (1998) Diallel crosses analysis of seed germination traits in sugar beet. Euphytica 103:259–263

Sadeghian SY, Yavari N (2004) Effect of water deficit stress on germination and early seedling growth in sugar beet. J Agron Crop Sci 190:138–144

Salekdeh GH, Komatsu S (2007) Crop proteomics: aim at sustainable agriculture of tomorrow. Proteomics 7:2976–2996

Sayfzadeh S, Habibi D, Fathollah Taleghani D, Kashani A, Vazan S, Sadre Qaen SH, Habib Khodaei A, Mashhadi A Boojar M, Rashidi M (2011) Response of antioxidant enzyme activities and root yield in sugar beet to drought stress. Int J Agric Biol 13(3):357–362

Schneider K, Schäfer-Pregl R, Borchardt DC, Salamini F (2002) Mapping QTLs for sucrose content, yield and quality in a sugar beet population fingerprinted by EST-related markers. Theor Appl Genet 104:1107–1113

Schneider K, Kulosa D, Soerensen TR, Heine SMHM, Durstewitz G, Polley A, Weber E, Jamsari Lein J, Hohmann U, Tahiro E, Weisshaar B, Schulz B, Koch G, Jung Cand Ganal M (2007) Analysis of DNA polymorphisms in sugar beet (*Beta vulgaris* L.) and development of an SNP-based map of expressed genes. Theor Appl Genet 115:601–615

Seaman J (2007) Mechanisms of salt tolerance in halophytes: can crop plant resistance to salinity be improved. APS 402 Dissertation. Candidate no: 000124971, pp 1–11

Serrano R, Culiañz-Macia FA, Moreno V (1999) Genetic engineering of salt and drought tolerance with yeast regulatory genes. Sci Hort 78:261–269

Setiawan A, Koch G, Barnes SR, Jung C (2000) Mapping quantitative trait loci (QTLs) for resistance to Cercospora leaf spot disease (*Cercospora beticola* Sacc.) in sugar beet (*Beta vulgaris* L.). Theor Appl Genet 100:1176–1182

Shaheen R, Hood-Nowotny RC (2008) Carbon isotope discrimination: potential for screening salinity tolerance in rice at the seedling stage using hydroponics. Plant Breed 124(3):220–224

Shannon MC (1985) Principles and strategies in breeding for higher salt tolerance. Plant Soil 89:227–241

Shannon MC (1996) New insights in plant breeding efforts for improved salt tolerance. Hort Tech 6(2):96–99

Sharma S, Villamor JG, Verslues PE (2011) Essential role of tissue-specific proline synthesis and catabolism in growth and redox balance at low water potential. Plant Physiol 157:292–304

Shoresh M, Spivak M, Bernstein N (2011) Involvement of calcium-mediated effects on ROS metabolism in the regulation of growth improvement under salinity. Free Radic Biol Med 51(6):1221–1234

Singer MA, Lindquist S (1998) Multiple effects of trehalose on protein folding in vitro and in vivo. Mol Cell 1:639–648

Singh RK (2006) Plant breeding course. PBGB, International Rice Research Institute (IRRI)

Singh RK, Redona E, Refuerzo L (2010) Varietal improvement for abiotic stress tolerance in crop plants: special reference to salinity in rice. In: Pareek A, Sopory SK, Bohnert HJ, Govindjee, N (eds). Abiotic stress adaptations in plants: physiological, molecular and genomic foundation. Springer, The Netherlands, pp 385–415

Smirnoff N (1998) Plant resistance to environmental stress. Curr Opin Biotechnol 9:214–219

Stevanato P, Trebbi D, Saccomani M (2010) Root traits and yield in sugar beet: identification of AFLP markers associated with root elongation rate. Euphytica 173:289–298

Stich B, Piepho HP, Schulz B, Melchinger AE (2008a) Multi-trait association mapping in sugar beet (*Beta vulgaris* L.). Theor Appl Genet 117:947–954

Stich B, Melchinger AE, Heckenberger M, Mohring J, Schechert A, Piepho HP (2008b) Association mapping in multiple segregating populations of sugar beet (*Beta vulgaris* L.). Theor Appl Genet 117:1167–1179

Subbarao GV, Johansen CMK (1999) Strategies and scope for improving salinity tolerance in crop plants. In: Pessarakli M (ed) Handbook of plant and crop stress. Marcel Dekker, New York, pp 1069–1082

Subbarao GV, Johansen C, Kumar Rao JVDK, Jana MK (1990) Salinity tolerance in F_1 hybrids of pigeonpea and a tolerant wild relative. Crop Sci 30:785–788

Tal M (1985) Genetics of salt tolerance in higher plants: theoretical and practical considerations. Plant Soil 89:199–226

Talei D, Valdiani A, Yusop MK, Abdullah MP (2013) Estimation of salt tolerance in *Andrographis paniculata* accessions using multiple regression model. Euphytica 189:147–160

Tester M, Davenport R (2003) Na^+ tolerance and Na^+ transport in higher plants. Ann Bot 91:503–527

Thomson MJ, Ocampo DM, Egdane J, Katimbang M, Singh RK, Gregorio G, Ismail M (2007) QTL mapping and marker assisted backcrossing for improved salinity tolerance in rice. Plant and animal genomes XV conference, 13–17 January 2007, San Diego CA

Tsialtas JT, Maslaris N (2009) Selective absorption of K^+ over Na^+ in sugar beet cultivars and its relationship with yield and quality in two contrasting environments of central Greece. J Agron Crop Sci 195(5):384–392

Turan S, Cornish K, Kumar S (2012) Salinity tolerance in plants: breeding and genetic engineering. Aust J Crop Sci 6(9):1337–1348

Uno Y, Kanechi M, Inagaki N, Sugimoto M, Maekawa S (1996) The evaluation of salt tolerance during germination and vegetative growth of asparagus, table beet and sea aster. J Japan Soc Hort Sci 65:579–585

Voznesenskaya EV, Koteyeva NK, Edwards GE, Ocampo G (2010) Revealing diversity in structural and biochemical forms of C_4 photosynthesis and a C_3–C_4 intermediate in genus *Portulaca* L. (*Portulacaceae*). J Exp Bot 61:3647–3662

Wakeel A, Asif AR, Pitann B, Schubert S (2011) Proteome analysis of sugar beet (*Beta vulgaris* L.) elucidates constitutive adaptation during the first phase of salt stress. J Plant Physiol 168(6):519–526

Weber WE, Borchardt DC, Koch G (1999) Combined linkage maps and QTLs in sugar beet (*Beta vulgaris* L.) from different populations. Plant Breed 118:193–204

Weber WE, Borchardt DC, Koch G (2000) Marker analysis for quantitative traits in sugar beet. Plant Breed 119:97–106

Würschum T, Maurer H, Schulz B, Möhring J, Reif J (2011) Genome-wide association mapping reveals epistasis and genetic interaction networks in sugar beet. Theor Appl Genet 123(1):109–118

Xia T, Apse MP, Aharona GS, Blumwald E (2002) Identification and characterization of a NaCl-inducible vacuolar Na^+/H^+ antiporter in *Beta vulgaris*. Physiol Plantarum 116:206–212

Yamada N, Promden W, Yamane K, Tamagake H, Hibino T, Tanaka Y, Takabe T (2009) Preferential accumulation of betaine uncoupled to choline monoxygenase in young leaves of sugar beet-importance of long-distance translocation of betaine under normal and salt-stressed conditions. J Plant Physiol 166(18):2058–2070

Yancey PH (1994) Compatible and counteracting solutes. In: Strange K (ed) Cellular and molecular physiology of cell volume regulation. CRC Press, Boca Raton, pp 81–109

Yang AF, Duan AG, Gu XF, Gao F, Zhang JR (2005) Efficient transformation of beet (*Beta vulgaris*) and production of plants with improved salt-tolerance. Plant Cell Tiss Org Culture 83:259–270

Yang L, Ma C, Wang L, Chen L, Li H (2012) Salt stress induced proteome and transcriptome changes in sugar beet monosomic addition line M14. J Plant Physiol 169(9):839–850

Yeo AR, Yeo ME, Flowers SA, Flowers TJ (1988) Screening of rice (*Oryza sativa* L.) genotypes for physiological characters contributing to salinity resistance and their relationship to overall performance. Theor Appl Genet 79:377–384

Zein FI, El-Yamani MS, Moustafa ATA, El-Abaseri MA (2002) Tolerance of some sugar beet cultivars to soil salinity. Egypt J Soil Sci 42(2):319–330

Zhang HX, Blumwald E (2001) Transgenic salt-tolerant tomato plants accumulate salt in foliage but not in fruit. Nature Biotechnol 19:765–768

Zhang N, Si H, Wen G, Du H, Liu B, Wang D (2011) Enhanced drought and salinity tolerance in transgenic potato plants with a BADH gene from spinach. Plant Biotechnol Rep 5:71–77

Zhu JK (2000) Genetic analysis of plant salt tolerance using *Arabidopsis*. Plant Physiol 124:941–948

Chapter 6
Genotypic Variation for Drought Tolerance in Wheat Plants

Jalal A. Aliyev and Irada M. Huseynova

Abstract Many natural disasters, including drought, result in crop production and quality losses worldwide and global climate change makes the situation worse. A wide range of strategies used to enhance the drought tolerance depend on the genetically determined plant capacity and sensitivity, as well as the intensity and duration of the stress. Understanding the physiological, biochemical, and molecular mechanisms of drought tolerance is very important in development of selection and breeding strategies. Among crop plants, wheat is the staple food for more than 35% of world population and is often grown in water-limited conditions. Wheat anti-drought study is of importance to worldwide wheat production and biological breeding. For this purpose, a rich gene fund was created, comprising thousands of wheat genotypes with contrasting photosynthetic traits, productivity, and tolerance to drought stress, and both selected from the ancient, aboriginal varieties of national selection and introduced from the world gene pool, particularly from Centro Internacional de Mejoramiento de Maíz y Trigo (CIMMYT), International Centre for Agricultural Research in Dry Areas (ICARDA), and other international centers. All these genotypes are grown in the field conditions on a wide area at the Absheron Experimental Station of the Research Institute of Crop Husbandry (Baku, Azerbaijan) under normal irrigation and drought. Numerous winter durum (*Triticum durum* Desf.) and bread (*Triticum aestivum* L.) wheat genotypes were the main target of the research. The main parameters for selection of these genotypes were grain yield, plant phenotypic features (stem height, area and architectonics of the leaf surface, etc.), duration of the vegetation, and other morphophysiological traits, as well as drought resistance. Random amplified polymorphic DNA (RAPD)-polymerase chain reaction (PCR) spectra were analyzed in 220 bread (*T. aestivum* L.) and 46

J. A. Aliyev (✉) · I. M. Huseynova
Institute of Botany, Azerbaijan National Academy of Sciences,
40 Badamdar Highway, AZ 1073 Baku, Azerbaijanrg
e-mail: aliyev-j@botany-az.org

J. A. Aliyev
Department of Plant Physiology and Biotechnology, Research Institute of Crop Husbandry,
Ministry of Agriculture of the Republic of Azerbaijan, AZ 1098 Baku, Azerbaijan

I. M. Huseynova
e-mail: huseynova-i@botany-az.org

P. Ahmad et al. (eds.), *Improvement of Crops in the Era of Climatic Changes*, Volume 2, DOI 10.1007/978-1-4614-8824-8_6,

durum (*T. durum* Desf.) wheat genotypes. Using primer P6 (TCGGCGGTTC), 920-bp fragment was revealed in 234 genotypes. Primer P7 (TCGGCGGTTC) produced a 750-bp band only in 203 genotypes. Results obtained by using both P6 and P7 primers match in 74 % of the analyzed genotypes: In 184 genotypes, specific loci were amplified by both markers. The obtained results show that these genotypes possess loci associated with drought tolerance. In 13 genotypes (total 5 %), none of the primers amplified specific fragments. So, the presented data could be used for monitoring the environmental stresses in field-grown plants and help in selection of stress-resistant varieties.

Keywords Drought · Wheat genotypes · Gene pool · RAPD markers · PCR

1 Introduction

World food production is primarily restricted by the impact of the environment. It is very difficult to find areas "without stress," where crops can reach their maximum yield potential. Abiotic environmental factors are considered the main source (71 %) of yield reduction (Boyer 1982; Khan et al. 2013). Different in quantity and intensity, abiotic factors affect the plants in various ways, but certain quantity of each abiotic environmental factor is necessary for the optimal plant development. Any excess or deficit of the abiotic stress factors adversely affects the plant growth and productivity (Koyro et al. 2012). Drought is one of the most common external stresses affecting plant growth and development through changes in metabolism and gene expression (Aranjuelo et al. 2011). Global climate change leads to a higher probability of occurrence of dry years in the regions where drought was a rare event, and also to appearance of new areas with arid conditions that require urgent development of new crop varieties resistant and/or tolerant to drought conditions. Development of the drought-resistant high-yielding varieties is very important for the agriculture in our country and all over the world.

Drought effects on crop production have been investigated for decades. However, studying the drought response is still a challenge because of the complex and quantitative nature of the trait (Budak et al. 2013). Drought tolerance depends on the environmental interactions. Duration and severity of the dehydration stress and its dependence on other abiotic and biotic stress factors are also significant. One of the most effective methods for increasing the crop productivity is to increase the adaptation to growth conditions (Rashed et al. 2010). In this regard, studies on the development of adaptive breeding techniques based on the latest achievements of genetics and biotechnology are of particular relevance and allow expanding the range of variability of the existing genotypes and creating forms with new features and properties in a short period of time.

Wheat is a very valuable food crop, which occupies a leading place in grain balance for most of the world population. It is also one of the oldest and most important crops in Azerbaijan. It is strategically important foodstuff for about 2 billion people (36% of the world population; Rashed et al. 2010). Current UN projections indicate that world population could increase by 2.25 billion people from today's level, reaching 9.15 billion by 2050. At the global level, agricultural production and consumption will need to be raised by 70%, and food production in the developing world will need to *double,* which also makes it necessary to increase the wheat production all over the world (www.fao.org; www.populationinstitute.org). However, this task faces a number of problems (climate change, development of epiphytotics, erosion of arable land), overcoming of which requires breeding and genetic improvement of wheat. Its productivity depends on the implementation of genetically inherent yielding potential, as well as the impact of specific weather and climate conditions. Wheat production capacity suffers from fluctuation of yield from year to year and from site to site. Plant species differ in their sensitivity and response to decline in water potential caused by drought, low temperatures, or high salinity. Stress as a result of drought may occur at the beginning or at the end of the season or at the stages of grain filling and ripening. Furthermore, under the same conditions drought tolerance of various species, subspecies, and cultivars are different. This fact emphasizes the importance of genetic diversity as the major factor of drought. Drought-tolerant plants are the main objectives of the investigations. They can be used as sources of drought-related genes and gene regions and facilitate the improvement of modern crop varieties. *Aegilops tauschii,* which is more drought tolerant than *Triticum* and wild emmer wheat (*T. dicoccoides*), has drought tolerance characteristics lost during cultivation of modern lines (Budak et al. 2013).

Yield increase and wheat properties improvement through traditional breeding methods do not match the growth rate of the world population. Therefore, development of more effective approaches is required. Improving the productivity of wheat varieties in drought conditions is one of the important tasks in wheat-breeding programs.

Molecular breeding of plants is the basis for crop improvement in the twenty-first century (Moose and Mumm 2008; Fleury et al. 2010). Integration of advances in biotechnology, genomic research, and application of molecular markers in common plant-breeding practice created a basis for molecular breeding of plants—an interdisciplinary science revolutionizing the crop improvement. An important issue of food security that people currently face around the world led scientists to the genetic revolution through the progress in biotechnology (Ashraf 2010). The genetic revolution actually implies change of qualitative and quantitative traits in the organism by transferring the desired genes from one species to another. This strategy is called a transgenic approach. In contrast to classical breeding, transgenic approach allows including only specific cloned genes and restricts the transfer of undesirable genes from the donor organism. Pyramiding of genes with similar features could also be achieved through this approach. Rapid advancement of recombinant DNA technology and development of throughput and effective gene transfer led to an efficient transformation and development of transgenic lines of several crop species (Gosal et al. 2009). The transgenic approach is widely used around the world to improve the features, including resistance to abiotic and biotic stresses, in a number of crops (Ashraf et al. 2008). Plant responses to drought are complex, because they include a lot of genes with additive effects, so the prospects for improving the drought resistance in crops are not yet clear. Despite this fact, many efforts to develop transgenic lines of different crops demonstrating improved drought tolerance were made over the past few decades. Bioengineering plays a special role in development of the genes encoding the compatible organic osmolytes, plant growth regulators, antioxidants, heat shock, and late embryogenesis proteins, as well as transcription factors involved in gene expression (Manavalan et al. 2009; Ashraf 2010).

One of the main phenomena occurring in the cell during water shortage is extensive modification of gene expression (Liu et al. 2013; Savitri et al. 2013). Therefore, through a variety of molecular and biochemical approaches, key genes responsible for drought resistance were studied (Kirigwi et al. 2007; Xu et al. 2007; Miyama and Tada 2008; Wei et al. 2009). Regulation of expression of genes involved in stress tolerance is necessary for the improvement of this feature in plants. It is known that resistance to drought, as well as other abiotic stresses, is under the control of many minor genes (polygenes), which are expressed with superposition (Mohammadi et al. 2005; Lang and Buu 2008). Loci on chromosomes belonging to this type of genes are now referred to as quantitative trait loci (QTL). It is necessary to develop effective DNA markers in order to identify a number of QTL for different crops. The genetic variability of cultivated plants can be used in direct breeding, under in vitro or in vivo stress conditions, or in QTL (polygenes) mapping and subsequent breeding using markers. QTL mapping allows determining the location, number, and value of phenotypic effects and gene action patterns (Vinh and Paterson 2005). The role of polygenes in controlling the traits is widely assessed by traditional means, but the use of DNA markers and QTL mapping allowed to distinguish the complex traits conveniently (Humphreys and Humphreys 2005). For QTL analysis, phenotypic evaluation of a large number of plants belonging to populations segregated by a variety of genetic markers is performed, then a part or the entire population is genotyped and, finally, the corresponding statistical analysis is performed to accurately determine the loci controlling the trait (Asins 2002). QTL mapping by drought resistance was performed for a variety of crops, especially for maize, wheat, barley, cotton, sorghum, and rice (Quarrie et al. 1994; Sari-Gorla et al. 1999; Teulat et al. 1997; Saranga et al. 2001; Sanchez et al. 2002; Bernier et al. 2008). Root penetration ability (Uga et al. 2011), osmotic adjustment (Robin et al. 2003), grain yield and yield components (Xu et al. 2005), stay green (Jiang et al. 2004), canopy temperature, leaf rolling, and leaf drying (Yue et al. 2006) that represent QTLs for morphology and other traits have been reported. Recently, several yield QTLs have been identified in wheat through linkage analysis and association mapping. Most QTLs for drought tolerance in wheat have been determined through yield and yield-related measurements under water-limited conditions. However, yield and drought are complex traits that involve multiple loci and show genotype–environment interactions. So it is very difficult to describe yield with respect to water use, and its accurate phenotyping is a challenge since QTLs established in one environment may not be confirmed in the other.

QTL analysis and other subsequent study using molecular markers in wheat revealed that chromosomes 5B, 4B, and 7B carry important genes for drought tolerance. A QTL on chromosome 5B located between two markers (M51P65 and Psr136) showed positive correlation with drought tolerance. However, QTLs discovered on chromosome 4B and 7B (between M62P64d–Rht and M83P65d–M21P76n markers, respectively) are negatively correlated with drought tolerance (Rana et al. 2013). A marker located on chromosome 4A (Xwmc89) was found significantly associated with drought tolerance in tolerant wheat genotypes (Kirigwi et al. 2007). Several other reports on QTL in homologous Triticeae group 4 chromosomes highlight the importance of this chromosome group for drought stress tolerance.

Wheat genomic data are necessary when performing the analysis of drought response (Edae et al. 2013). However, thanks to the International Triticeae Mapping Initiative (ITMI) and International Wheat Genome Sequencing Consortium (IWGSC), whole genome sequencing of bread wheat is almost complete. The extensive reservoir of alleles in drought-tolerant wild germplasm has been investigated with the application of the whole-genome sequencing. "Omics" strategies are also used in drought research since osmotic stress response is regulated at the posttranscriptional and posttranslational levels (Budak et al. 2013). The ability to introduce only desired loci is the advantage of the transformation/selection strategies over the conventional and marker-assisted breeding (MAB). Breeding or molecular transformation of novel genes obtained from screening of wheat germplasms will contribute to the creation of high-yield wheat under drought conditions, balancing the production with the consumption of the increasing human population.

In recent years, molecular biology techniques have led to the development of DNA markers that have been effectively used to identify a number of features in various crops (Barakat et al. 2010). Effects caused by the environmental impact usually present in conventional breeding have been completely eliminated by using this approach. Through the creation of complete genetic linkage maps based on molecular markers, it became possible to map accurately the genes responsible for variability of complex quantitative traits in order to use them in various breeding programs. Molecular markers are a beneficial addition to the morphological and physiological characteristics of the varieties, as they are numerous, independent of the impact of tissues or environment, and allow to identify varieties in the early stages of plant development. The use of molecular markers in breeding will provide information about a particular feature at the early stages of development without waiting for the phenotypic expression of the feature, and will simplify testing the properties, for example, resistance to various diseases, requiring scrupulous evaluation using traditional investigation techniques (Ameen 2013). Molecular markers can be used for isolation and cloning of genes in order to study properties controlled by them and transferring these properties to other varieties (i.e., for genetic transformation). In addition, with the development of detailed molecular interaction maps, significant progress was reached in breeding procedures using markers that allowed pyramiding desirable properties to achieve significant improvements in drought resistance of crops (Lörz and Wenzel 2008). It is now possible to test the suitability of thousands of genomic regions of crop germplasm under limited water supply through MAB, which was not previously possible. Assessing the significance of breeding of each genome region, the breeder can combine genes of different nature using new methods which were impossible to do previously by means and methods of conventional breeding. The use of DNA markers is required to increase stress resistance due to complexity of the mechanism of tolerance to abiotic stress and problems in the selection by phenotype (Manavalan et al. 2009). We can list a number of DNA markers, such as restriction fragment length polymorphism (RFLPs), random amplified polymorphic DNA (RAPD), cleaved amplified polymorphic sequences (CAPS), insertion/deletion polymerase chain reaction markers (PCR indels), amplified fragment length polymorphism (AFLPs), simple sequence repeat (SSR, microsatellites), single nucleotide polymorphism (SNPs), and functional markers (FMs), currently used for studying the inheritance of stress tolerance.

RAPD technology is an important tool for rapid identification of markers associated with drought resistance and is very effective in determining the genetic variation among wheat genotypes (Iqbal et al. 2007; Ali et al. 2013). Marker-assisted selection programs (MAS programs) were improved specifically through RAPD markers (Wang et al. 1995; Penner et al. 1996; Abdel-Tawab et al. 1998). RAPD analysis may be used as an express method for detection of genetic polymorphism, and as a source for unique locus-specific markers. The aim of the study was to perform screening of wheat genotypes for drought resistance using RAPD markers and to identify the most valuable genotypes for the subsequent use in breeding.

2 Materials and Methods

2.1 Plant Material

The study used 266 genotypes of durum (*T. durum* Desf.) and bread (*T. aestivum* L.) wheat collected in the gene pool of the Research Institute of Crop Husbandry. Plants were grown in the field on a wide area at the Absheron Experimental Station of the Research Institute of Crop Husbandry (Baku, Azerbaijan) under normal irrigation and drought. The main parameters for selection of these genotypes were grain yield, plant phenotypic features (stem height, area and architectonics of the leaf surface, etc.), duration of the vegetation, and other morphophysiological traits, as well as drought resistance.

2.2 Plant DNA Isolation

DNA extraction was performed using cetyltrimethyl ammonium bromide (CTAB) method with some modifications (Murray and Thompson 1980). Fresh plant tissue was grinded in liquid nitrogen and suspended in 1,000 μL of CTAB extraction buffer (100-mM Tris–HCl, pH 8.0; 20-mM ethylenediaminetetraacetic acid (EDTA), pH 8.0; 1.4-mM NaCl; 40-mM β-mercaptoethanol), preheated in a water bath to 60 °C. Homogenization was completed by intense shaking on vortex. Then, 400 μL

Table 6.1 Nucleotide sequence of RAPD primers used for DNA amplification

Primers	5′–3′ sequence
P6	TCGGCGGTTC
P7	CTGCATCGTG

of chloroform (99.8 %) was added to each tube and the contents were gently mixed. The test tubes were placed in a water bath and incubated for 10 min at 60°C. After incubation, the tubes were centrifuged in a tabletop Eppendorf-type centrifuge (15,000 g) for 10 min at room temperature. After centrifugation, the supernatant was carefully isolated and transferred to clean tubes, and then 600 μL of cold isopropanol was added, mixed well, and left at room temperature for 3–5 min. At this stage, the disperse DNA precipitate can be observed. The tube contents were centrifuged at room temperature (15,000 g) for 10 min. The precipitate was washed several times with 70 % ethanol, dried in a thermostat at 56 °C for 5 min, and dissolved in Tris–EDTA (TE) buffer (10-mM Tris–HCl, pH 8.0; 1-mM EDTA). Samples were left overnight at 4 °C to completely dissolve the DNA in the buffer.

2.3 Determination of DNA Concentration

After dissolution of the DNA, its concentration was determined by optical density (OD) at 260 nm using the ULTROSPEC 3300 PRO spectrophotometer ("AMERSHAM", USA). Purity of genomic DNA was determined by A260/A280 absorbance ratio. Quality of the DNA was verified in a 0.8 % agarose gel stained with 10-mg/ml ethidium bromide in 1 × TBE buffer (Tris base, boric acid, EDTA). The gel was photographed under UV light using "gel documentation system UVITEK" (UK).

2.4 DNA Amplification

A PCR using RAPD marker was carried out using Williams method (Williams et al. 1990). DNA amplification was performed in a reaction mix of 25-μL volume containing 10x buffer; 20 ng of genomic DNA; 0.2-μM primer; 200 μM each of dATP, dCTP, dGTP, and dTTP; 2.5-mM $MgCl_2$; and 0.2 units of Taq polymerase in incubation buffer. Two decamer oligonucleotide primers P6 and P7 (Eurogentec S.A., Belgique) associated with drought resistance (Pakniyat and Tavakol 2007) were used for RAPD analysis. Nucleotide sequences of the primers are shown in Table 6.1. PCR was performed in the "Applied Biosystems 2720 Thermal Cycler" (Singapore) at the following conditions: 1cycle—4 min at 94 °C; 10 cycles—1 min at 94 °C, 1 min at 36 °C, and 1 min at 72 °C; 35 cycles—1 min at 94 °C, 1 min at 36.2 °C, and 1 min at 72 °C; the final elongation cycle was performed at 72 °C for 15 min and stored at 4 °C.

The reaction products were separated by electrophoresis in a 1.2 % agarose gel in the HR-2025-High Resolution ("IBI SCIENTIFIC", USA) horizontal electrophoresis device with the addition of ethidium bromide. Gels were documented using "gel documentation system UVITEK." To estimate the size of amplified fragments, 100-bp DNA Ladder Plus marker was used. Statistical analysis included construction of binary matrices for each of the primers, where the "presence" (+) or "absence" (−) of fragments with the same molecular weight on electrophoregram was specified.

3 Results and Discussion

For the identification of drought-resistance genes, 266 wheat genotypes were used. While 220 of them were bread wheat genotypes (*T. aestivum* L.), the remaining 46 were durum wheat genotypes (*T. durum* Desf.). Drought resistance was tested using RAPD-PCR analysis. For this purpose, decamer oligonucleotide RAPD primers P6 and P7, which are associated with drought resistance in accordance with publications, were used (Pakniyat and Tavakol 2007). Table 6.2 shows the results of PCR reactions performed with both markers. As seen in the table, P6 molecular marker (TCGGCGGTTC) gives a positive result in 234 genotypes (this is approximately 88 % of all genotypes used for this analysis). In detail, amplification fragments specific for P6 primer were identified in 194 genotypes of bread wheat and in 40 genotypes of durum wheat. Products were not amplified in 32 genotypes, that is only 12 % of the analyzed genotypes, and 26 samples among them were bread wheat; the remaining 6 were durum wheat.

Figure 6.1 shows electrophoretic profiles obtained by using this primer (figures are shown only for 50 genotypes). It is known that P6 primer produces a fragment of 920 bp. As seen from these profiles, amplification fragments of 920-bp region are well visualized only in 41 genotypes. Expected fragments were not amplified in sensitive to drought genotypes such as tetraploid Garagylchyg-2 and hexaploids Giymatli-2/17 and Gyrmyzy gul-1 (figures not shown), as well as in some genotypes considered to be medium resistant to drought. Genotypes of durum—Terter-2, Mirvari, and bread wheat—Renan, Fatima, Murov-2, Seba, and Podolyanka (Fig. 6.1) could serve as an example of the latter. It is noteworthy that the fragment associated with P6 primer is also not amplified in two tetraploid drought-resistant genotypes—Shirvan-5 and Kehreba.

Interesting results were also obtained during the RAPD-PCR analysis with a second primer—P7 (TCGGCGGTTC). Positive result during application of this molecular marker was observed in 203 genotypes, in other words, in 76 % of the used 266 *Triticum* samples (see Table 6.2). Of these, 87 % (176 genotypes) were bread wheat genotypes, whereas the remaining 13 % (27 genotypes) were durum wheat genotypes. Negative result was obtained for 24 % genotypes (i.e., in 63 samples, which is two times more in comparison with the first marker). Among them, 44 genotypes are hexaploids (bread wheat genotypes) and the remaining 19 are tetraploids (durum wheat genotypes).

Figure 6.2 reflects PCR profiles obtained by using RAPD marker P7. Amplification fragments are of 750-bp size. In 34 samples of 50 genotypes, the amplified

Table 6.2 Results of PCR analysis using RAPD markers P6 (5′ TCGGCGGTTC 3′) and P7 (5′ TCGGCGGTTC 3′)

No.	Wheat genotypes	Primers	
		P6	P7
Hexaploids (Triticum aestivum L.*)*			
1	Akinchi-84	+	+
2	Pirchahin-1	+	+
3	Guneshli	+	+
4	Dagdash	+	+
5	FARAN Dolc	+	+
6	Renan	−	−
7	Avreca	+	+
8	Pactole	+	+
9	38 IBWSN (129 №)	+	+
10	10 SAWVT (11 №)	+	+
11	13th FAWWON (117 №)	+	+
12	8th WWEERYT (32 №)	+	+
13	4th RWVT-LRCA (89 №)	+	+
14	14th FAWWON (86 №)	+	+
15	8th WON-SA (65 №)	+	+
16	9th WON-SA (27 №)	−	+
17	3 RBWYT (521 №)	+	+
18	3 RBWYT (510 №)	+	+
19	3 RBWYT (536 №)	+	+
20	3 RBWYT (518 №)	+	+
21	39 IBWSN (113 №)	+	+
22	14 SAWYT (49 №)	+	+
23	39 IBWSN (97 №)	+	−
24	11st IWWYT-R (9816 №)	+	−
25	S5	+	−
26	12nd IWWYT-IR 29	+	−
27	15th FAWWON-IR 30	+	−
28	T3	+	−
29	T2	+	+
30	FO3 № N 907	+	+
31	FO2 №3 N 107A	+	+
321	TH1	+	−
33	FO2 №206.1	+	−
34	DD №7	+	+
35	FO2 №2 N 9 (ART)	+	−
36	FOFO2 N7 N6	+	−
37	D8 №5	+	+
38	FO2 N8 N4	+	−
39	S4	+	−
40	AYT SIR 5002	+	+
41	AYT SIR 5081	+	+
42	15th FAWWON-SA 34	+	+
43	15th FAWWON-SA 29	+	+
44	12th IWWYT-IR (25)	+	+
45	11th IWWYT-SA (32)	+	+

Table 6.2 (continued)

No.	Wheat genotypes	Primers	
		P6	P7
46	11th IWWYT-SA (16)	+	+
47	16th FAWWON-SA (32)	+	+
48	16th FAWWON-SA (33)	+	+
49	16th FAWWON-SA (10)	+	+
50	16th FAWWON-SA (26)	+	−
51	16th FAWWON-SA (6)	+	+
52	16th FAWWON-SA (14)	+	+
53	16th FAWWON-IR (66)	+	+
54	16th FAWWON-IR (43)	+	+
55	83	−	−
56	DD 08 №1	+	+
57	DD 08 №3	−	−
58	FO2 №208	+	+
59	FO2 №308	+	−
60	FO2 №2 N308	+	−
61	S3	+	+
62	Selective SS	−	−
63	16th FAWWON-IR (32)	+	+
64	16th FAWWON-IR (44)	+	+
65	16th FAWWON-IR (51)	+	+
66	16th FAWWON-IR (75)	+	+
67	16th FAWWON-IR (11)	+	−
68	16th FAWWON-IR (12)	+	+
69	16th FAWWON-IR (10)	+	−
70	16th FAWWON-IR (19)	+	+
71	16th FAWWON-IR (23)	+	+
72	16th FAWWON-IR (29)	+	−
73	16th FAWWON-IR (21)	+	+
74	Mirbashir-128	+	+
75	Yegane	+	−
76	Zirve-80	+	−
77	Fatima	−	−
78	Aran	+	−
79	Azeri	+	−
80	Murov	+	−
81	Murov-2	−	−
82	Seba	−	−
83	Tereggi	+	−
84	Beyaz	+	+
85	Shafag	+	−
86	KSI-13	+	−
87	Pirshahin	+	+
88	Ugur	+	+
89	Perzivan-1	+	+
90	Perzivan-2	+	+
91	Sheki-1	+	+
92	16th FAWWON-IR (61)	+	+

Table 6.2 (continued)

No.	Wheat genotypes	Primers	
		P6	P7
93	16th FAWWON-IR (46)	+	+
94	16th FAWWON-IR (52)	+	+
95	16th FAWWON-IR (90)	+	+
96	16th FAWWON-IR (47)	+	+
97	29 ES WVT (7)	+	+
98	29 ES WVT (26)	+	−
99	29 ES WVT (38)	+	+
100	29 ES WVT (30)	−	+
101	16 SAWWVT (29)	+	+
102	16 SAWWVT (34)	+	+
103	S1	+	+
104	Giymatli-2/17	−	−
105	Gobustan	+	+
106	Nurlu-99	+	+
107	Gyrmyzy gul-1	−	+
108	Azamatli-95	+	+
109	Tale-38	+	+
110	Ruzi-84	+	+
111	12nd FAWWON №97 (130/21)	+	+
112	4th FEFWSN №50 (130/32)	+	+
113	30 ESWVT (20)	+	+
114	30 ESWVT (28)	+	+
115	3rd WWEERYT №10	+	+
116	3rd WWEERYT №21	+	+
117	1st WWEERYT №43	+	+
118	FO2 N7-A (dwarf)	+	+
119	2nd WWEERYT №4	+	+
120	2nd WWEERYT №19	+	+
121	9th FAWWON №74	+	+
122	11st FAWWON №22	+	+
123	Gizil bugda	+	+
124	9th FAWWON №9	+	+
125	2nd WWEERYT №13	+	+
126	RB WON (SAA) №75	+	+
127	4 WWEERYT №3	+	+
128	4 WWEERYT №45	+	+
129	Bezostaya-1	+	+
130	2nd FAWWEYT 9918	+	−
131	7th HRWSN 184/27	+	−
132	9th FAWWON №28	+	+
133	Canada 2	+	−
134	ATIF A.C.	+	+
135	17th FAWWON-IR (15)	+	+
136	17th FAWWON-IR (14)	−	+
137	17th FAWWON-IR (17)	+	+
138	17th FAWWON-IR (23)	+	+
139	17th FAWWON-IR (27)	+	+

Table 6.2 (continued)

No.	Wheat genotypes	Primers	
		P6	P7
140	17th FAWWON-IR (37)	+	+
141	17th FAWWON-IR (51)	–	+
142	17th FAWWON-IR (13)	+	+
143	17th FAWWON-IR (31)	–	+
144	17th FAWWON-SA (54)	–	+
145	13th IWWYT-IR (12)	+	+
146	13th IWWYT-IR (11)	+	+
147	13th IWWYT-IR (10)	+	+
148	13th IWWYT-IR (78)	+	+
149	13th IWWYT-IR (59)	+	+
150	13th IWWYT-IR (48)	+	+
151	13th IWWYT-IR (49)	+	+
152	13th IWWYT-IR (51)	+	+
153	Selective Sunbul kiz	+	+
154	S9	+	+
155	13th IWWYT-IR (17)	+	+
156	17th FAWWON-SA (33)	+	+
157	17th FAWWON-IR (124)	+	+
158	CWANA 10th SBW-ON (14)	+	+
159	CWANA 10th SBW-ON (41)	–	–
160	CWANA 10th SBW-ON (69)	+	+
161	CWANA 10th SBW-ON (68)	+	+
162	CWANA 10th SBW-ON (117)	+	+
163	CWANA-CA 10th IRSBWVT (5)	+	+
164	CWANA-CA 10th IRSBWVT (15)	+	+
165	CWANA-CA 10th IRSBWVT (6)	+	–
166	CWANA-CA 10th IRSBWVT (22)	+	+
167	CWANA-CA 10th IRSBWVT (1)	+	+
168	CWANA-CA 10th IRSBWVT (3)	+	+
169	CWANA-CA 10th IRSBWVT (18)	+	+
170	CWANA-CA 10th IRSBWVT (4)	+	+
171	CWANA-CA 10th IRSBWVT (16)	+	+
172	CWANA-CA 10th IRSBWVT (19)	+	+
173	CWANA-CA 10th IRSBWVT (9)	+	+
174	30 ESWVT (45)	+	+
175	Selective Sunbul ag	+	+
176	17th FAWON-SA (80)	+	+
177	13th IWWVT-IR (61)	+	+
178	Bogdanka	+	+
179	Kripsinka	+	+
180	42 IBWSN (51)	+	+
181	42 IBWSN (169)	+	+
182	42 IBWSN (159)	+	+
183	42 IBWSN (34)	+	+
184	42 IBWSN (118)	+	–
185	42 IBWSN (148)	+	+
186	42 IBWSN (24)	–	+

Table 6.2 (continued)

No.	Wheat genotypes	Primers	
		P6	P7
187	42 IBWSN (157)	+	+
188	42 IBWSN (23)	−	+
189	42 IBWSN (166)	−	+
190	42 IBWSN (170)	−	+
191	42 IBWSN (129)	−	+
192	42 IBWSN (158)	+	+
193	42 IBWSN (163)	+	+
194	42 IBWSN (165)	−	−
195	42 IBWSN (173)	+	+
196	42 IBWSN (115)	+	+
197	42 IBWSN (12)	+	+
198	42 IBWSN (13)	+	+
199	42 IBWSN (17)	+	+
200	42 IBWSN (18)	+	+
201	42 IBWSN (9)	−	−
202	17th FAWWON-IR (112)	+	+
203	17th FAWWON-IR (54)	+	−
204	17th FAWWON-IR (58)	+	+
205	17th FAWWON-IR (53)	+	+
206	17th FAWWON-IR (59)	+	+
207	17th FAWWON-IR (94)	+	+
208	17th FAWWON-IR (78)	+	+
209	17th FAWWON-IR (79)	+	+
210	17th FAWWON-IR (88)	+	+
211	17th FAWWON-IR (90)	+	+
212	13th IWWVT-IR (53)	+	+
213	13th IWWVT-IR (57)	−	+
214	13th IWWVT-IR (54)	−	+
215	CWANA 10th SBW-ON (1)	+	+
216	CWANA 10th SBW-ON (82)	+	+
217	CWANA 10th SBW-ON (13)	+	+
218	Polsha	+	+
219	Podolyanka	−	+
220	Mironovskaya	+	+
Tetraploids (Triticum durum Desf.*)*			
221	Vugar	+	+
222	Shiraslan-23	+	+
223	Alinja-84	+	−
224	Terter	+	−
225	Sherg	+	−
226	Tigre	+	+
227	Terter-2	−	+
228	Garabag	+	+
229	Yagut	+	+
230	Turan	+	+
231	Mirbashir-50	+	+

Table 6.2 (continued)

No.	Wheat genotypes	Primers	
		P6	P7
232	Shirvan-3	+	+
233	Shirvan-5	−	+
234	Mugan	+	+
235	Ag bugda	+	+
236	Kehreba	−	+
237	Mirvari	−	+
238	Salt tolerant Mapping pop (F7)	+	−
239	ID VT 06-DTA (6)	+	−
240	S2	−	−
241	Barakatli-95	+	−
242	Garagylchyg-2	−	−
243	Gyrmyzy bugda	+	+
244	Kollektivnaya 77	+	+
245	RB WON (DA) №93	+	+
246	DON-D №56	+	+
247	Fadda 98	+	+
248	Gioaza	+	−
249	Beltago	+	−
250	DYT-MSA №3	+	+
251	DYT-MTA (60 №29) 1	+	−
252	Polonicum	+	−
253	2nd WDEEVT №32	+	+
254	DYT-MCA №4	+	−
255	BDME	+	+
256	RBWHADTON (4)	+	+
257	RBWHADTON (41)	+	−
258	33rd IDON-MD (71)	+	−
259	CWANA-CA 10th DSBWVT (20)	+	+
260	CWANA-CA 10th DSBWVT (8)	+	+
261	CWANA-CA 10th DSBWVT (25)	+	+
262	33rd IDON-MD (105)	+	+
263	33th IDON-MD (19)	+	−
264	33rd IDYT-MD (7)	+	−
265	33rd IDYT-MD (15)	+	−
266	33rd IDYT-MD (28)	+	−

+ Presence of the locus, − Absence of the locus

locus can be clearly seen in the 750-bp region. However, among genotypes in which this fragment is not visualized, bread wheat genotype Zirve-80 can be noted. In field conditions, this genotype is evaluated as drought resistant. Another drought-resistant genotype, where locus associated with P7 marker was not amplified, is durum wheat genotype Barakatli-95 (figure not shown).

And again, these phenomena can be explained by certain mutations in the regions complementary to the primer. Among drought-sensitive genotypes, such as durum wheat genotype Garagylchyg-2 and bread wheat genotype Giymatli-2/17,

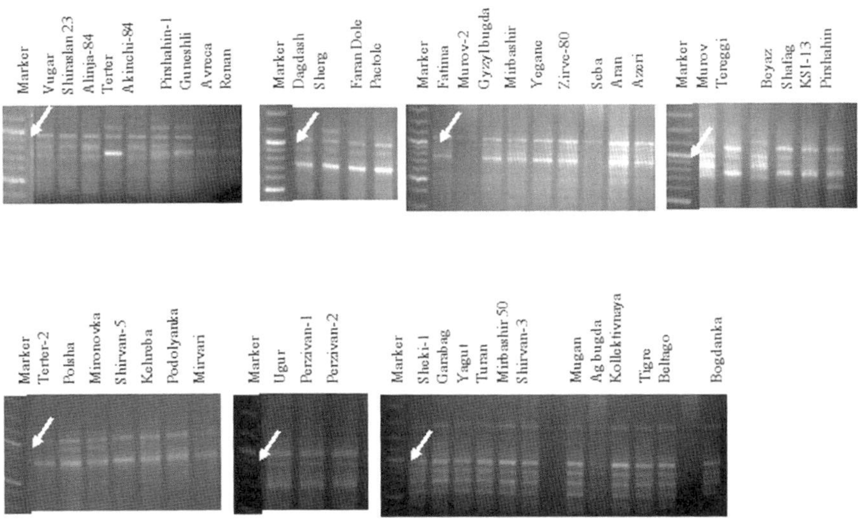

Fig. 6.1 RAPD profiles of wheat genotypes induced by P6 primer. *Arrow* indicates the zone of 920 bp; M (molecular weight marker), 100 bp

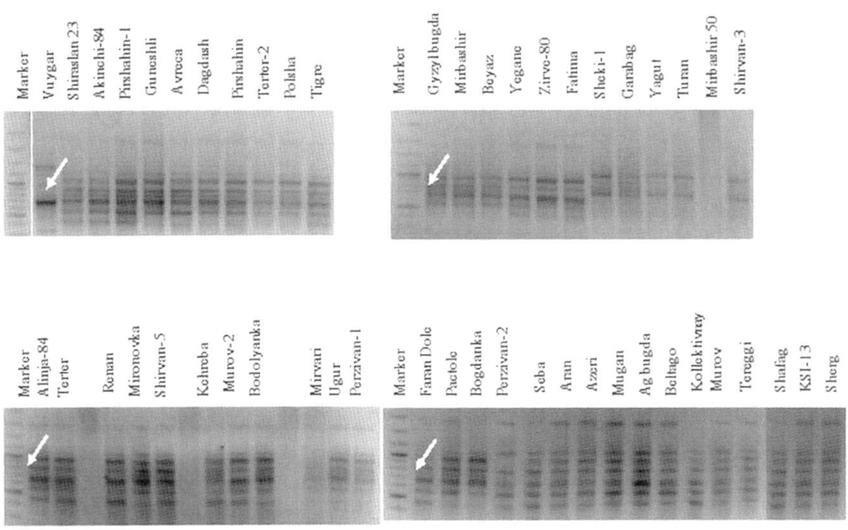

Fig. 6.2 RAPD profiles of wheat genotypes induced by P7 primer. *Arrow* indicates the zone of 750 bp; M (molecular weight marker), 100 bp

locus associated with drought resistance was not amplified. The next exception was bread wheat genotype Gyrmyzy gul-1 which possesses resistance locus, but demonstrates sensitivity to water stress in field conditions.

Comparative analysis of PCR profiles obtained with both RAPD markers was conducted. Results obtained using molecular markers P6 and P7 coincide approximately in 74 % of all tested genotypes: In 184 genotypes, synthesis of characteristic fragments successfully occurred by applying both markers, indicating that these genotypes have loci associated with drought resistance. Specific amplification fragments in 13 (approximately 5 %) of 266 used genotypes were not identified with any of the applied markers. These results confirm our data for genotypes Garagylchyg-2 and Giymatli-2/17 where biochemical and physiological parameters also indicate their sensitivity to drought (Huseynova et al. 2007). Meanwhile, among these genotypes, there are also genotypes medium resistant to drought, such as Renan, Fatima, Murov-2, and Seba.

Mismatch results during PCR analyses with different molecular markers are also observed. In 7 % of genotypes, fragments specific for P6 marker in the 920-bp region were not synthesized, but on the contrary fragments were amplified in the 750-bp region specific for P7 marker. Four of them are tetraploid genotypes: Terter-2 and Mirvari that are medium resistant and Shirvan-5 and Kehreba that are resistant to drought. The remaining 15 genotypes are hexaploid: 9th WON-SA (27), 29 ES WVT (30), Gyrmyzy gul-1, 17th FAWWON-SA (14), 17th FAWWON-SA (31), 17th FAWWON-SA (51), 17th FAWWON-SA (54), 42 IBWSN (24), 42 IBWSN (129), 42 IBWSN (170), 42 IBWSN (166), 42 IBWSN (23), 13th IWWVT-IR (57), 13th IWWVT-IR (54), and Podolyanka. It is known that Gyrmyzy gul-1 is drought sensitive and Podolyanka is medium resistant to drought. Response to drought in the remaining genotypes in field conditions has not yet been investigated. Approximately in 19 % of genotypes, the amplification products specific for P6 marker were absent, and on the contrary synthesis of PCR profiles specific for P7 was successful. Of these genotypes, 33 are bread wheat genotypes, the remaining 17 are durum wheat genotypes. For some of these genotypes, their response to water shortage in field conditions was investigated. For example, it is known that bread wheat genotypes Yegane, Aran, Azeri, Murov, Tereggi, Shafaq, and KSI-13 are medium resistant to drought. The exception is Zirve-80 genotype resistant to water shortage. Among durum wheat genotypes where the results obtained using different markers did not coincide, there are genotypes medium resistant to drought, such as Alinja-84, Terter, Sharg, as well as resistant genotype-Barakatli-95.

4 Conclusions and Future Perspective

Despite the latest research advances in the field of biotechnology that revived interest in the targeted breeding for drought resistance and the use of new tools of genomics to improve drought resistance of crops (Moose and Mumm 2008), nowadays it has already been recognized that the problem of adaptation syndrome complexity in drought conditions can only be solved through an integrated approach that combines physiological analysis of drought avoidance and specific patterns of drought resistance in plants using molecular and genetic techniques. Therefore, in addition to recent rapid achievements in genomics, a better understanding of physiological mechanisms of drought response will contribute to advances in the field of genetic improvement of the drought resistance of crops (Fleury et al. 2010). This approach will increase the efficiency of improving drought resistance of wheat (Barakat et al. 2010).

Thus, we can conclude that the results obtained by PCR analysis of wheat genotypes contrasting in drought resistance could be used for further wheat drought tolerance studies at the molecular and genetic levels.

Acknowledgements This research was financially supported by the Science Development Foundation under the president of the Republic of Azerbaijan (EIF-2010-1(1)-40/24-M-20).

References

Abdel-Tawab FM, Stuber CW, Rashed MA, Bahieldin A, Atta AH (1998) Genome mapping and detection of QTLs linked with salt tolerance in maize. Proc Int Conf Mol Genet 1:230–248

Ali A, Ali N, Ali I, Adnan M, Ullah N, Swati ZA (2013) Morphological and genetic diversity of Pakistani wheat germplasm under drought stress. Int J Adv Res Tech 2(5):186–193

Ameen TE (2013) Molecular markers for drought tolerance in bread wheat. Afr J Biotechnol 12(21):3148–3152

Aranjuelo I, Molero G, Erice G, Avice JC, Nogués S (2011) Plant physiology and proteomics reveals the leaf response to drought in alfalfa (*Medicago sativa* L.). J Exp Bot 62:111–123

Ashraf M (2010) Inducing drought tolerance in plants: recent advances. Biotechnol Adv 28:169–183

Ashraf M, Athar HR, Harris PJC, Kwon TR (2008) Some prospective strategies for improving crop salt tolerance. Adv Agron 97:45–110

Asins MJ (2002) Present and future of quantitative trait locus analysis in plant breeding. Plant Breed 121:281–291

Barakat MN, Al-Doss AA, Moustafa KA, Ahmed EI, Elshafei AA (2010) Morphological and molecular characterization of Saudi wheat genotypes under drought stress. J Food, Agric Environ 8(1):220–228

Bernier J, Kumar A, Serraj R, Spaner D, Atlin G (2008) Review: breeding upland rice for drought resistance. J Sci Food Agric 88:927–939

Boyer JS (1982) Plant productivity and environment. Science 218:443–448

Budak H, Kantar M, Kurtoglu KY (2013) Drought tolerance in modern and wild wheat. Sci World J 2013:548246. http://www.hindawi.com/journals/tswj/2013/548246/. vol. 2013, Article ID 548246, 16 p, doi:10.1155/2013/548246, accessed 28.05.2014

Edae EA, Byrne PF, Manmathan H, Haley SD, Moragues M, Lopes MS, Reynolds MP (2013) Association mapping and nucleotide sequence variation in five drought tolerance: candidate the genes in spring wheat. Plant Genome 6(2). https://www.crops.org/publications. 13 p, DOI: 10.3835/plantgenome2013.04.001, accessed 28.05.2014

Fleury D, Jefferies S, Kuchel H, Langridge P (2010) Genetic and genomic tools to improve drought tolerance in wheat. J Exp Bot 61:3211–3222

Gosal SS, Wani SH, Kang MS (2009) Biotechnology and drought tolerance. J Crop Improv 23:19–54

Humphreys MO, Humphreys MW (2005) Breeding for stress resistance: general principles. In: Ashraf M, Harris PJC (eds) Abiotic stresses: plant resistance through breeding and molecular approaches. Haworth, New York, pp 19–46

Huseynova IM, Suleimanov SY, Aliyev JA (2007) Structural-functional state of thylakoid membranes of wheat genotypes under water stress. Biochim Biophys Acta 1767(6):869–875

Iqbal A, Khan AS, Khan IA, Awan FS, Ahmad A, Khan AA (2007) Study of genetic divergence among wheat genotypes through random amplified polymorphic DNA. Genet Mol Res 6:476–481

Jiang GH, He YQ, Xu CG, Li X, Zhang Q (2004) The genetic basis of stay-green in rice analysed in a population of doubled haploid lines derived from an indica by japonica cross. Theor Appl Genet 108:688–698

Khan MA, Iqbal M, Akram M, Ahmad M, Hassan MW, Jamil M (2013) Recent advances in molecular tool development for drought tolerance breeding in cereal crops: a review. Zemdirbyste-Agric 100(3):325–334

Kirigwi FM, Ginkel M, Brown-Guedira G, Gill BS, Paulsen GM, Fritz AK (2007) Markers associated with a QTL for grain yield in wheat under drought. Mol Breed 20:401–413

Koyro HW, Ahmad P, Geissler N (2012) Abiotic stress responses in plants: an overview. In: Ahmad P, Prasad MNV (eds) Environmental adaptations and stress tolerance of plants in the era of climate change. Springer Science + Business Media, USA, pp 1-28, DOI 10.1007/978-1-4614-0815-4_1

Lang NT, Buu CB (2008) Fine mapping for drought tolerance in rice (*Oryza sativa* L). Omonrice 16:9–15

Liu P, Xu Z-S, Lu P-P, Hu D, Chen M, Li L-C, Ma Z-Y (2013) A wheat PI4K gene whose product possesses threonine auto phosphorylation activity confers tolerance to drought and salt in *Arabidopsis*. J Exp Bot 64(10):2915–2927

Lörz H, Wenzel G (eds) (2008) Molecular marker systems in plant breeding and crop improvement. In: Nagata T, Lörz H, Widholm JM (eds) Biotechnology in agriculture and forestry. Springer-Verlag, Berlin, p 478

Manavalan LP, Guttikonda SK, Phan Tran LS, Nguyen HT (2009) Physiological and molecular approaches to improve drought resistance in soybean. Plant Cell Physiol 50(7):1260–1276

Miyama M, Tada Y (2008) Transcriptional and physiological study of the response of Burma mangrove (*Bruguiera gymnorhiza*) to salt and osmotic stress. Plant Mol Biol 68:119–129

Mohammadi M, Taleei A, Zeinali H, Naghavi MR, Ceccarelli S, Grando S, Baum M (2005) QTL analysis for phenologic traits in doubled haploid population of barley. Int J Agric Biol 7(5):820–823

Moose SP, Mumm RH (2008) Molecular plant breeding as the foundation for 21st century crop improvement. Plant Physiol 147:969–977

Murray MG, Thompson WF (1980) Rapid isolation of high molecular weight plant DNA. Nucleic Acids Res 8:4321–4325

Pakniyat H, Tavakol E (2007) RAPD markers associated with drought tolerance in bread wheat (*Triticum aestivum* L). Pak J Biol Sci 10:3237–3239

Penner GA, Lee SJ, Bezte LJ, Ugali E (1996) Rapid RAPD screening of plant DNA using dot blot hybridization. Mol Breed 2:7–10

Quarrie SA, Gulli M, Calestani C, Steed A, Marmiroli N (1994) Location of a gene regulating drought-induced abscisic acid production on the long arm of chromosome 5A of wheat. Theor Appl Genet 89:794–800

Rana RM, Rehman SU, Ahmed J, Bilal M (2013) A comprehensive overview of recent advances in drought stress tolerance research in wheat (*Triticum aestivum* L.). Asian J Agric Biol 1(1):9–37

Rashed MA, Sabry SRS, Atta AH, Mostafa AM (2010) Development of RAPD markers associated with drought tolerance in bread wheat (*Triticum aestivum*). Egypt J Genet Cytol 39:131–142

Robin S, Pathan MS, Courtois B, Lafitte R, Carandang S, Lanceras S, Amante M, Nguyen HT, Li Z (2003) Mapping osmotic adjustment in an advanced backcross inbred population of rice. Theor Appl Genet 107:1288–1296

Sanchez AC, Subudhi PK, Rosenow DT, Nguyen HT (2002) Mapping QTLs associated with drought resistance in sorghum (*Sorghum bicolor* (L.) Moench). Plant Mol Biol 48:713–726

Saranga Y, Menz M, Jiang CX, Wright RJ, Yakir D, Paterson AH (2001) Genomic dissection of genotype × environment interactions conferring adaptation of cotton to arid conditions. Gen Res 11:1988–1995

Sari-Gorla M, Krajewski P, Difonzo N, Villa M, Frova C (1999) Genetic analysis of drought tolerance in maize by molecular markers II Plant height and flowering. Theor Appl Genet 99:289–295

Savitri ES, Basuki N, Aini N, Arumingtyas EL (2013) Identification and characterization drought tolerance of gene *LEA-D*11 soybean (*Glycine max* L. Merril) based on PCR-sequencing. Am J Mol Biol 3:32–37

Teulat B, Monneveux P, Wery J, Borriès C, Souyris I, Charrier A (1997) Relationships between relative water content and growth parameters in barley: a QTL study. New Phytol 137:99–107

Uga Y, Okuno K, Yano M (2011) Dro1, a major QTL involved in deep rooting of rice under upland field conditions. J Exp Bot 62(8):2485–2494

Vinh NT, Paterson AH (2005) Genome mapping and its implication for stress resistance in plants. In: Ashraf M, Harris PJC (eds) Abiotic stresses: plant resistance through breeding and molecular approaches. Haworth, New York, pp 109–124

Wang RR, Chen CJ, Joppa LR (1995) Production and identification of chromosome specific RAPD markers for Langdon durum wheat disomic substitution lines. Crop Sci 35:886–888

Wei B, Jing R, Wang C, Chen J, Mao X, Chang X, Jia J (2009) Dreb1 genes in wheat (*Triticum aestivum* L.): development of functional markers and gene mapping based on SNPs. Mol Breed 23:13–22

Williams JG, Kubelik KJ, Livak JA, Tingey SV (1990) DNA polymorphisms amplified by arbitrary primers are useful genetic markers. Nucleic Acids Res 18:6531–6535

Xu JL, Lafitte HR, Gao YM, Fu BY, Torres R, Li ZK (2005) QTLs for drought escape and tolerance identified in a set of random introgression lines of rice. Theor Appl Genet 111:1642–1650

Xu ZS, Xia LQ, Chen M, Cheng XG, Zhang RY, Li LC, Zhao YX, Lu Y, Ni ZY, Liu L, Qiu ZG, Ma YZ (2007) Isolation and molecular characterization of the *Triticum aestivum* L. ethylene-responsive factor 1 (TaERF1) that increases multiple stress tolerance. Plant Mol Biol 65:719–732

Yue B, Xue W, Xiong L, Yu X, Luo L, Cui K, Lin D, Xing Y, Zhang Q (2006) Genetic basis of drought resistance at reproductive stage in rice: separation of drought tolerance from drought avoidance. Genetics 172:1213–1228

Chapter 7
Soil Contaminants: Sources, Effects, and Approaches for Remediation

Ghulam Murtaza, Behzad Murtaza, Nabeel Khan Niazi and Muhammad Sabir

Abstract The contamination of soils with various inorganic and organic contaminants led to the degradation of large expenses of urban and arable lands throughout the world. The presence of toxic contaminants poses a significant health risk to humans and other ecological systems. Scattered literature is harnessed to critically review the various natural and anthropogenic sources and potential hazards and to identify the best possible remediation strategies for a number of contaminants, mainly those inorganic in nature such as arsenic (As), cadmium (Cd), chromium (Cr), copper (Cu), lead (Pb), mercury (Hg), nickel (Ni), and zinc (Zn) commonly found in the contaminated soils. The remediation methods including chemical and phytoremediation techniques are discussed in this chapter. Chemical remediation methods such as immobilization, soil washing, and vitrification are relatively expansive and hazardous to the environment, and are not suitable for large-scale soil remediation activities. Conversely, phytoremediation has emerged as an environmentally friendly and feasible technology for restoration of contaminated soils, but very limited efforts have been directed to demonstrate this technology under field conditions. Remediation of heavy metal-contaminated soils is necessary to reduce the associated risks, make the land resource available for agricultural production, enhance food security, and scale down land tenure problems arising from changes in the land-use pattern.

Keywords Contaminant · Heavy metal(loid) · Health risk · Remediation · Phytoremediation

---

G. Murtaza (✉) · B. Murtaza · N. Khan Niazi · M. Sabir
Institute of Soil and Environmental Sciences, University of Agriculture, Faisalabad-38040, Pakistan
e-mail: gmurtazauaf@gmail.com

B. Murtaza
e-mail: behzadmurtaza@yahoo.com

N. Khan Niazi
e-mail: nabeelkniazi@gmail.com

M. Sabir
e-mail: cmsuaf@gmail.com

P. Ahmad et al. (eds.), *Improvement of Crops in the Era of Climatic Changes*, Volume 2,
DOI 10.1007/978-1-4614-8824-8_7,

1 Introduction

A material which does not exist in nature but is introduced into the soil through natural processes and anthropogenic activities, and hence affects the composition of the soil, is known as a contaminant. It can be categorized as chemical, biological, physical, or radiological substances that, in sufficient concentrations, can adversely affect the living organisms through soil, water, air, and food. A contaminant becomes a pollutant when it exerts detrimental effects, e.g., chlorine gas does not occur in the atmosphere, so it is a contaminant, but when it is released into the atmosphere through human-induced activity, it becomes a pollutant because of its harmful effects on humans and animals. Contaminants are released into the soils through natural and anthropogenic sources which are described next.

Naturally, contaminants are added into the soils through volcanic eruptions and weathering of soil parent material (Garrett 2000). Natural sources of metals include mineral dust which makes up a significant portion of aerosols (Artinano et al. 2003; Ikem et al. 2008). Major trace elements in crustal dust in the order of abundance are iron (Fe), manganese (Mn), zinc (Zn), lead (Pb), vanadium (V), chromium (Cr), nickel (Ni), copper (Cu), cobalt (Co), mercury (Hg), and cadmium (Cd), but these abundances vary greatly over the earth's surface (Garrett 2000). Concentrations of metals have also been found to increase with decreasing particle size, especially for highly weathered and wind-eroded soils (Meza-Figueroa and Maier 2009). Asian dust storm events contribute high levels of Cu, Zn, and Pb concentrations, showing that such events can be important in geochemical cycling of trace elements (Ma et al. 2001). Other natural contributors of metals on parent material include volcanic (Dias and Edwards 2003; Garrett 2000), biogenic (Artaxo and Hansson 1995), and forest fire emissions (Andreae and Merlet 2001; Karl et al. 2007).

Anthropogenic activities are the major source of contaminants to the soils and water. Major anthropogenic activities include the use of pesticides, herbicides, and fertilizers in agriculture; mining and smelting activities; leather tanning industries; and heavy mechanical industries. These activities add various organic and inorganic contaminants to the soils, which make the soil unsafe for humans, animals, plants, and microbes.

Each of these sources can be divided into two categories, i.e., point sources and nonpoint sources, on the basis of their origin and modes of contamination. The term "point source" means any discernible, confined, and discrete conveyance, including, but not limited to, any pipe, ditch, channel, tunnel, conduit, well, discrete fissure, container, rolling stock, concentrated animal feeding operation, or vessel or other floating craft, from which contaminants are or may be discharged (Fig. 7.1a, b). This term does not include agricultural storm water discharges and return flows from irrigated agriculture. The major point sources of contaminants released into the soils include untreated industrial effluent, sewage plants, effluent treatment plants, effluents from agricultural farm buildings, solid waste disposal sites, application of plant growth regulators, and pesticides manufacturing sites.

The contaminated runoff (also known as nonpoint source of contamination) from agriculture, urban development, forestry, recreational boating and marinas, hydro-

Fig. 7.1 The discharge of industrial effluents (**a**) and smoke (**b**); point sources of contamination

Fig. 7.2 The runoff of water from a contaminated land and nonpoint sources of contamination (**a**, **b**)

modification, and wetlands activities is the leading cause of water pollution in waters (Fig. 7.2a, b). The nonpoint sources are mainly related to erosion of virgin lands, forests and other natural vegetation, and in situ weathering of rocks or minerals in soils. These also include unscientific application of pesticides, herbicides, fungicides, fertilizers, cultivation of marginal sloping lands, construction sites, transportation, strip mining, and piling of dust and litter on impervious surface in urban areas.

The point and nonpoint sources of contamination in the rural and urban territories need separate consideration and assessment approaches. General differences between the point and nonpoint sources of contaminants are briefly summarized in Table 7.1.

2 Classification of Contaminants

2.1 Biological Contaminants

Several biological contaminants are found in sewage water, sewage sludge, and other solid wastes such as municipal wastes coming from both point and nonpoint

Table 7.1 Point and nonpoint sources of soil contaminants. (Ghafoor et al. 2012)

Sr. No.	Point sources	Nonpoint sources
1	Steady flow; variability ranges less than 1 order of magnitude	Highly dynamic among intermittent intervals; variability is severalfold of one
2	Severe impact during long flows over summer season due to concentration effect	Severe impact during or post storms and/or floods
3	Enter the water bodies at identifiable points	Point of entry is difficult to define
4	Mainly chemical oxygen demand (COD), biological oxygen demand (BOD), dissolved oxygen (DO), suspended particles, and nutrients are of importance	Sediment load, toxic/hazardous material/ions, DO, and pH are of significance

sources (Neumann et al. 2002), for example, *Campylobacter* species, *Clostridium perfringens, Escherichia coli, Salmonella* species, *Shigella,* viruses (polio virus, rotavirus), parasitic protozoa (*Cryptosporidium parvum* oocysts, *Entamoeba histolytica, Giardia lamblia* cysts), and Helminths (*Ascaris* species, *Ancylostoma* species, *Trichuris* species). In addition, hookworms are also found in raw sewage sludge. These microbes cause a number of diseases such as typhoid, many types of dysentery (bacillary, amoebic), gastroenteritis, tuberculosis, diarrhea, anemia, vomiting, and ascariasis (Gerardi and Zimmerman 2004). Some of these ailments could lead toward epidemics. However, *Salmonella* bacteria and *Shigella dysenteriae* are the most dangerous and commonly found microbes in human excreta which cause typhoid and dysentery, respectively (Yang et al. 2005; Zychlinsky and Sansonetti 1997). In developing countries including Pakistan, microbe-oriented contamination is usually very dangerous since one drainage/sewerage system carries all types of wastes.

2.2 Organic Contaminants

The major group of chemicals includes halogenated aliphatic and aromatic hydrocarbons including benzenes and petroleum hydrocarbons, polynuclear hydrocarbons (PAHs), organochlorine pesticides, polychlorinated biphenyles (PCBs), phthalate esters, oil, and grease which are contributed from a number of point and nonpoint sources.

In the past 50 years, production of synthetic organic chemicals for industrial and domestic uses has increased enormously. For example, between 1991 and 1995, production of HCOOH increased from 186 to 205 (1,000 t) (Weissermel 2008). As a consequence, the occurrence and concentration of organic contaminants in effluents, sewage, and biosolids have also increased. The presence and level of organic contaminants depend greatly on the quality of the wastewater, the different local point sources, the physicochemical properties of particular organic compounds, and operational parameters of the wastewater treatment plant. Concentrations of organic contaminants are generally greater in industrial sewage than in domestic effluent (Smith 2000). More than 300 organic chemicals from a diverse range of classes of compounds have been identified, and their concentrations vary from the pictograms per kilogram to grams per kilogram level (Jacobs et al. 1987; Smith 2000). Although water treatment plants were primarily designed to remove organic matter, and the mechanisms of degradation of bulk organic components are well studied and understood, the processes by which synthetic organics are degraded have received relatively little study. An organic contaminant could undergo a number of processes including (1) sorption to solid surfaces, (2) volatilization, (3) chemical degradation, and (4) biodegradation. Generally, the more hydrophobic a compound is, the more susceptible it will be to accumulation onto sewage sludge particles.

Although there is a scarcity of data on the behavior of organic contaminants during the water treatment and sludge digestion processes, some generalizations can be made (Rogers 1996; Weissermel 2008). For example, molecules with highly branched hydrocarbon chains are generally less susceptible to biodegradation than unbranched compounds, and short chains are not as quickly degraded as long chains. In addition, unsaturated aliphatic compounds are generally more susceptible to degradation than saturated analogs. Due to their low water solubility and high lipophilicity, they are generally believed to partition into sludges during sedimentation (Bhandari and Xia 2005). Nonetheless, the fate of these chemicals during wastewater treatment processes is not well characterized. Many are not readily degradable under anaerobic conditions and therefore persist in biosolids following anaerobic digestion. There is very limited information in the literature on the behavior and fate of biosolids-borne organic contaminants in agricultural soils. However, the literature available suggests that some are rapidly lost following land application, since they are readily degraded under aerobic soil conditions and lost by volatilization or photolysis. Some are, however, known or thought to be only slowly degraded (e.g., organotins, brominated flame retardants, and high molecular mass PAHs), while some are highly persistent, e.g., dioxins and organochlorines (Haynes et al. 2009).

Uptake of organic contaminants from soil into plants is characteristically low, and the main pathway is via volatilization from the soil surface, followed by uptake of contaminated air by aboveground vegetation. Accumulation of organics into grazing animals could occur through ingestion of biosolids adhering to vegetation directly after application and by direct soil/biosolids ingestion. Many organics are metabolized in animals, but halogenated biphenyls, chlorinated pesticides and hydrocarbons, and polychlorinated dibenzo-p-dioxins and furans (PCDD/Fs) are of primary concern since they are resistant to degradation and tend to bioconcentrate in animal fat and animal products (e.g., milk). The European Union has proposed limit values for several organic contaminants/groups of contaminants in biosolids, although the relevance of these is not yet fully understood. A better understanding of the reactions and fate of organic contaminants in soils is required before more rigorous regulations can be drafted.

2.3 Inorganic Contaminants

A variety of contaminants from point and nonpoint sources belong to this group, e.g., heavy metal(loid)s in the form of their compounds, ion pair or ion complex, poisonous gases, and products generated as a result of their interactions with other constituents in soil. The major inorganic contaminants are heavy metals / metalloids (e.g., Cu, Zn, Cd, Pb, Ni, Cr, As) which mainly originate from discharges from industry and from domestic grey water, including leaching from Cu and Pb pipes and Zn from domestic products (skin creams, deodorants, etc.). Heavy metals enter the sludge during primary treatment through their association with/adsorption to sedimenting particles and during secondary treatment through adsorption to bacterial cell walls and/or accumulation into bacterial cells. Concerns regarding the heavy metal loads in sewage sludge/biosolids have resulted in guidelines and regulations being developed in many parts of the world which are usually based on maximum allowable metal concentration limits in biosolids and/or the allowable loading limits of metals added in biosolids to the soil. A limitation of these approaches is that they consider total rather than biologically active (extractable) concentrations of heavy metals in soils. Where agronomic rates are used (2–8 Mg ha^{-1}), heavy metal toxicities limiting crop growth in biosolids-amended soils are very rare (Krogmann et al. 1999; Murtaza et al. 2011). Nevertheless, a major concern is that heavy metals may accumulate in edible portions of plants and subsequently enter the food chain and have toxic effects on grazing animals and/or humans that ingest them (Cobb et al. 2000). For the most part, heavy metals are not readily translocated to the aboveground edible portions of crops, so toxicities from ingestion of food crops are not likely under current regulations. The main potential pathway for accumulation of heavy metals into the meat of grazing animals is via direct soil/biosolids ingestion, and substantial accumulation is unlikely under adequate grazing management. There is a growing body of evidence that biosolids-induced heavy metal accumulation in soils can have negative effects on soil microbial/biochemical activity, but the significance/importance of this has yet to be fully understood (Haynes et al. 2009).

3 Biological Toxicity of Contaminants

3.1 Organic Contaminants

Soil contamination with organic contaminants is a serious threat to environment including humans, animals, plants, and microbes. A large quantity of organic chemicals used in agriculture and urban areas has shown adverse health effects upon exposure to contaminated soils. The contamination of soils with these organic chemicals is linked to various industrial activities: coal burning, motor vehicle emissions, waste incineration, waste dumping, and oil spills. In agricultural areas, farmers have been using an ever-increasing amount of organic chemicals as pesticides that leads to widespread environmental damage.

The immensity of the soil has led to the dilution of these contaminants, and most of the contaminants remain on or near the surface of the soil unless they have been moved by the action of water, organisms, or mechanical mixing (Abou Seeda et al. 2005). This dilution has reduced the toxicity of these contaminants, but the unknown factor is the action of the soil, its chemistry, and the combined action of all the microorganisms, plants, and invertebrates that live in the soil. This biological action, combined with influences of the soil components, has the potential of creating new metabolites and chemicals. Toxicologists need to expand their studies to include the persistent organic contaminants in order to estimate their bioaccumulation in living organisms (plants, animals, microbes, and humans; Vallack et al. 1998; Van der Oost et al. 2003).

3.2 Inorganic Contaminants

Both point and nonpoint sources are responsible for the release of certain toxic elements known especially as heavy metal(loid)s in the soil. The most commonly encountered metal(loid) contaminants include Cd, As, Ni, Pb, Cr, Co, Cu, Mn, Fe, Zn, and Mo. These heavy metal(loid)s originate from a variety of sources; for instance, industrial manufacturing; sewage sludge application; and composts chiefly from urban solid wastes, burned fuel, household wastes, agrochemicals (pesticides, fertilizers), radionuclides, mining and smelting operations, and tanning activities are some of the major anthropogenic sources of heavy metal(loid) entry into the environment (Murtaza et al. 2010). Natural processes, such as volcanic eruptions, weathering of parent material, and hydrothermal release of metal(loid)s, also contribute to soil contamination with heavy metal(loid)s (Garrett 2000).

3.2.1 Cadmium

Cadmium is a naturally occurring metal, but it is usually present in the environment combined with chlorine, oxygen, and sulfur. Both short- and long-term exposure to Cd causes serious health problems. Cadmium poisoning is an occupational hazard which is associated with industrial processes such as metal plating and the manufacturing of Ni–Cd batteries, pigments, plastics, and other synthetics. Primarily, inhalation in industries is the major route of Cd entry into human body. Inhalation of Cd-containing fumes can result initially in metal fume fever but may progress to chemical pneumonitis, pulmonary edema, and death (Bakand et al. 2007). Human exposures to Cd are primarily the result of burning of fossil fuels and municipal wastes. There are notable instances of Cd toxicity as a result of long-term consumption of Cd in food and water. Up to World War II, Japanese mining operations contaminated the Jinzu River with Cd and traces of other toxic metals. Consequently, Cd accumulated in rice crops grown along the riverbanks downstream of mines. The local communities eating contaminated rice developed itai-itai disease and renal abnormalities, including proteinuria and glucosuria (Nogawa et al. 2004). Cadmium is one of the six substances that have been banned by the European Union Restriction on Hazardous Substances (RoHS) directive. The directive imposed ban on using certain Cd-derived hazardous substances in electronics. Cadmium and several Cd-containing compounds are known carcinogens and can induce many types of cancers (Hecht 2003).

It is reported that Cd toxicity to human body is caused by Zn-binding proteins, particularly proteins that contain Zn-finger protein structures. Zn and Cd are in the same group of periodic table, contain the same common oxidation state (2^+), and upon ionization acquire almost the same size. Owing to these similarities, Cd can replace Zn in many biological processes, e.g., systems that contain softer ligands such as sulfur. Cadmium can bind up to ten times more strongly than Zn in certain biological systems, although these replacements are rare. The major toxicity symptoms in humans include vomiting, nausea, stomach cramps, diarrhea, kidney damage and disease, fragile bones, and lung damage causing chest pain or shortness of breath.

Tobacco smoking is the most important single source of Cd for general population. It has been estimated that ~10% of Cd contents of a cigarette are inhaled through smoking. The absorption of Cd from lungs is much more effective and efficient than that from the gut, and as much as 50% of the Cd inhaled via cigarette smoke may be absorbed (Elinder et al. 1983; Galażyn-Sidorczuk et al. 2008). On an average, smokers have four to five times higher blood Cd and two to three times higher kidney Cd concentrations than nonsmokers (Jarup 2003). The Environmental Protection Agency (EPA) has classified Cd as a human carcinogen.

3.2.2 Chromium

Chromium enters air, water, and soil mostly as chromite (Cr(III)) and chromate (Cr(VI)) ions; the latter is highly toxic and mobile in soil environment. In air, Cr compounds are present mostly in the form of fine dust particles which settle over land and water surfaces. It is strongly adsorbed onto soil particles, and a very small quantity gets dissolved in water, which could move into deeper soils or groundwater.

Intake of high levels of Cr (VI) through inhalation could cause irritation to nose, resulting in runny nose, nosebleeds, ulcer, and holes in the nasal septum. Ingesting large amount of Cr(VI) usually causes stomach upsets, stomach ulcer, convulsions, kidney and liver damage, and even death. Direct contact of skin with certain Cr(VI) compounds may cause skin ulcers. Chromate is a very strong oxidizing agent, which reacts quickly to form complexes. The main reason for Cr(VI) toxicity is that one of the reduction products of Cr(VI) is Cr(V) which is a known carcinogen and could lodge in any tissue to form cancerous growths. In human body, acidity and action of enzymes with Cr(VI) could promote the formation of Cr(V) in small quantities. The only place where Cr(V) is likely to accumulate is in some of the fine capillaries in kidneys, intestines, or lungs. During the passage out, Cr(VI) will continue to oxidize anything it can, leaving deposits of the relatively safe Cr(III) and completely unsafe Cr(V) behind. The World Health Organization (WHO), department of health and human services (DHHS), and EPA have maintained that Cr(VI) is a human carcinogen. Boonyapookana et al. (2002) reported that the concentration of Cd and Cr in *Wolffia globosa* (duckweed) significantly increased with an increase in exposure time and concentration in nutrient solution. The effects of Cd and Cr on the biomass production and total chlorophyll contents in *W. globosa* significantly decreased with exposure time and metal concentration in the growth medium. In a greenhouse experiment, Jamal et al. (2009) demonstrated that there was no effect of Al and Cr (each alone or in combined treatment) on seed germination and dry biomass of wheat. They reported that the root, shoot, and seedling length of both the wheat varieties significantly decreased compared to those for the control.

3.2.3 Arsenic

Arsenic is a metalloid and is present in group VA of the periodic table of elements. According to the International Agency for Research on Cancer and World Health Organization, As is classified as class 1 human carcinogen and a highly toxic element. Arsenic exists in organic (monomethyl arsenic acid, dimethyl arsenic acid) and inorganic forms in the soil and water environments. Inorganic As species, mainly arsenate (As(V)) and arsenite (As(III)), are toxic and highly mobile forms of As in soil; As(III) is 60 times more mobile and toxic than As(V). Arsenate prevails in the oxidized conditions while As(III) prevails under the reduced soil environments in the normal soil pH range (pH 4–8; Madsen 2011; Niazi et al. 2011a; Niazi et al. 2011b; Smith 2000).

In small doses, As generally causes nausea, vomiting, and diarrhea, but in high doses it could cause abnormal heartbeat, damage to blood vessels, and a feeling of pins and needles in hands and feet. The symptoms of As toxicity can lead to formation of small corns or warts on palms of hands and soles of feet. Arsenic also creates malfunctioning in plant species, for example, As(V) can disrupt the phosphate uptake pathway of plants resulting in nutritional imbalance and ultimately plant death. This is attributed to the chemical similarities in As(V) and phosphate. After entering into soil, the fate of As depends on the soil pH, redox potential, and presence of Fe and Al oxides, since they control the availability and speciation of As in the contaminated soil environments.

The massive As contamination of groundwater has been reported in several countries throughout the world including the USA, Brazil, Canada, and in many countries in Southeast Asia region, such as Bangladesh, India, China, Sri Lanka, and Pakistan. More than 100 million people have been affected by ingestion of As-contaminated drinking water and food (mainly rice) in Southeast Asia region; the worst scenario is in Bangladesh, where 64 districts are badly hit by As poisoning. In these As-affected areas, As concentration is significantly higher than the safe limit of As (0.01 mg L^{-1}) in water set by WHO. The As-containing groundwater is widely used for the irrigation of land used to produce food crops (e.g., rice, vegetables) which led to an increase in As concentration up to 60 mg kg^{-1} in several Bangladeshi soils.

3.2.4 Cobalt

In small amounts, Co is essential to many living organisms including humans. The Co concentration between 0.13 and 0.30 mg kg^{-1} in soils markedly improves the health of grazing animals. It is a central component of the vitamin cobalamin (vitamin B_{12}). Although Co is an essential element for life in minute amounts, it has similar mutagenic and carcinogenic effects at higher levels. In 1966, addition of Co compounds to stabilize beer foam in Canada led to cardiomyopathy, which is known as beer drinker's cardiomyopathy. Powdered Co in metal form is a fire hazard. After Ni and Cr, Co is a major cause of contact dermatitis (Basketter et al. 2003).

Phytotoxicity and bioavailability of Co in barley (*Hordeum vulgare* L.), oil seed rap (*Brassica napus* L.), and tomato (*Lycopersicon esculentum* L.) shoot growth in the ten soils varying widely in soil properties using a standardized shoot biomass assay were investigated. The effective concentration (EC) of added Co causing 50% inhibition ranged from 40 to 1,708 mg kg^{-1}, from 7 to 966 mg kg^{-1}, and from 7 to 733 mg kg^{-1} for barley, oil seed rape, and tomato, respectively, representing 43-, 138-, and 105-fold variation among soils. The EC_{50} based on Co in soil solution varied less among soils (4–15 fold) than that based on the total added Co, suggesting that solubility of Co is a key factor influencing its toxicity to plants.

In another study, cauliflower (*B. oleracea* L.) was grown in sand with complete nutrition (control) and at 0.5 mM each of Co, Cr, and Cu (Chatterjee and Chatterjee 2000). Visible phytotoxic symptoms of Co appeared first, which were most pronounced compared to Cu and Cr. Excess of each heavy metal decreased the biomass, shoot Fe, chlorophyll a and b contents, and protein, and the activity of catalase in leaves remained in order Co>Cu>Cr.

3.2.5 Nickel

The ingestion of Ni and its soluble compounds should be <0.05 mg cm^{-3} in Ni equivalents per 40-h workweek. The NiS as fumes or dust and many other compounds are carcinogenic (Kasprzak et al. 2003). In addition, Ni as Ni-carbonyl $[Ni(Co)_4]$ is a highly toxic gas. Sensitized individuals usually show skin allergy to Ni, which clinically is a form of dermatitis. Sensitivity to Ni may also be present in patients with pompholyx. Nickel is an important cause of contact allergy, partially owing to its use in jewellery for pierced ears (Thyssen and Menni 2009). Nickel allergies affecting pierced ears are often marked by itchy and/or red skin (Magaye and Zhao 2012).

3.2.6 Mercury

Several fish species have natural potential to concentrate Hg in bodies mostly as methylmercury which is an extremely toxic organic compound for humans. Species of fish that are high on the food chain such as shark, swordfish, king mackerel, albacore tuna, and tilefish are included in table food, but they all contain generally higher Hg than others. Since Hg and methylmercury are fat soluble, they primarily accumulate in the viscera in meager amount in the muscle tissues. When this fish is consumed by a predator, the Hg level is bioaccumulated. Thus, fish species common in food acquire ten times more body Hg than the species they consume. Mercury poisoning happened in Minamata, Japan, which later came to be known as Minamata disease.

Treatments of seeds of rice cultivars (Ratna and IR36) with 10^{-5} and 10^{-4} M $PbCl_2$and $HgCl_2$ decreased germination, germination index (GI), shoot and root length, tolerance index (TI), vigor index (VI), and dry mass of shoot and root but increased percentage difference in germination and phytotoxicity in both the cultivars from control. It was concluded that phytotoxic effect of Hg was greater than Pb at identical concentrations and that cv. IR36 proved to be more tolerant to these metals than Ratna (Bera et al. 2005). Among the monitoring indices examined, TI, VI, and percent toxicity seemed good biological methods.

3.2.7 Lead

The level of Pb in drinking water is <0.015 mg L^{-1} in the USA and <0.01 mg L^{-1} in California, but both are above its natural level, i.e., a reflection of anthropogenic contamination (Ponder et al. 2000). The body of an adult human contains 2 ppm Pb with a range of 1.4–5.7 ppm. Its 90 % contents are in bones at 20–40 ppm. It could enter the body through breathing air but is mostly consumed through food, beverages, drugs, supplements, and many other items ingested.

In low traffic areas, Pb in soils ranges from 30 to 50 ppm, which increases up to 300 ppm in high-traffic zones in the USA. In dry shoots of several plants from these areas, Pb was reported to be in the range of 0.4–4.0 ppm, except one grass sample that contained 7.5 ppm Pb. Shoots of this grass from the countryside had high Pb (0.9 ppm), which was considerably lower than in the urban zone plants. The shoots for other plant species from rural areas had Pb<0.1 ppm. It could be concluded that high-traffic city zones contained considerably more Pb in soils than that in rural areas. In a study, Ghafoor et al. (2008a) reported clear clues that plants and soils near high-traffic roads contained more Pb than those away-from-roads counterparts. High correlation between Pb in soils and that in human blood serum is commonly reported. It has been concluded that when soil Pb increases to 500 ppm (50 times higher than uncontaminated soils), the Pb in blood of children will increase significantly. Such effects are mainly caused in young children (<5 year) through ingestion of leaded paints, chipping of buildings, discharge from motor vehicles, and industrial activities while playing outdoors in Pb-contaminated areas. Although Pb in air is decreasing with efforts to discourage its addition in gasoline as antiknock agent, its pollution is common due to many other sources, like industrial activity, use of sewage and sludge, and continued use of leaded soil in some countries. The highest Pb concentration, up to >10 µg m^{-3}, is reported near Pb smelters and along urban freeways where leaded gas is used in vehicles. Pb in air at Kyoto was 0.075 $\mu g\ m^{-3}$, whereas the polluted city of Manila reported that it ranged from 0.3–1.1 $\mu g\ m^{-3}$ in urban areas to 1/2 to 1/3 of that amount in surrounding rural areas (0.15–0.30 $\mu g\ m^{-3}$) and 0.01 $\mu g\ m^{-3}$ in areas far away from traffic. Even with relatively lower Pb in rural air, it was estimated that natural Pb in air during prehistoric times was lower (0.01 $\mu g\ m^{-3}$).

4 Technologies for Remediation of Heavy Metal(loid)-Contaminated Soils

4.1 Chemical Remediation

4.1.1 Immobilization

Immobilization technologies are designed to reduce the mobility of contaminants by changing the physical or leaching characteristics of the contaminated matrix. Mobility is usually decreased by physically restricting contact between the contaminant and the surrounding groundwater or by chemically altering the contaminant to make it more stable with respect to dissolution in groundwater. The aqueous- and solid-phase chemistry of metals is conducive to immobilization by these techniques. A variety of methods are available for immobilization of metal contaminants, including those that use chemical reagents and/or thermal treatment to physically bind the contaminated soil or sludge. Most immobilization technologies can be performed *ex situ* or in situ. The in situ processes are preferred due to the lower labor and energy requirements, but implementation of in situ processes will depend on specific site conditions.

4.1.2 Solidification/Stabilization

Solidification/stabilization, also referred to as waste fixation, reduces the mobility of hazardous substances and contaminants in the environment through both physical and chemical means (Sherwood and Qualls 2001). Stabilization generally refers to the process that reduces the risk posed by a waste by converting the contaminant into a less soluble or immobile form (Anderson and Mitchell 2003). The in situ stabilization and solidification method involves three main components: (1) a means of mixing the contaminated soil in place, (2) reagent storage, preparation, and feed system, and (3) a means to deliver the reagents to the soil-mixing zone (Nyer 2010). The in situ and *ex situ* stabilization/solidification methods are usually applied to soils contaminated by heavy metals and other inorganic compounds. However, stabilization of soils that contain low levels of organic constituents is feasible even for volatile organics (Druss 2002). Most stabilization/solidification technologies have limited effectiveness against organics and pesticides, except for asphalt batching and vitrification which destroy most organic contaminants (Raag 2000; Wilk 2003).

4.1.3 Asphalt Batching

Asphalt batching is a stabilization/solidification method for treating hydrocarbon-contaminated soils, which incorporates petroleum-laden soils into hot asphalt mixtures as a partial substitute for stone aggregate; the mixture can be utilized for paving. This process involves excavation of the contaminated soils followed by an initial thermal treatment and incorporation of the treated soil into an aggregate for asphalt. During the incorporation process, heating of the mixture results in the volatilization of the more volatile hydrocarbon constituents (Asante-Duah 1996; Khan et al. 2004). The remaining compounds are incorporated into an asphalt matrix during cooling, thereby limiting constituent migration. After it is given sufficient time to set and cure, the resulting solid asphalt now has the waste uniformly distributed throughout it and is impermeable to water (Alpaslan and Yukselen 2002).

4.1.4 Vitrification

Vitrification or the use of molten glass is a method of stabilization/solidification that uses a powerful source of energy to melt soil or other earthen materials at extremely high temperatures (1,600–2,000 °C), immobilizing most inorganics and destroying organic contaminants by pyrolysis (Acar et al. 1995; Ghosh and Singh 2005). During this process, the majority of contaminants initially present in the soil are volatilized, while the remaining contaminants are converted into chemically inert products, stable glass, and crystalline products (Asante-Duah 1996; Khan et al. 2004). The high temperatures destroy any organic constituents, resulting in few by-products. Inorganics such as heavy metals and radionuclides are actually incorporated into a glass structure which is generally strong, durable, and resistant to leaching (Dermatas and Meng 2003).

4.1.5 Chemical Oxidation

The use of chemical oxidation technology as a form of in situ soil and groundwater remediation has been practiced for at least two decades. Early practitioners realized the potential for harnessing chemical oxidation processes employed in wastewater treatment to rapidly destroy organic contaminants in soil and groundwater. Valuable contributions have been made by researchers in academia as well as practitioners in the field regarding specific oxidative chemistries, the interactions of the oxidative species with native materials, and the requirements for successful delivery. Further, it is now clear that chemical oxidation is best coupled with accelerated bioremediation for more successful site management.

Chemical oxidation has been used to successfully remove significant contaminant mass from soils and groundwater at numerous sites (Mulligan et al. 2001). Using a variety of oxidants, such as hydrogen peroxide, permanganate, persulfate, percarbonate, and ozone, success has been achieved with majority of the contaminants of concern

(COCs). Hydrogen peroxide has been shown to be effective for petroleum-based COCs and permanganate for chlorinated solvents (Silva et al. 2011). Most researchers and practitioners believe that technology has the ability to rapidly reduce large masses of contamination; however, the promise of chemical oxidation to rapidly and completely degrade target contaminants in situ has been overstated. Site owners or responsible parties have been disappointed, as chemical oxidation has failed to meet their expectations of reaching low (ppb) contaminant concentrations within a short period of time. The requirement for direct oxidant–contaminant interaction, matrix interaction effects, contaminant desorption, plume distribution, and a range of other factors is now seen as inhibiting the simplistic view of complete and rapid in situ treatment.

4.1.6 Soil flushing

In situ soil flushing is an innovative remediation technology that floods contaminated soils with a solution that moves the contaminants to an area where they can be removed (Logsdon et al. 2002; Otterpohl 2002). Soil flushing is accomplished by passing an extraction fluid through the soil using an injection or infiltration process. Contaminated groundwater and extraction fluids are captured and pumped to the surface using standard groundwater extraction wells. Recovered groundwater and extraction fluids with the adsorbed contaminants may need treatment to meet the appropriate discharge standards before being recycled or released to local, publicly owned, wastewater treatment works or receiving streams (Otterpohl 2002; Raag 2000; Son et al. 2003). Soil flushing applies to all types of soil contaminants and is generally used in conjunction with other remediation technologies such as activated carbon, biodegradation, and pump and treat (Boulding 1996; Di Palma et al. 2003). Since soil flushing is conducted in situ, it reduces the need for excavation, handling, or transportation of hazardous substances. Juhasz et al. (2003) have discussed in situ remediation of dichlorodiphenyltrichloroethane (DDT)-contaminated soil using a two-phase cosolvent-flushing fungal biosorption process.

5 Phytoremediation

Phytoremediation is the use of plants to clean up the heavy metal(loid)-contaminated soils. This process takes advantage of the ability of plants to extract, accumulate, and/or degrade constituents that are present in soil and water environments (Alkorta et al. 2004). All plants extract necessary components, including nutrients and heavy metals from the contaminated soil. Some plants are referred to as hyperaccumulators as they have the ability to store large amounts of heavy metals which they do not use in their metabolic functions. Plants have also been known to take up various organics and either degrade or process them for use in physiological processes (Khan et al. 2004; Vouillamoz and Milke 2001). Barter (1999) has presented a good overview of phytoremediation of contaminated soils. It mainly outlines the five basic phytoremediation techniques: rhizofiltration, phytoextraction, phytotransformation, phytostimulation, and phytostabilization.

5.1 Phytoextraction

Phytoextraction is the most commonly recognized of all the phytoremediation technologies, and is the focus of research nowadays. The phytoextraction process involves the use of plants to facilitate the removal of metals from contaminated soils (Niazi et al. 2012). The metal-accumulating plants are grown in the contaminated soils and are established using agricultural practices. The roots of plants absorb metals from soil and translocate them to shoots. If metal availability in soils is inadequate for sufficient plant uptake, chelates, acidifying agents, or mineral nutrients (such as phosphate to enhance arsenate availability) may be used to release them into soil solution (Sabir et al. 2008; Saifullah et al. 2010). After sufficient plant growth and metal accumulation, the aboveground portion of plant is harvested, which permanently removes metals from the site. There is a secondary contamination threat from the harvested metal-containing plant biomass. Some experts suggest that the incineration of harvested plant dramatically decreases the volume of plant material requiring disposal to a dump site (Kumar et al. 1995; McGrath and Zhao 2003). In some cases, valuable metals can be extracted from the metal-rich ash and can serve as a source of revenue, thereby offsetting the expense of remediation (Cunningham and Berti 2000). Also, the harvested huge amount of biomass can be used as a raw material for the production of biofuel to produce bioenergy. Phytoextraction should be viewed as a long-term remediation effort, requiring many cropping cycles to reduce metal concentration to acceptable levels (Kumar et al. 1995; McGrath and Zhao 2003).

The time required for phytoremediation depends on the type and extent of metal contamination, the length of growing season, and the efficiency of metal-removing plants. Kertulis-Tartar et al. (2006) conducted a field experiment for the phytoremediation of an As-contaminated site historically contaminated with copper–chromium–arsenate application in FL, USA. They used *Pteris vittata* for the phytoextraction of As from soil with As concentration in *P. vittata* fronds more than 2,000 mg kg^{-1} dry weight. The authors suggested that ~8 years would be required to decrease the concentration of acid-extractable As in soil from a mean value of 82–40 mg kg^{-1} As, the limit set by EPA. In another field study, Niazi et al. (2012) used the two As-hyperaccumulating fern species, *P. vittata* and *Pityrogramma calomelanos* var. *austroamericana,* for the remediation of an As-contaminated disused cattle dip site in northern NSW, Australia. The authors estimated that up to 155 and 400 years would be required by *P. calomelanos* var. *austroamericana* and *P. vittata*, respectively, to bring down the total As content (313–1,902 mg kg^{-1}, at 0–40-cm depth) below the Ecological Investigation Level (EIL) value of 20 mg kg^{-1} As in Australia. These differences in remediation times were due to variations in the fern species performance for extraction of As from soil at the site.

Phytoextraction technology is suitable for the remediation of large areas of land that are contaminated at shallow depth with low-to-moderate levels of heavy metal(loid)s (Kumar et al. 1995; Luo et al. 2005). Many factors determine the effectiveness of phytoextraction in remediating metal-contaminated soils (Blaylock and Huang 2000). Soil metals should be bioavailable or subject to absorption by plant roots. The land should be relatively free of obstacles such as fallen trees or boulders and have an acceptable topography to allow for normal cultivation practices with agriculture equipment. As a plant-based technology, the success of phytoextraction is inherently dependent upon several plant characteristics. The two most important characters include the ability to produce large quantities of biomass and the ability to accumulate large quantities of metal(loid)s in shoots (McGrath and Zhao 2003). Ebbs et al. (1997) reported that *B. juncea* is more effective for Zn removal from soil than *Thlaspi caerulescens,* a known hyperaccumulator of Zn. This advantage is due primarily to the fact that *B. juncea* produces ten times more biomass than *T. caerulescens*. Plants being considered for phytoextraction must be tolerant to the targeted metal(s) and be efficient at translocating them from roots to harvestable portion of plants (Blaylock and Huang 2000). Other desirable plant characteristics include the ability to tolerate stressed soil conditions (i.e., soil pH, salinity, sodicity, soil structure, water content), production of dense root system, ease of care and establishment, and few disease and insect problems. Although some plants show promise for phytoextraction, there is no plant which possesses all of these traits. Finding the perfect plant is the focus of plant breeding and genetic engineering research efforts.

5.2 Phytostabilization

Sometimes there is no immediate effort to clean metal-contaminated sites, either because the responsible companies no longer exist or because the sites are not on high-priority list of remediation agenda (Berti and Cunningham 2000). The traditional means to decrease metal toxicity is by in situ inactivation, a remediation technique that employs the use of soil amendments to immobilize or fix metals in soils. Although metal migration from soil to plants is minimized, soils are often subject to erosion and still pose an exposure risk to humans and other animals. Phytostabilization, also known as phytorestoration, is a plant-based remediation technology that stabilizes contaminants and prevents exposure pathways via wind and water erosion; provides hydraulic control, which suppresses the vertical migration of contaminants into groundwater; and physically and chemically immobilizes contaminants by sorption to roots or complexation with root exudates (Berti and Cunningham 2000; Schnoor 2000). This technique is actually a modified version of the in situ inactivation in which the function of plants is secondary to amend soils. Unlike phytoextraction, the goal of phytostabilization in not to remove metals from soils but to stabilize them and decrease risk to the human health and environment.

The phytostabilization process is explained in detail by Bolan et al. (2011) and Berti and Cunnigham (2000). Before planting, the contaminated soil is plowed to prepare a seed bed and to incorporate lime, fertilizer, or other amendments (Ghafoor et al. 2008b) for inactivating metal contaminants. The amendments applied to soil are required to fix metals rapidly following incorporation and chemical alteration, preferably long lasting if not permanent. The most promising fertilizers are phosphate fertilizers, organic matter (OM), biosolids, iron or manganese oxyhydroxides, natural or artificial clay minerals, or mixture of these amendments. Plants chosen for phytostabilization should be poor translators of metal contaminants to aboveground plant tissues that could be consumed by humans or animals. The lack of appreciable metals in shoots also eliminates the necessity of treating harvested shoot residue as hazardous waste (Brunner et al. 2008; Flathman and Lanza 1998). Selected plant should be easy to establish and care for, grow quickly, have dense canopies and root systems, and be tolerant to metal contaminants and other site conditions which may limit plant growth. The research of Bradshaw (1997) led to the development of two cultivars of *Agrostis tenuis* Sibth and one of *Festuca rubra* L which are now commercially available for phytostabilization of Pb-, Zn-, and Cu-contaminated soils.

Phytostabilization is most effective at sites having fine-textured soils with high OM contents, but is suitable for treating a wide range of soils in contaminated areas (Berti and Cunningham 2000; Bolan et al. 2011). However, some highly contaminated sites are not suitable for phytostabilization because plant growth and survival are not possible (Berti and Cunningham 2000). At sites which support plant growth, fight manager must be concerned with the migration of contaminated plant residues off-site or disease and insect problems which limit the longevity of plants (Schnoor 2000). Phytostabilization has advantages over other soil remediation practices in that it is less expensive, less environmentally evasive, easy to implement, and offers aesthetic value (Berti and Cunningham 2000; Schnoor 2000). When decontamination strategies are in practice, because of the size of contaminated area or lack of funds, phytostabilization has advantages. It may also serve as an interim strategy to decrease risk at sites where complications delay selections of the most appropriate technique for a site.

5.3 Phytovolatilization

Some metal contaminants like As, Hg, and Se do exist as gaseous species in environment. Recently, scientists have searched for naturally occurring or genetically modified plants that are capable of absorbing elemental forms of such metal from soils, biologically converting them to gaseous species within the plant and releasing them into the atmosphere. This process is called phytovolatilization—the most controversial of all the phytoremediation technologies. Mercury, As, Cr, CN, and Se are toxic (Chapman et al. 2003; Suszcynsky and Shann 1995), and there is doubt about whether the volatilization of these elements into the atmosphere is safe (Vara Prasad and de Oliveira Freitas 2003; Watanabe 1997). Selenium phytovolatilization has been given the most attention (McGrath et al. 2002) because this element is a serious problem in many parts of the world (Brooks et al. 1998; Tagmount et al. 2002). However, there have been considerable efforts to insert bacterial Hg ion reductase genes into plants for the purpose of Hg phytovolatilization (Rugh 2001). Although there have been no efforts to genetically engineer plants which volatilize As, it is likely that researchers will pursue this possibility in future. According to Brooks et al. (1998), the release of volatile Se compounds from higher plants was first reported by Lewis et al. (1966). Terry et al. (1992) and Panwar (2010) reported that members of the Brassicaceae family are capable of releasing up to 40 g Se ha^{-1} day^{-1} as various gaseous compounds. Some aquatic plants such as cattails are also good for Se phytovolatilization (Pilon-Smits 2005). Unlike plants that are being used for Se volatilization, those which volatilize Hg are genetically modified organism.

Arabidopsis thaliana L. and tobacco have been genetically modified with bacterial organomercurial lyase (Mer B) and mercuric reductase (Mer A) genes (Rugh 2001). These plants absorb elemental Hg(II) and methylmercury (MeHg) from soil and release volatile Hg (O) from leaves into the atmosphere (Bizily et al. 2000; Heaton et al. 1998). Volatile Se compounds, such as dimethylselenide, are 1/600 to 1/500 as toxic as inorganic form of Se found in soil (Desouza et al. 2000). The volatilization of Se and Hg is also a permanent site solution, because the inorganic forms of these elements are removed and gases are not likely to redeposit at or near the site (Heaton et al. 1998; Pilon-Smits 2005). Further, sites that utilize this technology may not require much management after the original planting. This remediation method has the added benefits of minimal site disturbance, less erosion, and no need for disposal of contaminated plant material (Heaton et al. 1998; Rugh 2001). Heaton et al. (1998) suggest that the addition of HgO into the atmosphere would not contribute significantly to the atmospheric pole. However, those who support this technique also agreed that phytovolatilization would not be wise for sites near population centers or at places with unique metrological conditions that promote the rapid depositions of volatile compounds (Heaton et al. 1998; Rugh 2001).

5.4 Phytostimulation

Soil microbes have been found suitable to enhance the bioavailability of metals for phytoextraction and increase the phytoremediation potential of plants by complementing the process in many ways. Soil microbes may degrade organic contaminants into simpler organic compounds and supply them as nutrient to plants for enhanced phytoremediation of the contaminated sites. Microbial activity in the rhizosphere of plants is severalfold higher than in the bulk soil. The population of microflora present in the rhizosphere is much higher than that in the soil with less vegetation. This is due to the presence of nutrients from the root exudates of plants (Suthersan and Payne 2004). A typical microbial population present in the rhizosphere per gram of air-dried soil comprises bacteria (5×10^6), actinomycetes (9×10^5), and fungi (2×10^3; Schnoor 2000). The rhizosphere in plants is divided into two general areas: the inner rhizosphere at the root surface and the outer rhizosphere immediately adjacent to soil. The microbial population is larger in the inner zone where the root exudates are concentrated. A large variety of exudates are secreted from roots in the form of sugars, amino acids, and essential vitamins.

Root exudates may also include acetates, esters, and benzene derivatives. Enzymes are also present in the rhizosphere and this may act as substrates for the microbial population. Enzymes secreted by plant roots or the microbial community in the rhizosphere comprise esterases and different oxidoreductases (phenoloxidases and peroxidases). Plant peroxidases are exuded by some members of Fabaceae, Gramineae, and Solanaceae. White radish (*Raphanus sativus*) and horseradish (*Armoracia rusticana*) secrete "peroxidase," while the aquatic green algae Nitella and Chara secrete "laccase" (Tavares-Dias 2006).

Chemolithotrophic bacteria have been shown to enhance metal availability. Several strains of *Bacillus* and *Pseudomonas* have been reported to increase Cd accumulation by *B. juncea*. Naturally occurring Rhizobacteria were found to promote Se and Hg bioaccumulation in plants growing in wetlands (Singh et al. 2011). These microbes grow much better if organic manures are added to the soil. Uptake of hydrophobic xenobiotics of larger size can be facilitated by primary microbial biodegradations in the rhizosphere. The hydrophobic persistent organic contaminants like PCDD/Fs and 2,3,7,8-tetrachlorodibenzo-pdioxin (TCDD), with present log K_{ow} value above 4, have been reported to be taken up by roots and transported to shoots in *Cucurbita pepo* (Singh et al. 2011).

Root exudates, in the form of compounds structurally analogous to the chemicals utilized by microorganisms, would enhance the activity of the bacteria present in the rhizosphere, accelerating the degradation process. Certain phenolic compounds in the root exudates have been found to support growth of PCB-degrading bacteria and serve as structural analogs for PCB degradation (Zhao et al. 2001). Sometimes, inoculation of plant soil by microbes enhances the phyto-degradation of organic contaminants present in the soil. Degradation of "pentachlorophenol" (PCP) was accelerated by the prosomillet after the soil in which it was growing was inoculated with *Pseudomonas* strain SR 3.

5.5 Phytotransformation

Several inorganic and organic contaminants, once absorbed inside the root, may become biochemically bound to cellular tissues (biotransformed) in forms that are biologically inert or less active (Vara Prasad and de Oliveira Freitas 2003; Watanabe 1997). In many cases, plants have the ability to metabolize organic contaminants by phytotransformation and conjugation reactions followed by compartmentalization of products in their tissues. Poplar trees (*Populus* species) have been found to transform trichloroethylene in soil and groundwater.

Plants possess the necessary mechanism to detoxify the cyanides from mining wastes resulting from gold (Au) and silver (Ag) mining. Cyanide (ammonium thiocyanate) is the leach reagent of choice for Au and Ag extraction (Sinha et al. 2010). During several metabolic functions, plants are confronted with cyanide as a by-product of metabolism. This occurs particularly during the synthesis of "ethylene" in mature tissues, where "hydrogen cyanide" (HCN) is formed as a by-product. Consequently, vascular plants have evolved effective strategies for detoxifying the toxic cyanide with the aid of enzymes. Those identified with cyanide detoxification are *Salix* spp. and *Sorghum* spp. Plants only survive cyanide exposure up to the doses they can eliminate. The cyanide-detoxifying enzyme system is "beta-cyano-alanine synthase," which connects free cyanide and cysteine to cyanoalanine. The final metabolic product is "asparagines," a nontoxic essential amino acid in plants (Trapp et al. 2003).

5.6 *Rhizofiltration*

Metal contaminants in industrial wastewater and groundwater are most commonly removed by precipitation or flocculation, followed by sedimentation and disposal of the resulting sludge (Ensley 2000). A promising alternative to this conventional clean up method is rhizofiltration. The process involves raising plants hydroponically and transplanting them into metal-polluted waters where plants absorb and concentrate metals in their roots and shoots. Root exudates and changes in rhizosphere pH induced precipitation of metals onto root surface. As these become saturated with metal contaminants, roots or whole plants are harvested for disposal (Brunner et al. 2008; Flathman and Lanza 1998). Most researchers believe that plants for phytoremediation should accumulate metals only in roots (Flathman and Lanza 1998). Dushenkov et al. (1995) explained that the translocation of metals to shoots would decrease the efficiency of rhizofiltration by increasing the amount of contaminated plant residue needing disposal. In contrast, Zhu et al. (1999) suggested that the effectiveness of this process could be increased by using plants with ability to absorb and translocate metals within the plant. Despite such differences, it is apparent that proper plant selection is the key to ensuring the success of rhizofiltration as water cleanup strategies.

Raskin and Ensley (2000) described characteristics of an ideal plant for rhizofiltration. Plants should be able to accumulate and tolerate a significant amount of target metal(s) in conjunction with easy handling, low maintenance cost, and a minimum of secondary waste requiring disposal. It is also desirable that plants could produce significant root biomass or root surface area. Several aquatic species having ability to remove metals from water include water hyacinth, pennwort, and duckweed (Kamal et al. 2004). However, these plants have limited potential for rhizofiltration, because they are not efficient for metal removal, a result of their small and slow-growing roots (Dushenkov et al. 1995; Jadia and Fulekar 2009). These authors also point out that the high water contents of aquatic plants complicate their drying, composting, or incineration. Despite limitations, Zhu et al. (1999) indicated that water hyacinth was effective for removing trace elements from wastewater. Terrestrial plants are thought to be more suitable for rhizofiltration because these produce more substantial and often fibrous root system with large surface area for metal sorption. Sunflower and Indian mustard are the most promising terrestrial candidates for metal removal from water. The roots of Indian sunflower remove lead (Dushenkov et al. 1995; Vara Prasad and de Oliveira Freitas 2003), U, Cs^{137}, and Sr^{90} (Dushenkov et al. 1997) from hydroponic solutions.

Rhizofiltration is a cost-competitive technology in the treatment of surface water or groundwater containing low, but significant, concentrations of metals like Cr, Pb, and Zn (Ensley 2000; Kumar et al. 1995; Verma et al. 2006). The commercialization of this technology is driven by economics as well as such technical advantages as applicability to many problem metals, ability to treat high volumes, lesser need for toxic chemicals, decreased volume of secondary waste, possibility of recycling, and likelihood of regulatory and public acceptance (Dushenkov et al. 1995; Khan et al. 2004; Kumar et al. 1995). However, application of this plant-based technology may be more challenging and acceptable to failure than other methods or similar cost. The production of hydroponically grown transplants and the maintenance of successful hydroponic system in a field will require the expertise of qualified personnel, and the facilities and specialized equipment required can increase overhead cost. Perhaps the greatest benefit of this remediation method is related to positive public perception. The use of the plants at a site where contamination exists conveys the ideas of cleanliness and progress to the public that would have normally been perceived as polluted.

6 Conclusion and Future Perspectives

Background knowledge of the sources, chemistry, and potential health and environmental risks of toxic heavy metal(loid)s from contaminated soils is imperative for the selection of appropriate remedial options. The remediation of soils contaminated by heavy metal(loid)s is vital to reduce the associated health risk, make the land resources available for agricultural production, enhance food security, and scale down land tenure problems. Chemical remediation methods discussed earlier in this chapter are relatively expansive as well as hazardous to the environment. On the other hand, phytoremediation using metal(loid)-accumulating plant species has emerged as an environmentally friendly and feasible technology for cleaning up heavy metal(loid)-contaminated soils but have not been demonstrated practically under field conditions with the exception of few field trials. Application of phytoremediation (particularly phytoextraction and phytostabilization) to the field scale requires attention by the scientists and commercial/government organizations, so the new plant species should be explored for their phytoremediation potential. Although long remediation time is the main obstacle in the implementation of phytoremediation, there is a dire need for future research on the methods or approaches to enhance the plant biomass production and metal(loid) removal ability of plants from contaminated soils.

References

Abou Seeda M, Zaghloul A, AbdEl-Galil A (2005) The role of cement dust in chemical remediation of sludge treated soil. ASS Univ Bull Environ Res 8:98–105

Acar YB, Gale RJ, Alshawabkeh AN, Marks RE, Puppala S, Bricka M, Parker R (1995) Electrokinetic remediation: basics and technology status. J Hazard Mater 40:117–137

Alkorta I, Hernandez-Allica J, Becerril J, Amezaga I, Albizu I, Onaindia M, Garbisu C (2004) Chelate-enhanced phytoremediation of soils polluted with heavy metals. Rev Environ Sci Biotech 3:55–70

Alpaslan B, Yukselen MA (2002) Remediation of lead contaminated soils by stabilization/solidification. Water Air Soil Pollut 133:253–263

Anderson A, Mitchell P (2003) Treatment of mercury-contaminated soil, mine waste and sludge using silica micro-encapsulation. TMS annual meeting, extraction and processing division, March 2–6, 2003, San Diego, CA, pp 265–274

Andreae MO, Merlet P (2001) Emission of trace gases and aerosols from biomass burning. Glob Biogeochem cycles 15:955–966

Artaxo P, Hansson H-C (1995) Size distribution of biogenic aerosol particles from the Amazon Basin. Atmos Environ 29:393–402

Artinano B, Salvador P, Alonso D, Querol X, Alastuey A (2003) Anthropogenic and natural influence on the PM(10) and PM(2.5) aerosol in Madrid (Spain). Analysis of high concentration episodes. Environ Pollut 125:453–465

Asante-Duah DK (1996) Managing contaminated sites: problem diagnosis and development of site restoration. Wiley, Chichester

Bakand S, Winder C, Hayes A (2007) Comparative in vitro cytotoxicity assessment of selected gaseous compounds in human alveolar epithelial cells. Toxicol Vitro 21:1341–1347

Barter M (1999) Phytoremediation-an overview. J New Engl Water Environ Assoc 33:158–164

Basketter DA, Angelini G, Ingber A, Kern PS, Menne T (2003) Nickel, chromium and cobalt in consumer products: revisiting safe levels in the new millennium. Contact Dermat 49:1–7

Bera AK, Bera A, Samadrita BR (2005) Impact of heavy metal pollution in plants. In: Bose B, Hemantaranajan A (eds) Developments in physiology biochemistry and molecular biology of plants, vol 1. New India Publishing Agency, New Delhi, pp 105–124

Berti WR, Cunningham SD (2000) Phytostabilization of metals. Phytoremediation of toxic metals: using plants to clean-up the environment. Wiley, New York, pp 71–88

Bhandari A, Xia K (2005) Hazardous organic chemicals in biosolids recycled as soil amendments. In: Water pollution. Springer Berlin, pp 217–239

Bizily SP, Rugh CL, Meagher RB (2000) Phytodetoxification of hazardous organomercurials by genetically engineered plants. Nat Biotechnol 18:213–217

Blaylock MJ, Huang JW (2000) Phytoextraction of metals. Phytoremediation of toxic metals: using plants to clean-up the environment. Wiley, New York, pp 53–70

Bolan NS, Park JH, Robinson B, Naidu R, Huh KY (2011) Phytostabilization: a green approach to contaminant containment. Adv Agron 112:145–204

Boonyapookana B, Upatham ES, Kruatrachue M, Pokethitiyook P, Singhakaew S (2002) Phytoaccumulation and phytotoxicity of cadmium and chromium in duckweed *Wolffia globosa*. Int J Phytoremed 4:87–100

Boulding R (1996) EPA environmental assessment sourcebook. CRC Press, Chelsea

Bradshaw A (1997) Restoration of mined lands-using natural processes. Ecol Eng 8:255–269

Brooks RR, Chambers MF, Nicks LJ, Robinson BH (1998) Phytomining. Trends Plant Sci 3:359–362

Brunner I, Luster J, Gunthardt-Goerg MS, Frey B (2008) Heavy metal accumulation and phytostabilisation potential of tree fine roots in a contaminated soil. Environ Pollut 152:559–568

Chapman PM, Wang F, Janssen CR, Goulet RR, Kamunde CN (2003) Conducting ecological risk assessments of inorganic metals and metalloids: current status. Hum Ecol Risk Assess 9:641–697

Chatterjee J, Chatterjee C (2000) Phytotoxicity of cobalt, chromium and copper in cauliflower. Environ Pollut 109:69–74

Cobb GP, Sands K, Waters M, Wixson BG, Dorward-King E (2000) Accumulation of heavy metals by vegetables grown in mine wastes. Environ Toxicol Chem 19:600–607

Cunningham SD, Berti WR (2000) Phytoextraction and phytostabilization: technical, economic and regulatory considerations of the soil-lead issue. In: Terry N, Banuelos G (eds) Phytoremediation of contaminated soils and waters. CRC Press LLC, Boca Raton, pp 359–375

Dermatas D, Meng X (2003) Utilization of fly ash for stabilization/solidification of heavy metal contaminated soils. Eng Geol 70:377–394

Desouza M, Pilon-Smits E, Terry N (2000) The physiology and biochemistry of selenium volatilization by plants. Phytoremediation of toxic metals: using plants to clean-up the environment. Wiley, New York, pp 171–190

Di Palma L, Ferrantelli P, Merli C, Biancifiori F (2003) Recovery of EDTA and metal precipitation from soil flushing solutions. J Hazard Mater 103:153–168

Dias G, Edwards G (2003) Differentiating natural and anthropogenic sources of metals to the environment. Hum Ecol Risk Assess 9:699–721

Druss DL (2002) Guidelines for design and installation of soil-cement stabilization. Geotech Special Publ 1:527–539

Dushenkov V, Kumar PN, Motto H, Raskin I (1995) Rhizofiltration: the use of plants to remove heavy metals from aqueous streams. Environ Sci Technol 29:1239–1245

Dushenkov S, Kapulnik Y, Blaylock M, Sorochisky B, Raskin I, Ensley B (1997) Phytoremediation: a novel approach to an old problem. Stud Environ Sci 66:563–572

Ebbs S, Lasat M, Brady D, Cornish J, Gordon R, Kochian L (1997) Phytoextraction of cadmium and zinc from a contaminated soil. J Environ Qual 26:1424–1430

Elinder CG, Kjellström T, Lind B, Linnman L, Piscator M, Sundstedt K (1983) Cadmium exposure from smoking cigarettes: variations with time and country where purchased. Environ Res 32:220–227

Ensley BD (2000) Rationale for use of phytoremediation. In: Raskin I, Ensley BD (eds) Phytoremediation of toxic metals using plants to clean up the environment. Wiley, New York, pp 3–11

Flathman PE, Lanza GR (1998) Phytoremediation: current views on an emerging green technology. J soil contam 7:415–432

Galażyn-Sidorczuk M, Brzoska MM, Moniuszko-Jakoniuk J (2008) Estimation of Polish cigarettes contamination with cadmium and lead, and exposure to these metals via smoking. Environ Monit Assess 137:481–493

Garrett RG (2000) Natural sources of metals to the environment. Hum Ecol Risk Assess 6:945–963

Gerardi MH, Zimmerman MC (2004) Wastewater pathogens. Wiley.com

Ghafoor A, Murtaza G, Ahmad B, Boers TM (2008a) Evaluation of amelioration treatments and economic aspects of using saline-sodic water for rice and wheat production on salt-affected soils under arid land conditions. Irrig Drain 57:424–434

Ghafoor A, Rehman MZ, Ghafoor A, Murtaza G, Sabir M (2008b) Fractionation and availability of cadmium to wheat as affected by inorganic amendments. Int J Agr Biol 10:469–474

Ghafoor A, Murtaza G, Rehman MZ, Sabir M, Ahmad HR, Saifullah (2012) Environment pollution: types, sources & management. Allied Book Centre, Lahore

Ghosh M, Singh S (2005) A review on phytoremediation of heavy metals and utilization of it's by products. Appl Ecol Environ Res 3:1–18

Haynes R, Murtaza G, Naidu R (2009) Inorganic and organic constituents and contaminants of biosolids: implications for land application. Adv Agron 104:165–267

Heaton AC, Rugh CL, Wang N-j, Meagher RB (1998) Phytoremediation of mercury-and methylmercury-polluted soils using genetically engineered plants. J soil contam 7:497–509

Hecht SS (2003) Tobacco carcinogens, their biomarkers and tobacco-induced cancer. Nat Rev Can 3:733–744

Ikem A, Campbell M, Nyirakabibi I, Garth J (2008) Baseline concentrations of trace elements in residential soils from Southeastern Missouri. Environ Monit Assess 140:69–81

Jacobs LW, O'Connor G, Overcash M, Zabik M, Rygiewicz P (1987) Effects of trace organics in sewage sludges on soil-plant systems and assessing their risk to humans. Lewis, Chelsea
Jadia CD, Fulekar M (2009) Phytoremediation of heavy metals: recent techniques. Afr J Biotechnol 11:664–691
Jamal A, McKenzie K, Clark M (2009) The impact of health information technology on the quality of medical and health care: a systematic review. HIM J 38:26–37
Jarup L (2003) Hazards of heavy metal contamination. Br Med Bull 68:167–182
Juhasz AL, Smith E, Smith J, Naidu R (2003) In situ remediation of DDT-contaminated soil using a two-phase cosolvent flushing-fungal biosorption process. Water Air Soil Pollut 147:263–274
Kamal M, Ghaly A, Mahmoud N, Cote R (2004) Phytoaccumulation of heavy metals by aquatic plants. Environ Int 29:1029–1039
Karl T, Guenther A, Yokelson RJ, Greenberg J, Potosnak M, Blake DR, Artaxo P (2007) The tropical forest and fire emissions experiment: emission, chemistry, and transport of biogenic volatile organic compounds in the lower atmosphere over Amazonia. J Geophys Res 112:D18302. doi:10.1029/2007JD008539
Kasprzak KS, Sunderman FW Jr, Salnikow K (2003) Nickel carcinogenesis. Mutat Res-Fund Mol M 533:67–97
Kertulis-Tartar G, Ma L, Tu C, Chirenje T (2006) Phytoremediation of an arsenic-contaminated site using Pteris vittata L.: a two-year study. Int J Phytoremediat 8:311–322
Khan FI, Husain T, Hejazi R (2004) An overview and analysis of site remediation technologies. J Environ Manage 71:95–122
Krogmann U, Boyles LS, Bamka WJ, Chaiprapat S, Martel CJ (1999) Biosolids and sludge management. Water Environ Res 71:692–714
Kumar PN, Dushenkov V, Motto H, Raskin I (1995) Phytoextraction: the use of plants to remove heavy metals from soils. Environ Sci Technol 29:1232–1238
Lewis B, Johnson C, Delwiche C (1966) Release of volatile selenium compounds by plants. Collection procedures and preliminary observations. J Agric Food Chem 14:638–640
Logsdon S, Keller K, Moorman T (2002) Measured and predicted solute leaching from multiple undisturbed soil columns. Soil Sci Soc Am J 66:686–695
Luo C, Shen Z, Li X (2005) Enhanced phytoextraction of Cu, Pb, Zn and Cd with EDTA and EDDS. Chemosphere 59:1–11
Ma C-J, Kasahara M, Höller R, Kamiya T (2001) Characteristics of single particles sampled in Japan during the Asian dust-storm period. Atmos Environ 35:2707–2714
Madsen EL (2011) Microorganisms and their roles in fundamental biogeochemical cycles. Curr Opin Biotechnol 22:456–464
Magaye R, Zhao J (2012) Recent progress in studies of metallic nickel & nickel-based nanoparticles' genotoxicity and carcinogenicity. Environ Toxicol Pharmacol 34(3):644–650
McGrath SP, Zhao F-J (2003) Phytoextraction of metals and metalloids from contaminated soils. Curr Opin Biotechnol 14:277–282
McGrath S, Zhao J, Lombi E (2002) Phytoremediation of metals, metalloids, and radionuclides. Adv Agron 75:1–56
Meza-Figueroa D, Maier RM (2009) The impact of unconfined mine tailings in residential areas from a mining town in a semi-arid environment: Nacozari, Sonora, Mexico. Chemosphere 77:140–147
Mulligan C, Yong R, Gibbs B (2001) Remediation technologies for metal-contaminated soils and groundwater: an evaluation. Eng Geol 60:193–207
Murtaza G, Ghafoor A, Qadir M, Owens G, Aziz M, Zia M (2010) Disposal and use of sewage on agricultural lands in Pakistan: a review. Pedosphere 20:23–34
Murtaza G, Haynes RJ, Naidu R, Belyaeva ON, Kim K-R, Lamb DT, Bolan NS (2011) Natural attenuation of Zn, Cu, Pb and Cd in three biosolids-amended soils of contrasting pH measured using rhizon pore water samplers. Water Air Soil Pollut 221:351–363
Neumann M, Schulz R, Schäfer K, Müller W, Mannheller W, Liess M (2002) The significance of entry routes as point and non-point sources of pesticides in small streams. Water Res 36:835–842

Niazi NK, Singh B, Shah P (2011a) Arsenic speciation and phytoavailability in contaminated soils using a sequential extraction procedure and XANES spectroscopy. Environ Sci Technol 45:7135–7142

Niazi NK, Singh B, Van Zwieten L, Kachenko AG (2011b) Phytoremediation potential of *Pityrogramma calomelanos* var. *austroamericana* and *Pteris vittata* L. grown at a highly variable arsenic contaminated site. Int J Phytoremed 13:912–932

Niazi NK, Singh B, Van Zwieten L, Kachenko AG (2012) Phytoremediation of an arsenic-contaminated site using *Pteris vittata* L. and *Pityrogramma calomelanos* var. *austroamericana*: a long-term study. Environ Sci Pollut Res 19:3506–3515

Nogawa K, Kobayashi E, Okubo Y, Suwazono Y (2004) Environmental cadmium exposure, adverse effects and preventive measures in Japan. BioMetals 17:581–587

Nyer EK (2010) *In situ* treatment technology. CRC Press, Florida

Otterpohl R (2002) Options for alternative types of sewerage and treatment systems directed to improvement of the overall performance. Water Sci Technol 45:149–158

Panwar B, Marton L, Kadar I, Anton A, Nemeth T (2010) Phytoremediation: a novel green technology to restore soil health. Acta Agron Hung 58:443–458

Pilon-Smits E (2005) Phytoremediation. Annu Rev Plant Biol 56:15–39

Ponder SM, Darab JG, Mallouk TE (2000) Remediation of Cr (VI) and Pb (II) aqueous solutions using supported, nanoscale zero-valent iron. Environ Sci Technol 34:2564–2569

Raag (2000) Evaluation of Risk Based Corrective Action Model, Remediation Alternative Assessment Group, Memorial University of Newfoundland, St John's, NF, Canada

Raskin I, Ensley BD (2000) Phytoremediation of toxic metals. Wiley, New York

Rogers HR (1996) Sources, behaviour and fate of organic contaminants during sewage treatment and in sewage sludges. Sci Total Environ 185:3–26

Rugh CL (2001) Mercury detoxification with transgenic plants and other biotechnological breakthroughs for phytoremediation in vitro cell development. Biol-Plant 37:321–325

Sabir M, Ghafoor A, Saifullah, Rehman M, Murtaza G (2008) Effect of organic amendments and incubation time on extractability of Ni and other metals from contaminated soils. Pak J Agric Sci 45:18–24

Saifullah, Zia MH, Meers E, Ghafoor A, Murtaza G, Sabir M, Zia-ur-Rehman M, Tack F (2010) Chemically enhanced phytoextraction of Pb by wheat in texturally different soils. Chemosphere 79:652–658

Schnoor JL (2000) Phytostabilization of metals using hybrid poplar trees. Phytoremediation of toxic metals: using plants to clean-up the environment. Wiley, New York, pp 133–150

Sherwood LJ, Qualls RG (2001) Stability of phosphorus within a wetland soil following ferric chloride treatment to control eutrophication. Environ Sci Technol 35:4126–4131

Silva WS, Lapis AA, Suarez PA, Neto BA (2011) Enzyme-mediated epoxidation of methyl oleate supported by imidazolium-based ionic liquids. J Mol Catal B-Enzym 68:98–103

Singh J, Upadhyay SK, Pathak RK, Gupta V (2011) Accumulation of heavy metals in soil and paddy crop (*Oryza sativa*), irrigated with water of Ramgarh Lake, Gorakhpur, UP, India. Toxicol Environ Chem 93:462–473

Sinha RK, Valani D, Sinha S, Singh S, Herat S (2010) bioremediation of contaminated sites: a low-cost nature's biotechnology for environmental clean up by versatile microbes, plants & earthworms. In: Faerber T, Herzog J (eds) Solid waste management and environmental remediation. Nova Science, USA, pp 1–72

Smith S (2000) Are controls on organic contaminants necessary to protect the environment when sewage sludge is used in agriculture? Prog Environ Sci 2:129–146

Son A-J, Shin K-H, Lee J-U, Kim K-W (2003) Chemical and ecotoxicity assessment of PAH-contaminated soils remediated by enhanced soil flushing. Environ Eng Sci 20:197–206

Suszcynsky EM, Shann JR (1995) Phytotoxicity and accumulation of mercury in tobacco subjected to different exposure routes. Environ Toxicol Chem 14:61–67

Suthersan SS, Payne FC (2004) In situ remediation engineering. CRC Press, Florida

Tagmount A, Berken A, Terry N (2002) An essential role of S-Adenosyl-l-Methionine: l-MethionineS-Methyltransferase in selenium volatilization by plants. Methylation of selenomethionine

to Selenium-Methyl-l-Selenium-Methionine, the precursor of volatile selenium. Plant Physiol 130:847–856
Tavares-Dias M (2006) A morphological and cytochemical study of erythrocytes, thrombocytes and leukocytes in four freshwater teleosts. J Fish Biol 68:1822–1833
Terry N, Carlson C, Raab T, Zayed AM (1992) Rates of selenium volatilization among crop species. J Environ Qual 21:341–344
Thyssen JP, Menni T (2009) Metal Allergyî: a review on exposures, penetration, genetics, prevalence, and clinical implications. Chem Res Toxicol 23:309–318
Trapp S, Larsen M, Pirandello A, Danquah-Boakye J (2003) Feasibility of cyanide elimination using plants. Eur J Mineral Proc Environ Prot 3:128–137
Vallack HW, Bakker DJ, Brandt I, Brostrom-Lunden E, Brouwer A, Bull KR, Gough C, Guardans R, Holoubek I, Jansson B (1998) Controlling persistent organic pollutants-what next? Environ Toxicol Pharmacol 6:143–175
Van der Oost R, Beyer J, Vermeulen NP (2003) Fish bioaccumulation and biomarkers in environmental risk assessment: a review. Environ Toxicol Pharmacol 13:57–149
Vara Prasad MN, de Oliveira Freitas HM (2003) Metal hyperaccumulation in plants: biodiversity prospecting for phytoremediation technology. Electron J Biotechnol 6:285–321
Verma P, George K, Singh H, Singh S, Juwarkar A, Singh R (2006) Modeling rhizofiltration: heavy-metal uptake by plant roots. Environ Model Assess 11:387–394
Vouillamoz J, Milke M (2001) Effect of compost in phytoremediation of diesel-contaminated soils. Water Sci Technol 43:291–295
Watanabe ME (1997) Phytoremediation on the brink of commericialization. Environ Sci Technol 31:182A–186A
Weissermel K (2008) Industrial organic chemistry. Wiley.com
Wilk C (2003) Solidification/stabilization treatment: principles and practice. Air and Waste Management Association's Magazine for Environmental Managers July, 31–37
Yang F, Yang J, Zhang X, Chen L, Jiang Y, Yan Y, Tang X, Wang J, Xiong Z, Dong J (2005) Genome dynamics and diversity of Shigella species, the etiologic agents of bacillary dysentery. Nucleic Acids Res 33:6445–6458
Zhao JS, Ward OP, Lubicki P, Cross JD, Huck P (2001) Process for degradation of nitrobenzene: combining electron beam irradiation with biotransformation. Biotechnol Bioeng 73:306–312
Zhu Y, Zayed A, Qian J, De Souza M, Terry N (1999) Phytoaccumulation of trace elements by wetland plants: II. Water hyacinth. J Environ Qual 28:339–344
Zychlinsky A, Sansonetti P (1997) Perspectives series: host/pathogen interactions. Apoptosis in bacterial pathogenesis. J Clin Invest 100:493–495

Chapter 8
Role of Macronutrients in Plant Growth and Acclimation: Recent Advances and Future Prospective

Durgesh Kumar Tripathi, Vijay Pratap Singh, Devendra Kumar Chauhan, Sheo Mohan Prasad and Nawal Kishor Dubey

Abstract Macronutrients play a very important role in plant growth and development. Their functions range from being structural units to redox-sensitive agents. Generally, application of macronutrient increases yield, growth, and quality of crops. In the recent years, however, plant physiologists, biotechnologists, and ecophysiologists have been working to investigate various other blind features of these minerals and their future prospective, because nutrients are involved in every step of plant life. Every macronutrient has its own character, and is therefore involved in different metabolic processes of plant life. Herein, this chapter deals with the recent progress made in discovering the roles of macronutrients in plant growth and acclimation process as well as future prospective of elemental research in plants.

Keywords Macronutrients · Growth · Acclimation · Stresses

D. K. Chauhan (✉)
Department of Botany, University of Allahabad, Allahabad 211002, India
e-mail: dkchauhanau@ahoo.com

D. K. Tripathi · N. K. Dubey
Department of Botany, Center of Advanced Study, Banaras Hindu University, Varanasi 221005, India
dktripathiau@gmail.com
nkdubeybhu@gmail.com

V. P. Singh
Govt. Ramanuj Pratap Singhdev Post Graduate College, Baikunthpur, Korea, Chhattisgarh 497335, India
e-mail: vijaypratap.au@gmail.com

S. M. Prasad
Department of Botany, Ranjan Plant Physiology and Biochemistry Laboratory, University of Allahabad, Allahabad 211002, India
profsmprasad@gmail.com

P. Ahmad et al. (eds.), *Improvement of Crops in the Era of Climatic Changes,* Volume 2,
DOI 10.1007/978-1-4614-8824-8_8, © Springer Science+Business Media New York 2014

1 Introduction

The study related to plant mineral nutrients is a subject of marvelous curiosity and issue of immense importance among the plant physiologists, molecular biologists, and agronomists. Therefore, much research has been focused on this aspect to tackle problems related to the mineral nutrition of plants. Various studies demonstrated that plant nutrients are either chemical elements or generally compounds which are essential for growth, development, and better yield while also playing counter roles in external activities and metabolism of plants (Vitousek 1982; Alam 1999; Subbarao et al. 2003; He and Yang 2007). Studies related to nutrient's essentialities for plants were started from sixteenth and seventeenth centuries by renowned chemists Van Helmont, Boyle, Glauber, and Mayow (Street and Opik 1970). The ninetieth century saw an emergence of agricultural chemistry and witnessed the use of artificial fertilizer for improving the crop quality and yield. Furthermore, with the progress of water and sand culture techniques, appreciation of elements which are essential for plants was also discovered, and the essentialities of nitrogen (N), phosphors (P), sulfur (S), potassium (K), calcium (Ca), magnesium (Mg), and iron (Fe) for the plants were identified (Street and Opik 1970). Through the systematic development of elemental research, it has been shown that some plants complete their life cycle in the presence of various elements; however, in their absence, plants were incapable. Therefore, Epstein (1972) has classified elemental requirement in two ways, i.e., essential and nonessential elements. Figure 8.1 clearly reveals the abundance percentage of different elements with respect to plants, soil, and ocean; however, Fig. 8.2 depicts the pathway of mineral translocation from soil to roots and from roots to upper parts of the plants. Table 8.1 describes the basic information of elements regarding their weight percentage, symbol, major forms, order of their abundance on the earth's crust, etc. (http://hyperphysics).

Studies suggested that there are 20 elements which are regarded as essential or beneficial for the survival of plants; however, essentialities of 14 elements for plants are also a matter of enormous debate among the scientists working on elemental dynamics (Marschner 1995; Mengel et al. 2001). Furthermore, carbon, hydrogen, and oxygen are regarded as three major elements which are taken up through both air and water, while remaining elements accumulate through the plant roots from the soil. On the basis of elements' quantity and their requirement on the earth crust, they have been classified in many other ways such as macronutrients, i.e., N, P, K, Ca, and Mg, and micronutrients or trace nutrients which consist of B, Cl, Cu, Fe, Mo, Mn, Ni, Na, and Zn. Studies suggested that these elements enhance the growth and yields, besides performing different dynamic functions to protect the internal or external integrity of plant life (White and Brown 2010). Besides this, some elements like silicon (Si) and cobalt (Co) are characterized as beneficial elements but have not been considered as essential for all plants, nonetheless reaching the criteria of essential elements for some members of Poaceae and Cyperaceae (Tripathi et al. 2011, 2012b; Epstein 1999; Pilon-Smits et al. 2009; Barker and Pilbeam 2010).

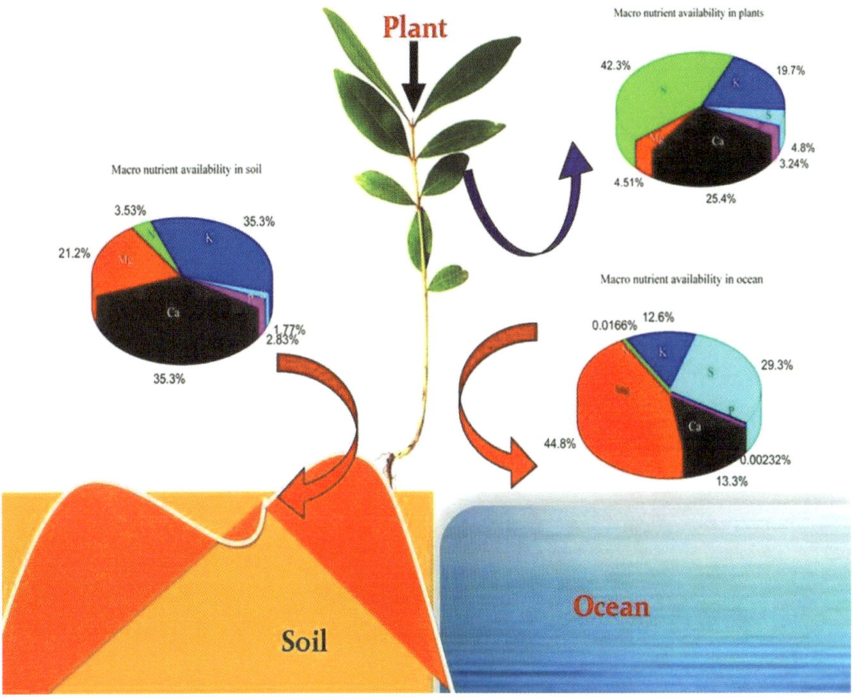

Fig. 8.1 Percent availability of macronutrients in soil, plant, and ocean

The Human body is known to require 22 mineral elements for sustainability of good health, and these mineralsare involved in different activities of human body; conversely, an appropriate diet is a major source of their availability (White and Broadley 2005a; White et al. 2009). It has been mentioned in the reports of Welch and Graham (2002) and White and Broadley (2005a) that 6 billion population of the world (above 60 %) are malnourished. On the other hand, it has also been noticed that, mainly in developed and developing countries, Ca, Mg, and Cu deficiencies are very common, and this has happened because of low amount of minerals in the natural sources such as crops, foods, water, etc.

From the beginning of nutrient evolution in various science disciplines, most of the works have been devoted to know the importance of elements in plants, animals, and ecosystems. However, in recent years, various studies related to the mechanism of elements' translocation from plant roots and subsequent distribution within the plant cells and their molecular dynamics have been well investigated (Sharma and Dubey 2005; Karley and White 2009; Miller et al. 2009; Miwa et al. 2009; Puig and Pen˜arrubia 2009; White and Broadley 2009; White and Brown 2010; Tripathi et al. 2011, 2012). Therefore, in this chapter we have summarized the progress and future prospective of macronutrients in plant science research.

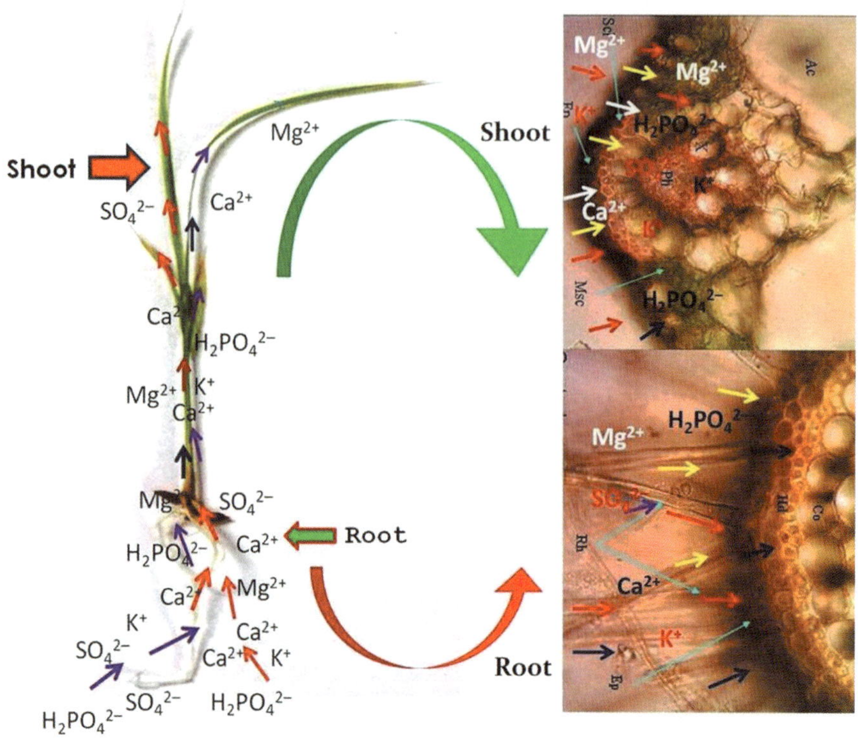

Fig. 8.2 Schematic diagram showing absorption of different macronutrients via plant's roots

Table 8.1 Some of the macronutrients with their order of abundance on the earth crust, major form of abundance, and weight % in the earth crust

Element	Symbols	Order of abundance in the earth crust	Discovered by	Year	Major form of abundance	Weight % in the earth crust
Calcium	Ca	Fifth	Humphry Davy	1808	Ca^{2+}	4.1
Magnesium	Mg	Eighth	Joseph Black	1755	Mg^{2+}	2.3
			Humphry Davy	1808		
Potassium	K	Ninth	Humphry Davy	1807	K^{+}	2.1
Phosphorous	P	Eleventh	Hennig Brand	1669	$H_2PO_4^{2-}$	0.1
			Antoine Lavoisier	1977		
Sulfur	S	Fifteenth	Antoine Lavoisier	1777	SO_4^{2-}	0.03
Nitrogen	N	Seventh	Daniel Rutherford	1772	NO_3^{-}	0.002

2 Macronutrients

Macronutrientsbasically performa counter role in various metabolic processes of plants and human beings, and are therefore required in large quantities for their survival. On the basis of their functions, macronutrients have been classified into two groups: primary macronutrients, i.e., N, P, and K, and secondary macronutrients, i.e., Ca, Mg, and S (Ryan et al. 2001; García et al. 2003; Rowley et al. 2012; Morgan and Connolly 2013). These primary and secondary macronutrients play significant role during the entire plant life by performing various beneficial activities in plant metabolism as well as protecting plants from various abiotic and biotic stresses including the stresses of heavy metals, drought, heat, UV radiations, and from diseases and insect pest attacks (Shanker and Venkateswarlu 2011; Rowley et al. 2012; Morgan and Connolly 2013). These macronutrients also help to increase the yield, growth, and quality of various crops (Morgan and Connolly 2013). Moreover, in recent years, plant physiologists, biotechnologists, and eco-physiologists have been working to investigate various other blind features of these minerals and discuss their future prospective because of nutrients involvement at each step of plant life. Every macronutrient has its own unique character, and is therefore involved in different metabolic processes of plant life.

2.1 *Primary Macronutrients*

2.1.1 Nitrogen

Nitrogen is the 15th group element of the periodic table, which belongs to the P block, and was first identified by Daniel Rutherford in 1772, while Jean-Antoine Chaptal was the first man who gave its name as "nitrogen" in 1790 (http://en.wikipedia.org). Generally, it is found in the form of a colorless and tasteless gas which makes up 78.09% by quantity of the earth's atmosphere (Keeney and Nelson 1982; Fontana and Zickerman 2010; Alvarez et al. 2012). Nitrogen is required for plants in the greatest amount, which comprises about 1.5–2.0% of plant dry matter, besides approximately 16% of total plant protein (Frink et al. 1999; Craig Jr 2002; Chen et al. 2003; Lima et al. 2007; Alvarez et al. 2012). It has been suggested that in our entire solar system, N is probably the seventh abundant element and most essential component of all existing cells (Frink et al. 1999; Craig Jr 2002; Alvarez et al. 2012). Nitrogen is also regarded as the essential component of all proteinsand enzymes and further performs in various metabolic processes of energy transformation (Street and Kidder 1997). Therefore, sufficient amount of N availability in plants is required, because it is one of the major key factors of crop production (Nadeem et al. 2013). Studies reveled that nitrogen is also an essential constituent of chlorophylls, which is closely associated with photosynthetic process (Nursu'aidah et al. 2014). Furthermore, it facilitates to improve the fruit and seed production along with hasty plant growth, quality of forage crops, and leaf (Mengel and Kirkby 1987; Marschner 2011). In natural ecosystem and agriculture,

N occurs in many forms such as nitrate, nitrite, ammonium, amino acids, etc; however, it is taken up by plants from various sources like fertilizers, air, water, rain, and some other sources in molecular form (N_2) directly (Bernhard 2010; Khajuria and Kanae 2013). It is converted into various obtainable forms by several soil organisms. It has been shown that in the past 50 years, owing to the systemic use of N in the form of fertilizer, crop production has almost doubled (Shaviv 2001). However, it is also true that due to the industrial pollution and severe climate change, the natural N cycle has considerably been in trouble (Cleemputa and Boeckxa 2013).

Vidal and Gutierrez (2008) and Lošák et al. (2010) confirmed that under various patterns of N supply, plants have shown elaborate reaction in relation to physiological and morphological levels to regulate their development and growth (Kiba et al. 2011). Additionally, Lošák et al. (2010) again reported the consequences of N on metabolisms of essential and nonessential amino acids of maize crops and declared that 240 kg N/ha application reduced most of the 17 amino acids analyzed in the grain of maize compared to 0 and 120 kg N/ha treatments, and concluded that the application of N in high amount showed a considerable outcome on the maximal accumulation of isoleucine, valine, histidine, leucine, cysteine, phenylalanine, and alanine. Similarly, Kandi et al. (2012) have reported that concentration of methionine and lysine amino acids, which are required for human body and cannot be created biologically, were affected in different varieties of potato under different levels of N, which significantly increased. Effect of N supply on the modification of plant hormonal status was also studied and systematically reported by Glare (2001). Kiba et al. (2011) and Pavlíková et al. (2012) proposed that phytohormones like abscisic acid (ABA), indole-3-acetic acid (IAA), and cytokinins (CK) were strongly connected to nitrogen signaling. Additionally, Takei et al. (2001, 2002) and Cline et al. (2006) reported that N availability is closely connected with the cytokinins in many plant species. Results also suggested that cytokinins' metabolism and translocation were adjusted by N nutritional status in plants (Takei et al. 2001; Sakakibara et al. 2010; Pavlíková et al. 2012). Different N pathways; its fixation; uptake involving genes; genetic, biochemical, and ecological aspects; and future prospects in sustainable agricultural development have been well studied earlier (Simpson 1983; Jamiesona and Semenov 2000; Martre et al. 2003; Zhu et al. 2006; Foulkes et al. 2009; Hirel et al. 2011; Alvarez et al. 2012; Sharma et al. 2013).

2.1.2 Phosphorus

Phosphorus belongs to P block elements of periodic table and is considered as a nonmetallic element, first indentified by Hennig Brand in 1669; however, Antoine Lavoisier acknowledged phosphorus in the form of element in 1777 (Seaborg 1980; Luminaris 2005; http://en.wikipedia.org/wiki/Phosphorus). It is found as white P and red P; however, due to its soaring reactivity, it is never available in the form of free constituent on the earth. Maximally, its compounds are used as fertilizers; however, It also is used frequently in the form of detergents, pesticides, and matches (Diskowski and Hofmann 2005). In cell membranes of the plant, it is abundantly present in the form of phosphate, where it plays vital roles in being the constituent of DNA, RNA, and ATP (Brown and Weselby 2010; http://en.wikipedia.org/wiki/Phosphorus). Thus, it is regarded as an essential component for growth and development of plants; however, its availability in soil is often low, and therefore its high amount in the form of organic phosphate is exogenously used to attain high crop yields (Huang et al. 2011). Among crops, barley (*Hordeum vulgare*) and wheat (*Triticum aestivum*) plants use around 46 % of P in the form of fertilizers (Huang et al. 2011; Hackenberg et al. 2013). But wide-ranging fertilization in the crop fields with extensive P leads to fast exhaustion of nonrenewable P wealth, and therefore contributes to environmental pollution; however, its deficiency in soil causes significant reduction in crop yields (Elliott et al. 1997; Vance et al. 2003; Gahoonia and Nielsen 2004; Huang et al. 2011).

Studies related to various aspects of P utilization, its uptake, and its effect on diverse metabolic processes are the main research scopes for scientists working on agronomical and agricultural issues (Ballantyne 2009). Jia et al. (2011) proposed that phosphate transporters (PTs) are generally involved in uptake and translocation of their inorganic form (Pi) from the soil; however, biological belongings and physiological function of PTs are still lacking. In this context, Jia et al. (2011) demonstrated a PT gene *OsPht1* of family Pht1, which is keenly involved in P homeostasis in rice. Furthermore, Hammond et al. (2011) stated that regulatory hot spots are linked to plant gene appearance under different soil P availability in *Brassica rapa*. In addition, Huang et al. (2011) have shown that P utilization efficiency can be correlated with expression of low-affinity PTs and non-coding RNA, IPS1 in barley. Further, Postma and Lynch (2011) reported that suboptimal availability of P in root cortical aerenchyma improves the growth of maize plants. Due to a modification of cell division or cell elongation factors, leaf growth depression under P deficiency is well recognized (Chiera et al. 2002; Assuero et al. 2004; Kavanova et al. 2006).

Additionally, Kavanová et al. (2006) reported that lack of phosphorus content causes adecline in cell division and cell elongation in the leaves of various grasses. It has been well reported that P is a necessary component of photosynthetic processes which are systematically implicated in creation of sugars, oils, and starches and which further helps in the conversion of solar energy into chemical energy, proper plant maturation, and withstanding stress. It also promotes the rapid growth of plants as well as root systems; therefore, research related to P availability and its consumption is a need of further research from agronomical, agro-economical, as well as agricultural points of view.

2.1.3 Potassium

Potassium (K), which is regarded as one of the most valuable elements for the growth and development of plants, is alkaline in nature, and the term K is related to the Neo-Latin world kalium. It was first introduced and isolated by Humphry Davy in 1807 (Holden 2001). Potassium is a necessary and extremely mobile macronutrient in plants that is abundantly present in young parts of the plants. However, Fernando et al. (1992) reported that cytosol contained its highest amount, i.e., 30–

50 mMm, while 20 mM was found in vascular region of the cells. Mengel (2007) suggested that K plays a major role as a cationic inorganic element in the plants and is therefore regarded as an essential element to all plant life, and plants cannot survive without its presence. It is linked to many physiological processes which help in improving photosynthesis, enzyme activation, water relations, assimilates, transportation, as well as plant growth and development (Pettigrew 2008; Lidon and Cebola 2012; Zlatev and Lidon 2012). Helal and Mengel (1979) revealed that if K^+supply is poor in plants, protein synthesis will be inhibited and, therefore, forerunners of proteins like amino acids, amides, and nitrate could be accumulated. Conversely, Leigh and Wyn Jones (1986) suggested that it also endorses the uptake of amino acids and is also very important for protein synthesis. Furthermore, Anonymous (1998) explained the essentiality of K in plants during protein synthesis, i.e., transcription and translation, and suggested that these two processes would not be possible without ample K supply to the plants (Wakeel et al. 2011). It has also been observed that during the entire protein synthesis process, potassium is frequently associated with binding of transfer RNA (tRNA) to ribosomes (Wakeel et al. 2011).

Pfluger and Mengel (1972) reported that nicotinamide adenine dinucleotide phosphate (NADPH) plays a key function in photosynthesis, whereas K is responsible for the activation of biosynthesis of this coenzyme. Plant cell membranes are relatively permeable to K ions because of the existence of different selective K ion channels, and are therefore involved in control of various processes of the cell (Chrispeels et al. 1999; Yamaguchi and Blumwald 2005). Chrispeels et al. (1999) stated that K is an active constituent of plant metabolic system, which dynamically entails transportation of metabolites, nutrients, phytohormones, and water through xylem and phloem, while deficiency of K ions in plants severely affects the transportation mechanisms. It also helps in stabilization of pH (7–7.5) for proper enzymatic reactions as well as diffusion of various organic anions and other compounds within the plant (Mengel and Kirkby 1980). It has been shown by Anonymous (1998) that the process of opening and closing of stomata in most of the plants is dependent upon K ions, because their movement between the guard cells of stomata checks the loss of water, and therefore helps in minimizing the risk of drought stress. Potassium plays a valuable role in minimizing the worldwide agriculture and agronomical problems like heavy metal stress: Cd, Cr, and Al (Shaviv 2001).

2.2 *Secondary Macronutrients*

2.2.1 Calcium

Among the list of all available elements, calcium is found to be the fifth most plentiful element on the earth crust, and is also regarded as the fifth most ample liquefied ion in ocean (Krebs 2006). First discovered by Humphry Davy in 1808, calcium belongs to the second group and S block in the periodic table (Enghag 2008). It is one of the most essential elements for all living organisms and is required particularly in the form of calcium ions (Ca^{2+}) that helps as well as participate in many cellular processes (Marschner 1995; White and Broadley 2003). It is important to every plant for their growth and development and is involved in activating the enzymes, inducing water movement and salt balance in plant cells, and also activating K to control the process of opening and closing of stomata (Hepler 2005). It is also required for cell growth, division, elongation, and various essential biological functions (Berridge et al. 2000; Hirschi 2004). Calcium boosts the nutrient uptake, improves the plant tissue's resistance, makes cell wall stronger, and contributes to normal root system development (Berridge et al. 2000; Hirschi 2004). Hepler (2005) noticed that Ca is an essential regulator of plant growth and development and that deficiency in plants causes yellow coloration and black spots on leaves. Further, when deposited in plant tissues, it is immobile. For this, cells have developed several mechanisms for strongly regulating calcium ion (Ca^{2+}) fluxes as well as different Ca pools, and thus it is keenly transported from the cytosol into the mitochondria, endoplasmic reticulum, cell walls, plastids, and vacuoles (Bush 1993; Harper 2001; Pittman and Hirschi 2003; Volk et al. 2004). It has been reported that Ca ions alleviate ionic stress in plants, and various feasible mechanisms have been briefly described by which Ca ions are able to prevent the subsequent damage (Plieth 2005). Kochian (1995) reported that, due to protective behavior of Ca ions, the plants were protected from the Al toxicity.

Additionally, Rodriguez-Serrano et al. (2009) also reported similar observations in which Ca influenced the Cd toxicity in pea plants. Moreover, various studies also revealed that Ca protects plants from various biotic and abiotic stresses through its various advantageous signaling channels such as Ca^{2+}/H^{+} antiporters, Ca^{2+}-ATPases, etc. (Hong-Bo et al. 2008; Kudla et al. 2010) Channels of Ca porous ion are ultimately accountable for drought stress signal transduction (Bush 1995; Norelli and Miller 2004; Andjelkovic and Thompson 2006; Ahn and Suh 2007; Hong-Bo et al. 2008; Morgan et al. 2013).

Recently, Yang et al. (2013) suggested that *SlSRs* gene plays diverse roles in response to specific stress signals, and this *SlSRs* gene may act as a coordinator(s) connecting Ca-mediated signaling with other stress signal transduction pathways during fruit ripening and storage. Furthermore, Li et al. (2013) also discovered that under NaCl stress in tobacco, a gene, HSP, successfully expressed the Ca signaling pathways. Several other studies based on Ca response against the stresses suggested that many transcription factors like SR/CAMTA perform greater role against multiple abiotic and biotic stresses, preferably drought, cold, pathogens, and wounding, as well as ethylene, auxin, MeJA, and SA (Reddy et al. 2000; Yang and Poovaiah 2000; Bouche et al. 2002; Yang and Poovaiah 2002; Galon et al. 2010; Reddy et al. 2011; Yang et al. 2013).

2.2.2 Magnesium

Magnesium is a very common element which is found in all living beings on earth; among the comparative list of abundance, it is the eighth most abundant element on the earth crust and ninth in the universe (Ash 2005; Housecroft and Sharpe 2008; Luft 2012). In 1755, Joseph Black was the first scientist who introduced magnesium, though it was first isolated by Humphry Davy in 1808,and the term Mg comes from the Greek word magnesia. It performs several advantageous functions in plants and animals, and is known as one of the essential nutrient elements for the survival. However, with reference to plants, its amount has been found to be 0.2–0.4 % of plant dry matter and its requirement for preeminent plant growth is 1.5–3.5 g kg^{-1} in the vegetative parts (Hopkins and Hüner 2011; Marschner 2012; Chen and Ma 2013). Magnesium is a central atom of chlorophyll and therefore plays a major role in plant photosynthesis, and thus its deficiency degrades the chlorophyll content and leaves become yellowish in color, which is known as chlorosis; however, an adequate supply of Mg makes the plant healthy (Ding et al. 2008; Hermans et al. 2010). Shabala and Hariadi (2005) proposed that low or excess levels of Mg contents in plants may serve diverse impact on photosynthesis. As it is a movable element in plants, chlorophyll of the plants is first decreased in old leaves and the remaining amount of Mg in old leaves is transferred to younger leaves (Hermans et al. 2010). Shaul (2002) also reported that even minute differences in Mg level may affect the various important chloroplast enzymes. Studies suggested that Mg is an active constituent of electron transportation chain; therefore, during the entire process of electron transport chain of the chloroplast, Mg has a significant responsibility. Furthermore, due to the presence of appropriate level of Mg content, the action of antioxidative enzymes and the content of antioxidant molecules were reported to be increased in pepper, maize, bean, mulberry, and *Mentha pulegium* (Cakmak and Marschner 1992; Cakmak 1994; Candan and Tarhan 2003; Tewari et al. 2004, 2006; Anza et al. 2005, Ding et al. 2008; Waraich et al. 2012). Tewari et al. (2006) further suggested that Mg deficiency increases the oxidative stress in mulberry plants by enhancing the generation of ROS and triggering distinct redox changes in the cellular metabolism, and with its sufficiency, activation of antioxidant machinery, including induction of distinct superoxide dismutase (SOD) isoforms, takes place. A study carried out by Ding et al. (2008) clearly revealed that low level of Mg in rice plants was negatively associated with the concentration of Malondialdehyde (MDA) and three antioxidative enzymes; however, exogenous supply of Mg and K in rice plants showed significant interactive effects in shoot biomass, yield, chlorophyll content, photosynthetic rate, and the activities of SOD, catalase (CAT), Peroxidase(POD), and MDA contents (Ding et al. 2008).

Chen et al. (2012) reported that Mg alleviates stress in rice, and this behavior is closely associated with Mg transporter. Further, Chen and Ma (2013) stated that Mg alleviates Al toxicity through functioning of Mg transporters, which are accountable for its adequate translocation, distribution, and uptake in rice plants. Magnesium alleviates heavy metal stresses in other plant species which has also been well studied (Ryan et al. 1994; Tan et al. 1992; Silva et al. 2001; Watanabe and Okada 2005; Yang et al. 2007; Chou et al. 2011; Chen and Ma 2013). Another efficient role of Mg in disease resistance management in different plant species has also been recognized (Sugawara et al. 1998; Rogan et al. 2000; Jones and Huber 2007; Dordas 2009), and in this reference, recently, Jones and Huber (2007) successfully referred the facts that an adequate supply of Mg may control various harmful plant diseases. However, previously, it has been revealed that Mg successfully controlled root rot diseases caused by *Rhizoctonia solani* in bean plant (Bateman 1965). Besides, Thomas and Orellana (1964) reported effective role of Mg against the leaf spot diseases caused by *Botrytis* spp. in Castor bean; however, early blight disease caused by *Alternaria solani* inpotato has also been controlled by Mg addition (Panthee and Chen 2010).

2.2.3 Sulfur

Sulfur (S) is the ninth richest element on the earth's crust, which is naturally found in the form of pure sulfide and sulfate minerals (Khan and Mazid 2011). It is a P block element having the atomic weight of 32.06 and has been discovered in China before 2000 BC and later documented as an element in 1777 by Antoine Lavoisier. It is known as the most beneficial element for all living organisms and performs various dynamic roles for growth, development, and survival of plant life. Therefore, for maximum production, it is regarded as an essential plant nutrient necessary for all crop plants. Generally, plants take sulfur in the form of sulfate (SO_4^{2-}) which is verymobile in soil and recognized as the fourth most necessary element for the plants after N, P, and K (Jamal et al. 2010). Various studies have recognized the importance of S in plants and its possible role in agriculture development (Jamal et al. 2005, 2006a, b, c, 2009, 2010; Scherer 2009; Jamal et al. 2010). Major role of S has been differently recognized, i.e., it plays a crucial role in the synthesis of chlorophyll, proteins, seeds oil content, as well as amino acids methionine and cysteine (Tandon 1986; Jamal et al. 2005, 2006a, 2009; Jamal et al. 2010). Further, activities of various other essential elements like P, N, Mg, and Ca are closely connected with S deposition in plants, and the interaction of these minerals with S is beneficial for crop improvement. Therefore, in these aspects, various studies have been conducted. In the process of protein synthesis, S and N play a key role, while translocation of these two mineral nutrients in plants is highly interrelated and the relationship between these two elements has been well studied (Zhao et al. 1993; McGrath and Zhao 1996; Jamal et al. 2006a, 2010). Moreover, benefits of S and N interaction in the yield of plants has been well studied in many plants such as sunflower, soybean, mustard, and groundnut (McGrath and Zhao 1996; Ahmad et al. 1998, 1999; Ahmad and Abdin 2000; Fazili et al. 2010a, b, Verma and Swarankar 1986; Jamal et al. 2010). Lopez-Jurado and Hunnway (1985) and Jamal et al. (2010) reported that S is a main component of nitrogen fixation in legumes; however, it has also been shown that addition of S extensively improved the nitrogen fixation, growth, and yield of plants. All et al. (2002) reported alleviative effect of S on plant growth under saline conditions and the response of S application on K/Na in sunflower. Further, S has also been reported to alleviate the Cd toxicity in various plant species, i.e., *Arabidopsis thaliana, B. juncea, Nicotiana tabacum, T. aestivum,* and *B. campestris* (Zhu et al. 1999a, b; Harada et al. 2001; Harada et al. 2002; Ashraf and Harris 2004 Sarry et al. 2006; Khan et al. 2007; Anjum et al. 2008; Gill and Tuteja 2011).

Role of S in other aspects cannot be overruled, because S-rich protein may improve the plant defense mechanisms against the pathogens, since its related compounds are closely connected to biotic stress resistance (Rausch and Wachter 2005; Hell and Kruse 2007; Hell and Hillebrand 2008). Additionally, Cooper and Williams (2004) have described the role of S as an induced antifungal material in plant protection, while Williams et al. (2002) reported the role of S and thiol (S-containing group) accumulation in tomato plants against the fungal pathogen stress. Momose and Iwahashi (2001) accounted for the genes of S metabolism pathway in *Saccharomyces cerevisiae* under Cd interaction (Gill and Tuteja 2011).

Besides, two S-regulated genes *Sultr1* and *Sultr2,* which encode two sulfate transporters, have also been found to be affected due to the Cd treatment (Takahashi et al. 2000; Herbette et al. 2006; El Kassis et al. 2007; Khan et al. 2007; Pootakham et al. 2010; Gill and Tuteja 2011).

3 Conclusions and Future Perspective

In conclusion, macronutrients play a very important role in plant growth and development, and thus influence every stage of plant life. However, excess and/or deficit of macronutrients may adversely affect the overall growth and performance of plants. Therefore, the cellular status of an element must be tightly regulated. Report of Welch and Graham (2002) and White and Broadley (2005a) revealed that 6 billion population of the world (above 60%) are malnourished. Further, it has also been noticed that in developed and developing countries, Ca, Mg, and Cu deficiencies are very common. This condition has been developed due to low amount of minerals in sources such as crops, foods, water, etc. The condition warrants plant scientists to optimize methods in order to maintain the amount of essential nutrients in desired quantity, while simultaneously keeping the toxic element at the least level. This biofortification of crops with desired macronutrient may help to abolish mineral deficiency. Besides this, by managing the amount of particular macronutrient, the unwanted loss of crop productivity due to other stresses will also be minimized, which could help to feed the growing population.

References

Ahmad A, Abdin MZ (2000) Interactive effect of Nitrogen and Sulfur on the oil and protein contents and on the fatty acid profiles of oil in the seeds of rapeseed (*Brassica campestris* L.) and mustard (*Brassica juncea* L. Czern and Coss). J Agronomy Crop Science 183:1–6

Ahmad A, Abraham G, Gandotra N, Abrol YP, Abdin MZ (1998) Interactive effect of nitrogen and sulfur on growth and yield of rapeseed-mustard (*Brassica juncea* L. Czern. and Coss. and *Brassica campestris* L.) genotypes. J Agron Crop Sci 181:193–199

Ahmad A, Abrol YP, Abdin MZ (1999) Effect of split application of sulfur and nitrogen on growth and yield attributes of *Brassica* genotypes differing in time of flowering. Can J Plant Sci 79:175–180

Ahn IP, Suh SC (2007) Calcium restores prepenetration morphogenesis abolished by methylglyoxal-bis-guanyl hydrazone in *Cochliobolus miyabeanus* infecting rice. Phytopathology 97(3):331–337

Alam SM (1999) Nutrient uptake by plants under stress conditions. In: Pessarakli M (ed) Handbook of plant and crop stress, vol 3. Marcel Dekker, New York, pp 285–313

All A, Sahm M, Niazi BH (2002) Role of sulphur for potassium/sodium ratio in sunflower under saline conditions. Helia 25(37):69–78

Alvarez JM, Vidal EA, Gutiérrez RA (2012) Integration of local and systemic signaling pathways for plant N responses. Curr Opin Plant Biol 15(2):185–191

Andjelkovic V, Thompson R (2006) Changes in gene expression in maize kernel in response to water and salt stress. Plant Cell Rep 25:71–79

Anjum NA, Umar S, Ahmad A, Iqbal M, Nafees NA (2008) Sulphur protects mustard (*Brassica campestris* L.) from cadmium toxicity by improving leaf ascorbate and glutathione: sulphur protects mustard from cadmium toxicity. Plant Growth Regul 54:271–279

Anonymous (1998) Functions of potassium in plants. Better Crops 82:4–5

Anza M, Riga P, Garbisu C (2005) Time course of antioxidant responses of *Capsicum annuum* subjected to progressive magnesium deficiency. Ann Appl Biol 146:123–134

Ash R (2005) The Top 10 of everything 2006: the ultimate book of lists. Dk Publishing, pp 1–256, ISBN 0-7566-1321–3, United Kingdom

Ashraf MPJC, Harris PJC (2004) Potential biochemical indicators of salinity tolerance in plants. Plant Sci 166(1):3–16

Assuero SG, Mollier A, Pellerin S (2004) The decrease in growth of phosphorus-deficient maize leaves is related to a lower cell production. Plant Cell Environ 27:887–895

Ballantyne P (2009) CGIAR science forum. Suppl Crop Sci 50:1–192

Barker AV, Pilbeam DJ (2010) Handbook of plant nutrition. CRC Press 117:1–605

Bateman DF (1965) Discussion of the soil environment. In: Baker KF, Snyder WC (eds) Ecology of soil-borne plant pathogens: prelude to biological control. Univ California Press, Berkeley p 139

Berridge MJ, Lipp P, Bootman MD (2000) The versatility and universality of calcium signalling. Nat Rev Mol Cell Biol 1:11–21

Bernhard A (2010) The nitrogen cycle: processes, players, and human impact. Nature Educ Knowl 2(2):12

Bouche N, Scharlat A, Snedden W, Bouchez D, Fromm H (2002) A novel family of calmodulin-binding transcription activators in multicellular organisms. J Biol Chem 277:21851–21861

Brown D, Weselby C (2010) NASA-funded research discovers life built with toxic chemical. NASA Feature, posted on NASA December, 2. http://www.nasa.gov/

Bush DS (1993) Regulation of cytosolic calcium in plants. Plant Physiol 103:7–13

Bush DS (1995) Calcium regulation in plant cells and its role in signaling. Ann Rev Plant Physiol 46:95–122

Cakmak I (1994) Activity of ascorbate-dependent H_2O_2-scavenging enzymes and leaf chlorosis are enhanced in magnesium and potassium-deficient leaves, but not in phosphorus-deficient leaves. J Exp Bot 45:1259–1266

Cakmak I, Marschner H (1992) Magnesium deficiency and high light intensity enhance activities of superoxide dismutase, ascorbate peroxidase and glutathione reductase in bean leaves. Plant Physiol 98:1222–1227

Candan N, Tarhan L (2003) Relationship among chlorophyll-carotenoid content, antioxidant enzyme activities and lipid peroxidation levels by Mg^{2+} deficiency in the *Mentha pulegium* leaves. Plant Physiol Biochem 41:35–40

Chen ZC, Ma JF (2013) Magnesium transporters and their role in Al tolerance in plants. Plant Soil 368:51–56

Chen ZC, Yamaji N, Motoyama R, Nagamura Y, Ma JF (2012) Up-regulation of a magnesium transporter gene *OsMGT1* is required for conferring aluminum tolerance in rice. Plant Physiol 159:1624–1633

Chen QS, Yi KK, Huang G, Wang XB, Liu FY, Wu YR, Wu P (2003) Cloning and expression pattern analysis of nitrogen-starvation-induced genes in rice. Acta Botanica Sinica-Chinese Edition 45(8):974–980

Chiera J, Thomas J, Rufty T (2002) Leaf initiation and development in soybean under phosphorus stress. J Exp Bot 53:473–481

Chrispeels MJ, Crawford NM, Schroeder JI (1999) Proteins for transport of water and mineral nutrients across the membranes of plant cells. Plant Cell 11:661–676

Chou TS, Chao YY, Huang WD, Hong CY, Kao CH (2011) Effect of magnesium deficiency on antioxidant status and cadmium toxicity in rice seedlings. J Plant Physiol 168:1021–1030

Cleemputa OV, Boeckxa P (2013) Nitrogen. Encyclopedia of environmental management. doi:10.1081/E-EEM-120001574

Cline MG, Thangavelu M, Dong K-II (2006) A possible role of cytokinin in mediating long-distance nitrogen signaling in the promotion of sylleptic branching in hybrid poplar. J Plant Physiol 163:684–688
Cooper RM, Williams JS (2004) Elemental sulphur as an induced antifungal substance in plant defence. J Exp Bot 55(404):1947–1953
Craig CC Jr (2002) Nitrogen use efficiency of cotton following corn in rotation and foliar fertilization of cotton using leaf blade analysis. Doctoral Dissertation, Mississippi State University, pp 1–128
Ding YC, Chang CR, Luo W, Wu YS, Ren XL, Wang P, XU GH (2008) High potassium aggravates the oxidative stress induced by magnesium deficiency in rice leaves. Pedosphere 18(3):316–327
Diskowski H, Hofmann T (2005) Phosphorus in Ullmann's encyclopedia of industrial chemistry. Wiley-VCH, Weinheim. doi:10.1002/14356007.a19-505
Dordas C (2009) Role of nutrients in controlling plant diseases in sustainable agriculture: a review. In: Sustainable agriculture, Springer, Netherlands, pp 443–460
Elliott DE, Reuter DJ, Reddy GD, Abbott R (1997) Phosphorus nutrition of spring wheat (*Triticum aestivum* L.). Effects of phosphorus supply on plant symptoms, yield, components of yield, and plant phosphorus uptake. Aust J Agric Res 455:855–867
El Kassis E, Cathala N, Rouached H, Fourcroy P, Berthomieu P, Terry N, Davidian JC (2007) Characterization of a selenate-resistant Arabidopsis mutant. Root growth as a potential target for selenate toxicity. Plant Physiol 143(3):1231–1241
Enghag P (2008) Encyclopedia of the elements: technical data-history-processing-applications. Wiley, University of Michigan, United States 1–1309
Epstein E (1972) Mineral Nutrition of Plants: Principle and Perspectives. John Wiley and Sons, New York
Epstein E (1999) Silicon. Annu Rev Plant Phys Plant Mol Bio. 50:641–664
Fazili IS, Masoodi M, Ahmad S, Jamal A, Khan JS, Abdin MZ (2010a) Oil biosynthesis and its related variables in developing seeds of mustard (*Brassica juncea* L.) as influenced by sulphur fertilization. J Crop Sci Biotechnol 13:39–46
Fazili IS, Masoodi M, Ahmad S, Jamal A, Khan JS, Abdin MZ (2010b) Interactive effect of sulfur and nitrogen on growth and yield attributes of oilseed crops (*Brassica campestris* L. and *Eruca sativa* Mill.) differing in yield potential. J Plant Nutr 33:1216–1228
Fernando M, Mehroke J, Glass ADM (1992) De novo synthesis of plasma membrane and tonoplast polypetides of barley roots during short-term K deprivation: in search of high-affinity K transport system. Plant Physiol 100:1269–1276
Fontana PM, Zickerman CP (2010) Mitigating the environmental footprint of towed streamer seismic surveys. In 72nd EAGE conference and exhibition. http://www.earthdoc.org/. Accessed 14 June 2010
Foulkes MJ, Hawkesford MJ, Barraclough PB, Holdsworth MJ, Kerr S, Kightley S, Shewry PR (2009) Identifying traits to improve the nitrogen economy of wheat: recent advances and future prospects. Field Crop Res 114(3):329–342
Frink CR, Waggoner PE, Ausubel JH (1999) Nitrogen fertilizer: retrospect and prospect. Proc Natl Acad Sci U S A 96:1175–1180
Gahoonia TS, Nielsen NE (2004) Barley genotypes with long root hairs sustain high grain yields in low-P field. Plant Soil 262:55–62
Galon Y, Aloni R, Nachmias D, Snir O, Feldmesser E, Scrase-Field S, Boyce JM, Bouche N, Knight MR, Fromm H (2010) Calmodulin-binding transcription activator 1 mediates auxin signaling and responds to stresses in *Arabidopsis*. Planta 232:165–178
García AG, Dourado-Neto D, del VBasantaM, Ovejero RFL, Favarin JL (2003) Logistic rice model for dry matter and nutrient uptake. Sci Agricola 60(3):481–488
Gill SS, Tuteja N (2011) Cadmium stress tolerance in crop plants: probing the role of sulfur. Plant Signal Behav 6(2):215–222

Glare TR, O'Callaghan M, Malone LA, Burgess EP (2001) Summary of current scientific awareness of the effect of genetically modified organisms on the natural environment. Rep Minist Environ 1–188

Hackenberg M, Huang PJ, Huang CY, Shi BJ, Gustafson P, Langridge P (2013) A comprehensive expression profile of MicroRNAs and other classes of non-coding small RNAs in barley under phosphorous-deficient and-sufficient conditions. DNA Res 20(2):109–125

Hammond JP, Mayes S, Bowen HC, Graham NS, Hayden RM, Love CG, Broadley MR (2011) Regulatory hotspots are associated with plant gene expression under varying soil phosphorus supply in *Brassica rapa*. Plant Physiol 156(3):1230–1241

Harada E, Eui Choi Y, Tsuchisaka A, Obata H, Sano H (2001) Transgenic tobacco plants expressing a rice cysteine synthase gene are tolerant to toxic levels of cadmium. J Plant Physiol 158:655–661

Harada E, Yamaguchi Y, Koizumi N, Sano H (2002) Cadmium stress induces production of thiol compounds and transcripts for enzymes involved in sulfur assimilation pathway in *Arabidopsis*. J Plant Physiol 159:445–448

Harper JF (2001) Dissecting calcium oscillators in plant cells. Trends Plant Sci 6:395–397

He ZL, Yang XE (2007) Role of soil rhizobacteria in phytoremediation of heavy metal contaminated soils. J Zhejiang Univ Sci 8(3):192–207

Helal HM, Mengel K (1979) Nitrogen metabolism of young barley plants as affected by NaCl salinity and potassium. Plant Soil 51:457–462

Hell R, Kruse C (2007) Sulfur in biotic interactions of plants in their environment. In: Hawkesford MJ, de Kok LJ (eds) Sulfur in plants: an ecological perspective. Dordrecht, Springer, pp 197–224

Hell R, Hillebrand H (2008) Evaluation of future developments in agrobiotechnology: the potential roles of protein nitrogen and sulfur for better crop plants. Landbauforsch Volk 58:91–96

Hepler PK (2005) Calcium: a central regulator of plant growth and development. Plant Cell 17(8):2142–2155

Herbette S, Taconnat L, Hugouvieux V, Piette L, Magniette MLM, Cuine S, Auroy P, Richaud P, Forestier C, Bourguignon J, Renou JP, Vavasseur A, Leonhardt N (2006) Genome-wide transcriptome profiling of the early cadmium response of *Arabidopsis* roots and shoots. Biochimie 88:1751–1765

Hermans C, Vuylsteke M, Coppens F, Cristescu SM, Harren FJ, Inzé D, Verbruggen N (2010) Systems analysis of the responses to long-term magnesium deficiency and restoration in *Arabidopsis thaliana*. New Phytol 187(1):132–144

Hirel B, Tétu T, Lea PJ, Dubois F (2011) Improving nitrogen use efficiency in crops for sustainable agriculture. Sustainability 3(9):1452–1485

Hirschi KD (2004) The calcium conundrum. Both versatile nutrient and specific signal. Plant Physiol 136:2438–2442

Holden NE (2001) History of the origin of the chemical elements and their discoverers. National nuclear data center (NNDC). http://www.nndc.bnl.gov/content/elements. Accessed 12 March 2004

Hong-Bo S, Li-Ye C, Ming-An S, Shi-Qing L, Ji-Cheng Y (2008) Bioengineering plant resistance to abiotic stresses by the global calcium signal system. Biotechnology advances 26(6):503–510

Hopkins William G, Hüner Norman PA (2011) Introduction to plant physiology, 4th ed. Wiley, pp 1–528, ISBN:978-0-470-46142-6

Housecroft CE, Sharpe AG (2008) Inorganic chemistry, 3rd ed. Prentice Hall, pp 305–306, ISBN:978-0131755536 http://en.wikipedia.org

Huang CY, Shirley N, Genc Y, Shi B, Langridge P (2011) Phosphate utilization efficiency correlates with expression of low-affinity phosphate transporters and noncoding RNA, IPS1, in barley. Plant physiol 156(3):1217–1229

Jamal A, Fazli IS, Ahmad S, Abdin MZ, Yun SJ (2005) Effect of sulphur and nitrogen application on growth characteristics, seed and oil yield of soybean cultivars. Korean J Crop Sci 50(5):340–345

Jamal A, Fazli IS, Ahmad S, Abdin MZ (2006a) Interactive effect of nitrogen and sulphur on yield and quality of groundnut (*Arachis hypogea* L.). Korean J Crop Sci 51(6):519–522

Jamal A, Fazli IS, Ahmad S, Kim KT, Oh DG, Abdin MZ (2006b) Effect of sulfur on nitrate reductase and ATP sulfurylase activities in groundnut (*Arachis hypogea* L.). J Plant Biol 49(6):513–517

Jamal A, Fazli IS, Ahmad S, Abdin MZ, Yun SJ (2006c) Effect of nitrogen and sulphur application on nitrate reductase and ATP- sulphurylase activities in soybean. Korean J Crop Sci 51(4):298–302

Jamal A, Ko K, Kim HS, Cho YK, Joung H, Ko K (2009) Role of genetic factors and environmental conditions in recombinant protein production for plant molecular biofarming. Biotechnol Adv 27:914–923

Jamal A, Moon YS, Abdin MZ (2010a) Enzyme activity assessment of peanut (*Arachis hypogea*) under slow-release sulphur fertilization. Aus J Crop Sci 4(3):169–174

Jamal A, Moon YS, Abdin MZ (2010b) Sulphur-a general overview and interaction with nitrogen. Aust J Crop Sci 4(7):523–529

Jamieson PD, Semenov MA (2000) Modeling nitrogen uptake and redistribution in wheat. Field Crop Res 68(1):21–29

Jia H, Ren H, Gu M, Zhao J, Sun S, Zhang X, Xu G (2011) The phosphate transporter gene *OsPht1* is involved in phosphate homeostasis in rice. Plant Physiol 156(3):1164–1175

Jones JB, Huber DM (2007) Magnesium and plant disease. In: Datnoff LE, Elmer WH, Huber DM (eds) Mineral nutrition and plant disease. American Phytopathological Society (APS Press), St Paul, pp 95–100, ISBN-13:978-0890543467

Kandi MAS, Tobeh A, Golipouri A, Godehkahriz SJ, Rastgar Z (2012) Concentration changes of lysine and methionine amino acids in potatoes varieties affected by different levels of Nitrogen fertilizer. Tech J Eng Appl Sci 2 (4):93–96

Karley AJ, White PJ (2009) Moving cationic minerals to edible tissues: potassium, magnesium, calcium. Curr Opin Plant Biol 12:291–298

Kavanová M, Lattanzi FA, Grimold AA, Schnyder H (2006) Phosphorus deficiency decreases cell division and elongation in grass leaves. Plant physiol 141(2):766–775

Kavanova´ M, Grimoldi AA, Lattanzi FA, Schnyder H (2006) Phosphorus nutrition and mycorrhiza effects on grass leaf growth. P status- and size mediated effects on growth zone kinematics. Plant Cell Environ 29:511–520

Keeney DR, Nelson DW (1982) Nitrogen-inorganic forms. Methods of soil analysis, part 2, chemical and microbiological properties, Soil Science Society of America Book No 5. 643–698 ISBN-13:978-0891188100

Khajuria A, Kanae S (2013) Potential and Use of Nitrate in Agricultural Purposes. J Water Res Prot 5:529–533

Khan TA, Mazid M (2011) Nutritional significance of sulphur in pulse cropping system. Biol Med 3(2):114–133

Khan NA, Samiullah, Singh S, Nazar R (2007) Activities of antioxidative enzymes, sulphur assimilation, photosynthetic activity and growth of wheat (*Triticum aestivum*) cultivars differing in yield potential under cadmium stress. J Agro Crop Sci 193:435–444

Kiba T, Kudo T, Kojima M, Sakakibara H (2011) Hormonal control of nitrogen acquisition: Roles of auxin, abscisic acid, and cytokinin. J Exp Bot 62:1399–1409

Kochian LV (1995) Cellular mechanisms of aluminum toxicity and resistance in plants. Annu Rev Plant Physiol Plant Mol Biol 46:237–260

Krebs RE (2006) The history and use of our earth's chemical elements: a reference guide. Greenwood Publishing Group 1-376 ISBN-13:978-0313301230

Kudla J, Batistič O, Hashimoto K (2010) Calcium signals: the lead currency of plant information processing. The Plant Cell Online, 22(3):541–563

Leigh RA, Wyn Jones RG (1986) Cellular compartmentation in plant nutrition: The selective cytoplasm and the promiscuous vacuole. Adv Plant Nutr 2:249–279
Li M, Guo S, Xu Y, Meng Q, Li G, Yang X (2013) Glycine betaine-mediated potentiation of HSP gene expression involves calcium signaling pathways in tobacco exposed to NaCl stress. Physiol Plant 150:63–75
Lidon ZZ, Cebola F (2012) An overview on drought induced changes in plant growth, water relations and photosynthesis. Emir J Food Agr 24(1):57–72
Lima PS, Rodrigues VLP, de Medeiros JF, de Aquino BF, da Silva J (2007) Yield and quality of melon fruits as a response to the application of nitrogen and potassium doses. Revista Caatinga 20(2)
Lopez-Jurado G, Hannway DB (1985) Sulphur nutrition effects on dinitrogen fixation of seedling alfalfa. J Plant Nutr 8:1103–1121
Losak T, Hlusek J, Filipcik R, Pospisilova L, Manasek J, Prokes K, Orosz F (2010) Effect of nitrogen fertilization on metabolisms of essential and non-essential amino acids in field-grown grain maize (*Zea mays* L.). Plant Soil Environ 56(12):574–579
Luft FC (2012) Whither magnesium? Clin Kidney J 5(Suppl 1):i1–i2
Luminaris M (2005) Phosphorus-from discovery to commodity. Indian J Chem Tech 12:108–122
Marschner H (1995) Mineral nutrition of higher plants, 2nd ed. London, Academic, pp 1–889, ISBN:978-0-12-473542-473542
Marschner H (2011) Marschner's mineral nutrition of higher plants, vol 89. Academic, London, pp 1–651, ISBN:9780123849052
Marschner H (2012) Mineral nutrition of higher plants, 3rd ed. Academic, London, pp 1–672, ISBN-13:978-0123849052
Martre P, Porte JR, Jamieson PD, Triboï E (2003) Modeling grain nitrogen accumulation and protein composition to understand the sink/source regulations of nitrogen remobilization for wheat. Plant Physiol 133(4):1959–1967
McGrath SP, Zhao FJ (1996) Sulphur uptake, yield response and the interactions between N and S in winter oilseed rape (*Brassica napus* L.). J Agric Sci 126:53–62
Mengel K (2007) Potassium. In: Barker AV, Pilbeam DJ (eds) Handbook of plant nutrition. Taylor & Francis, Boca Ratan, pp 91–120
Mengel K, Kirkby EA (1980) Potassium in crop production. Adv Agron 33:59–110
Mengel K, Kirkby EA (1987) Principles of plant nutrition, vol 73. International Potash Institute, Bern, pp 588–594, ISBN:3906535037
Mengel K, Kirkby EA, Kosegarten H, Appel T (2001) Principles of plant nutrition, vol 5. Springer Netherlands, pp 1–849
Miller AJ, Shen Q, Xu G (2009) Freeways in the plant: transporters for N, P and S and their regulation. Curr Opin Plant Biol 12:284–290
Miwa K, Kamiya T, Fujiwara T (2009) Homeostasis of the structurally important micronutrients, B and Si. Curr Opin Plant Biol 12:307–311
Momose Y, Iwahashi H (2001) Bioassay of cadmium using a DNA microarray: genome-wide expression patterns of *Saccharomyces cerevisiae* response to cadmium. Environ Toxicol Chem 20:2353–2360
Morgan JB, Connolly EL (2013) Plant-soil interactions: nutrient uptake. Nat Educ Knowl 4(8):2
Morgan SH, Lindberg S, Mühling KH (2013) Calcium supply effects on wheat cultivars differing in salt resistance with special reference to leaf cytosol ion homeostasis. Physiol Plant 149:321–328
Nadeem SM, Ahmadb M, Zahir ZA, Javaid A, Ashraf M (2013) The role of mycorrhizae and plant growth promoting rhizobacteria 3 (PGPR) in improving crop productivity under stressful environments. Biotechnol Adv. http://dx.doi.org/10.1016/j.biotechadv.2013.12.005
Norelli JL, Miller SS (2004) Effect of prohexadione-calcium dose level on shoot growth and fire blight in young apple trees. Plant Dis 88(10):1099–1106
Nursu'aidah H, Motior MR, Nazia AM, Islam MA (2014) Growth and photosynthetic responses of long bean (Vigna unguiculata) and mung bean (*Vigna radiata*) response to fertilization. J Anim Plant Sci 24(2):573–578

Panthee DR, Chen F (2010) Genomics of fungal disease resistance in tomato. Curr Genomics 11:30–39

Pavlíková D, Neuberg M, Zizkova E, Motyka V, Pavlík M (2012) Interactions between nitrogen nutrition and phytohormone levels in *Festulolium* plants. Plant Soil Environ 58:367–372

Pettigrew WT (2008) Potassium influences on yield and quality production for maize, wheat, soybean and cotton. Physiol Plant 133(4):670–681

Pfluger R, Mengel K (1972) The photochemical activity of chloroplasts obtained from plants with different potassium nutrition. Plant Soil 36:417–425

Pilon-Smits EA, Quinn CF, Tapken W, Malagoli M, Schiavon M (2009) Physiological functions of beneficial elements. Curr Opin Plant Biol 12(3):267–274

Pittman JK, Hirschi KD (2003) Don't shoot the (second) messenger: endomembrane transporters and binding proteins modulate cytosolic Ca^{2+} levels. Curr Opin Plant Biol 6:257–262

Plieth C (2005) Calcium: just another regulator in the machinery of life? Ann Bot 96(1):1–8

Postma JA, Lynch JP (2011) Root cortical aerenchyma enhances the growth of maize on soils with suboptimal availability of nitrogen, phosphorus, and potassium. Plant Physiol 156(3):1190–1201

Pootakham W, Gonzalez-BallesterD, Grossman AR (2010) Identification and regulation of plasma membrane sulfate transporters in Chlamydomonas. Plant Physiol 153(4):1653–1668

Puig S, Pen˜arrubia L (2009) Placing metal micronutrients in context: transport and distribution in plants. Curr Opin Plant Biol 12:299–306

Rausch T, Wachter A (2005) Sulfur metabolism: a versatile platform for launching defence operations. Trends Plant Sci 10:503–509

Ryan J, Estefan G, Rashid A (2001) Soil and plant analysis laboratory manual, Interaction Center for Agricultural Research in the Dry Areas (ICARDA), 2nd ed. Aleppo, Syria, pp 1–18

Reddy ASN, Reddy VS, Golovkin M (2000) A calmodulin binding protein from *Arabidopsis* is induced by ethylene and contains a DNA-binding motif. Biochem Biophys Res Commun 279:762–769

Reddy ASN, Ali GS, Celesnik H, Day IS (2011) Coping with stresses: roles of calcium- and calcium/calmodulin-regulated gene expression. Plant Cell 23:2010–2032

Rodriguez-Serrano M, Romero-Puertas MC, Pazmino DM, Testillano PS, Risueno MC, del Rio LA, Sandalio LM (2009) Cellular response of pea plants to cadmium toxicity: cross talk between reactive oxygen species, nitric oxide, and calcium. Plant Physiology 150:229–243

Rogan GJ, Bookout JT, Duncan DR, Fuchs RL, Lavrik PB, Love SL, Mueth M, Olson T, Owens ED, Raymond PJ, Zalewski J (2000) Compositional analysis of tubers from insect and virus resistant potato plants. J Agric Food Chem 48:5936–5945

Rowley S, Cardon G, Black B (2012) Macronutrient management for Utah Orchards. USU Extension Publication Horticulture/Fruit/2012-01pr

Ryan PR, Kinraide TB, Kochian LV (1994) Al^{3+} -Ca^{2+} interactions in aluminum rhizotoxicity. Planta 192:98–102

Sakakibara H, Takei K, Hirose N (2010) Interactions between nitrogen and cytokinin in the regulation of metabolism and development. Trends Plant Sci 9:440–448

Sarry JE, Kuhn L, Ducruix C, Lafaye A, Junot C, Hugouvieux V, Jourdain A, Bastien O, Fievet JB, Vailhen D, Amekraz B, Moulin C, Ezan E, Garin J, Bourguignon J (2006) The early responses of *Arabidopsis thaliana* cells to cadmium exposure explored by protein and metabolite profiling analyses. Proteomics 7:2180–2198

Scherer HW (2009) Sulfur in soils. J Plant Nutr Soil Sci 172:326–335

Seaborg GT (1980) The new elements: much progress has been made in the discovery and understanding of new elements in the last forty years, but there are still opportunities to increase this understanding in the future. Am Sci 68(3):279–289

Shabala S, Hariadi Y (2005) Effects of magnesium availability on the activity of plasma membrane ion transporters and light-induced responses from broad bean leaf mesophyll. Planta 221:56–65

Shanker AK, Venkateswarlu B (2011) Abiotic stress in plants-mechanisms and adaptations. Tech Publisher, pp 1–428, ISBN 978-953-307-394-1

Sharma P, Dubey RS (2005) Lead toxicity in plants. Brazilian J Plant Physiol 17(1):35–52

Sharma SB, Sayyed RZ, Trivedi MH, Gobi TA (2013) Phosphate solubilizing microbes: sustainable approach for managing phosphorus deficiency in agricultural soils. Springer Plus 2:587
Shaul O (2002) Magnesium transport and function in plants: the tip of the iceberg. BioMetals 15:309–323
Shaviv A (2001) Advances in controlled-release fertilizers. Adv Agron 71:1–49
Silva IR, Smyth TJ, Carter TE, Rufty TW (2001) Altered aluminum root elongation inhibition in soybean genotypes in the presence of magnesium. Plant Soil 230:223–230
Simpson RJ, Lambers H, Dalling MJ (1983) Nitrogen redistribution during grain growth in wheat (*Triticum aestivum* L.) IV. Development of a quantitative model of the translocation of nitrogen to the grain. Plant Physiol 71(1):7–14
Street JJ, Kidder G (1997) Soils and Plant nutrition, corporative extension service, vol 8. Institute of Food and Agriculture Sciences, University of Florida. SL, pp 1–4
Street, HE, Öpik, H (1970) The physiology of flowering plants: their growth and development. Edward Arnold (Publishers) Ltd
Subbarao GV, Ito O, Berry WL, Wheeler RM (2003) Sodium-a functional plant nutrient. Crit Rev Plant Sci 22(5):391–416
Sugawara K, Singh UP, Kobayashi K, Ogoshi A (1998) Scanning electron microscopical observation and X-ray microanalysis of *Erysiphe pisi* infected leaves of pea (*Pisum sativum* L.). Phytopathol Z 146:223–229
Takahashi H, Watanabe-Takahashi A, Smith F, Blake-Kalff M, Hawkesford M, Daito K (2000) The role of three functional sulfate transporters involved in uptake and translocation of sulphate in *Arabidopsis thaliana*. Plant J 23:171–182
Takei K, Sakakibara H, Taniguchi M, Sugiyama T (2001) Nitrogen-dependent accumulation of cytokinins in root and thetranslocation to leaf: implication of cytokinin species thatinduces gene expression of maize response regulator. Plant Cell Physiol 42:85–93
Takei K, Takahashi T, Sugiyama T, Yamaya T, Sakakibara H (2002) Multiple routes communicating nitrogen availability from roots to shoots: a signal transduction pathway mediated by cytokinin. J Exp Bot 53:971–977
Tan K, Keltjens WG, Findenegg GR (1992) Aluminium toxicity with sorghum genotypes in nutrient solutions and its amelioration by magnesium. J Plant Nutr Soil Sci 155:81–86
Tandon HLS (1986) Sulfur research and agricultural production in India 2nd edition. Fertilizer development and consultation organization, New Delhi, p 76
Tewari RK, Kumar P, Tewari N, Srivastava S, Sharma PN (2004) Macronutrient deficiencies and differential antioxidant responses-influence on the activity and expression of superoxide dismutase in maize. Plant Sci 166:687–694
Tewari RK, Kumar P, Sharma PN (2006) Magnesium deficiency induced oxidative stress and antioxidant responses in mulberry plants. Sci Hortic 108:7–14
Thomas CA, Orellana RG (1964) Phenols and pectin in relation to browning and maceration of castor bean capsules by *Botrytis*. J Phytopathol 50:359–366
Tripathi DK, Kumar R, Chauhan DK, Rai AK, Bicanic D (2011) Laser-induced breakdown spectroscopy for the study of the pattern of silicon deposition in leaves of *Saccharum* species. Instrum Sci Technol 39(6):510–521
Tripathi DK, Kumar R, Pathak AK, Chauhan DK, Rai AK (2012a) Laser-induced breakdown spectroscopy and phytolith analysis: An approach to study the deposition and distribution pattern of silicon in different parts of Wheat (*Triticum aestivum* L.). Plant Agric Res 1(4):352–361
Tripathi DK, Singh VP, Kumar D, Chauhan DK (2012b) Impact of exogenous silicon addition on chromium uptake, growth, mineral elements, oxidative stress, antioxidant capacity, and leaf and root structures in rice seedlings exposed to hexavalent chromium. Acta Physiol Plant 34(1):279–289
Tripathi DK, Singh VP, Kumar D, Chauhan DK (2012c) Rice seedlings under cadmium stress: effect of silicon on growth, cadmium uptake, oxidative stress, antioxidant capacity and root and leaf structures. Chem Ecol 28(3):281–291
Vance CP, Uhde-Stone C, Allan DL (2003) Phosphorus acquisition and use: critical adaptations by plants for securing a nonrenewable resource. New Phytol 157:423–447

Verma MM, Swarnkar KC (1986) Response of linseed to nitrogen, phosphorous and sulfur application in sandy loam soil. Indian Agric 30:223–228
Vidal EA, Gutiérrez RA (2008) A systems view of nitrogen nutrient and metabolite responses in Arabidopsis. Curr Opin Plant Biol 11(5):521–529
Vitousek P (1982) Nutrient cycling and nutrient use efficiency. Am Nat 119:553–572
Volk GM, Goss LJ, Franceschi VR (2004) Calcium channels are involved in calcium oxalate crystal formation in specialized cells of *Pistia stratiotes* L. Ann Bot 93(6):741–753
Wakeel A, Farooq M, Qadir M, Schubert S (2011) Potassium substitution by sodium in plants. Crit Rev Plant Sci 30(4):401–413
Waraich EA, Ahmad R, Halim A, Aziz T (2012) Alleviation of temperature stress by nutrient management in crop plants: a review. J Soil Sci Plant Nutr 12(2):221–244
Watanabe T, Okada K (2005) Interactive effects of Al, Ca and other cations on root elongation of rice cultivars under low pH. Ann Bot 95:379–385
Welch RM, Graham RD (2002) Breeding crops for enhanced micronutrient content. Plant Soil 245(1):205–214
White PJ, Broadley MR (2003) Calcium in plants. Ann Bot 92(4):487–511
White PJ, Broadley MR (2005) Biofortifying crops with essential mineral elements. Trends Plant Sci 10(12):586–593
White PJ, Broadley MR (2009) Biofortification of crops with seven mineral elements often lacking in human diets-iron, zinc, copper, calcium, magnesium, selenium and iodine. New Phytol 182:49–84
White PJ, Brown PH (2010) Plant nutrition for sustainable development and global health. Ann Bot 105(7):1073–1080
White PJ, Bradshaw JE, Finlay M, Dale B, Ramsay G, Hammond JP, Broadley MR (2009) Relationships between yield and mineral concentrations in potato tubers. Hortic Sci 44(1):6–11
Williams JS, Hall SA, Hawkesford MJ, Beale MH, Cooper RM (2002) Elemental sulfur and thiol accumulation in tomato and defense against a fungal vascular pathogen. Plant Physiol 128(1):150–159
Yamaguchi T, Blumwald E (2005) Developing salt-tolerant crop plants: challenges and opportunities. Trends Plant Sci 10:615–620
Yang TB, Poovaiah BW (2000) An early ethylene up-regulated gene encoding a calmodulin-binding protein involved in plant senescence and death. J Biol Chem 275:38467–38473
Yang TB, Poovaiah BW (2002) A calmodulin-binding/CGCG box DNA-binding protein family involved in multiple signaling pathways in plants. J Biol Chem 277:45049–45058
Yang JL, You JF, Li YY, Wu P, Zheng SJ (2007) Magnesium enhances aluminum-induced citrate secretion in rice bean roots (*Vigna umbellata*) by restoring plasma membrane H^+-ATPase activity. Plant Cell Physiol 48:66–73
Yang T, Peng H, Whitaker BD, Jurick WM (2013) Differential expression of calcium/calmodulin-regulated SISRs in response to abiotic and biotic stresses in tomato fruit. Physiol Plant 148:445–455
Zhao FJ, Evans EJ, Bilsborrow PE, Syers JK (1993) Influence of S and N on seed yield and quality of low glucosinolate oilseed rape (*Brassica nepus* L.). J Sci Food Agric 63:29–37
Zhu GB, Peng YZ, Wu SY, Ma B, Wang YY (2006) Study on optimal operation of step feed biological nitrogen removal process. China Water Wastewater 22(21):1–5
Zhu YL, Pilon-Smits EAH, Jouanin L, Terry N (1999a) Cadmium tolerance and accumulation in Indian mustard is enhanced by overexpressing γ-glutamylcysteine synthetase. Plant Physiol 121:1169–1177
Zhu YL, Pilon-Smits EAH, Jouanin L, Terry N (1999b) Over expression of glutathione synthetase in Indian mustard enhances cadmium accumulation and tolerance. Plant Physiol 119:73–80
Zlatev Z, Lidon FC (2012) An overview on drought induced changes in plant growth, water relations and photosynthesis. Emirates J Food Agri 24(1)

Chapter 9
Mutation Breeding: A Novel Technique for Genetic Improvement of Pulse Crops Particularly Chickpea (*Cicer arietinum* L.)

Mohd Rafiq Wani, Mohd Imran Kozgar, Nasya Tomlekova, Samiullah Khan, Alvina Gul Kazi, Subzar Ahmad Sheikh and Parvaiz Ahmad

Abstract Breeding of pulses, especially chickpea, by exploiting genetic diversity using conventional methods has been practiced in the past. Nevertheless, these methods at present are inadequate for making any significant breakthrough to handle the world's ever-increasing food demand. In this bizarre scenario, induced mutations have emerged as big relief, and are largely exploited for developing improved high-yielding crop varieties and for discovering desired genes that control important agronomical traits. Gene mutation, leading to the quality advancement of well-adapted existing varieties, has been the pedestal for germplasm improvement.

M. R. Wani (✉) · S. A. Sheikh
Department of Botany, Government Degree College (Boys), Anantnag 192102,
Jammu and Kashmir, India
e-mail: botanyrafiq@gmail.com;
sheikhsubzar00@gmail.com

M. I. Kozgar
KERB Biotech, Bemina, Srinagar 190018, Jammu and Kashmir, India
e-mail: m.i.kozgar@gmail.com

N. Tomlekova
Molecular Biology Laboratory, Department of Breeding,
Variety Maintenance and Introduction, Maritsa Vegetable Crops Research Institute,
Plovdiv, Bulgaria
e-mail: nasia.tomlekova@gmail.com

S. Khan
Mutation Breeding Laboratory, Department of Botany, Aligarh Muslim University,
Aligarh, Uttar Pradesh 202002, India
e-mail: khan_drsami@yahoo.co.in

A. G. Kazi
Atta Ur Rehman School of Applied Biosciences, National University of Sciences and
Technology, Islamabad, Pakistan
e-mail: alvina_gul@yahoo.com

P. Ahmad
Department of Botany, S.P. College, Srinagar 190001,
Jammu and Kashmir, India
e-mail: parvaizbot@yahoo.com

P. Ahmad et al. (eds.), *Improvement of Crops in the Era of Climatic Changes*, Volume 2,
DOI 10.1007/978-1-4614-8824-8_9,

Pulses are more prone to biotic and abiotic stresses as compared to cereals. As a result, there is a serious need to develop such varieties having high resistance to the above-mentioned stresses. During the past eight and a half decades, around 3,139 improved crop varieties have been released globally derived either as direct mutants or from their progenies (http://mvgs.iaea.org). Vast numbers of these varieties like cereals, pulses, oil crops, root and tuber crops and ornamentals have been released in developing countries for cultivation, including India, resulting in massive economic impact. Lately, mutagenesis has received an immense impel for its use in a newfangled promising technique known as targeting induced local lesions in genomes. With the unfolding of novel biological fields such as genomics, functional genomics, bioinformatics and the emergence of other technologies based on these sciences, there has been an increased surge in induced mutations within the scientific community. The knowledge of functional and basic genetics of model legume crops will benefit chickpea breeders to comprehend that marker-assisted selection has great potential to develop biotic and abiotic stress-resistant varieties. The basic understanding of genes, which direct major agronomical traits, is essential for plant breeders to frame apposite approaches and execute them in breeding programmes for promising results. In this era, with growing human population, hunger ghosts are haunting millions of people all around. Under these circumstances, the salvaging step lies in tailoring better crop varieties embedded with superior proteins, minerals and high yield. Mutagenic agents, physical as well as chemical, are used to induce mutations and generate variations from which desired mutants may be selected. However, basic information vis-à-vis effectiveness of various mutagens and their possible role in generating polygenic variability is meagre among pulses in general and chickpea in particular. Hence, the present review condenses various facets of contemporary knowledge for pulse crop varietal improvement, particularly chickpea, through induced mutagenesis with special thrust on qualitative as well as yield-attributing traits.

Keywords Mutagenesis · Morphological mutations · Quantitative traits · Proteins and minerals · Anti-nutritional factors · Chickpea and pulses

1 Introduction

Chickpea or Bengal gram comprises 43 taxa, of which 9 are annual (including the cultivated one, *Cicer arietinum* L.), 33 perennial and 1 unspecified (Van der Maesen 1987; Goyal et al. 2011). The cultivated chickpeas are categorized into two distinct groups namely *microsperma* or Desi and *macrosperma* or Kabuli (Van der Maesen 1972; Moreno and Cubero 1978; Toker et al. 2011). The Kabuli-type chickpea has large seeds with salmon white testa, while the seed of Desi-type chickpea is less than half the size of Kabuli type and has a thicker seed coat that is brown to yellow in colour. The Desi type is genetically diverse as compared to Kabuli-type chickpea (Moreno and Cubero 1978) and has been clearly separated by molecular diversity analyses (Iruela et al. 2002). The Kabuli type has been grown traditionally in the Mediterranean basin and Central Asia, while the Desi type has been mainly produced in the Indian subcontinent, East Africa and Central Asia. Polymorphism has been reported between accessions from *Cicer arietinum* and the wild species *Cicer reticulatum* (Udupa et al. 1993). Contrary to cultivated chickpea, wild *Cicer* species have been shown to be resistant for different biotic and abiotic stresses (Abbo et al. 2007; Sharma et al. 2006; Canci and Toker 2009). Natural mutation and selections from Desi-type chickpea have resulted in the development of Kabuli-type varieties (Moreno and Cubero 1978; Salimath et al. 1984; Jana and Singh 1993). Chickpea is one of the main grain legumes of the Indian subcontinent and forms a major source of dietary protein for the predominantly vegetarian population of India (Fig. 9.1). Its protein quality is better than other legume crops such as pigeon pea, black gram and green gram (Kaur and Singh 2005).

2 Botanical Descriptions, Biosystematics and Cytogenetics

Genus *Cicer* belongs to the family Papilionaceae (Fabaceae). The plant has a deep taproot provided with nodules. The stem is mostly erect and green. Leaves are stipulate and imparipinnately compound, usually small, leaflets in each leaf which are arranged on a rachis with a small petiole. All external surfaces of the plant, with the exception of corolla, are covered by glandular hairs. Flowers are pedicellate, bisexual with papilionaceous corolla and borne singly in axillary racemes with ten stamens (9+1). The ovary is monocarpellary, unilocular, 1–2 ovules and superior with a terminal slightly bent style and blunt stigma. Pistil and anthers usually remain inside the keel. Pollination takes place before the opening of the bud. Thus, self-pollination is the rule (Auckland and Van der Maesen 1980; Singh and Ibrahim 1990). The fruit is an inflated pod with 1–2 seeds. The seed surface may be wrinkled or smooth. Germination is hypogeal (Khan 2011).

The *Cicer* belongs to tribe Cicereae (Iruela et al. 2002; Parveen 2006; Goyal et al. 2011). However, some researchers argue that it belongs to tribe Vicieae (Singh et al. 1997). Chickpea (*Cicer arietinum* L.) is a diploid legume crop known to have the somatic chromosome number of $2n=16$ (Mercy et al. 1974; Ahmad and Godward 1980; Singh and Singh 1997) with genome size 1C=740 Mbp (million base pair). There are also reports of $2n=14$ chromosome number, but such plants of *Cicer* are very rare and may not be able to maintain themselves in nature (Singh et al. 1997). The wild species of *Cicer* have largely remained unutilized due to crossability barrier, and most studies on wild *Cicer* species have focussed on annual species because of various difficulties associated with propagation of perennial species (Gowda and Gaur 2004).

Fig. 9.1 Chickpea seeds and pod bearing branch of chickpea. (Source: science.howstuffworks.com)

3 Economic Importance and Basic Research Constraints

Pulses hold a demanding place among the Indian agricultural system because of their precious qualities. They play a fundamental role in overcoming the protein malnutrition in developing countries like India, where a majority of the population is vegetarian. The amino acid composition of pulse proteins is such that a mixed diet of cereal and pulse has a superior biological value than either of the foods alone. The indications about positive correlation of atherosclerosis with diet rich in saturated fatty acids and reported decrease in blood cholesterol level with inclusion of pulses have led to a growing realization of substituting a part of animal protein by vegetable protein (Chandra and Ali 1986). Pulses are also important because they perform relatively well even under poor management and low soil fertility. They have the ability to fix atmospheric nitrogen in the soil through symbiosis with rhizobia. Pulses offer the most important means of increasing agricultural production in dry land because of their low water requirement, which constitute about 40 % of the total cropped area of our country. Despite having high economic importance, its yield did not witness much appreciation during the recent past. It is argued that one of the main reasons of failure to accomplish a real productivity breakthrough in chickpea is the lack of genetic variability. India has witnessed a decreasing trend in per capita availability of pulses in recent years. The problem of dwindling per capita availability can be unravelled from rapid enhancement in production levels by fully exploiting the yield potential of existing varieties through genetic manipulation.

The data for 2011–2012 indicate that pulses occupy an area of 10.84 million hectares (first advance estimate released on 14.09.2011 by Directorate of Economics and Statistics, Department of Agriculture and Cooperation, Ministry of Agriculture, Government of India) and produce 18.24 million tonnes with an average yield of 694 kg/ha (Table 9.1). In India, chickpea was grown over an area of 8.31 million hectares with the production of 7.58 million tonnes in 2011–2012 (Table 9.1). The average yield of 912 kg/ha is less and is not adequate to meet the increasing demand. For breaking yield plateau in chickpea, the development of high-yielding varieties coupled with appropriate growth habit is essential. Low productivity can be ascribed to narrow genetic base, low-yielding potential of the varieties and their susceptibility to diseases. In India, chickpea is mainly grown in drier areas as they are best suited for its production. The main chickpea-producing states in India are Madhya Pradesh, Uttar Pradesh, Rajasthan, Maharashtra and Andhra Pradesh (Goyal et al. 2011). With the instigation of All India Pulse Improvement Project in cooperation with various premier agricultural institutes, an extensive multidisciplinary approach for the advancement of pulse crops in India started in 1965; nevertheless, the focal emphasis was primarily on the production aspect. The key constraints of pulses production in the country are the presence of genotypes with undesirable characters like low yield, low harvest index, indeterminate growth habit and staggered flowering (Kumar et al. 2011). Since pulse crops have been cultivated under rainfed conditions in marginal lands with low input management, the genotypes of

Table 9.1 All India area, production and yield of total pulses and chickpea. (Source: Directorate of Economics and Statistics, Department of Agriculture and Cooperation, Ministry of Agriculture, Government of India)

Periods	Area (million ha)		Production (million tonnes)		Yield (kg.ha-1)	
From sowing to harvesting	Total pulses	Chickpea	Total pulses	Chickpea	Total pulses	Chickpea
2000–2001	20.35	5.19	11.08	3.86	544	744
2001–2002	22.01	6.42	13.37	5.47	607	853
2002–2003	20.50	5.91	11.13	4.24	543	717
2003–2004	23.46	7.05	14.91	5.72	635	811
2004–2005	22.76	6.71	13.13	5.47	577	815
2005–2006	22.39	6.93	13.39	5.60	598	808
2006–2007	23.19	7.49	14.20	6.33	612	845
2007–2008	23.63	7.54	14.76	5.75	625	762
2008–2009	22.09	7.89	14.57	7.06	659	895
2009–2010	23.28	8.17	14.66	7.48	630	915
2010–2011	26.28	9.21	18.09	8.25	689	896
2011–2012	*10.84*[a]	*8.31*	*18.24*	*7.58*	*694*	*912*

[a]First advance estimate only released on 14.09.2011

different pulses have adapted themselves merely for survival rationale, thus fixing their genetic potential at lower yield levels. The pulses, in general, are quite ineffective in fully exploiting the solar and nutrient energy and are also poor in partitioning the photosynthates between vegetative parts and grains. A lot of dry matter goes for production of the stalk. Accordingly, the harvest index is poor. Photosynthetic rate and the activity of photosynthetic enzymes also decline after the commencement of pod setting. The nodules usually crumble; besides, the conversion of carbohydrates and proteins also utilizes some of the energy produced, leading to stumpy productivity. Most of the main pulse crops have an indeterminate growth habit and increased cost of cultivation and fail to do well in farmer's fields in terms of productivity (Kumar et al. 2011). The lower productivity is associated with the risk of crop failure due to several biotic and abiotic stresses. Global climatic change is now a reality and a major challenge for agricultural production systems. No extensive efforts have been made to genetically improve the crops, barring a few efforts to identify the important morphological descriptors and develop advanced breeding lines to meet the region-specific adaptation (Kumar et al. 2011).

The Indian Agricultural Research Institute (IARI) in New Delhi, Bhabha Atomic Research Centre (BARC) in Mumbai, Tamil Nadu Agricultural University (TNAU) in Coimbatore and the National Botanical Research Institute (NBRI) in Lucknow are some of the pioneer research centres in India which are strongly committed to mutation breeding for several crops and have contributed substantially to the development and release of a large number of mutant varieties. The breeding potential of a crop plant lies in its ability to exploit the existing variability through selection or created variability through hybridization or using spontaneous mutation. However, in pulses, the genetic variability has been decreased due to natural selection; hence,

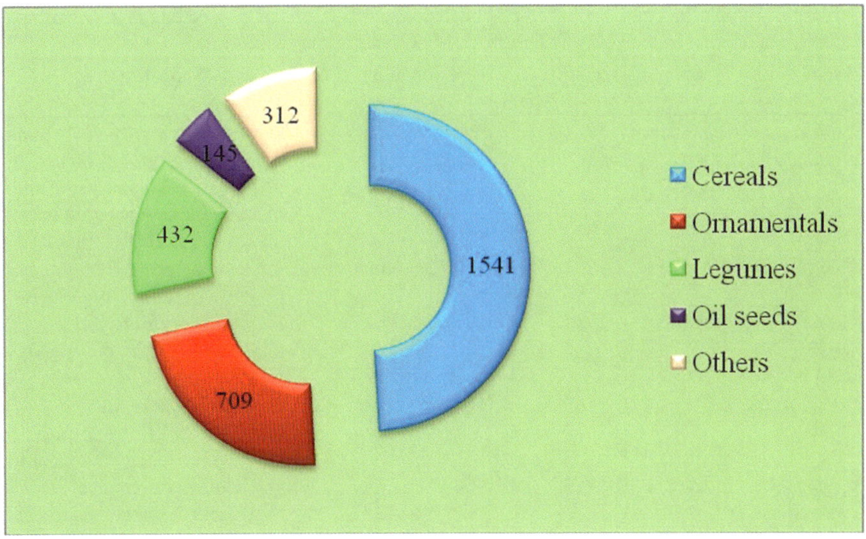

Fig. 9.2 Number of mutant varieties released in the world. (Source: Joint FAO/IAEA, Vienna Mutant Variety Database (MVD); http://mvgs.iaea.org)

conventional breeding methods are not much fruitful. Induced variability, which can be most useful in breeding programmes, is generally obtained using mutagens that are known to react with particular bases of a DNA molecule. Any agent which can change the base sequence either in resting state or during subsequent DNA metabolism has the potential to produce a mutant or changed organism (Solanki et al. 2011). Interest in induced mutagenesis has been revived in recent years due to the fact that mutant organisms are an indispensable tool for the science of genetics.

During the past 85 years, around 3,139 mutant varieties have been officially registered/released either as direct mutants or from their progenies for commercial cultivation in the world. The major contribution is from cereals (1,541) followed by ornamentals (709). Globally, so far 432 legume varieties have been released and are in commercial cultivation (Fig. 9.2). These are soybean (170), groundnut (72), mungbean (36), chickpea (21), lentil (13), cowpea (12), urdbean (9), pigeon pea (7) and others (92) (http://mvgs.iaea.org). Chickpea is generally grown on marginal lands, and several biotic and abiotic factors such as drought, heat, salinity, cold and insects and diseases constrain its productivity (Toker and Cagirgan 2004). To overcome such stresses restricting yield, genetic variation available in the germplasm collection are used in the development of resistant varieties. An efficient tool to induce such desirable variability is mainly mutagenesis. The details of some improved varieties of chickpea developed through induced mutations are given in Table 9.2. Though the work on mutations started long back, the first mutation-based variety M 699 (Hyprosola) was released in 1981 in Bangladesh which was found to mature early and be high yielding. In India, the four high-yielding, *Ascochyta* blight-resistant and wilt disease-resistant chickpea mutant varieties, viz. Pusa 408

Table 9.2 Details of chickpea varieties developed through mutation breeding. (Sources: Joint FAO/IAEA, Vienna Mutant Variety Database, MVD; http://mvgs.iaea.org)

Mutant variety name	Country	Year of registration	Developed by	Main improved attributes
Hyprosola (M-699)	Bangladesh	1981	200-Gy gamma rays	10 days early maturity, more pods, higher harvest index, higher planting density and 19 % higher yield compared to initial variety
CM72	Pakistan	1983	150-Gy gamma rays	Resistance to chickpea blight (*Ascochyta rabiei*), high yield
Kiran (RSG-2)	India	1984	Fast neutrons	Erect plant type, more pods, high yield, early maturity, salt tolerance
Pusa 408 (Ajay)	India	1985	600-Gy gamma rays on seeds	High yield, blight resistance, semi-erect, 140–155 days to maturity and plant architecture
Pusa 413 (Atul)	India	1985	600-Gy gamma rays on seeds	High yield, wilt resistance, moderate blight resistance, resistance to stunt virus, foot rot, root rot, semi-erect, higher branching number, more than two grains/pod, 130–140 days to maturity
Pusa 417 (Girnar)	India	1985	600-Gy gamma rays on seeds	High yield, short, semi-erect, profusely branched, high pod number, 110–130 days to maturity, wilt resistance, moderate resistance to stunt virus, collar root, foot rot, root rot, low pod borer and nematode damage
NIFA-88 (CM-1918)	Pakistan	1990	100-Gy gamma rays on seeds	Moderate resistance to *Ascochyta rabiei*, 2-week earlier maturity, high yield (15–20 %), higher nitrogen fixation than the initial
Line 3	Egypt	1992	50-Gy gamma rays + 0.025 % EMS	High yield
CM-88	Pakistan	1994	100-Gy gamma rays	Resistance to *Ascochyta*, resistance to *Fusarium* and high yield
NIFA-95	Pakistan	1995	200-Gy gamma rays	Resistance to bacterial blight
CM-98	Pakistan	1998	300-Gy gamma rays	Resistance to *Ascochyta* blight and *Fusarium* wilt
Binasola-2[a]	Bangladesh	1998	NA	High yield, maximum 2.2 tons.ha^{-1} (average 1.6 tons.ha^{-1}), 50–60-cm-taller plants, larger seed size than Hyprosola, 120–130 days initial maturity period, suitable for cultivation in dry areas particularly in Barind tract

Table 9.2 (continued)

Mutant variety name	Country	Year of registration	Developed by	Main improved attributes
CM 2000	Pakistan	2000	150-Gy gamma rays	High yield and resistance to diseases
Hassan-2K	Pakistan	2000	450-Gy gamma rays on seeds	High yield, resistance to blight and wilt, higher protein content (24 %)
Binasola-3	Bangladesh	2001	200-Gy gamma rays on dry seeds and selections in later generation	Early maturity, erect plant type, rough seed coat and larger seed size than exotic genotype G-97 (initial)
Binasola-4	Bangladesh	2001	Hybridization with one ICRISAT line K-850 and mutant variety Hyprosola	Higher seed yield than parents, medium seed size, bright seed coat colour
Borina[b]	Bulgaria	2004	150-Gy gamma rays	Morphological mutations
Strandja[b]	Bulgaria	2004	50-Gy gamma rays + 0.05 % EMS	Morphological mutations
BGM 547	India	2005	Gamma rays	High yield, bold grain size, attractive golden brown colour, moderate resistance to wilt, root rot, stunt and *Helicoverpa armigera*
Pusa 547	India	2006	600-Gy gamma rays	High yield, good cooking quality, tolerance to *Fusarium* wilt, stunt virus and root rot
TAEK-SAGEL	Turkey	2006	150-Gy gamma rays	early maturity (95–100 days), higher yield capacity (180–220 kg.da^{-1}), higher seed protein (22–25 %), higher first pot height (20–25 cm), 100-seed weight (42–48 g), cooking time (35–40 min.), *Ascochyta* blight (*Ascochyta rabiei*) resistance and better quality than the control variety
THAL-2006	Pakistan	2006	Hybridization with mutant line CM82/87	Tolerance to blight, tolerance to moisture stress and bold seed size
CM-2008	Pakistan	2008	0.2 % EMS	Seed size, resistance to wilt and high yield

[a]Bangladesh Institute of Nuclear Agriculture, BINA; http://www.bina.gov.bd/
[b]Tomlekova 2010

(Ajay), Pusa 413 (Atul), Pusa 417 (Girnar) and Pusa 547, developed at Indian Agriculture Research Institute, New Delhi, and released by the Indian government for commercial cultivation, are the first examples of direct use of induced mutagenesis in chickpea. The mutant variety Pusa 547, released in 2006 for farmer's cultivation, has high yield, attractive bold seeds, thin testa, good cooking quality, high tolerance to wilt and root rot complex and performs well as a late-sown crop (Kharkwal and Shu 2009).

4 Food Value

Chickpea is valued for its highly nutritive seeds that have protein content ranging from 19% to 20%. The protein quality of chickpea is considered to be better than other pulses. Additional nutritional attributes of chickpea include 60.65% carbohydrate, 6.04% fat, 17.4% total dietary fibre, 10.7% total sugar and 2.48% of ash (Table 9.3). The mineral profile consists of potassium (875 mg/100 g), phosphorous (366 mg/100 g), magnesium (115 mg/100 g), calcium (105 mg/100 g), sodium (24 mg/100 mg), iron (6.24 mg/100 g), zinc (3.43 mg/100 g) and manganese (2.204 mg/100 g). Chickpea is primarily used for human consumption and only a small proportion is used for livestock feed. Since chickpeas are high in fibre and protein and have a low glycemic index, they help in controlling human body weight. Including approximately 175 mL of chickpeas in daily diet can help in lowering low-density lipoprotein cholesterol levels, which reduces the risk of fatal cardiac diseases. Chickpea seeds are eaten fresh as green vegetables, fried, roasted, broiled in snack foods and condiments, and their flour is used in soups/salads and to make bread (Parveen 2006). High-protein chickpea flour is gluten free and can also be used to make pakoras or can even serve as a substitute for eggs in baked products. Chickpea is also known for its use in herbal medicine and cosmetics and is an excellent source of important vitamins such as riboflavin, niacin, thiamin, folate and vitamin A precursor β-carotene. Like other pulses, chickpea seeds also contain anti-nutritional factors which can be reduced or eliminated by various cooking techniques. Moreover, chickpea has numerous prospective health paybacks and, in amalgamation with other pulses and cereals, could have beneficial effects on certain worrisome human diseases such as CVD, type 2 diabetes, digestive disorders and some cancers. By and large, chickpea is an important pulse crop with a diverse array of potential nutritional and health benefits (Jukanti et al. 2012).

5 Constraints for Pulses Production

Leguminous crop species cultivated for grain production and well suited for monoculture, viz. poor lodging resistance, possess nitrogen-fixing ability, which is considered an advantage during natural evolution under prevailing inter-species

Table 9.3 Composition of mature raw seeds of chickpea valuable for daily diet. (Source: NDB No.16056, nutrient values and weights are for edible portion, USDA National Nutrient Database for Standard Reference, Release 21, 2008)

Nutrient substances	Units	Value per 100 g	Non-nutrient/bioactive substances	Units	Value per 100 g
Energy	kcal	364	Water	g	11.53
Protein	g	19.30	Ash	g	2.48
Total lipid (fat)	g	6.04	Fibre, total dietary	g	17.4
Carbohydrate	g	60.65	*Minerals*		
Sugar, total	g	10.70	Calcium (Ca)	mg	105
Lipids			Iron (Fe)	mg	6.24
Fatty acids, total saturated	g	0.626	Magnesium (Mg)	mg	115
Fatty acids, total monosaturated	g	1.358	Phosphorus (P)	mg	366
Fatty acids, total polyunsaturated	g	2.694	Potassium (K)	mg	875
Phytosterols	mg	35	Sodium (Na)	mg	24
Amino acids			Zinc (Zn)	mg	3.43
Tryptophan	g	0.185	Copper (Cu)	mg	0.847
Threonine	g	0.716	Manganese (Mn)	mg	2.204
Isoleucine	g	0.828	Selenium (Se)	mkg	8.2
Leucine	g	1.374			
Lysine	g	1.291	*Vitamins*		
Methionine	g	0.253	Vitamin C	mg	4.0
Cysteine	g	0.259	Thiamine	mg	0.477
Phenylalanine	g	1.034	Riboflavin	mg	0.212
Tyrosine	g	0.479	Niacin	mg	1.541
Valine	g	0.809	Pantothenic acid	mg	1.588
Arginine	g	1.819	Vitamin B-6	mg	0.535
Histidine	g	0.531	Folate, total	mkg	557
Alanine	g	0.828	Folate, food	mkg	557
Aspartic acid	g	2.270	Choline, total	mkg	95.2
Glutamic acid	g	3.375	Carotene, beta	mkg	40
Glycine	g	0.803	Vitamin E (alpha-tocopherol)	mg	0.82
Proline	g	0.797	Vitamin K (phylloquinone)	mkg	9.0
Serine	g	0.973			

competition. During the past two decades, significant breakthroughs have not been achieved in improving the yield of pulses; hence, they have been pushed largely to marginal areas (Singh et al. 2011). The enhancement and sustainability of the production of pulses is of paramount importance for meeting food needs in the future. In recent years, global climate change is becoming a major concern worldwide as it has affected the productivity of number of crops adversely. The associated effects of high temperature results in terminal moisture stress in rabi pulses and early onset of flowers or pods in kharif pulses. It has been observed that instead of normal flowering, most of the genotypes of pulses have shown early flowering and pod development, but poor pod filling resulted in low yields. Pulses are more prone to biotic and abiotic stresses as compared to cereals. The other constraints limiting higher productivity of pulses are low genetic potential, low and unstable yield and narrow genetic base. Hence, there is a pressing need to develop varieties having resistance to biotic and abiotic stresses. A major constraint to abiotic factors is lack of understanding of complex genetic basis and the difficulty in efficiently combining favourable alleles into desirable agronomic base, which has led to limited success until now. Consequently, there is a persistent need to not only dissect physical basis of stress environments but also dissect the genetic tolerance into its components. Germplasm conferring resistance on insect pests or diseases has not been easy to find in crop gene pools, or, even if available, it is technically difficult to access recent developments in characterization against various stresses. Techniques for efficient screening of germplasm, marker-assisted selection (MAS) and mutation breeding have offered the possibility to breed high-yielding varieties with tolerance to such stresses (Singh et al. 2011).

6 Mutagenesis

Hugo de Vries in 1900 used the term "mutation" to describe phenotypic changes that are inheritable. Thus, a mutation can briefly be defined as a sudden heritable change in the DNA of a living cell, not caused by the common phenomenon of genetic segregation or recombination. Mutations are the ultimate sources of genetic variation. They provide the raw material upon which other factors of evolution act, and therefore all new species ultimately arise from mutation. In the last decade, mutation breeding is no longer used only as a tool for crop improvement of traditional traits, e.g. yield, nutritional quality, resistance to diseases and pests, but more frequently for diversified uses of crop end products (Shu 2005). Once gene(s) for resistance to a particular disease or stress cannot be found in the available gene pool, plant breeders have no obvious choice, but to attempt mutation induction. Mutagenesis is predominantly carried out in autogamous diploid or allopolyploid crops and ornamentals, but to some extent also in allogamous crops in which, however, more methodological difficulties have to be conquered. In those crops, which cannot be used in cross-breeding because of their sterility, the induction of mutations is the only way to increase the genetic variability of the species within a short period of time. The main handicaps of mutation breeding are the negative selection value and the pleiotropic action of mutant genes selected. If a mutant shows a positive character which might be of interest for plant breeding, it shows very often, one or several negative traits making it useless for breeding purposes. Genetic variation among genotypes and relationship between major yield-contributing traits are of vital importance to a breeding programme, which aims to produce important varieties in pulse crops such as chickpea. Earlier, this programme primarily involved various options for creating genetic variability among pulses, the major ones being the introgression of useful genes from related wild species, development of ideal plant type, exploitation of hybrid vigour through cytoplasmic male sterility (CMS)-based hybrids and incorporation of multiple and broad-based resistance against the key stresses.

The discovery of X-rays by Wilhelm Roentgen in 1895 led to the application of X-rays for inducing mutations in *Drosophila melanogaster* by Muller (1927) and in barley by Stadler (1928). This technique later became the most important tool for locating genes on chromosomes, studying gene structure, gene expression and regulation and for exploring genomes (Solanki et al. 2011). *Nicotiana tabacum* was the first crop in which the first commercial mutant variety "chlorine type" was induced. Later, with the development of international coordination and availability of financial assistance by FAO/IAEA to research workers/organizations, systematic work on mutation breeding started (van Harten 1998), and now plant breeders are being encouraged to employ mutation breeding as one of the "peaceful" uses of atomic energy (Solanki et al. 2011). A chemical such as mustard gas was found to be highly mutagenic (Auerbach and Robson 1946) and since then a number of chemical agents have been discovered that can increase the frequency of artificially induced mutations. Despite the availability of a range of mutagenic agents, it is still difficult to direct the induction of mutation with the desired expression of characters. In general, ionizing radiations such as X-rays and gamma rays are preferred because of their easy application, good penetration, reproducibility, high mutation frequency and less disposal problems. Among the chemical mutagens, the most commonly used are ethylmethane sulphonate (EMS), diethyl sulphate (DES), *N*-Nitroso-*N*-methylurea (NMU), *N*-Nitroso-*N*-ethylurea (NEU) and ethylene imine (EI) all of which belong to a special class of alkylating agents. All these chemicals react with DNA by alkylating the phosphate groups and also the purine and pyrimidine bases. The dose of a mutagen applied is an important consideration in any mutagenesis programme. The lethal dose-50 (LD_{50}) gives an idea about the optimum dose of the mutagens. Optimum dose produces the maximum mutations with minimum hazards. An overdose of mutagens will kill too many plants, while low dose will produce too less mutation spectrum and frequency. The mutagenic dose mainly depends upon the concentration, duration of treatment and temperature during the treatment. Pre-soaking, pH of solution, metallic ions, carrier agents, post washing, post drying and storage of seeds are the modifying factors for the mutagenic effect (Solanki et al. 2011). The physical and chemical mutagens cause three types of effects, i.e. physiological damage, gene mutations and chromosomal aberrations (Swaminathan 1968). Gene and chromosomal mutations may be transferred from M_1 to succeeding generations; however, physiological effects are generally restricted to the M_1 generation.

Mutation induction with radiation was the most frequently used method to develop direct mutant varieties, as improvement by acclimatization, selection and hybridization have proved to be time consuming and laborious with limited genetic variation (Yaqoob and Rashid 2001). Gene mutations occur spontaneously as errors during DNA replication. Mostly these errors are repaired; however, some may pass to the subsequent cell division and establish in plant offsprings. Artificial induction of mutations by ionizing radiation dates back to the beginning of the twentieth century. But it took many years to establish that such changes could be beneficial for plant breeding. Hence, crop improvement using classical mutagenesis is now well standardized and as a result, new methods of radiation treatment, as well as chemical agents with mutagenic properties are serving as invaluable tools for augmentation of the genetic variation in crops to circumvent bottleneck conditions. An extensive programme on mutation breeding in chickpea was initiated by Kharkwal (1979) and since then, several mutants with agronomically useful traits have been induced in chickpea (Toker et al. 2005). These include mutants for improved yield and proteins, early maturity, root nodulation, erect plant type, determinate growth, compact growth habit and resistance to *Ascochyta* blight, *Fusarium* wilt, root rots, nematodes, etc. The progress in breeding for resistance to insect pests in pulse crops has been limited mainly due to nonavailability of sources with high level of resistance and the lack of effective screening techniques (Gaur and Chaturvedi 2004). *Ascochyta* blight is the most widespread and economically destructive disease of chickpea. Consequently, mutants have been induced in chickpea having resistance to *Ascochyta* blight (Omar and Singh 1995), *Fusarium* wilt (Bhatnagar et al. 1979), pod borer (Shaikh 1983) and nematodes (Bhatnagar et al. 1985). At least 19 chickpea varieties have been developed in Asia through induced mutation breeding, which include nine varieties (CM-72, NIFA-88, CM-88, NIFA-95, CM-98, CM 2000, Hassan-2K, THAL-2006 and CM-2008) released in Pakistan, six varieties (RSG-2, Pusa 408, Pusa 413, Pusa 417, BGM 547 and Pusa 547) released in India and four varieties namely M-699, Binasola-2, Binasola-3 and Binasola-4, released in Bangladesh (Table 9.2).

Before starting the mutation breeding programmes, the following should be taken into consideration: (1) Dominant mutations occur at very low frequencies and can be selected in M_1 (Micke and Donini 1993). Mutations are predominantly recessive and can be selected in the second generation (Salimath et al. 2007; Toker et al. 2007; Toker et al. 2011). Selection for polygenic traits should be initiated at individual plant progenies in M_3 (Micke and Donini 1993; van Harten 1998; Toker et al. 2011). (2) Mutations are beneficial with very low frequencies, whilst mutagenic treatments by and large reduce germination, growth rate, vigour, pollen and ovule fertility in living organisms (Toker et al. 2011). (3) Mutations are randomly induced and they might occur in any gene(s). However, some gene(s) can be more induced than others (Toker et al. 2011). (4) Mutations can be recurrent. The same gene(s) in a crop plant species may be induced again. (5) Mutations have generally pleiotropic effects due to closely linked gene(s) (Singh 2005; Salimath et al. 2007; Toker et al. 2007; Toker et al. 2011). The choice of a parent material is crucial to achieve well-defined objectives through mutation breeding. The purpose of mutation breeding is to improve the locally adapted varieties by creating new alleles not available in germplasm collections (Solanki et al. 2011). The variety selected for mutagenesis should be one of the best varieties released recently. Since genotypes give differential response to mutagens, two or more varieties must be taken for mutagenesis. It is always beneficial to select a well-adapted high-yielding variety for improving one or two specific traits (Anonymous 1977).

All plant parts can be treated with a particular type of mutagen. The common practice is to treat seeds, but in vegetatively propagated crops, whole plants, dormant cuttings, bulbs, tubers and corns can also be treated. Soaking of seeds in water before treatment gives good results. The seed is the most commonly treated biological material because it can tolerate physical conditions which generally are tolerated only by nonliving molecules (Solanki et al. 2011). Through mutation breeding, a lot of variability can be created in a crop species which subsequently can be used for germplasm enhancement as well as for developing new varieties. Another major advantage of mutation breeding is the time required for the development of a new variety as compared to hybridization. Generally, 11–12 years are needed to develop a new variety through hybridization, whereas it takes only 8–9 years in case of mutation breeding (Brock 1977). Mutations may induce both qualitative as well as quantitative variation comparatively in a shorter period of time by altering alleles at known loci as well as at previously unknown loci, besides altering linkage groups (Konzak et al. 1977). With the newfangled impulse in plant mutation research from basic mutational studies to modern reverse genetics, breeders at present are able to exploit mutation techniques more sophisticatedly than ever before dreamed achievable.

7 Chlorophyll Mutations, Effectiveness and Efficiency of Mutagens

Chlorophyll mutations, in general, are considered as the measure to access the effectiveness of treatments of various mutagens. Gustafsson (1940) grouped chlorophyll mutations into albina, xantha, viridis, chlorina, striata, tigrina and maculata classes. Numerous authors have so for reported the incidence of different types of chlorophyll mutations such as albina, xantha, chlorina, viridis, virescent, tigrina, etc. in M_2 generation following mutagenic treatments (Waghmare 2001; Toker and Cagirgan 2004; Yamaguchi et al. 2006; Khan and Tyagi 2009; Arulbalachandran and Mullainathan 2009; Wani et al. 2011b; Girija and Dhanavel 2013a). Chemical mutagens induce higher frequency of chlorophyll mutations as compared to radiations in chickpea (Kharkwal 1998b) and grasspea (Waghmare and Mehra 2001). Jurgan et al. (1985) irradiated the dry seeds of winged bean by different doses of gamma rays and detected a chimaeric plant with lighter green and variegated leaves in mericlinal form in M_1 population. The frequency of chlorophyll variegation reduced in subsequent generations and the mutant tissues in leaves were eliminated. Toker and Cagirgan (2004) noticed different chlorophyll-deficient mutants (viridis, xantha, albino) in chickpea and concluded that the frequency of chlorophyll mutations in M_1 corresponded to the occurrence of morphological mutants in M_2. Effective doses of mutagen varied from genotype to genotype. Klu (2000) irradiated the dry seeds of winged bean with 150 and 250 Gy of gamma rays to induce mutations. High degree of chimerism was detected in M_2 which was segregated. Chlorophyll mutations were detected from M_1 progeny and their frequency was decreased in M_2 population. Khan et al. (2005b) observed five different types of chlorophyll mutations, viz.

albina, chlorina, tigrina, viridis and xantha in chickpea when treated with EMS, SA (sodium azide) and hydrazine hydrate. They concluded that lower and moderate concentrations of EMS gave the higher frequency of chlorophyll mutations, while no such trend was noticed with the other two mutagens. Singh and Singh (2007) reported four different types of chlorophyll mutations, i.e. albina, xantha, chlorina and viridis in mungbean. They reported that the frequency of chlorophyll mutations was higher in the population treated with EMS. Dadke and Kothekar (2005) reported a chlorophyll mutant (xantha) after gamma irradiation. Interestingly, the crude protein and extractable protein contents of the said mutant were found to be higher than the control. They incorporated this xantha mutant in an inter-mutant hybridization programme in order to know its genetics, and it was found to be recessive in nature, true breeding and segregating according to the Mendelian principle.

The worth of a mutagen depends on both its effectiveness as well as efficiency. Mutagenic effectiveness defines mutagen dose to the mutational events, while mutagenic efficiency is the production of desirable changes that are free from associations with undesirable genetic alterations. This is generally measured by the proportion of the mutation frequency in relation to damages associated to mutagenic treatments such as height reduction, chromosome breakages, sterility, lethality, etc. (Sikder et al. 2013). The use of mutagens in crop improvement helps to understand the mechanism of mutation induction and quantify the frequency as well as the pattern of changes in different selected plants by mutagens. The ability of these mutagens to enter the cell of living organisms and to interact with DNA produces the general toxic effects. Thus, their effects are mainly due to the direct interaction between the mutagen and the DNA molecules (Al-Qurainy and Khan 2009).

Kaul (1989) put in plain words that the efficiency of each mutagen is dependent upon the retrieval of types and frequency of mutations induced. As per his findings, the most enviable mutagen is the one that is least damaging and a high mutation yielder. The response of the biological system to physical, in addition to chemical mutagens, is influenced to a varying degree by numerous biological, environmental and chemical factors. These factors amend the effectiveness and efficiency of different mutagens greatly (Savin et al. 1968; Nilan et al. 1973; Kodym and Afza 2003). Kharkwal (1998a) while studying the comparative effectiveness and efficiency of gamma rays, fast neutrons, NMU and EMS on three chickpea varieties (two Desi and one Kabuli) established that NMU was the most potent mutagen, EMS the least efficient, while gamma rays showed least effectiveness. Mutagenic effectiveness and efficiency were found to be increased at lower doses of the mutagens. Chemical mutagens, on the whole, have been found to be more efficient than gamma rays (Kharkwal 1998a). The values of efficiency provide an idea of the extent and type of damage caused by the mutagen in question. Sharma and Sharma (1979) and Gautam et al. (1992) have reported that mutagenic efficiency increased with increase in dose of the mutagens, but Khan and Siddiqui (1993) in mungbean and Khan (1999) in black gram reported higher mutagenic efficiency at lower doses. At lower concentrations, the higher efficiency of mutagenic agents is presumably due to the fact that biological damage, viz. seedling injury and pollen sterility, increases at a faster rate with increasing mutagenic doses (Konzak et al. 1965).

8 Mutations Affecting Morphology of Crop Plants

Phenotypically, mutations are classified into two groups by Gaul (1964). The first of these is macro-mutations or large mutations. These are easily noticeable and morphologically distinct at the individual plant level and are further divided into two groups: (1) trans-specific macro-mutations, which occur very seldom and greatly reduce fertility and (2) intra-specific macro-mutations, which have been the most frequently selected. In macro-mutations, traits that are already known within the limits of the species are usually changed. The second of phenotypic mutations are micro-mutations or small mutations which are divided into two groups: (1) manifest micro-mutations and (2) cryptic micro-mutations. These result in a small effect that can be detected in a group of plants either by eye inspection or by statistical measurements (Toker et al. 2011). Plant architecture is recognized as one of the important traits for higher grain productivity in legumes. Morphological mutations affecting plant attributes like growth, leaf size and number, habit, plant architecture and floral organization in many cases prove to be promising from the breeder's point of view. In chickpea, morphological mutations have been isolated for leaf shape (Pundir and Reddy 1998; Gaur and Gour 2003; Toker and Cagirgan 2004), flower colour and structure (Khosh-Khui and Niknejad 1971; Knight 1993; Khan et al. 2004b) and size and colour of seeds (Phadnis 1978; Muehlbauer and Singh 1987; Gaur and Gour 2001). Flower colour mutants can be exploited as genetic markers in different breeding experiments (Datta and Sengupta 2002; Atta et al. 2003). Glabrous mutants were observed in an EMS-mutated population of chickpea (Pundir and Reddy 1989). The EMS induces mutants with brachytic growth (compact growth), erect growth habit, thick and sturdy stem, short internodal and interleaflet distances and few tertiary and later-order branches which may be useful in ideotype breeding (Gaur et al. 2008). In chickpea, the spectrum of mutations affecting major genes was found to be high with NMU as compared to EMS and gamma rays (Kharkwal 2000). The mutants obtained were miniature, dwarf, compact and upright resembling the ideotypes that are needed in chickpea varietal development. Dwarf mutants occur widely in different plant species (Khan et al. 2004b). Dwarfness may be due to reduced internode length or internode number or both (Sjodin 1971). With respect to quality parameters, mutagenic treatments with gamma rays increase the degree of softness of seed, thus improving the cooking quality (Graham et al. 2002).

The presence of more than one mutation in a single plant was termed as "multiple mutation" by Sharma (1969). Multiple mutations have been reported earlier by Kharkwal (1999) in chickpea. Toker and Cagirgan (2004) irradiated the seeds of chickpea with different doses of gamma rays and reported the maximum viable mutations at 200 Gy of gamma rays. According to Sharma (1969), agents with higher mutagenic efficiency induce more multiple mutations, and such mutations may accumulate several desirable characters within one plant. The frequency of morphological mutations has been found to increase with increasing doses of the mutagen (Thakur and Sethi 1995). The morphological mutants may be exploited in hybridization programmes to transfer some of the useful traits to other high-yielding varieties of crop plants. Morphological mutants might be the result of pleiotropic effects of mutated genes or chromosomal aberrations or gene mutations. Such mutant types were found to be under the influence of polygenes (Konzak et al. 1969). Khan et al. (2005a) reported a bushy mutant in mungbean after seed treatment with 0.02 % MMS (methylmethane sulphonate). The mutant had profuse branching with short height, which gave it a bush-like manifestation. The yield and yield components were found to have been adversely affected. However, the flowering and maturity period showed no significant difference from that of the parental variety. Charry and Bhalla (1988) isolated sterile mutants in pigeonpea (*Cajanus cajan*) and reported that sterility is governed by a single recessive allele and can be used in the development of composite crosses and in evolutionary breeding methods. A wide range of morphological mutants were identified by Wani et al. (2011a) in two varieties of mungbean. The mutants involved traits affecting plant height, growth habit, seed and pod. The frequency of morphological mutants differed in different mutagenic treatments and varieties. The highest mutation frequency was noticed in the EMS-treated population, whereas the lowest frequency was observed in SA treatments. Var. NM-1 displayed the broader spectrum and frequency of morphological mutations than var. PDM-11. Variation in the colour of seed testa and seed size was observed by Khadke and Kothekar (2011) and Girija et al. (2013) in moth bean and cowpea, respectively. The seed coat colour is affected by various genetic factors like pigmentation factor, pigment complementary and modifying factors (Moh 1971). The variations in seed coats were also reported by Sharma (1969), Rayyon (1995) and Gaikward (2002) in different plant systems.

Klu et al. (1989) studied radiation-induced mutations for improved seed quality in winged bean (*Psophocarpus tetragonolobus*). Veeresh et al. (1995) irradiated the dry seeds of winged bean var. "chimbu" with 10, 15, 20, 25, 30 and 35 kR of gamma rays and reported various types of morphological abnormalities (leaves) in plants after irradiation. Manjaya and Nandanwar (2007) irradiated the dry seeds of soybean var. JS 80–21 with 250 Gy gamma rays to induce genetic variability for early maturity, improved plant architecture and higher seed yield. A large number of mutants affecting morphological characters were identified and characterized. They include dwarf, crinkled leaf, pink flower colour and good pod-bearing mutants. Their four mutants, viz. M-21, M-76, M-91 and M-107 had a good number of pods. In addition, the mutant M-107 exhibited pleiotropic effect with alteration in more than one character having dwarf stature with small leaves, more branches, dark green foliage and late maturity.

9 Induction of Polygenic Variability

Chickpea, being a self-pollinated crop, exhibits natural cross-pollination ranging between 0 and 1 % (Singh 1987). This nature of *Cicer* and its sexual incompatibility with most of the wild genotypes in natural interspecific crosses have been responsible for the limited exploitation of wide hybridization (Saxena and Singh 1987).

Under such scarce natural variability, induced mutagenesis supplements the existing meagre germplasm variability. Gaul (1965) emphasized that micro-mutations are useful in plant breeding for two reasons: (1) They occur more frequently than macro-mutations and (2) they often do not adversely affect the viability as compared to macro-mutations because of minute physiological changes which are less drastic. Several workers have so far reported encouraging results on the induction of useful quantitative variability in chickpea. These include mutants for high yield, high protein content, early maturity, root nodulation, erect plant type, determinate growth, compact growth habit and resistance to *Ascochyta* blight, etc. (Gowda and Gaur 2004; Khan et al. 2004a; Canci et al. 2004; Cagirgan and Toker 2004; Kozgar et al. 2012; Wani et al. 2013). An extensive programme on mutation breeding in chickpea was initiated by Kharkwal (1979) with special reference to yield improvement. He reported nine mutants (six Desi and three Kabuli) which had high and consistent performance up to M_7 generations in multilocational trials.

Haq and Shakoor (1980) irradiated the seeds of *Cicer* with gamma rays and found a mutant which was resistant to blight disease. Khatoon and Bhalla (1986) used magnetic fields in combination with NMU in chickpea. They reported that the combination treatments can be very useful in holding back some of the mutants which would have otherwise been lost due to lethality of higher doses. Large variability has been exploited in the cultivated species for morphological traits (Gowda and Gaur 2004; Khan et al. 2004a; Girija and Dhanavel 2013b). It has been suggested that the high level of phenotypic variability seen for morphological traits could be because of the expression of a limited number of mutant loci, as a single mutation can have marked influence on the plant traits (Gaur and Gour 2003). Few early-maturing mutants were selected from M_3 population of lentil by Khan et al. (2006a). Early-maturing chickpea mutant L-84 was obtained through mutation breeding by Shamsuzzaman et al. (2005). The mutant strain matured significantly earlier by about 2 weeks than its parent variety Binasola-2 and had highest seed yield. Gamma ray-induced early fruiting has been reported by Jugran et al. (1998) in winged bean. In addition to significant early maturity (45 days) in flowering and fruiting, changes in stem and calyx colour from green to violet blue also occurred in the mutant. They concluded that the genetic changes were induced by gamma rays at several independent loci. The flowering and fruiting mutants have practical importance from the breeding point of view (Banerji 2011). A reduction in plant height was reported by Khan et al. (2006b) in EMS and SA-treated population of mungbean. Subba Rao (1988) has attributed the plant growth depression to slow rate of cell division, decreased amylase activity and increased peroxidase activity. The reduction in mean plant height was also reported by Bajaj et al. (1970) in *Phaseolus vulgaris*. On the other hand, Singh et al. (2000) reported an increase in plant height after treatments with gamma rays and EMS in *Vigna mungo*. Use of mutations for obtaining early-maturing varieties has been a frequent breeding objective (Micke 1979). Wang et al. (2003a) in soybean, Shamsuzzaman et al. (2005) and Wani et al. (2013) in chickpea also reported a significant reduction in days to maturity after mutagenic treatments. Wani et al. (2011c) appraised the increasing extent of genetic variability for quantitative traits in M_3 mutants of green gram following mutagenesis with EMS, HZ and

SA. Considerable increase in mean values for fertile branches per plant, pods per plant and total plant yield was discerned among the mutant lines. Contrarily, there are few reports of increasing variance coupled with more or less unaffected mean values after mutagenic treatments (Sharma 1986).

Seed yield in pulses is a complex trait and is influenced by many other quantitative traits like fertile branches per plant, pods per plant, seeds per pod and 100-seed weight (Wani et al. 2012). In *Vigna mungo,* a reduction in mean seed yield per plant in M_2 and M_3 generations has been reported by Singh et al. (2000). Waghmare and Mehra (2000) achieved considerably increased mean seed yield in M_3 after gamma rays and EMS treatments in *Lathyrus sativus.* Bhatia and Swaminathan (1962) concluded that the mean of the irradiated population, where no selection has been applied with regard to specific character under study, tends to go down in comparison to the control. Srivastava et al. (2011) suggested that selection could be applied at interfamily and intrafamily levels to detect some promising mutant families having high mean and coefficients of variation for various characters. In chickpea, different researchers reported augmented genetic variability for different agronomic characters in mutagen-treated populations (Nerker and Mote 1978; Barshile et al. 2009; Wani et al. 2013). Opinions differ regarding the direction of mutations. Gaul (1965) and Aastveit (1966) hold the view that induced polygenic mutations do not follow any particular direction, but occur at random. According to Goud (1967), polygenic mutations always follow a particular direction opposite to the previous history of selection. The response of an unselected character will, however, depend not only on its previous selection history but also on whether it is genetically correlated with the selected character (Brock 1967). Gregory (1965) held another view regarding the problem of shift in mean. According to him, mutations with very small phenotypic effect will occur with high frequency and will have an equal probability of being positive or negative in their effect. He further added that with the removal of mutations of drastic effect, the variance would be reduced in correlated character but largely unaffected in genetically non-correlated characters. The means for both the instances would, however, be approximately equal to the control population. There is a big discrepancy among the opinions of researchers as to which generation is actually appropriate for selection of quantitative traits. Gaul (1964) is of the opinion that selection for quantitative traits should be delayed until M_3 or later generations following mutagenic treatments. But many other workers have contrarily proposed that effective selection for polygenic traits should be done in early generations even in M_2 (Sneep 1977; Shakoor and Haq 1980; Kharkwal 1983; Sheeba et al. 2003).

The estimates of genotypic coefficient of variation and heritability of various quantitative traits are essential (Kaul and Garg 1979; Sakin and Yildirim 2004; Khan and Wani 2005; Bhogal et al. 2013) since they indicate the degree of stability to environmental fluctuations and the potential transmissibility of a trait from generation to generation. Heritability is of interest to the plant breeder as an index of transmissibility (Ibrahim and Sharaan 1974; Bhogal et al. 2013). Wani et al. (2012) reported relatively higher estimated heritability (broad sense) for various yield-attributing traits in M_4 generation of chickpea. Lower heritability for yield has been reported by earlier workers in pigeonpea (Srivastava and Singh 1993). The disparity in results could be because heritability is a property not only of a character but also of the population, environment and the circumstances to which the genotypes are subjected to. Genetic advance confers the degree of stability and genetic progress for a particular trait under apposite selection and as a result carries much impact in self-pollinated crops like chickpea. Heritability together with genetic advance is more reliable in envisaging the effect of selection than the heritability alone (Johnson et al. 1955), because of the fact that heritability estimates are prone to certain estimation errors and genotype–environment interactions (Lin et al. 1979).

In a mutation-breeding programme, there is an insistent need to embark studies on correlation coefficients in a routine manner in addition to the estimation of genetic variability among various indispensable yield and yield-contributing traits (Aastveit and Gaul 1967). The correlation among yield-contributing traits in a population is a composite of the effects of selection, gene linkage and pleiotropy (Sehrawat et al. 1996). The usefulness of mutations in weakening, strengthening or altering character association has been reported earlier (Kaul and Garg 1982; Yadav et al. 2002; Khan and Wani 2005). If the nature of selection practiced in the control and treated population is same, any difference in correlation coefficient between the two populations will be due to the effect of mutagens or altered pleiotropic effects of newly mutated genes. However, according to Gottschalk (1987), climatic factors can also influence a pleiotropic pattern positively or negatively. Such alterations in correlation among various traits may be utilized to enhance the rate of selection response in quantitative traits.

10 Seed Protein and Mineral Contents

Seed protein content is considered to be a complex character of a crop controlled by many genes located on several chromosomes (Frey 1977; Konzak et al. 1978; Coffman and Juliano 1979). The bulk of proteins in legume seeds are comprised of salt-soluble globulins, or storage proteins which are synthesized during seed development, stored in protein bodies and hydrolysed during germination to provide nitrogen and carbon skeletons for developing seedlings (Wang et al. 2003b). Seed protein showed a nonsignificant negative correlation with yield in different mutant lines of mungbean (Khan and Wani 2006). Similar negative correlation between yield and seed protein content has been reported earlier (Bliss et al. 1973; Karjalainen and Kortet 1987; Khan and Wani 2005). Khan and Claydon (1975) determined the role of induced mutations in the improvement of winged bean by isolating such mutants which had a potential new source of proteins. Protein content is influenced by the interaction of gene(s) and environmental factor(s) as has been reported in chickpea (Singh and Ibrahim 1990), mungbean (Ignacimuthu and Babu 1989) and pea (Gottschalk and Wolf 1983). Divergent reports are available vis-à-vis success extent of induced mutations for high grain yield coupled with high protein content of the mutants. Few researchers opined that high protein content is intricate to merge with high yield, since the two traits divulge more or less negative correlation (Abo-Hegazi 1980; Khan and Wani 2005; Parveen 2006). However, high-yielding mutants coupled with high protein contents were reported by various workers in the past and in recent times (Misra et al. 1973; Shaikh et al. 1982; Olejniczak 1986; Borah and Goswami 1995; Kharkwal 1998c; Kalia et al. 2000; Naik et al. 2002; Hiremath et al. 2010; Wani et al. 2012).

A good amount of variability with regard to soluble seed protein content was perceptibly noticed in viable mutants of moth bean by Khadke and Kothekar (2011). The highest protein content was detected in dwarf with erect habit mutants, while the lowest protein content of defatted seed powder was observed in tall mutants. Gottschalk and Muller (1970) proposed that the improvement in protein content and composition can be achieved by genetic manipulation, while Narahari and Bhatia (1975) suggested the pivotal role of mutation breeding for improving the quality of proteins. Several attempts in the past have been made to induce variations for protein quality and quantity using macro-mutations and micro-mutations (Gottschalk and Muller 1970; Singh and Chaturvedi 1980). Dadke and Kothekar (2005) made an extensive effort to accomplish the qualitative improvement of winged bean pertaining to its crucial biochemical features like seed proteins and seed trypsin inhibitors (TIs). Excellent variability has been gained in developing and characterizing the low TI-containing lines of winged bean through mutation breeding. Such lines can be released as new varieties for large-scale commercialization (Banerji 2011).

Grain legumes have also been reported to be good sources of essential minerals required by man, though their concentrations may fluctuate owing to both genetic and environmental factors. Mukhopadhyay (2000) studied the molecular evolutionary history of winged bean alpha-chemotrypsin inhibitor and carried out modelling of its high structural analysis. Altered seed mineral profiles have been identified in pea mutants. The degenerative leaves (*dgl*) mutation bestow an unrestrained hyperaccumulation of Fe into vegetative tissues and transport surplus Fe to seeds (Marentes and Grusak 1998). These mutants have been used to realize several aspects of whole-plant Fe homeostasis and the worth of phloem transport in seed Fe deliverance (Grusak 2000). In those areas, where legumes comprise an important constituent of human diet, mineral deficiencies, particularly Fe and Zn, seem to be prevailing. Efforts are needed to understand the mechanism of how minerals move from soil, through the plant and into the developing seeds has gained a great deal of curiosity recently (Grusak 2002). However, biological utilization of inherent nutrients of legumes is limited owing to the presence of various anti-nutritional substances including phytate, which is the main inhibitor of iron and zinc absorption (Sandberg 2002; Tajoddin et al. 2011). The poor bioavailability of iron not only retards the normal brain development in infants but also affects the success rate of pregnancies. Zinc deficiency prevents normal growth of a child and greatly weakens the immune system, resulting in more infections (Ann-Sofic Sandberg 2002).

11 Anti-nutritional Factors

Although legumes are an important source of dietary proteins and starch for humans, their acceptability and utilization have been limited owing to some anti-nutritional substances such as TIs, phytate, tannins, etc. (Sharma et al. 2013). TIs and tannins inhibit the digestibility of proteins and starch, whereas phytic acid reduces the bioavailability of some essential minerals, viz. iron, zinc, etc. (Rehman and Shah 2001). TIs were first reported by Read and Hass (1938). Later on, Bowman (1946) purified them and Kunitz (1945) isolated them in crystalline form. Plant protease inhibitors are quite stable molecules and are often resistant to heat, pH extremes and proteolysis (Ryan 1981). The inhibitors of trypsin and chemotrypsin have been implicated in reducing protein digestibility and in pancreatic hypertrophy (Liener 1976). Protease inhibitors have a significant role in developing pest and pathogen resistance in plants. The TI in its inactivated form has been considered to comprise a good source of sulphur-containing amino acids, in view of a rich content of cysteine amino acid in such proteins (Kakade et al. 1969).

TIs are widespread anti-nutritional substances which block trypsin activity and thereby reduce the digestibility of proteins. Kothekar et al. (1996) studied the effects of gamma irradiation with particular reference to induction of low trypsin and chymotrypsin inhibitor mutants in winged bean. Dhadke and Kothekar (2005) reported low TI content in gamma rays induced two mutants (early maturing and flat pod) of winged bean. The substantial (40–60%) reduction in the antitryptic activity in these mutants clearly indicates the role of induced mutations in enhancing the nutritional quality of seed proteins. Various factors such as amylase inhibitors, phytate, etc. are supposed to affect the starch digestibility in legumes (Yadav and Khetarpaul 1994). Reduced digestibility of starch lowers the glucose release into the blood stream, which is beneficial for diabetic patients (Zia-ul-Haq et al. 2007). However, digestibility of starch can be improved through heat treatments, e.g. cooking, roasting and autoclaving. Tannins are polymeric flavonoids that comprise a small part of the broad and diverse group of phenolic compounds produced by plants as secondary metabolites (Diaz et al. 2010). Tannins have been reported to inhibit digestive enzymes and thereby lower the digestibility of most important nutrients especially proteins and starch (Khattab and Arntfield 2009). Tannins also inhibit the utilization of nutrients through astringency and enzyme inhibition. As phenolics, tannins are water soluble and may be eliminated by thermal and hydrothermal processing treatments. Anti-nutritional factors should not pose a problem to humans if the seeds are correctly processed. Taking into consideration the overall nutrient and proximate composition analysis, the chickpea varieties can become an economic and alternative protein source that could alleviate protein malnutrition in developing countries, including India, in addition of improving the nutritional status of functional foods in developed countries.

12 Conclusions and Future Perspective

Globally, food security has witnessed a major deterioration in the past few years; food costs are mounting brusquely and poor people are threatened with serious malnutrition. With population explosion, the demand for food is enormously on the rise, while natural resources are depleting with every passing day. Erratic rainfalls, impetuous drought conditions, excessive floods, etc., often related to climate change, further exacerbate the miseries by deteriorating the crop production conditions. Under these circumstances, it is imperative that the yield potential of the crop plants has to be significantly increased to combat the aggravating food security situation. Induced mutations have the ability to increase the rate of domestication of many underexploited species of legume crops that may be potentially useful as a source of food, forage and industrial raw materials. It is striking that a huge number of mutant varieties have been developed and widely cultivated in developing countries, hence greatly improving their food security. In recent years, induced mutations, as a tool, have been gaining momentum in the field of plant molecular biology to identify, isolate and study the structure and function of such genes which are actually of imperative use in breeding studies. The fundamental knowledge of genes which control core agronomical and quality traits is vital for plant breeders to frame appropriate strategies and ingeniously implement them in breeding programmes for prosperous results.

The massive advent of plant molecular biology is anticipated to provide a sound solution to further increase the food production. Recently, mutation techniques have also been integrated with other molecular technologies, such as molecular marker or high-throughput mutation-screening techniques, and are becoming more powerful and effective in breeding crop varieties. In future, sequencing of mutated genes and development of molecular markers will receive more attention for pulse improvement. Varieties with improved efficiency vis-à-vis yield, early maturity, uptake of micronutrients, tolerance to abiotic stresses like drought, cold, salinity and resistance to biotic stresses such as disease and insect pests can be developed using mutation breeding and MAS. Therefore, induced mutations will continue to play a significant role for improving world food security in coming years and decades. Since chickpea is a nutritious legume in human diet, its high-yielding mutants coupled with higher protein and mineral contents would assume substantial economic importance. Efforts are on to develop transgenic plants of chickpea which are resistant to gram pod borer (*Helicoverpa armigera*) using *Bt* crystal protein genes. Besides use of *Bt* genes, genes of plant origin like lectins, protease and amylase inhibitors also hold great promise. In addition, the reverse-genetic strategies like targeting induced local lesions in genomes that target lesions to specific genes are anticipated to speed up the process of gene function analysis and the efficiency of mutation breeding for better crop future.

References

Aastveit K (1966) Use of induced barley mutants in a cross breeding programme. Mutations in plant breeding. IAEA, Vienna, pp 7–14

Aastveit K, Gaul H (1967) Variation and selection of micromutants. Rad Bot 1:353–361

Abbo S, Reddenm RJ, Yadav SS (2007) Utilization of wild relatives. In: Yadav SS, Redden B, Chen W, Sharma B (eds) Chickpea breeding and management. CAB International, Wallingford, pp 338–354

Abo-Hegazi AMT (1980) Seed protein and other characters in M_4 generation of chickpea. Indian J Genet 40(1):122–126

Ahmad S, Godward MBE (1980) Cytological studies on the cultivars of *Cicer arietinum* L. from Pakistan. Caryologia 33:55–68
Al-Qurainy F, Khan S (2009) Mutagenic effects of sodium azide and its application in crop improvement. World App Sci J 6(12):1589–1601
Anonymous (1977) Manual on mutation breeding, 2nd edn. IAEA, Tech Rep Ser No. 119, Vienna
Arulbalachandran D, Mullainathan L (2009) Chlorophyll and morphological mutants of blackgram (*Vigna mungo* (L.) Hepper) derived by gamma rays and EMS. J Phytol 1(4): 236–241
Atta BM, Ahsanul HM, Shah TM, Sadiq M, Mahmudul H, Syed H (2003) Induced flower colour mutations in chickpea. Int Chickpea Pigeonpea Newsl 10:6–7
Auckland AK, Van der Maesen LJG (1980) Chickpea. In: Fehr WR, Hadley HH (eds) Hybridization of crop plants. American Society of Agronomy and Crop Science Society of America, Madison, pp 249–259
Auerbach C, Robson JM (1946) Chemical production of mutations. Nature 157:302
Bajaj YPS, Saettler AW, Adams MW (1970) Gamma radiation studies on seeds, seedlings and callus tissue cultures of *Phaseolus vulgaris* L. Rad Bot 10:119–124
Banerji BK (2011) Mutation breeding in winged bean (*Psophocarpus tetragonolobus* (L.) D.C.)-a review. In: Khan S, Kozgar MI (eds) Breeding of pulse crops. Kalyani Publishers, Ludhiana, pp 22–33
Bhatia CR, Swaminathan MS (1962) Induced polygenic variability in bread wheat and its bearing on selection procedure. Z Pflanzenzucht 48:317–328
Bhatnagar CP, Luthra JP, Kumar A, Avasthi AK (1979) An early high yielding mutant in gram (*Cicer arietinum* L.). Proceedings of symposium on the role of induced mutations in crop improvement. September 13–16, Hyderabad, India
Bhatnagar CP, Handa DK, Mishra A (1985) Prospects of utilizing mutation breeding in inducing nematode resistance in chickpea. Int Chickpea Pigeonpea Newsl 12:5
Bhogal NS, Chauhan JS, Singh M, Singh KH, Meena ML, Meena DR (2013) Genetic variability for micro and macro nutrients in Indian mustard (*Brassica juncea* L.) under two environments. Indian J Genet 73(2): 177–181
Bliss FA, Barker LN, Franckowiak JO, Hall TC (1973) Genetic and environmental variation on seed yield, yield components and seed protein quantity and quality of cowpea. Crop Sci 13:656–660
Borah SP, Goswami BC (1995) Induced high protein mutants with renewed kernel protein distribution and altered amino acid composition in rice. Proceedings of symposium Genetic Research and Education: Current trends and the next fifty years. New Delhi, vol III, pp 1239–1247
Bowman DE (1946) Differentiation of soybean antitryptic factors. Proc Soc Exptl Biol Med 57:547
Brock RD (1977) Prospects and perspectives in mutation breeding. In: Muhammed A, Askel R, von Borstel RC (eds) Genetic diversity in plants. Plenum Press, New York, pp 117–132
Brock RD (1967) Quantitative variation in *Arabidopsis thaliana* induced by ionizing radiations. Rad Bot 7: 193–203
Cagirgan MI, Toker C (2004) Selection of chickpea mutants for cold tolerance and ascochyta blight resistance. Int Chickpea Pigeonpea Newsl 11:14–16
Canci H, Toker C (2009) Evaluation of annual wild *Cicer* species for drought and heat resistance under field conditions. Genet Resour Crop Evol 56:1–6
Canci H, Toker C, Cagirgan MI (2004) Preliminary screening for nodulation in chickpea mutants. Int Chickpea Pigeonpea Newsl 11:13–14
Chandra S, Ali M (1986) Recent achievement in pulse production. Tech Bull DPR, Kanpur
Chary SN, Bhalla JK (1988) EMS induced male sterile mutant in pigeonpea (*Cajanus cajan* (L.) Millsp.). Indian J Genet 48(3):303–304
Coffman WR, Juliano BO (1979) Seed protein improvement in rice. In: Seed protein improvement in cereals and grain legumes II. IAEA, Vienna, pp 261–276
Dadke RD, Kothekar VS (2005) Mutation breeding: a novel technique for quality improvement of winged bean (*Psophocarpus tetragonolobus* (L.) D.C.). In: Datta SK (eds) Role of classical mutation breeding in crop improvement. Daya Publishing House, New Delhi, pp 119–131

Datta AK, Sengupta K (2002) Induced viable macromutants in coriander (*Coriandrum sativum* L.). Indian J Genet 62(3):273–274

Diaz AM, Caldas GV, Blair MW (2010) Concentration of condensed tannin and anthocyanin in common bean seed coat. Food Res Int 43:595–601

Frey KJ (1977) Protein of oats. Z Pflanzenzucht 78:185–215

Gaikward NB (2002) Genetic improvement in lentil (*Lens culinaris* Medik) through mutagenesis. Ph D Thesis, Dr. Babasaheb Ambedkar University, Aurangabad, India

Gaul H (1964) Mutations in plant breeding. Rad Bot 4:155–232

Gaul H (1965) The concept of macro and micromutations in barley. Rad Bot (Suppl) 5:407–428

Gaur PM, Chaturvedi SK (2004) Genetic options for managing biotic stresses in pulse crops. In: Masood A, Singh BB, Kumar S, Dhar V (eds) Pulses in new perspective. Indian Society of Pulses Research and Development, IIPR, Kanpur

Gaur PM, Gour VK (2001) A gene inhibiting flower colour in chickpea (*Cicer arietinum* L.). Indian J Genet 61(1):41–44

Gaur PM, Gour VK (2003) Broad few leaflets and outwardly curved wings: two new mutants of chickpea. Plant Breed 122:192–194

Gaur PM, Krishnamurthy L, Kashiwagi J (2008) Improving drought avoidance root traits in chickpea (*Cicer arietinum* L.). Current status of research at ICRISAT. Plant Prod Sci 11:3–11

Gautam AS, Sood KC, Richaria AK (1992) Mutagenic effectiveness and efficiency of gamma rays, ethylmethane sulphonate and their synergistic effects in black gram (*Vigna mungo* L.). Cytologia 57:85–89

Girija M, Dhanavel D (2013a) Induced chlorophyll mutation in cowpea (*Vigna unguiculata* L. Walp.). Int J Cur Tr Res 2(1):136–140

Girija M, Dhanavel D (2013b) Effect of gamma rays on quantitative traits of cowpea in M_1 generation. Int J Res Biol Sci 3(2):84–87

Girija M, Dhanavel D, Gnanamurthy S (2013) Gamma rays and EMS induced flower color and seed mutants in cowpea (*Vigna unguiculata* L. Walp). Adv Appl Sci Res 4(2):134–139

Gottschalk W (1987) The genetic basis of variation. In: Improving vegetatively propagated crops. Academic, London, pp 317–334

Gottschalk W, Muller H (1970) Monogenic alterations of seed protein content and protein pattern in X-ray induced *Pisum* mutants. In: Improving plant protein by nuclear techniques. IAEA, Vienna, p 238

Gottschalk W, Wolf G (1983) Induced mutations in plant breeding. Springer, Heidelberg, p 238

Goud JV (1967) Induced mutations in bread wheat. Indian J Genet 27:40–45

Gowda CLL, Gour PM (2004) Global scenario of chickpea research—present status and future thrusts. In: Ali M, Singh BB, Kumar S, Dhar V (eds) Pulses in new perspective. Indian Society of Pulses Research and Development, IIPR, Kanpur, pp 1–22

Goyal S, Kozgar MI, Fatma S, Khan S (2011) Genetic improvement of black gram, chickpea and fababean through conventional and advanced approaches. In: Khan S, Kozgar MI (eds) Breeding of pulse crops. Kalyani Publishers, Ludhiana, pp 149–174

Graham JA, Panozzo JF, Lim PC, Brouwer JB (2002) Effects of gamma irradiation on physical and chemical properties of chickpeas (*Cicer arietinum*). J Sci Food Agric 82:14

Gregory WC (1965) Mutation frequency, magnitude of changes and the probability of improvement in adaptation. In: The use of induced mutations in plant breeding. FAO/IAEA, Tech Meet Rome, pp 429–441

Grusak MA (2000) Strategies for improving the iron nutritional quality of seed crops: lessons learned from the study of unique iron-hyperaccumulating pea mutants. Pisum Genet 32:1–5

Grusak MA (2002) Enhancing mineral content in plant food products. J Am Coll Nutr 21:178–183

Gustafsson A (1940) A mutation system of the chlorophyll apparatus. Lund Univ Arsskr NF Avd 2 36:1–40

Haq MA, Shakoor A (1980) Use of induced mutations for improving resistance against Ascochyta blight in chickpea and yellow mosaic virus in mungbean. In: Induced mutations for improvement of grain legume production. IAEA, Vienna, pp 63–67

Hiremath C, Nadaf HL, Ganapathi M, Praveen KB (2010) Induced mutants in groundnut (*Arachis hypogaea* L.). Karnataka J Agric Sci 23(2):327–329
Ibrahim AF, Sharaan AN (1974) Studies on certain early barley mutants in M_3 and M_4 generations after seed irradiation with gamma rays. Z Pflanzenzucht 73:44–57
Ignacimuthu S, Babu CR (1989) Induced variation in protein quantity and quality in the wild and cultivated urd and mungbean. Indian J Genet 49(2):173–181
Iruela M, Rubio J, Cubero JI, Gil J, Millan T (2002) Phylogenetic analysis in the genus *Cicer* and cultivated chickpea using RAPD and ISSR markers. Theor Appl Genet 104:634–651
Jana S, Singh KB (1993) Evidence of geographical divergence in kabuli chickpea from germplasm evaluation data. Crop Sci 33:626–632
Johnson HW, Robinson HF, Comstock RE (1955) Estimates of genetic and environmental variability in soybeans. Agron J 47:314–318
Jukanti AK, Gaur PM, Gowda CL, Chibbar RN (2012) Nutritional quality and health benefits of chickpea (*Cicer arietinum* L.)-a review. Br J Nutr 108(Suppl 1):S11–S26. doi:10.1017/S0007114512000797
Jurgan MH, Nath P, Banerji BK (1985) Induction of mutation in winged bean (*Psophocarpus tetragonolobus* (L.) D.C.) by gamma rays. Himal Chem Pharm Bull 2(1):58
Jurgan MH, Banerji BK, Datta SK (1998) Gamma ray induced early fruiting mutant in winged bean. In: National symposium on advances in legume research in India. Abstract No. 12, December 15–17, NBRI, Lucknow, p 81
Kakade ML, Simons NR, Liener IE (1969) An evaluation of natural vs synthetic substrates for measuring the antitryptic activity of soybean samples. Cereal Chem 46:518
Kalia CS, Kharkwal MC, Singh MP (2000) Recovery of desirable mutations in wheat. Indian J Genet 60(4):465–470
Karjalainen R, Kortet S (1987) Environmental and genetic variation in protein content of peas under northern growing conditions and breeding implications. J Agric Sci 59:1–9
Kaul MLH (1989) Mutation research in dilemma. In: Bir SS, Saggo MIS (eds) Perspectives in plant sciences in India. Today and Tomorrow Printers and Publishers, New Delhi, pp 93–108
Kaul MLH, Garg R (1979) Population performance and genetic parameters of some promising pea lines. Pisum Newsl 11:15–16
Kaul MLH, Garg R (1982) Radiation genetic studies in garden pea XIII. Genetic variability, interrelationships and path analysis in protein rich genotypes. Biol Zbl 101:271–282
Kaur M, Singh N (2005) Studies on functional, thermal and pasting properties of flours from different chickpea (*Cicer arietinum* L.) cultivars. Food Chem 91:403–411
Khadke SG, Kothekar VS (2011) Genetic improvement of moth bean (*Vigna aconitifolia* (Jacq.) Marechal) through mutation breeding. In: Khan S, Kozgar MI (eds) Breeding of pulse crops. Kalyani Publishers, Ludhiana, pp 34–54
Khan MN (1999) Mutagenic effectiveness and efficiency of EMS, gamma rays and their combinations in black gram (*Vigna mungo* (L.) Hepper). Adv Plant Sci 12(1):203–205
Khan S (2011) Impact of mutation breeding in chickpea- a review. In: Khan S, Kozgar MI (eds) Breeding of pulse crops. Kalyani Publishers, Ludhiana, pp 126–138
Khan TN, Claydon A (1975) Role of induced mutation in the improvement of potential new source of protein-winged bean (*Psophocarpus tetragonolobus*). Panel proceeding series (IAEA), Research coordination meeting of the seed protein improvement programme 3, Hahnenklee, Germany, 5 May/FAO, Vienna, Austria
Khan S, Siddiqui BA (1993) Chlorophyll mutations in (*Vigna radiata* (L.) Wilczek) II. Mutagenic effectiveness and efficiency of chemical mutagens. Pak J Bot 25(2):161–166
Khan MH, Tyagi SD (2009) Studies on induction of chlorophyll mutations in soybean (*Glycine max* (L.) Merrill). Front Agric China 3(3):253–258
Khan S, Wani MR (2005) Genetic variability and correlations studies in chickpea mutants. J Cytol Genet 6:155–160
Khan S, Wani MR (2006) Induced mutations for yield contributing traits in green gram. Int J Agric Biol 8(4):528–530

Khan S, Wani MR, Parveen K (2004a) Induced genetic variability for quantitative traits in *Vigna radiata* (L.) Wilczek. Pak J Bot 36(4):845–850

Khan S, Wani MR, Bhat MD, Parveen K (2004b) Induction of morphological mutants in chickpea. Int Chickpea Pigeonpea Newsl 11:6–7

Khan S, Wani MR, Parveen K (2005a) An induced bushy mutant in mungbean. Mutat Breed Newsl Rev 1:10

Khan S, Wani MR, Bhat MD, Parveen K (2005b) Induced chlorophyll mutations in chickpea (*Cicer arietinum* L.). Int J Agric Biol 7(5):764–767

Khan S, Wani MR, Parveen K (2006a) Sodium azide induced high yielding early mutant in lentil. Agric Sci Digest 26(1):65–66

Khan S, Wani MR, Parveen K (2006b) Quantitative variability in mungbean induced by chemical mutagens. Legume Res 29(2):143–145

Kharkwal MC (1979) Mutation breeding in chickpea in India. In: Symposium on the role of induced mutations in plant breeding. Osmania University, Hyderabad, pp 177–183

Kharkwal MC (1983) New selection technique for micromutations. Abstract of the XV International Congress of Genetics, New Delhi, India, p 298

Kharkwal MC (1998a) Induced mutations in chickpea (*Cicer arietinum* L.) I. Comparative mutagenic effectiveness and efficiency of physical and chemical mutagens. Indian J Genet 58(2):159–167

Kharkwal MC (1998b) Induced mutations in chickpea (*Cicer arietinum* L.) II. Frequency and spectrum of chlorophyll mutations. Indian J Genet 58(4):465–474

Kharkwal MC (1998c) Induced mutations for improvement of protein in chickpea (*Cicer arietinum* L.). Indian J Genet 58(1):61–68

Kharkwal MC (1999) Induced mutations in chickpea (*Cicer arietinum* L.) III. Frequency and spectrum of viable mutations. Indian J Genet 59(4):451–464

Kharkwal MC (2000) Induced mutations in chickpea (*Cicer arietinum* L.) IV. Types of macromutations induced. Indian J Genet 60:305–320

Kharkwal MC, Shu QY (2009) The Role of induced mutations in world food security. In: Shu QY (ed) Induced plant mutations in the genomics era. Food and Agriculture Organization of the United Nations, Rome, pp 33–38

Khatoon A, Bhalla JK (1986) Induced mutation in chickpea. Proc Bot Res (Abst):31

Khattab RY, Arntfield SD (2009) Nutritional quality of legume seeds as affected by some physical treatments II. Antinutr Factors LWT 42:1113–1118

Khosh-Khui M, Niknejad M (1971) Heritability and number of genes controlling flower size in chickpeas. J Heredity 62:209–210

Klu GYP, Quaynor-Addy M, Dinku E, Dikumwin E (1989) Induced mutations in winged bean in Ghana. Mut Breed Newslett 34:15–16

Klu GYP (2000) Induced mutations for accelerated domestication-a case study of winged bean (*Psophocarpus tetragonolobus* (L.) D.C.). West Afr J Appl Ecol 1:47–52

Knight EJ (1993) Fasciation in chickpea: genetics and evaluation. Euphytica 69:163–166

Kodym A, Afza R (2003) Physical and chemical mutagenesis. Methods Mol Biol 236:189–203

Konzak CF, Nilan RA, Wagner J, Foster RJ (1965) Efficient chemical mutagenesis. Rad Bot 5(Suppl):49–70

Konzak CF, Woo SC, Dickey J (1969) An induced semi-dwarf plant height mutation in spring wheat. Wheat Inf Serv 28:10–12

Konzak CF, Nilan RA, Kleinhofs A (1977) Artificial mutagenesis as an aid in overcoming genetic vulnerability of crop plants. In: Muhammed A, Aksel R, Von Borstel RC (eds) Genetic diversity in plants. Plenum, New York, pp 163–177

Konzak CF, Mung NV, Warner RL, Rubenthaler GL, Finney PL (1978) Advances in technology and in genetic information for breeding improvements in wheat protein potentials. Seed protein improvement by nuclear techniques, IAEA, Vienna, pp 519–531

Kothekar VS, Harsulkar AM, Khandelwal AR (1996) Low trypsin and chymotrypsin inhibitor mutants in winged bean (*Psophocarpus tetragonolobus* (L.) D.C.). J Sci Food Agric 71:137–140

Kozgar MI, Khan S, Wani MR (2012) Variability and correlations studies for total iron and manganese contents of chickpea (*Cicer arietinum* L.) high yielding mutants. Am J Food Technol 7(7):437–444
Kumar Y, Ghosh J, Haider ZA (2011) Strategies for yield improvement in pulses. In: Khan S, Kozgar MI (eds) Breeding of pulse crops. Kalyani Publishers, Ludhiana, pp 175–191
Kunitz M (1945) Crystallisation of trypsin inhibitors from soybean. Science 101:668–669
Liener IE (1976) Annual review. Plant Physiol 27:291
Lin CY, Pevzner I, Friars GW (1979) Experimental investigation of errors of heritability estimates in index selection. Can J Genet Cytol 21:303–308
Manjaya JG, Nandanwar RS (2007) Genetic improvement of soybean variety JS 80-21 through induced mutations. Plant Mutat Rep 1(3):36–40
Marentes E, Grusak MA (1998) Iron transport and storage within the seed coat and embryo of developing seeds of pea (*Pisum sativum* L.). Seed Sci Res 8:367–375
Mercy ST, Kakar SN, Chowdhury JB (1974) Cytological studies in three species of the genus *Cicer*. Cytologia 39:383–390
Micke A (1979) Crop improvement by induced mutations. Use of mutation induction to alter the ontogenic pattern of crop plants. Gamma Field Symp 18:1–23
Micke A, Donini B (1993) Induced mutations. In: Hayward MD, Bosemark NO, Ramagosa I (eds) Plant breeding-principles and prospects. Chapman and Hall, London, pp 53–62
Misra RN, Jachuck PJ, Sampath S (1973) Induced mutations for practical utility in rice. International symposium on use of isotopes and radiations in agriculture and animal husbandry, New Delhi, pp 19–71
Moh CC (1971) Mutation breeding in seed coat colour of bean (*Phaseolus vulgaris* L.). Euphytica 20:119–125
Moreno M, Cubero JI (1978) Variation in *Cicer arietinum* L. Euphytica 27:465–485
Muehlbauer FJ, Singh KB (1987) Genetics of chickpea. In: Saxena MC, Singh KB (eds) The chickpea. CAB International, United Kingdom, pp 99–126
Mukhopadhyay D (2000) The molecular evolutionary history of a winged bean alpha-chemotrypsin inhibitor and modelling of its mutational high structural analysis. J Mol Evol 50(3):214–223
Muller HJ (1927) Artificial transmutation of gene. Science 66:84
Naik BS, Singh B, Kole C (2002) A promising mungbean (*Vigna radiata* (L.) Wilczek) genotype with high protein content and seed yield. Indian J Genet 62(4):342–344
Narhari P, Bhatia CR (1975) Induced mutations for increasing protein content in wheat and rice. In: Breeding for seed protein improvement using nuclear techniques. IAEA, Vienna, pp 23–29
Nerker YS, Mote SB (1978) Induced mutagenesis in bengal gram (*Cicer arietinum* L.) II.Micromutations. J Maharashtra Agric Univ 3:180–183
Nilan RA, Sideris EG, Kleinhofs A, Sander C, Konzak CF (1973) Azide-a potent mutagen. Mut Res 17:142–144
Olejniczak J (1986) Characteristics of early and protein mutants in maize (*Zea mays* L.). In: Prasad AB (ed) Mutagenesis: basic and applied. Print House, Lucknow, pp 149–158
Omar M, Singh KB (1995) Development of early mutants with resistance to Ascochyta blight or leaf minor. Int Chickpea Pigeonpea Newsl 2:10–11
Parveen K (2006) Mutation studies with early generation selection for polygenic variability in chickpea (*Cicer arietinum* L.). Ph D Thesis, Aligarh Muslim University, Aligarh, India
Phadnis BA (1978) Inheritance of seed coat colour in bengal gram. Indian J Genet 38(1):13–19
Pundir RPS, Reddy GV (1989) Induction, genetics and possible use of glabrousnes in chickpea. Euphytica 42:141–144
Pundir RPS, Reddy GV (1998) Two new traits-open flower and small leaf in chickpea (*Cicer arietinum* L.). Euphytica 102:357–361
Rayyan A (1995) Mutagenesis and tissue culture studies in *Vigna mungo* (L.) Hepper. Ph D Thesis, Osmania University, Hyderabad, India
Read JW, Haas LW (1938) Studies on the baking quality of flour as affected by certain enzyme actions V. Further studies concerning potassium bromate and enzyme activity. Cereal Chem 15:59–68

Rehman ZU, Shah WH (2001) Tannin contents and protein digestibility of black grams (*Vigna mungo*) after soaking and cooking. Plant Foods Hum Nutr 56:265–273

Ryan CA (1981) Proteinase inhibitors. In: Marcus A (ed) The biochemistry of plants, proteins and nucleic acids, vol 6, Academic, New York, pp 351

Sakin MA, Yildirim A (2004) Induced mutations for yield and its components in durum wheat (*Triticum durum* Desf.). Food Agric Environ 2(1):285–290

Salimath PM, Bahl PN, Mehra RB (1984) Genetic diversity in chickpea (*Cicer arietinum* L.). Z Pflanzenzucht 92:52–60

Salimath PM, Toker C, Sandhu JS, Kumar J, Suma B, Yadav SS, Bahl PN (2007) Conventional breeding methods. In: Yadav SS, Redden R, Chen W, Sharma B (eds) Chickpea breeding and management. CAB International, Wallingford, pp 369–390

Sandberg AS (2002) Bioavailability of minerals in legumes. Br J Nutr 88(3):281–285

Savin VN, Swaminathan MS, Sharma B (1968) Enhancement of chemically induced mutation frequency in barley through alteration in the presoaking of the seeds. Mut Res 6:101–107

Sehrawat KD, Singh LN, Sharma SK (1996) Effect of mutagens on the association between various traits in siratro (*Macroptilium atropurpureum* (D.C.) URB). Indian J Genet 56(1):117–120

Shaikh MAQ (1983) Genetic improvement of food legumes in developing countries by mutation induction. In: Summerfield RJ (eds) World crops: cool season food legumes. Kluwer Academic, Dordrecht

Shaikh MAQ, Ahmed ZU, Majid MA, Wadud MA (1982) A high yielding and high protein mutant of chickpea (*Cicer arietinum* L.) derived through mutation breeding. Env Exp Bot 22(4):483–489

Shakoor A, Haq MA (1980) Improvement of plant architecture in chickpea and mungbean. In: Induced mutations for improvement of grain legume production I. IAEA, Vienna, pp 59–62

Shamsuzzaman KM, Islam MM, Begum M, Subhan A (2005) Development of an early maturing chickpea variety, BINASOLA-3. Mut Breed Newsl Rev 1:8–9

Sharma B (1969) Chemically induced mutations in cowpea (*Vigna sinensis* L. Savi). Curr Sci 38:520–521

Sharma B (1986) Increasing the efficiency of mutagenesis for micromutations by early generation selection. Indian J Genet 46(1):88–100

Sharma SK, Sharma B (1979) Mutagenic effectiveness and efficiency of gamma rays and N-nitroso-N-methyl urea in lentil. Indian J Genet 39:516–520

Sharma HC, Bhagwat MP, Pampapathy G, Sharma JP, Ridsdill-Smith TJ (2006) Perenial wild relatives of chickpea as potential sources of resistance to *Helicoverpa armigera*. Genet Resour Crop Evol 53:131–138

Sharma S, Yadav N, Singh A, Kumar R (2013) Nutritional and antinutritional profile of newly developed chickpea (*Cicer arietinum* L.) varieties. Int Food Res J 20(2):805–810

Sheeba A, Ibrahim SM, Yogameenakshi P, Babu S (2003) Effect of mutagens on quantitative traits in M_2 generation in sesame (*Sesamum indicum* L.). Indian J Genet 63(2):173–174

Shu QY (2005) To our readers. Mutat Breed Newsl Rev 1:1–2

Sikder S, Biswas P, Hazra P, Akhtar S, Chattopadhyay A, Badigannavar AM, D'Souza SF (2013) Induction of mutation in tomato (*Solanum lycopersicum* L.) by gamma irradiation and EMS. Indian J Genet 73(4):392–399

Singh KB (1987) Chickpea breeding. In: Saxena MC, Singh KB (eds) Common Wealth Agricultural Bureau. Int Willingford, Oxon, pp 127–162

Singh BD (2005) Plant breeding: principles and methods. Kalyani Publishers, Ludhiana

Singh VP, Chaturvedi SN (1980) The productivity of some mutants of the mungbean (*Vigna radiata* (L.) Wilczek) I. Bigger grain size and high yielding mutants. J Cytol Genet 15:152–155

Singh KB, Ibrahim H (1990) Hybridization techniques in chickpea. ICARDA, Aleppo

Singh KB, Singh O (1997) Prospects of creating higher yield potential in chickpea. In: Asthana AN, Ali M (eds) Recent advances in pulses research. Indian Society of Pulses Research and Development, IIPR, Kanpur

Singh AK, Singh RM (2007) Mutagenic effectiveness and efficiency of gamma rays, ethylmethane sulphonate and their synergistic effects in mungbean (*Vigna radiata* (L.) Wilczek). Crop Res 34:198–202
Singh VP, Chaturvedi SN, Srivastava A (1997) Genetic impropvement in pulse crops through mutation breeding. In: Siddiqui BA, Khan S (eds) Plant breeding advances and *in vitro* culture. CBS Publishers and Distributors, New Delhi, pp 25–42
Singh VP, Singh M, Lal JP (2000) Gamma ray and EMS induced genetic variability for quantitative traits in urdbean (*Vigna mungo* (L.) Hepper). Indian J Genet 60(1):89–96
Singh AK, Gupta RK, Singh IP (2011) Role of induced mutations for the improvement of pulse crops with special reference of mungbean. In: Khan S, Kozgar MI (eds) Breeding of pulse crops. Kalyani Publishers, Ludhiana, pp 104–125
Sjodin J (1971) Induced morphological variations in *Vicia faba* L. Hereditas 67:155–180
Sneep J (1977) Selection for yield and early generation of self-fertilizing crops. Euphytica 26:27–30
Solanki RK, Gill RK, Verma P, Singh S (2011) Mutation breeding in pulses: an overview. In: Khan S, Kozgar MI (eds) Breeding of pulse crops. Kalyani Publishers, Ludhiana, pp 85–103
Srivastava A, Singh VP (1993) Induced polygenic variability in pigeonpea (*Cajanus cajan* (L.) Millsp.). J Indian Bot Soc 72:281–284
Srivastava P, Marker S, Pandey P, Tiwari DK (2011) Mutagenic effects of sodium azide on the growth and yield characteristics in wheat (*Triticum aestivum* (L.) em. Thell.). Asian J Plant Sci 10(3):190–201
Stadler LJ (1928) Mutation in barley induced by X-rays and radium. Science 69:186–187
Subba Rao K (1988) Gamma rays induced morphological and physiological variations in *Cicer arietinum* L. Indian J Bot 11(1):29–32
Swaminathan MS (1968) Mutation breeding. Proc XII Int Cong Genet, Tokyo, Japan, 3:327–347
Tajoddin M, Shinde M, Lalitha J (2011) Phytic acid and mineral contents of some mungbean cultivars. J Food Legumes 24(2):163–164
Thakur JR, Sethi GS (1995) Comparative mutagenicity of gamma rays, ethylmethane sulphonate and sodium azide in barley (*Hordeum vulgare* L.). Crop Res 9:350–357
Toker C, Cagirgan MI (2004) Spectrum and frequency of induced mutations in chickpea. Int Chickpea Pigeonpea Newsl 11:8–10
Toker C, Uzun B, Canci H, Ceylan FO (2005) Effects of gamma irradiation on the shoot length of *Cicer* seeds. Radiat Phys Chem 73:265–367
Toker C, Yadav SS, Solanki IS (2007) Mutation breeding. In: Yadav SS, McNeil D, Stevenson PC (eds) Lentil-an ancient crop for modern times. Springer, Dordrecht, pp 209–214
Toker C, Uzun B, Inci NE (2011) Mutation breeding in chickpea. In: Khan S, Kozgar MI (eds) Breeding of pulse crops. Kalyani Publishers, Ludhiana, pp 1–21
Tomlekova NB (2010) Induced mutagenesis for crop improvement in Bulgaria. Plant Mutat Rep 2(2):1–32
Udupa SM, Sharma A, Sharma RP, Rai PA (1993) Narrow genetic variability in *Cicer arietinum* L. as revealed by RFLP analysis. J Plant Biochem Biotechnol 2:83–86
Van der Maesen LJG (1972) In: Veenmam H, Zonen Q (eds) *Cicer* L., a monograph of the genus, with special references to the chickpea (*Cicer arietinum* L.), its ecology and cultivation. Mededlingen Landbouwhogeschool (Communication Agricultural University), Wageningen, p 342
Van der Maesen LJG (1987) Origin, history and taxonomy of chickpea. In: Saxena MC, Singh KB (eds) The chickpea. CAB International, Wallingford, pp 11–34
van Harten AM (1998) Mutation breeding theory and practical applications. Cambridge University Press, Cambridge
Veeresh LC, Shivashankar G, Hittalmani S (1995) Effect of seed irradiation on some plant characteristics of winged bean. Mysore J Agric Sci 29(1):1–4
Waghmare VN (2001) Induced mutations in grasspea (*Lathyrus sativus* L.) II. Frequency and spectrum of chlorophyll mutations. Ad Plant Sci 14(I):249–253
Waghmare VN, Mehra RB (2000) Induced genetic variability for quantitative characters in grasspea (*Lathyrus sativus* L.). Indian J Genet 60(1):81–87

Waghmare VN, Mehra RB (2001) Induced chlorophyll mutations, mutagenic effectiveness and efficiency in *Lathyrus sativus* L. Indian J Genet 61(1):53–56

Wang LZ, Wang L, Zhao RJ, Pei YL, Fu YQ, Yan QS, Li Q (2003a) Combining radiation mutation techniques with biotechnology for soybean breeding. In: Improvement of new, traditional and industrial crops by induced mutations and related biotechnology. IAEA, Vienna, pp 107–115

Wang TL, Domoney C, Hedley CL, Casey R, Grusak MA (2003b) Can we improve the nutritional quality of legume seeds. Plant Physiol 131:886–891

Wani MR, Khan S, Kozgar MI, Goyal S (2011a) Induction of morphological mutants in mungbean (*Vigna radiata* (L.) Wilczek.) through chemical mutagens. The Nucleus 48(3):243–247

Wani MR, Khan S, Kozgar MI (2011b) Induced chlorophyll mutations I. Mutagenic effectiveness and efficiency of EMS, HZ and SA in mungbean. Front Agric China 5(4):514–518

Wani MR, Khan S, Kozgar MI (2011c) An assessment of high yielding M_3 mutants of green gram (*Vigna radiata* (L) Wilczek). Romanian J Biol 56(1):29–36

Wani MR, Lone MA, Sheikh SA, Dar MS, Tak MA, Ahmad P, Khan S (2012) Induction and assessment of genetic variability for yield and yield contributing traits of chickpea (*Cicer arietinum* L.). J Plant Genomics 2(1):28–33

Wani MR, Kozgar MI, Khan S, Dar NA (2013) Induction of genetic variability through artificial mutagenesis in chickpea (*Cicer arietinum* L.). Thai J Agric Sci 46(3):141–147

Yadav S, Khetarpaul N (1994) Indigenous legume fermentation: effect of some anti-nutrients and *in vitro* digestibility of starch and protein. Food Chem 50:403–406

Yadav VS, Singh D, Yadav SS, Kumar J (2002) Correlation and path analysis in chickpea. Indian J Pulses Res 15(1):19–22

Yamaguchi H, Morishita T, Degi K, Tanaka A, Shikazono N, Hase Y (2006) Effect of carbon-ion beams irradiation on mutation induction in rice. Plant Mut Rep 1(1):25–27

Yaqoob M, Rashid A (2001) Induced mutation studies in some mungbean (*Vigna radiata* (L.) Wilczek) cultivars. Online J Biol Sci 1(9):805–808

Zia-Ul-Haq M, Shahid I, Shakeel A, Imran M, Niaz A, Bhamger MI (2007) Nutritional and compositional study of desi chickpea cultivars grown in Punjab, Pakistan. Food Chem 105:1357–1363

Chapter 10
Organic Farming: The Return to Nature

Saima Siddique, Madeeha Hamid, Ameema Tariq, and Alvina Gul Kazi

Abstract Organic farming is a modern and a sustainable form of agriculture that provides consumers fresh natural farm products. Organic farming works in synchronization with nature rather than against it. This objective is achieved by using techniques to improve crop yields without harming the natural environment as well as the people who live and work in it. Organic agriculture offers an exclusive amalgamation of environment-friendly practices, which require low external inputs, thereby contributing to increased food availability. Organic farming has a very positive influence especially on birds, insects, weeds, wildlife, and soil flora and fauna. Conventional farming is capital intensive, which requires more manufactured inputs and energy as compared to knowledge- and labor-intensive organic farming. Organic agriculture uses energy more competently than conventional agriculture. As compared to conventional agriculture, organic farming produces cost-effective food products, free of synthetic fertilizers and pesticides. It also provides employment opportunities and economic benefits to local communities. The methods utilized in organic farming are more costly and labor intensive, but prove to be more cost effective in the long run. Since organic agriculture supplies more greenhouse gases in the soil, the farmers across the globe can solve the climate disaster by switching to organic methods. In addition, organic agriculture has the potential to address food security issues. Enough evidence is available to prove that organic crops are a better source of nutrients than their corresponding conventional forms. Organic systems give higher animal immunity and increased disease resistance to plants, with 50% less mycotoxins in crops and a persistent shelf life. Organic

A. G. Kazi (✉) · S. Siddique · M. Hamid · A. Tariq
Atta-ur-Rahman School of Applied Biosciences, National University of Sciences and Technology, H-12, Islamabad, Pakistan
e-mail: alvina_gul@yahoo.com

S. Siddique
e-mail: symacidiq16@yahoo.com

M. Hamid
e-mail: madeeha@asia.com

A. Tariq
e-mail: ameematariq@hotmail.com

P. Ahmad et al. (eds.), *Improvement of Crops in the Era of Climatic Changes,* Volume 2,
DOI 10.1007/978-1-4614-8824-8_10,

foods have more plant secondary metabolites, higher micronutrient content, and more conjugated fatty acids for better human health, including lower incidences of noncommunicable diseases. Organic agriculture merges modernism, custom, and science to manage the shared surroundings encouraging fair relationship and high quality of life for everyone involved.

Keywords Organic farming · Conventional farming · Climate change · Pesticides · Biofertilizer

1 Introduction

Organic farming has engrossed much attention in current decades as a way to maintain farming production. At the same time, it has played an important role in dealing with the environmental harms rooted in traditional agricultural techniques. Organic farming not only produces fine and healthy food products but also improves the fertility and quality of soil (Isaacs 2012). Funtilana has described organic agriculture as: "Organic Agriculture is giving back to the environment what was taken from it" (Singh et al. 2012). Organic food is grown and processed without using any synthetic fertilizers or pesticides (insecticides, herbicides, and/or fungicides), plant growth regulators, such as hormones, livestock antibiotics and GM organisms, and human sewage sludge (John 2011).

Biological pesticides produced from natural sources can be used in the production of organic food. Some preventative measures have to be applied prior to adopting the latest technologies for organic agriculture due to considerable risks of unpredictable tools such as genetic engineering (Tuomisto et al. 2012). Organic farming is based on ecological cycles, and procedures therefore diminish the use of exterior contributions. This minimizes resource utilization of the farms and restricts nutrient heaps in the system. It lessens the danger of phosphorus and nitrogen eutrophication and evades overfertilization. Organic livestock farming is based on environment-friendly production, maintaining animals in good health, realizing animal benefit values thereby generating yields of high class. Organic livestock farming meets the demands of the rising number of consumers (Sundrum 2012).

Organic agriculture requires low external inputs, thereby contributing to increased food availability. It is a system based on generating food with negligible destruction to ecosystems, humans, or animals. Nevertheless, reviewers argue that organic farming may have lesser yields and would consequently require additional land to generate a similar quantity of food as conventional systems do, resulting in biodiversity loss and extensive deforestation, therefore undermining the ecological benefits of organic procedures (Seufert 2012). The expected boost of human world population, above 2 billion by the middle of the century, spots the requirement of an increase in agriculture to deal with the high demands of feed, food, biofuels, and fiber to meet the needs of the population. It, in addition, plays a decisive role in bringing additional ecosystem services, for instance, those that guard the excellence of the environment. Organic agriculture tackles a lot of traits and makes use of the biological regulation methods to put back external input, protecting biodiversity at the same time.

According to the Codex Alimentarius Commission, "organic agriculture is a holistic production management system that avoids use of synthetic fertilizers, pesticides and genetically modified organisms, minimizes pollution of air, soil and water, and optimizes health and productivity of interdependent communities of plants, animals and people." Organic farming is a production scheme, which mainly prohibits or avoids the utilization of artificial pesticides, fertilizers, livestock feed additives, and growth regulators. The objectives of environmental, financial, and social sustainability are the fundamentals of organic farming. The major features include protecting long-lasting fertility of soil by preserving organic matter level, nitrogen self-sufficiency through the use of biological nitrogen fixation and legumes, careful mechanical intervention, fostering soil biological activity, successful recycling of organic materials including livestock wastes and crop residues, and pest control relying mainly on crop rotation, diversity, natural predators, resistant varieties, and organic manuring. A huge emphasis is made on preserving the soil fertility by returning all the wastes to it primarily through compost to reduce the gap between nitrogen, phosphorous, and potassium (NPK) addition and its removal from the soil.

1.1 Food and Agriculture Organization and Organic Agriculture

As defined by Food and Agriculture Organization (FAO), "Organic farming is environmental friendly ecosystem management in which use of all kinds of synthetic input is eliminated." In March 1999, organic agriculture was officially included into FAO's agenda as a way to support sustainable progress. The interest of FAO in organic agriculture stems from its potential to contribute to rural development and global food security. FAO member countries in the International Conference on Organic Agriculture and Food Security in Rome (2007) emphasized the organic agriculture's capability to create a more stable food supply, increased access to food in rural areas, and maintenance of natural resources (Morgera et al. 2012).

2 Background

Organic farming is a type of agriculture practiced by early farmers for thousands of years. A full organic food production system is one of the most flexible and oldest agroecosystems. Inorganic methods were introduced by Industrial Revolution with severe side effects. An organic revolution was started in Central Europe in the mid-1920s by Rudolf Steiner. He created biodynamic agriculture system, which is considered to be an old version of organic agriculture system. Organic agriculture was developed by Albert Howard in the 1940s in England as an independent field. Though organic farming is primitive in its widest sense, Sir Albert Howard started the post Industrial Revolution organic movement, for which he is known as the "father of organic farming." Since then, the production of organic food has moved from small experimental garden plots to outsized and huge farms with their products sold under a unique organic label. Modern organic farming, from its start until now, has contributed only a small part of the total agricultural output. The increase in ecological knowledge in the general population has altered the former supply-driven movement into a demand-driven movement. Many farm products are produced according to conventional methods in the developing countries, but these methods, although similar to organic farming, are not yet licensed. In some cases, economic reasons have forced the farmers of the developing world to transform (John 2007).

3 Problems Caused by Chemical Agriculture

3.1 Environmental Problems

The current intensive agriculture system causes many problems, including:

- Artificial herbicides and fertilizers are easily washed away from the soil, polluting lakes, rivers, and water courses
- The long-term use of artificial fertilizers results in soils with low organic matter content, which is prone to erosion by rain and wind
- Increased dependency on artificial fertilizers, which are required every year in greater amount to produce the same crop yield
- Artificial chemicals deteriorate the soil microorganisms resulting in poor structure of soil as well as decreased aeration and nutrient availability
- It becomes a great challenge to control pests and diseases as they become resistant to artificial pesticides. The number of natural pests decreases because of pesticide use and habitat loss
- Thinning of eggshells of birds
- Increased extinction of preying birds
- Adverse effects on wildlife
- Environmental imbalance (Deshmukh 2010)

3.2 Effects on Human Health

- Toxic residues cause skin reactions and allergic sensitization
- Unfavorable and adverse effects on nervous system, peripheral neuropathies, and impairment of nervous system
- Disturbance of endocrine system

- Carcinogenicity
- Suppression of immunity
- Prolonged exposure to small amount of pesticides residues in food commodities can lead to:
  i. Spontaneous abortions and increase in miscarriages
  ii. Initiation of early puberty in girls. Undesirable effects on male reproductive system
  iii. Effects on kidney, liver, and brain
  iv. Deformities and abnormalities in newborn babies (Deshmukh 2010).

4 Conventional Farming Versus Organic Farming

Organic agriculture system consists of management practices without using artificial fertilizers and high input of energy. This may affect storage of soil organic material in the long run (Leifeld 2012). Some of the analysis of existing data indicates that overall organic yields are usually less than traditional yields but under certain conditions (better management practices, specific crop types, and growing circumstances), yield of organic farming systems can be increased (Seufert et al. 2012).

5 Principles, Pillars, Components, and Objectives of Organic Farming

5.1 Principles of Organic Farming

The main principles of organic farming are as follows:

- To draw upon local resources and to work as much as possible within a closed system
- To try to preserve the long-term fertility of soil
- To avoid all forms of pollution resulting from different types of agricultural techniques
- To produce food products in sufficient quantity and of high nutritional quality
- To cut down use of fossil energy in agricultural practice
- To allow agricultural workers' to build up their potentialities as human beings and earn a living through their work
- To provide such conditions of life to livestock that fulfill their physiological needs (Deshmukh 2010)

5.2 Pillars of Organic Farming

The organic standards, market network, technology packages, and certification/ regulatory mechanism are the four pillars of organic farming (Deshmukh 2010).

5.3 Components of Pure Organic Farming

The components of pure organic farming are:

1. Make use of only organic planting for farming
2. No dependence on artificial inputs
3. Feeding soil, but not the crop plants
4. Implementing food safety practices, i.e., never using GM crops or products
5. Documentation of not only the end product but also the entire procedure and distribution chain, i.e., the process and not the produce is to be certified. The certification should keep in view the export
6. Use of artificial growth promoter and enhancers should be prohibited, even the use of organic fertilizers particularly those having some mutants; cow dung from intensive dairy farms using antibiotics, oil cakes, and bone meal should be strictly restricted (Deshmukh 2010)

5.4 Objectives of Organic Farming

The objectives of organic farming concisely expressed in the standard document of the international federation of organic agriculture (IFOAM) are as follows:

1. To work in collaboration with the natural systems rather than dominating them
2. To enhance the biological cycles contained within the farming system, which involves soil flora and fauna, microorganisms, plants, and animals
3. To preserve and enhance long-term fertility of the soil
4. To use renewable resources as much as possible within locally organized agricultural system
5. To work as much as possible, within a closed system with regard to nutrient elements and organic matter
6. To provide such conditions of life to all livestock to enable to act upon all aspects of their innate behavior
7. To evade all forms of pollution resulting from agricultural techniques
8. To uphold the genetic diversity of agricultural system and its surroundings including protection of wildlife habitats and plants
9. To allow agricultural workers' satisfaction and an adequate return from their work including a safe working environment
10. To consider wider and environmental impacts of the farming system (Deshmukh 2010)

6 Why Organic Farming

Organic farming helps to provide long-term benefits to people as well as the environment. Other environmental advantages of organic farming include supporting local food markets, increased biodiversity, improved soil quality and reduced pesticide pollution and packaging waste, and water usage (Ziesemer 2007). Besides its potential to alleviate poverty and economic gains, organic farming proves to be valuable in a number of ways. Some of its benefits are listed below.

6.1 Environmental Benefits

Organic farming protects the environment from harmful effects, which arise from the use of synthetic inputs, specially pesticides, fertilizers, and hormones (Kotschi and Muller-Samann 2004). Fertilizers and pesticides release dangerous toxic chemicals into soil and water (Theriault 2006). Some pesticides can cause harm to the environment or on direct exposure, they can prove to be toxic and dangerous to human health. Children are at a higher risk than adults to direct exposure, since the toxic effects of pesticides are often more severe in children than adults (Committee on Pesticides in the diets of Infants and Children 1993). Agriculture, without pesticides and chemical fertilizers, might deliver in a number of situations, but outputs would be less than traditional farming. Therefore, generating the massive amount of the worldwide diet will need agricultural methods together with the use of fertilizers (Gilbert 2012). Organic agriculture more or less constantly supports more biodiversity and usually has a positive environmental impact per unit of land. It does not essentially have a positive impact per unit of production. Organic cereals and milk, all produce elevated greenhouse gas (GHG) emissions per unit of product than their traditionally farmed counterparts. On the other hand, organic olives and beef have lower emissions in majority of the cases. On the whole, organic foodstuff requires less energy input, but extra land than the traditional products. Studies show that organic farming system provides greater biodiversity as compared to the traditional farming system due to decreased soil changes and chemical application (Nascimbene et al. 2012).

Organic farming has a positive and favorable influence especially on birds, insects, weeds, wildlife, and soil flora and fauna (Deshmukh 2010). All non-crop species exhibit partiality for organic farming system in terms of both diversity and abundance. An average of 30% more species reside in organic farms. Butterflies, birds, beetles, spiders, earthworms, mammals, soil microbes, and vegetation are particularly affected (Gabriel et al. 2006). The birds' number and species show higher density in organic farms (Deshmukh 2010).

Agro-biodiversity or agricultural biodiversity is a division of biodiversity, which consists of all shapes of life directly significant to agriculture, and can subsist equally in a farm and crossway farms. Ecologists have disagreed that at the farm level, a boost in on-farm variety and a range of overlying collection of species improve the level of agricultural biodiversity. This amplifies crop resilience and ecological firmness. The farming of a huge number of crops at the farm level, i.e., crop biodiversity, is a component of agricultural biodiversity, and generates differentiation in soil fauna, pests, predators, and weeds at the farm level. More significantly, crop biodiversity has been accounted to raise agricultural output through the control of pest infestation and replacement of agricultural soil, directing to better farm income constancy and security (Nastis 2013). Population density and biodiversity fitness are improved by the lack of pesticides and herbicides (Gabriel et al. 2006). Beneficial insects are attracted by weed species which, in turn, improve forage on weed pests and soil quality (van Elsen 2000). Soil-bound organisms often get a wide range of benefits because of the large number of bacterial organisms produced by natural fertilizers and experience a reduced intake of pesticides and herbicides (Hole et al. 2005). The risk of getting poor yield is reduced to a great extent in organic farming because it promotes biodiversity (Fließbach et al. 2007).

Organic farms are more capable of withstanding harsh weather conditions as compared to the conventional farms. Occasionally in drought conditions, their yield is 70–90 % more than conventional farms (Lotter 2003). Organic farms have been found to be more cost effective in the drier states of the USA because of their better drought performance. In addition, organic farms can endure hurricane damage much better and keep hold of 20–40 % more topsoil, thus incurring smaller economic losses as compared to their neighbors (Holt-Gimenez 2000). Hence, organic farming contributes positively to the reduction in soil, air, and groundwater pollution. Moreover, it is also an excellent solution to nitrate pollution. In addition, it improves soil fertility, structure, and soil fauna (Fan et al. 2005).

The organic matter is globally recognized to enhance soil fertility. In addition, improving the soil's chemical, physical, and organic matter has the prospective to add to climate change alleviation by impounding C from the atmosphere. The basic methods to guard organic elements inflowing the soil against decay are:

a. Selective preservation and production of resistant molecular preparation, structure, and association (biochemical methods)
b. Physical detachment from O_2, enzymes, decomposers, etc., by occlusion in aggregates (physical systems)
c. Chemical diffidence by intimate sorption (association) with mineral exteriors (chemical mechanisms)

An 18-year-long study was conducted on organic methods on nutrient-depleted soil, which showed that conventional methods are better for increasing soil fertility and yield in cold and temperate climates (Kirchmann 2007).

6.2 Economic Benefits and Profitability

Various studies have revealed that organic crops use 97 % less pesticides and yield 95–100 % higher along with 50 % lesser expenditure on energy and fertilizer. Hence, organic agriculture consumes zero pesticides and less energy (Mader et al. 2002). High prices that consumers disburse for organic products along with decreased cost of pesticide inputs and synthetic fertilizers add to increased profits. Organic farms have always been found to be more profitable as compared to the conventional farms (Lotter 2003). According to the FAO, "Organic farming is a pioneer to establish energy reducing practices by using organic principles. Organic principles, which emphasize farm-level self-sufficiency, incorporation of externalities and environmental stewardship, can be improved to form plans for limiting the use of fossil fuel-based energy in organic farming. Especially in the post-production handling sections, advancements done in order to decrease the consumption of energy can affect the traditional parallel sectors." In the majority of cases, 30–50 % less energy is consumed by organic farming as compared to the traditional farming:

- Organic agriculture typically uses energy more competently than conventional agriculture.
- Organic agriculture often needs about one-third additional manual labor hours as a substitution for energy-intensive inputs used in traditional agriculture (Ziesemer 2007). Modern chemical-dependent farming methods:
  - Lessen soil of nutrients
  - Demolish important soil microorganisms
  - Contribute to global climate change and desertification
  - Oversupply farmlands with toxic fertilizers, herbicides and pesticides, which then move into groundwater, rivers, lakes, and oceans

For example, numerous regions of Minnesota, which is the most important farmland, are now facing the problem of increased nitrogen in drinking water. Health risks of nitrogen include a potential correlation with cancer, in addition to reproductive and thyroid problems in both livestock and humans (Mercola 2013).

Organic agriculture is about 30 % more efficient to produce the same amount of food as compared to the traditional farming. Conventional farming is capital intensive, which requires more manufactured inputs and energy as compared to the traditional organic farming which is knowledge and labor intensive (Halberg et al. 2006). The system engages large skilled and semi-skilled/unskilled labor for various tasks to be performed (sowing, planting, cultivating, rearing, maintenance, aftercare, harvesting, cleaning, washing, grading, bar coding, labeling, packing, transporting, and marketing) in order to follow a strict code of "organic farming" (Pimentel 2006). Serious issues are being raised about the energy-intensive nature of these methods and their unpleasant outcomes on soil yield and environmental excellence. Organic agriculture is capable of supporting about three to 4 billion people (Trewavas 2001). In a study of 1,144 organic farms conducted in UK and Republic of Ireland, organic farms engaged more workers as compared to the conventional ones.

6.3 Health Benefits

Food for starving population, fiber for clothing, and feed for animals and even, in a number of cases, fuel for vehicles come from worldwide agriculture. Consequently, in the world's temperate climates, human agriculture has displaced 45 % of temperate forests, 50 % of savannas, and 70 % of grasslands. Agriculture is one of the main sources of GHG emissions; the most important cause of deforestation in the tropics and a recurrent basis of water pollution and nonrenewable groundwater mining. A number of farmers have turned to the organic methods. Such a kind of farming is destined to reduce human health and environmental impacts by evading the use of chemical pesticides, synthetic fertilizers, and antibiotic or hormone treatment for livestock. The use of industrial methods, predominantly synthetic nitrogen fertilizer, has fed the human population during the previous century (Biello 2012). Currently, there is no noticeable evidence of any health benefit of consuming organic over conventionally produced food products (Dangour et al. 2009).

Individual studies have taken into account a variety of potential impacts, including residues of pesticides. Pesticide residues provide a second channel for health effects. The organically produced vegetables and fruits are likely to contain less agrochemical residues than their conventionally grown alternatives (Magkos et al. 2006). Nitrate concentration might be less, but the potential health impact of nitrates is arguable. The users trust that organic products are healthier than traditionally grown products. Research has shown that organic products contain less nitrate content, because larger amount can cause cancer of the alimentary tract and methemoglobinemia in infants (Forman and Silverstein 2012). There is a decreased risk of eczema associated with consumption of organic milk, though no similar evidence was found in case of organic vegetables, fruits, or meat. The higher cost of organic products (ranging from 45–200 %) may limit the intake of the recommended five servings per day of fruits and vegetables, which reduce the risk of cancer and improve health irrespective of their source (Magkos et al. 2006).

The utilization of vegetables and fruits has been linked with lesser risk of chronic human health harms like hypertension, cancer, cardiovascular diseases, and diabetes type II because of their elevated phytochemicals. The health advantages of vegetables and fruits have so far been endorsed to the antioxidant characteristics of phytochemicals. The cell membrane lipid peroxidation (LPO) degree is found to be 60 % higher in organic tomatoes. The superoxide dismutase (SOD) activity is also radically higher in organic fruits. The organic tomato fruits under oxidative stress build up higher content of soluble solids as sugar, vitamin C, and phenolic compounds. These have smaller mass and size than the conventionally grown systems. In addition, they are also rich in soluble solids, phenolic compounds, and phytochemicals including vitamin C. In the past few decades, yield has been of greater importance as compared to micronutritional and gustative quality of plant products. This might be right for staple food, but the micronutritional and gustative qualities of vegetables and fruits hold more significance than the energy supply. Growers should not struggle to decrease stress in order to increase fruit size and yield, but should allow stress to a certain level to enhance product quality. Further research is needed to properly understand the relation between stress and oxidative stress and also between secondary metabolism and oxidative stress. In addition, more studies are required to understand the physiological mechanisms responsible for positive outcomes of organic farming on quality of fruit (Oliveira et al. 2013).

6.4 Social Benefits

Organic farming may have an important social effect on local communities. To start with, organic agriculture may provide employment opportunities to the local people. More manual labor is often required in organic agriculture to compensate for pesticide and synthetic fertilizer loss, thus providing more jobs in local communities. Commonly, the labor required to run an organic farm is 10–20 % higher as compared to the traditional farms. Organic farmers also expand their crops and widen their planting schedules throughout the year in order to enhance soil health and maintain biodiversity. This establishes year-round employment opportunities and may lessen the problems related to migrant labor. More job opportunities will increase the population of local communities and also halt migration to urban areas. Thus, organic agriculture can increase the local communities and support rural development. In order to stay competitive, farmers must adjust to the local conditions by managing land, labor, and resources so that the production can be increased. Farmers also depend on their neighbors to sustain certain principles in order to guarantee the reliability of their own water, soil, and air. Ties within the community are strengthened by association on these issues, leading to greater association among organic farmers and also partnerships. Cooperatives or organized groups can thus gain power in trade negotiations, gather their resources, and enjoy greater access to markets. There is some proof that increased collaboration results in new businesses among local communities and more active participation in local government. Consumer protection is another keystone of organic farming. The well-built regulatory frameworks, whereby the government verifies organic certifications, are essential for consumers to trust the food that they buy (Morgera et al. 2012).

7 Environment-Enhancing Agriculture

In actuality, a conventional farmer needs a farming model that can leave the environment better than before, the forest healthier, the land more diverse, wildlife more prolific, and soil more fertile resulting in clean water, clean air, and healthy animals and plants (Deshmukh 2010).

7.1 IFOAM's Definition of Organic Agriculture

The IFOAM definition of organic agriculture is based on:

1. The principle of health
2. The principle of ecology
3. The principle of fairness
4. The principle of care

a. The Principle of Health Organic agriculture should be able to maintain and improve the health of plant, soil, animal, human, and planet. The principle highlights that the health of individuals and communities is dependent upon the health of ecosystem and soil that produces healthy crops, which in turn foster the health of people and animals. Health is integrity and wholeness of living systems. In farming, processing, distribution, or consumption, the key role of organic agriculture is to sustain and enhance the health of ecosystem from the smallest organisms in the soil to the human beings.

b. The Principle of Ecology Organic agriculture should be based on ecological cycles and systems. It should emulate and help maintain them. The ecological balance is maintained by establishment of habitats, maintenance of agricultural and genetic diversity, and designing of farming system.

c. The Principle of Fairness According to the principle of fairness, organic agriculture should build on associations that assure equality with regard to the common life opportunities and environment. Environmental and natural resources utilized for production and consumption should be supervised in a way that is economically and socially just and held in trust for upcoming generations.

d. The Principle of Care Organic agriculture should be dealt with in a responsible and precautionary manner in order to guard the health of the present and future generations and for the well-being of environment. According to this principle, responsibilities and precautions are key concerns in development, management, and technology choices in organic agriculture. A policy paper has been issued by National Academy of Agriculture Sciences (NAAS), according to which synthetic pesticides can be avoided; however, complete exclusion of fertilizers may not be suitable under all conditions (Deshmukh 2010).

7.2 Environmental Impacts of Conventional and Organic Farming

Organic agriculture attempts to increase water and crop quality by removing external inputs, genetically modified (GM) crops, as well as pesticides and synthetic fertilizers. While there are a number of ways to achieve this outcome, the main point is that pesticides and synthetic fertilizers are not used. The main reason for the difference in global warming potential (GWP) is the manufacturing and transportation of artificial fertilizers. The chemical method to produce the fertilizer (the Haber-Bosch process) releases carbon dioxide as a side product, but organic agriculture uses on-farm compost as its fertilizer. Storing and moving compost still result in CH_3 emission, but the GHG emissions do not even come close to generating off-farm mineral fertilizers. Fossil fuels are usually the source of energy for the manufacturing and shipment of artificial pesticides. As compared to the drawbacks of organic fertilization, the GWP of organic pest control is still less than that of traditional methods (Akeenan 2011).

7.3 Potential of Organic Farming to Alleviate the Impact of Agriculture on Global Warming

One of the major threats for food security especially in the tropical countries is global warming. Rainfall irregularity and drought conditions are expected to get worse in many countries due to global warming. Extenuating the emissions of GHGs is thus an important challenge to improve food security. One way to achieve this is by reducing CO_2 emission due to combustion of fossil fuel, but it can also be done by farming (Moreau 2007). A new study shows that organic agriculture supplies more GHGs in the soil than conventional agriculture. Many farmers across the globe may help to solve the climate disaster by switching to organic methods, along with avoiding the use of pesticides (Isaacs 2012).

The outcome of Gattinger's meta-analysis shows an overall positive worldwide development towards an increase in organic farming (0.9 % of total world agricultural land or 37 million ha). The author's meta-analysis suggests that organic agriculture allows more carbon to be stored in the soil than traditional agriculture. Carbon is stored in the soil instead of heating up the atmosphere, which acts like a carbon sink. The researchers guess that by using a combination of rotating crops (mixed farming) and livestock, more carbon can be stored in organically farmed soil than any other farming system depending on the use of plant protection chemicals and artificial nitrogen fertilizer. Their results show that 0.37 Gt of carbon is being isolated per year globally (0.04 Gt of carbon in the USA, 0.03 Gt of carbon in Europe), thus offsetting 3 % of the existing total GHG emissions (2.3 % for the USA, 2.3 % for Europe), or 25 % of the total existing agricultural emissions (36 % for the USA, 23 % for Europe). The collective alleviation by organic agriculture up to 2030 would contribute 13 % to the collective reduction essential by then to stay on the trail to keep temperatures from rising 2 degrees Celsius above preindustrial levels by 2100. This meta-analysis gives a strong clue in favor of organic agriculture (Isaacs 2012).

At least 30 % of global warming is due to agriculture. Three gases are responsible for this, namely CH_4 (methane), N_2O (nitrous oxide), and CO_2 (carbon dioxide). Fertilizer industry mainly emits CO_2 from the machines used on the farm. A significant contributor for CO_2 emission by agriculture is deforestation. CH_4 emissions from livestock are mostly from enteric fermentation, but also from rice fields and manure. Soil (denitrification) is a major source of N_2O, and a minor one is fertilizer from animals.

a. ***CO_2 Emissions***

About half of the energy used in agriculture in developed countries is in the manufacturing of fertilizers (mostly nitrogen fertilizers). However, there is an increased amount of energy utilized by the fertilizers due to minor mechanization and less efficient use of fertilizer plants. Since organic agriculture does not use artificial nitrogen fertilizer, less energy is consumed than conventional agriculture, and therefore less CO_2 is emitted. In Europe, it has been evaluated that for key crops, organic farming uses per acre about half the energy used in traditional agriculture systems. Despite lower yield in organic agriculture, the amount of crops produced remains a significant factor. In European livestock production, the energy consumed in the production of 1 liter of organic milk is 25 % of conventional milk production. This is because organic cows are mainly grazing, whereas the feed of conventional ones is based on grain and soybean cake (Moreau 2007).

d. ***Carbon Sequestration***

It has been noted that the organic content in many areas of the world has decreased progressively due to the increase in agriculture based on deep plowing and artificial fertilizers. About 50 % of soil organic carbon has been degraded over the past 50–100 years of cultivation in the Great Plains of North America. About 7 million ha have less than 2 % organic content in France. Deforestation in tropical countries has led to a faster decline in the organic content. As proved by repeated and long-term trials, the organic matter increases (carbon content of soil increases) and is maintained by organic agriculture. This ability to impound carbon contributes to alleviate the contribution of agriculture to the greenhouse effect (Moreau 2007).

e. ***Nitrous Oxide (N_2O) Emissions***

Soil mainly emits nitrous oxide. The Intergovernmental Panel on Climate Change (IPCC) has evaluated that about 1.25 % of the amount of nitrogen applied as fertilizer is represented by these emissions. However, many factors are responsible for this percentage. As compared to conventional farming, data for emission of N_2O are less for organic farming. The amount of nitrogen applied normally is lower in organic than conventional agriculture. Likewise, emissions of N_2O increase radically when nitrogen fertilization exceeds the needs of the crop, which happens habitually in conventional farming. It can therefore be concluded that organic farming emits less N_2O than conventional agriculture (Moreau 2007).

f. ***Methane (CH_4) Emissions***

After N_2O emission, methane emission is the main contributor of global warming. There are three main origins of methane from agriculture: anaerobic fermentation of flooded crops (rice), fermentation of animal dejections, and enteric fermentation of ruminants. The production of methane per animal is about the same in conventional and organic breeding. But, the emission per kilo of meat or milk is more in intensive production. This increase is, at least moderately, compensated by better endurance of organic cows. In intensive systems, particularly in milk production, cows have a very short life, generally up to 5 years. Methane emission by fermentation of compost is less in organic than conventional breeding, because composting is aerobic, whereas storing compost (heaps or slurry) is mainly anaerobic.

In short, it can be said that organic farming has decreased the contribution of agriculture to global warming. It, therefore, leads to a stable food supply, which is threatened by the climate change. On the other hand, more research is required in order to estimate the extent of this alleviation and identify what improvements can be done in organic farming to increase it. Another important way to reduce the contribution of food production to global warming is to alter our eating habits, especially in the developed countries by reducing consumption of red meat (Moreau 2007).

8 Organic Foods

Some studies have revealed and also some consumers think that organic products are rich in flavor and nutrients. The organic products are sold at high prices as compared to the conventional food products. Since organic foods are grown via more labor-intensive production methods, they are sold at high prices.

8.1 Organic Certification

Organic certification is a certification procedure for manufacturers of organic products. In common, a few trades in food manufacture are certified including farmers, retailers, food processors, seed suppliers, and restaurants (Organic Certification 2013). Organic farmers are qualified to assure that their agricultural techniques comply with principles of organic manufacture to reassure consumers, retailers, and wholesalers that their products are really organic (Department of Agriculture and Food 2010). Necessities differ from country to country and normally engage a set of manufacturing principles for raising, storage, packaging, processing, and delivery that comprise:

- Human manure sludge fertilizers are not employed in feed of animals or development of plants
- Prohibition of artificial chemical inputs not on the National List of Allowed and Prohibited Substances (e.g., pesticides, food additives, antibiotics, fertilizer etc.), irradiation, GM organisms and the use of sewage mud
- A complete written record of manufacturing and sales proceedings (audit trail) should be kept

- Firm substantial separation of organic goods from noncertified goods
- Carrying out periodic on-site assessment

Organic certification deals with increasing international demand for organic food substances as well as guaranteeing quality, avoiding deception and supporting commerce of food. Such an official recognition was not compulsory in the early days of the organic association. With the passage of time, as organics have developed fame, the small-scale producers trade their goods straight to farmers' marketplace. Mostly customers are buying organic products through established channels, for example, supermarkets.

8.2 Hazard Analysis

A number of significant food safety hazards to consider in an organic farm may comprise:

1. Microbiological contagion from fish emulsion fertilizers or compost
2. Quality of water used for washing or processing the produce
3. Pest control methods
4. Contamination from exterior sources, for instance transport suppliers
5. Clean-down measures (Department of Agriculture and Food 2010)

9 Organic Agriculture and Food Security

Organic farming can add to food security. Eight hundred and 50 million people still starve for food, even though the worldwide food supply is sufficient. Additionally, the price of food has increased noticeably in the past 10 years, and there is less genetic assortment in our food because of traditional agricultural techniques. As a result, populations are more exposed to the threat of food scarcity. Organic farming has the ability to meet up these confronts. Organic farming involves access to food by dropping threats of disease, raising productivity and biodiversity over the lasting period, and giving means for limited manufacture and access to food. Sponsors for traditional agriculture disagree that organic agriculture reduces yield. Organic advocates, on the other hand, think that yield is identical to traditional farms over the extended period of time (Morgera et al. 2012). Organic agriculture has the following roles in food security:

- Mitigates acute starvation during food emergency situations through increased ecosystem stability and diversification
- Improves domestic nutrient intake and capability to buy food via sustainable growth and commercialization of small-scale agriculture
- Develops self-reliant food systems, especially at the domestic level

- Plays a role in intake of micronutrient and improved diets by the diversification of production and reintroduction of underutilized varieties

For poverty mitigation, organic agriculture:

- Contributes to sustainable rural livelihood, as it gives a better return on labor
- Offers employment opportunities, as it requires 30% more labor input per hectare
- Contributes to more common well-being, through nonexploitive work and fair salary that develop control on income
- Contributes to development of rural areas, as rural economies are revived

For environmental sustainability, organic farming:

- Avoids harm by increasing energy and nutrient recycling and efficiency of resource use
- Is a low-energy track food system, as it forbids the use of nitrogen fertilizers
- Decreases transaction and transport costs via community-supported food short-supply series
- Restores biodiversity and conserves ecological values (International Conference on Organic Agriculture and Food Security 2007)

For food sourcing, organic agriculture:

- Allows smallholders to compete with specialty foods and quality products
- Offers higher farm-gate prices that are a reflection of environmental stewardship and real production costs
- Helps re-localize food systems where the poor and hungry live
- Develops effervescent local food supplies that reduce import surges and food-import reliance

There are many links with organic agriculture practices, including issues such as legal protection, authority in the food supply chain, economic development opportunities, avoidance of noncompetitive practices, inexpensive technology, protection of consumers, expansion of local and regional markets, small producer's integration into markets, protection of agro-biodiversity and drinking-water quality, maintenance of ecosystem carrying capability for present and future generations, promotion of gardens at both home and schools, availability of nutritious and diverse food, and support of traditions on food-related matters.

It is emphasized particularly that production in agriculture should (in order of priority):

- Allow import not only for locally grown items
- Target local food needs in local markets
- Export high cost products

9.1 Organic Agriculture and Food Availability

There are many factors that affect food availability including fossil fuel crisis and water scarcity, globalization that threatens smallholder viability, and urbanization and loss of farms and farmers. The role of organic agriculture in food availability considers these issues, in terms of both food import capacity and agricultural output.

By converting global agriculture to organic without the use of nitrogen fertilizers and conversion of wildland to agriculture would lead to global agricultural supply of 2,640–4,380 kcal/person/day. Sustainable growth in developing countries through the practice of organic agriculture would lead to an increase in the production by 56 %. Organic and conventional yields are comparable on average, although yields increase when converting from low-input systems and decline when converting from high-input systems. A case study in Tigray, Ethiopia, reported double yields due to organic soil management. The main challenges in semiarid environments are soil management practices and livestock production, whereas in tropical humid ecosystems, it is crop diversification. Through an efficient use of natural resources locally, input availability is increased in organic systems. Organic farms utilize 33–56 % less energy per hectare. They also improve economic competence through savings on inputs, but are laborious. Nutrient use is increased through minimizing losses and recycling, but phosphorus is not easily accessible.

Organic urban gardens enhance the urban food supplies through short supply chains between consumers and growers. At community and household level, organic rural and urban markets and networks play an important role in improving food quality, quantity, and diversifying food availability. It is seen that organic agriculture is emerging robustly in domestic markets of some developing countries, such as Brazil, India, and China. The role played by the developed country consumers in triggering organic production in developing countries is known, but the ability of the poor to feed themselves is still an interrogation.

A challenge related to international markets is to create participatory networks, to bring the producers together, and to develop value chains based on reasonable trade and informed choices. The significance of food traceability is stressed to authorize consumers and producers, especially as organic farming is stepping into the mainstream. Nature and More are the examples of the commercially efficient systems which internalizes social and environmental costs in food prices. It is made clear that organic markets are not for an economic influence, but for an "aware elite" prepared to pay elevated prices provided the label is reliable.

There is a requirement for better agroecological science as well as need to understand factors that play an important role in risk alleviation. The complex environment–livestock interactions have been highlighted as an area for further research and improvement of organic principles. The importance of participatory guarantee systems is considered important to reduce the costs and authorize farming communities to distinguish their organic products in sale.

There is a need to assess organic foods and farms fundamentally as a whole, with multiple measurements on both efficiency and productivity. For comparing organic systems with other food systems, the adequate methodology must consider total agricultural outputs of multiple cropping systems, including yields and secondary goods such as straw, environmental services, such as carbon sequestration, total energy efficiency from the farm to postharvest handling and distribution, water saving, and soil fertility, and nonfood benefits derived from agricultural systems such as social equity and disease reduction (International Conference on Organic Agriculture and Food Security 2007).

9.2 Organic Agriculture and Access to Food

Organic agriculture system improves food access by increasing output, variety, and conservation of natural resources, by raising salaries and by decreasing risks for farmers. Enhancement also results by sharing knowledge among farmers. These benefits lead to poverty alleviation and a turnaround of migration from rural to urban areas. Policy requirements to advance food access include expanding fair-trade systems along the full value chain, increasing farmer's rights to local varieties, seeds, and biodiversity, strengthening the rights of indigenous farmers, and evaluating current emergency aid and procurement plans.

The investigation of several case studies on organic agriculture system in Africa, Asia, and Latin America suggests that the economic effects of converting to organic agriculture depend on the previous farming system. When converting from traditional extensive farming to organic farming, input cost reduces, while yield and income are likely to increase. On the other hand, when converting from intensive farming, yield and income tend to reduce. In both cases, input costs reduce and labor costs increase. However, there are other advantages of organic agriculture beyond the purely financial ones, such as conservation of natural resources, health protection, risk reduction, increased flexibility to adverse weather, and farmer authority through the attainment of knowledge and higher dependence on limited inputs. It is thought that policies aiming at facilitating the import of cheap foods and providing subsidies to farmers in developed countries have unfavorable effects on farmers in developing countries. On the other hand, in rural areas, access to food is enhanced by organic agriculture system. Organic agriculture system also tends to reorganize gender roles, with women participating more in homegrown foods, but care should be taken in the sharing of workload.

It is being noted that the conventional food prices are frequently altered by subsidies and do not reflect the full cost to society as a whole or the environment. The higher production does not necessarily translate into higher local access to food, but food must also be ethnically appropriate. The superior farm-gate food prices are needed for livelihood security, but market intelligence is needed to sustain benefits from improved income generation. Six thousand farmers are improving their income by 10–20 % from organic cotton in Madhya Pradesh, India, just due to market knowledge: a business model that builds networks and partnerships among retailers, spinning and processing companies, and farmers. As knowledge is very crucial in organic agriculture, farmer organizations have a significant role to play in training, extension, and technical assistance. A participatory internal control system can be established to reduce the cost of certification and group marketing.

The private sector should be the engine of growth for the organic supply system, whereas the governments can be more influential in building capacity and supporting research and providing legal and institutional environment. Extended political dedication by governments is deemed essential for the sustainable development of organic agriculture (International Conference on Organic Agriculture and Food Security 2007).

9.3 Organic Agriculture and Stability of Food Supply

The strength of food supply is challenged by climate change and interannual variability, trade reform impacts on commodity prices, and the erosion of natural resources and environmental services. Organic agriculture system is analyzed through the environmental constancy of organic agroecosystems. Preventive measures are mainly emphasized by organic agriculture system that results in an overall stability of the agroecosystem, especially of soils that have high levels of microbial biomass and increased soil organic matter. Organic soil structure results in better percolation and water drainage, and improves water retention (20–40 % more), thus decreasing the requirement of irrigation and increasing crop yield in drought periods. Mandatory crop rotation is a road to better ecological balance through rehabilitation of functional biodiversity and the use of adapted seeds/breeds.

9.4 Organic Agriculture and Food Utilization

The utilization of food is challenged by dietary problems, health concerns and rapid urbanization, global trans-boundary diseases and higher occurrence of contamination, food consumer demands for quality food, and changing trade patterns. The role of organic agriculture system is analyzed in terms of consumer health, food quality, postharvest handling, and food security. Numerous benefits of organic agriculture system depend on the establishment of an ecological balance among the soil, plants, and animals, not just on substituting synthetic pesticides and fertilizers with organic ones. This primary difference is particularly significant for farmers with little knowledge of organic agriculture.

It is expected by the consumers that the organic food must be safe and should be equal to or better than the conventional foods. Organic foods should have more plant secondary metabolites, higher micronutrient content, and more conjugated fatty acids to contribute to better human health, including lower incidences of noncommunicable diseases. Organic systems provide higher animal immunity and increased disease resistance of plants, with 50 % less mycotoxins in crops and a persistent shelf life. The restriction on synthetic input use has led to safer drinking water, due to reduced amount of phosphates and nitrates being leached and pesticide poisoning is also avoided (about 20,000 deaths per year are caused by conventional agriculture chemicals).

Rural environmental pollution in China has led to the need for health and environmental protection among consumers. Organic land has increased from 342,000 ha in 2003 (0.28% of total land) to 978,000 ha in 2005, while increasing the yearly income of local farmers by nine times. The example of China is very inspirational because it provides three distinct organic supply models. The first is commercially successful and involves suburban or semi-urban areas close to large rich cities in the Eastern provinces. These organic gardens, which give jobs to needy workers, sell to both domestic supermarkets (owned by big entrepreneurs) and international market. In the second model, usually the most successful, domestic farmers take the entire risk and are in charge of group training and certification. The third model, more insecure, involves poor farmers living in remote locations, encouraged by local research institutions or local environment defense boards.

The benefits of higher yields can be less than the cost of illness. More knowledge is required to understand the cost of organic agriculture on the nutritional quality of diet, both in developed and in developing countries. It is decided that organic foods should not only be evaluated in terms of "harmless," but also in terms of other life quality and fitness values. The significance of food cultures is considered important, including knowledge systems for food harvest, storage and preservation. The revival of indigenous knowledge and local system adaptation is highlighted for the pattern shift towards food security. For the revival of local food systems, governments should facilitate homegrown policies. A holistic view of food systems, ahead of productivity to include social, environmental, and health impacts, could solve the current irony in farming (International Conference on Organic Agriculture and Food Security 2007).

10 Organic Farming Methods

The overarching objective of organic farming as defined by IFOAM is as follows: Organic agriculture is a manufacture system that maintains the health of soil and people. It depends on biodiversity, ecological procedures, and cycles modified to confined circumstances. Organic agriculture merges modernism, custom, and science to do well to the shared surrounding and encourages just relationships and a high quality of life for each and every one involved (International Federation of Organic Agriculture Movement 2008). Miscellaneous "organic fertilizers," e.g., cattle dung compost, farmyard manure, MSW compost, sewage sludge, poultry droppings, plants bio-fertilizers, microbial inoculants, and earthworm vermicastings, are used for agriculture all over the world where farmers are unable to afford expensive chemical manures (Sinha and Herat 2012). Most studies reflect that organic farming could increase the quantity and variety of total fungi and total bacteria in soil (Wang et al. 2012).

10.1 Chemical Control

The pest problem cannot be solved by the pesticides. Insecticide use has amplified tenfold, whereas crop fatalities from pest spoil have doubled in the past 50 years. Three significant causes of why natural power is preferable to pesticide use are given below:

10.1.1 Safety For People

Synthetic pesticides can rapidly discover their mean into watercourses and food chains. This can generate health risks for humans. Human health can also be debilitated by population consuming food (especially vegetables and fruit) holding remains of pesticides sprayed on the produce. Around the world, there are an estimated 1 million cases of poisoning by pesticides each year. Approximately 20,000 of these result in death. The majority of the deaths occur in tropical countries, where chemical pesticides prohibited in the USA or Europe are still available.

10.1.2 Cost

As natural techniques do not involve purchasing matter from the exterior, using natural disease and pest control is frequently cheaper than applying chemical pesticides.

10.1.3 Safety for the Environment

Chemical pesticides can have a number of harmful effects on the environment. They are:

- Useful insects can be killed by chemical pesticides. The equilibrium between pests and helpful predators can be disturbed by only one spray.
- Artificial chemicals can reside in the bodies of animals and in the environment causing troubles for several years.
- Insect pests can rapidly turn out to be resistant to synthetic products and are no longer restricted over a small number of breeding cycles. This means that stronger chemicals or enlarged quantity are subsequently required, generating more environmental, health, and economic problems.

10.2 Natural Control

The organic farmer can manage pests and diseases in a number of ways by:

- Raising vigorous crops that experience fewer harm from pests and diseases
- Selecting crops with an innate resistance to particular pests and diseases. Local varieties are superior in resisting diseases and local pests as compared to introduced varieties

- Avoiding the phase when the pest does the majority harm by timely planting of crops
- Companion sowing with other crops (garlic, onion) to keep the pests away
- Picking or trapping pests from the crop
- Recognizing diseases and pests properly. This will stop the farmer from unintentionally eradicating helpful insects or wasting time. It is hence useful to know preferred host plants, breeding habits, predators of pests, and life cycles
- Using crop rotation to check a carryover of pests to the subsequent period and breaking pest cycles
- Providing a natural environment to promote natural predators that manage pests

10.3 Weed Management

Weed management in organic farming is not a simple job, mainly in regions where labor for hand weeding is not reasonably priced or is scarce. On the other hand, the standard must be similar as in any traditional cropping scheme, i.e., weed competition desires to be prohibited in turn to attain the highest crop yields. This essentially involves weeding with nonchemical substances, however, this has to be done accurately at the precise time to get rid of weeds during the supposed significant phase of weed competition. Organic schemes also need the use of precautionary techniques prior to raising the crop and to set up a sensible crop rotation. Decayed seedbed preparation to kill the weeds manually or mechanically is an extremely fine choice to hold up the launch of weed competition. The use of green manure and cover crops, as well as mounting soil fertility, may possibly assist to manage a number of weed varieties. The common techniques used to put off weed competition in organically raised crops are elevated seeding rates and companion cropping with small-seeded legumes and narrow seed spacing/cross seeding (Food and Agriculture Organization of United States 2013).

Organic weed management encourages weed inhibition, to a certain extent, by promoting phytotoxic effects on weeds and crop competition. Organic farmers integrate biological, cultural, chemical, physical, and mechanical strategy to control weeds without artificial herbicides. Organic principles say that a single crop cannot be raised in a similar place without a different, dominant crop. Thus, organic standards need rotation of annual crops. Crops with different life cycles depress weeds linked with a particular crop. The weed-suppressive cover crops are repeatedly incorporated in organic crop rotation. Organic farmers struggle to amplify soil organic material content, which can hold up microorganisms to demolish ordinary weed seeds.

Additional cultural procedures used to improve crop competitiveness and lessen weed force include high-density planting, tight row spacing, choice of viable crop ranges, and delayed planting into lukewarm soil to promote quick crop germination. Physical and mechanical weed control procedures used on organic farms can be generally classified as (Schonbeck 2010):

- Tillage—rotating the soil between crops to integrate crop remains and soil alterations; eliminate accessible weed development and set up a seedbed for planting
- Cultivation—upsetting the soil after seeding
- Cutting and Mowing—eliminating apex enlargement of weeds
- Thermal weeding and flame weeding—using heat to kill weeds
- Mulching—jamming weed appearance with plastic films, landscape fabric, or organic materials (Szykitka 2004)

For herbicidal use, a few naturally supplied chemicals are acceptable. These include corn gluten meal, certain formulations of acetic acid (concentrated vinegar), and essential oils. A small number of choosy bioherbicides based on fungal pathogens have also been built up (Schonbeck 2010).

10.4 Crop Rotation

Crop rotations are at the heart of organic agriculture and aid organic methods to shield our surrounding. They engage in altering the form of crop developed in one area on a usual basis. Organic farmers sow alternating groups of plants (brassica, cereals, roots, and legumes) to put in fertility and stop diseases and pests from strengthening. A few plants, like clover, provide nutrients to the soil, whereas potatoes and wheat consume nutrients. Rotations frequently comprise of a "rest" phase for individual plots or fields, where "green manure" or grass such as clover is grown for a season or further to add fertility. This is known as planting "ley." Although using crop rotation may sound old fashioned, it is a more effective and sophisticated method than chemicals. Using chemicals to fertilize the soil frequently merely supplies crops with three basic elements (nitrogen, potassium, and phosphorous) that they require to grow before supplying them with each and every nutrient that they require.

In addition to making sure that soil nutrients do not get exhausted, crop rotations check the buildup of diseases and pests, which assist organic producers to prevent the use of pesticides. When a farmer grows the similar crop in one field year after year (known as monoculture), the pests and diseases that attack the crop set up and get enlarged in number with time. Nonorganic producers rely on pesticides to deal with the pest and disease. On the contrary, organic farmers stay away from this by growing discontinuous crops that are susceptible to different diseases and pests every year, stopping several from getting established in the similar location.

Crop rotations have a lot of significant functions:

1. They assist to manage diseases and pests.
2. They help to preserve soil fertility.
3. They help out to maintain soil structure and soil organic matter levels.
4. They make certain that sufficient nutrients are accessible to different crops each year.

In general, organic farming decreases environmental pollution and liberate GHGs from food manufacture.

10.5 Soil Management

Plants require potassium, phosphorus, and nitrogen, in addition to micronutrients. Green manure and crop rotation facilitate to give nitrogen through legumes (the Fabaceae family), which fix nitrogen from the atmosphere through symbiosis with rhizobial bacteria. Intercropping, which is occasionally used for disease and insect management, can add soil nutrients, but the rivalry among the crop and the legume can be challenging, and a greater gapping among crop lines is necessary (Watson et al. 2002). Organic farmers use animal manure and developed fertilizers such as a variety of mineral powders like greensand, rock phosphate, and seed meal. Collectively, these techniques assist to manage erosion. In a number of cases, pH might have to be altered. Except in the USA, few compounds such as magnesium sulfate, aluminum sulfate, soluble boron products, and iron sulfate are permitted in organic farming.

Diverse farms with both crops and livestock can function as lay farms, whereby the land collects soil fertility through increasing nitrogen-fixing grasses such as *alfalfa* or white clover and raise cereals or cash crops as soon as soil fertility is established. Farms exclusive of livestock may discover it more complicated to preserve fertility and could depend on peripheral contributions, such as grain legumes and green manures. Horticultural farms raising vegetables and fruits, which function in confined circumstances, are frequently more dependent upon exterior inputs (Watson et al. 2002).

10.6 Controlling Other Organisms

Nematodes, arthropods (e.g., mites, insects), bacteria, and fungi are organisms other than weeds, which form the basis of problems on organic farms. A broad range of integrated pest management techniques are used by organic farmers to avoid diseases and pests. These comprise, however, are not restricted to, nutrient management and crop rotation, providing environment for advantageous organisms, sanitation to eradicate pest territory, crop protection using physical obstacles, crop diversification by companion planting or founding of polycultures, and collection of pest-resistant crops and animals. Organic farmers frequently rely on the use of advantageous organisms to diminish pest populations (biological pest control). Examples of advantageous insects include big-eyed bugs, minute pirate bugs, and ladybugs (which are likely to fly away).

Natural insecticides permitted to use on organic farms include *Pyrethrum* (a chrysanthemum extract), *Bacillus thuringiensis* (a bacterial toxin), *rotenone* (a legume root extract), *neem* (a tree extract), and *spinosad* (a bacterial metabolite). These pesticides are used by less than 10% of organic farmers regularly (Lotter 2003). These are, at times, called green pesticides, since they are usually considered to be environment-friendly and safer. *Pyrethrum* and *rotenone* are mostly controversial, since they perform by affecting the nervous system just like the majority of traditional insecticides (Pottorff 2010). Rotenone is awfully toxic to fish and is able to provoke symptoms similar to Parkinson's disease in mammals (Sheer et al. 2003). Naturally resulting fungicides acceptable for use on organic farms include the fungus *Trichoderma harzianum* and bacteria *B. pumilus* and *B. subtilis*. These are chiefly successful for diseases attacking roots.

10.7 Genetic Modification

A key feature of organic agriculture is the refusal of genetically engineered animals and plants. Though resistance to the utilization of every transgenic technology in organic agriculture is powerful, some agricultural researchers persist to support incorporation of transgenic technologies into organic agriculture as the most favorable way to sustainable farming, predominantly in the developing countries (Herrera-Estrella and Alvarez-Morales 2001). In the same way, a few organic growers question the justification behind prohibiting the use of genetically engineered seeds as they view this sort of biotechnology to be steady with organic standards.

11 Production of Inputs for Organic Farming

11.1 Composting

11.1.1 Production of Compost

Compost production is the major sustaining factor of organic farming. It is done by the interaction of biological materials and the microorganisms. It is a decomposition method that produces a stable product from the organic matter. A large amount of organic matter from the farms or kitchens can be recycled back to the soil as compost. The level of organic matter can be increased by the addition of compost with many beneficial effects, such as water retention capacity improvement in humus content, absorption capability, etc. Compost can also be used for bedding.

11.1.2 Classification of Composing

a. Aerobic Composting: Aerobic composting involves aerobic conditions necessary for the composting of large-scale agricultural and municipal wastes. Aerobic microbes control aerobic composting. Therefore, oxygen is crucial for this decomposition process. It is characterized by the absence of foul odors, short stabilization, and high temperature. The high temperature helps in destroying the pathogenic organism and the weed seeds.

b. Anaerobic Composting: It is controlled by anaerobic bacteria, which work in the absence of atmospheric oxygen. The method works at lower temperature characterized by longer stabilization time and production of foul odors. The major benefit of anaerobic composting is that minimum attention is required for the process and once the compost bed is established, little or no energy is needed.

11.1.3 Vermi Compost

Earthworms set up compost from livestock and farm wastes. Earthworms constantly nourish upon the organic residues and manufacture casts. Vermi compost is a term normally given to the casts produced by earthworm. Casts of earthworms function as a good quality supply of manure for rising crops because they are typically wealthy in organic matter and nutrients. Earthworms, which are exclusively appropriate for the preparation of vermi compost are *Eudrilus eugeniae, Eisenia fetida,* and *Perionyx excavatus*.

11.1.4 Merits of Vermi Compost

1. Helps in improving the pH of soil, i.e., salinity and alkalinity of the soil
2. Helps in removing some of the toxic components from the soil
3. Helps in providing macroelements and small dose of essential elements
4. Helps in reducing the translocation of chemical fertilizers especially nitrogenous fertilizers
5. Helps in increasing the fertility of soil
6. Helps in increasing the water absorption capacity of the loose soil and penetration in compact soil
7. Assists in dropping soil influx by nematodes
8. Vermi compost is also full of enzymes responsible for general growth of plants, hormones such as auxins and cytokinins, and microorganisms for raising the proportion of nitrogen
9. Helps out in dropping the infestation of termites

11.2 Green Manure

For growing crops, numerous green manure crops give adequate nitrogen and organic matter. In a time range between 45 and 60 days, these can supply 60–90 kg nitrogen. Green manuring moreover aids to accelerate the nutrient cycling process and put together accessible nutrients to the crops by supplying large amount of easily decomposable organic matter to the soil.

11.2.1 Benefits of Using Green Manure

1. With modest additional labor, a low-cost method of improving crop is presented by green manuring.
2. Green manures are particularly significant where not enough animal manure is obtainable, e.g., on farms.
3. They aid in increasing the soil fertility.
4. Green manures add on different nutrients.
5. By improving drainage and letting additional air into the soil, they advance soil structure.
6. Green manures facilitate sandy soil not to drain so quickly by assisting soil to grasp added water.
7. Since the roots go through deep in the soil and grasp the plant in position, green manures furthermore help out to prevent the soil from being carried away by wind (Deshmukh 2010).

Inorganic fertilizers used in traditional methods might not protect soil structure, but may cause large fluctuation in the ion concentration and pH of the soil solution. These can considerably decrease a number of soil faunal populations, particularly earthworms. Organic manuring applications tend to preserve soil structure and are not disrupting to the soil chemical surrounding. It contributes to manage microbial pathogens and supports populations of valuable soil fauna. Where inorganic fertilization procedures stop working to preserve soil organic substance intensity, and consequently soil structure, they restrain crop rooting and decrease the water preservation capability of the soil. The accessibility of micronutrients and retention of macronutrients are improved by organic manuring procedures. The use of extremely soluble inorganic fertilizers usually results in greater losses of macronutrients, which can injure crop roots and interrupt crop uptake of all nutrients. It may be due to nutrient imbalance.

Inorganic fertilization procedures, which harm soil structure, may limit the cycling of a number of crop nutrients, and speed up the loss of others. Nutrient cycling effectiveness in farming systems has significant implications for water resources. Adding up of phosphorus to surface water results in eutrophication. Phosphorus losses are likely to be higher for traditional than organic farms. This is due to soil erosion and removal or unintentional loss of livestock wastes. Contamination of groundwater with nitrates in a lot of areas is strongly associated with agricultural practices, but the comparative impacts of organic and conventional structures are hard to estimate (Hodges 2012).

12 Implications of Organic Farming

IFOAM has outlined some principles for organic farming which state the maintenance of ecology and also the avoidance of all pollution spreading practices. Production values and economic feasibilities are also incorporated as key points (Deshmukh 2010).

12.1 Soil Health

Soil is given more care by organic farming. Organic farming enhances the status of micronutrients. Green manuring, in addition to rock phosphate and vesicular-arbuscular mycorrhiza (VAM) fungi, enhances the uptake of macronutrients such as phosphorus and nitrogen *(mycorrhizal associations produced by Glomeromycotan fungi are known as arbuscular mycorrhizas or vesicular-arbuscular mycorrhizas)*. An increase in the organic matter of soil occurs, pH of soil becomes stable, and organic carbon enhances, which are important for good biological environment of soil. Organic manure speeds up the nutrition transformation processes. The enhancement of soil's physical properties occurs in terms of aggregation stability, porosity, nutrient retention, soil aeration, and water-holding ability (Deshmukh 2010).

12.2 Water Pollution

Water contamination can be eliminated by prohibiting the use of synthetic peptides in organic farming as compared to conventional farming. No use of antibiotic and food additives in livestock production may decrease the risk factors involved in conventional livestock (Deshmukh 2010). Factors related to losses in nitrate leaching are nitrate load in the soil, different activities of crop production, and the type of crop harvested. It is also determined by inclusion of organic matter and nitrogen demand by crops. The nitrogenous humus being incorporated in the soil must be equal to the demand of nitrogen by the crop. Water pollution can only be caused by mishandling, otherwise organic farming has very low risk factors (Deshmukh 2010).

12.3 Air Pollution

Air pollution in organic farming is less than that in the conventional farming due to the following reasons:

- Excess of nitrogen is less in organic system and is likely to emit little amount of gaseous nitrogen. Losses of ammonia can occur through management and spreading of farmyard manure (FYM).
- Organic farming forbids the use of many wastes, which may result in water, soil, and air pollution.

In organic farming, emission of carbon dioxide is lower due to prohibition of use of chemicals (Deshmukh 2010).

13 Present Status of Organic Farming

The requirement of safe and sound food, in competition to improved environmental consciousness, has resulted in a rising demand for organic stuff. In developed countries, there is a steadily growing market for organic products, driven by rising consumer awareness for health and environment, which offers farmers a chance to produce for the best price markets and hence an opportunity to augment their farm profitability and improve their livelihoods. Organic agriculture is seen as a sustainable method of attaining social, environmental, and economic goals by producing above-average income in contrast to similar conventional farm sizes, new job opportunities due to comparatively elevated labor inputs, and improved habitat circumstances as synthetic fertilizers and pesticides are not applied.

14 Causes of Low Adoption of Organic Farming

In organic farming, the foremost challenges are management of nutrients and increase of yield (Toumisto et al. 2012). Although adaptation to organic agriculture often comes with a decrease in crop yield, proponents of organic agriculture highlight the sustainability mainly due to improvement in organic material-associated soil excellence. Based on current research on methods driving soil organic material turnover, though, it somewhat comes into view that low-input agroecosystems may convert to lesser competence in terms of substrate use by heterotrophs which might have an effect on soil organic material storage space in the long run. A collection of field statistics verifies lower use competence in some organic soils and thus questions the claim of a largely sustainable use of the soil reserve in organic agricultural methods (Leifeld 2012).

15 Conclusion and Future Perspectives

Organic farmers depend greatly on versatile understanding of soil science and ecology. Modern organic farming methods are used to guarantee fertility and pest/weed control with conventional techniques of crop rotation. Organic farming has to be implicated as a feasible option compared to conventional approaches in agriculture. Thousands of farmers have transformed to this method as a result of a higher demand for organically developed foodstuff. Organic farming also involves access to food by dropping threats of disease, raising productivity and biodiversity over the lasting period, and giving means for limited manufacture and access to food. Other advantages of organic agriculture beyond the purely financial ones include conservation of natural resources, health protection, risk reduction, increased flexibility to adverse weather, and farmer authority through the attainment of knowledge and higher dependence on limited inputs. Sustainable earnings in organic farming would not only include safe food production but also shielding of natural environment and maintenance of limited assets. Today, the organic agriculture sector is one of the fastest growing food segments, thrust to which is provided by many factors like introduction of policies that prove to be encouraging for organic agriculture, taking away of government funding on agricultural inputs, controversial debate on genetic modification related to food safety and crisis aggravated by foot and mouth and mad cow diseases and dioxin-contaminated food. On the demand side, forceful marketing and promotion strategies of supermarkets and retailers have produced new marketing prospects in northern countries. Food retailing chains have played a significant role in promoting the market growth for organic food products.

References

Akeenan (2011) Can organic farming really reduce GHG level. www.verus-co2.com/blog/?cat=27. Accessed 25 July 2013 (Versus Carbon Neutral)

Biello D (2012) Will organic food fail to feed the world? www.scientificamerican.com/article.cfm?id=organic-farming-yields-and-feeding-the-world-under-climate-change. Accessed 25 July 2013 (Scientific American)

Committee on Pesticides in the Diets of Infants and Children (1993) Pesticides in the diets of infants and children. Committee on Life Sciences, National Research Council, National Academy Press, Washington DC. www.nap.edu/openbook.php?isbn=0309048753. Accessed 25 May 2013

Dangour DA, Aikenhead MA, Hayter MA, Allen DE, Lock DK, Uauy PR (2009) Comparison of putative health effects of organically and conventionally produced foodstuffs: a systematic review. Nutrition and Public Health Intervention Research Unit. London School of Hygiene and Tropical Medicine. www.food.gov.uk/multimedia/pdfs/organicreviewappendices.pdf. Accessed 25 May 2013

Department of Agriculture and Food, Australia (2010) www.agric.wa.gov.au/PC_94253.html. Accessed 25 May 2013

Deshmukh SM (2010) Organic farming: principles, prospects and problems. www.charoka.com/Organic-Farming-Principles-Prospects-and-Problems/b/60009. Accessed 25 April 2013

Fan T, Stewart B, Payne W, Yong W, Luo J (2005) Long-term fertilizer and water availability effects on cereal yield and soil chemical properties in northwest China. Soil Sci Soc America J 69:842–855

Fließbach A, Oberholzer HR, Gunst L, Mader P (2007) Soil organic matter and biological soil quality indicators after 21 years of organic and conventional farming. Ecosyst Environ 118:273–284

Food and Agriculture Organization of United States (2013) AGP—What is integrated weed management. www.cgiar.org/wp-content/uploads/2013/04/eco-efficiency_book.pdf. Accessed 27 July 2013

Forman J, Silverstein J (2012) Organic foods: health and environmental advantages and disadvantages. Pediatrics 130(5):25–79

Gabriel D, Roschewitz I, Tscharntke T, Thies C (2006) Beta diversity at different spatial scales: plant communities in organic and conventional agriculture. Ecol Appl 16(5):2011–2021

Gilbert N (2012) Organic farming is rarely enough. Nature.com. www.nature.com/news/organic-farming-is-rarely-enough-1.10519. Accessed 26 May 2013

Halberg N, Alroe H, Knudsen M (2006) Global development of organic agriculture: challenges and prospects. www.cabi.org/CABeBooks/default.aspx?site=107&page=45&LoadModule=PDFHier&BookID=282. Accessed 26 May 2013

Herrera-Estrella L, Alvarez-Morales A (2001) GM crops: hope for developing countries? EMBO Rep. 2001 April 15; 2(4):258–258. www.ncbi.nlm.nih.gov/pmc/articles/PMC1083872. Accessed 26 May 2013
Hole D, Perkins A, Wilson J, Alexander I, Grice P, Evans A (2005) Does organic farming benefit biodiversity? Biol Conserv 122:113–130
Holt-Gimenez E (2000) Hurricane mitch reveals benefits of sustainable farming techniques. http://www.panna.org/legacy/gpc/gpc_200012.10.3.02.dv.html. Accessed 26 May 2013 (PANNA)
International Conference on Organic Agriculture and Food Security (2007) 3–5 May, 2007, Rome, Italy. ftp://ftp.fao.org/docrep/fao/meeting/012/j9918e.pdf. Accessed 4 July 2014
International Federation of Organic Agriculture Movements (2008) Classic.ifaom.org/press/press/2008/20090915_IFAOM_Annual_Report_2008.php. Accessed 26 May 2013
Isaacs JR (2012) Organic farming keeps carbon out of the atmosphere. http://news.mongabay.com/2012/1128-isaacs-organic-carbon. Accessed 26 May 2013
John P (2007) China's organic revolution. J Org Syst 2(1):1–11. http://orgprints.org/10949/01/10949.pdf. Accessed 26 May 2013
John P (2011) Nanomaterials in food and agriculture: the big issue of small matter for organic food and farming. Proceedings of the third scientific conference of ISOFAR (International Society of Organic Agriculture Research), pp 96–99
Kirchmann H (2007) Comparison of long-term organic and conventional crop-livestock systems on a previously nutrient-depleted soil in Sweden. Agron J 99:960–972
Kotschi J, Muller-Samann K (2004) The role of organic agriculture in mitigating climate change: a scoping study. International Federation of Organic Agriculture Movements, Bonn. www.shop.ifaom.org/bookstore/download-preview/ClimateStudy_intro.pdf. Accessed 26 May 2013
Leifeld J (2012) How sustainable is organic farming. Agric Ecosyst Environ 150:121–122
Lotter DW (2003) Organic agriculture. J Sustain Agric 21(4). http://donlotter.net/lotter-organicag.pdf. Accessed 26 May 2013
Mader P, Fliessbach A, Dubois D, Gunst L, Fried P, Niggli U (2002) Soil fertility and biodiversity in organic farming. Science 296:1694–1697
Magkos F, Arvaniti F, Zampelas A (2006) Organic food: buying more safety or just peace of mind? A critical review of the literature. Crit Rev Food Sci Nutr 46(1):23–56. doi:10.1080/10408690490911846.PMID16403682
Mercola D (2013) How organic farming could release us from the curse of fertilizer. Healthimpactnews.com/2013/how-organic-farming-could-release-us-from-the-curse-of-fertilizer/. Accessed 14 May 2012
Moreau TL (2007) International conference on organic agriculture and food security. Rome, Italy, 3–5 May 2007
Morgera E, Caro CB, Duran GM (2012) Organic agriculture and the law. FAO, Rome
Nascimbene J, Marini L, Paoletti M (2012) Organic farming benefits local plant diversity in vine yard farms located in intensive agricultural landscapes. Environ Manage 49:1054–1060
Nastis SA (2013) Crop biodiversity repercussions of subsidized organic farming. Land Use Policy 32:1–380
Oliveira A, Moura C, Enéas GF, Claudia AM, Laurent U, Maria Raquel AM (2013) The impact of organic farming on quality of tomatoes is associated to increased oxidative stress during fruit development. PLoS ONE 8(2):e56354. doi:10.1371/journal.pone.0056354
Organic Certification (2013) www.michiganorganic.msu.edu/michiganorganic/organic-certification. Accessed 14 May 2012
Pimentel D (2006) Impacts of organic farming on the efficiency of energy use in agriculture. The Organic Center, Cornell University, Ithaca
Pottorff L (2010) Some pesticides permitted in organic gardening. Denver. www.colostate.edu/Depts/CoopExt/4DMG/VegFruit/organic.htm. Accessed 14 May 2013
Schonbeck M (2010) An organic weed control toolbox. Virginia Association for Biological Farming
Seufert V, Ramankutty N, Foley JA (2012) Comparing the yields of organic and conventional agriculture. Nature 485:229–232
Sheer TB, Betarbet R, Testa CM, Seo BB, Richardson JR, Kim JH et al (2003) Mechanism of toxicity in rotenone models of Parkinson's disease. J Neurosci 23(34):10756–10764

Singh S, George R, Prafull B (2012) Organic farming practices of plains and hills farmers and their extent of compliance with National Program for Organic Production (NPOP) guidelines. J Agric Sci 3:55–62

Sinha R, Herat S (2012) Organic farming: producing chemical-free, nutritive and protective food for the society while also protecting the farm soil by earthworms and vermicompost—reviving the dreams of sir Charles Darwin. Agric Sci Res J 2:217–239

Sundrum A (2012) Organic livestock farming: a critical review. Livest Prod Sci 67(3):207–215

Szykitka W (2004) The big book of self-reliant living: advice and information on just about everything you need to know to live on planet earth. The Lyons Press, Guilford

Theriault M (2006) Growing demand for organic produce helps young farmers. The Bellingham Herald. www.truth-out.org/archive/item/60536:growign-demand-for-organic-produce-helps-young-farmers. Accessed 14 May 2012

Toumisto H, Hodge I, Riordan P, Macdonald D (2012) Does organic farming reduce environmental impacts—A meta analysis of European research. J Environ Manage 112:309–320

Trewavas A (2001) Urban myths of organic farming. Nature 410:409–410

van Elsen T (2000) Species diversity as a task for organic agriculture in Europe. Agric Ecosyst Environ 77:101–109

Wang S, Li Z, Fan G (2012) Soil quality and microbes in organic and conventional farming systems. Afr J Microbiol Res 6:5077–5085

Watson CA, Atkinson D, Gosling P, Jackson LR, Rayn FW (2002) Managing soil fertility in organic farming systems. Soil Use Manage 18:239–247. doi:10.1111/j.1475-2743.2002.tb00265.x. http://orgprints.org/8060/. Accessed 14 May 2012

Yadav SK, Babu S, Yadav MK, Singh K, Yadav GS, Pal S (2013) A review of organic farming for sustainable agriculture in Northern India. Int J Agron. http://dx.doi.org/10.1155/2013/718145. Accessed 14 May 2012

Ziesemer J (2007) Energy use in organic food systems. Natural Resources Management and Environment Department, Food and Agriculture Organization of the United Nations, Rome

Chapter 11
The Role of Cytological Aberrations in Crop Improvement Through Induced Mutagenesis

Mohammad Imran Kozgar, Shaheen Hussain, Mohd Rafiq Wani and Samiullah Khan

Abstract Mutation breeding for the past few decades has been pivotal in developing such high-yielding varieties which greatly help to feed the ever-growing global human population. Although induced mutations have great relevance for growing superior plant types in different crops plant, most of them are lethal or semi-lethal and do not have any practical values possibly due to the doses monitored and/or mutagens employed. Thus, to administer successful mutagenesis, selection of efficient mutagen and treatment is a basic prerequisite. The cytological analysis with respect to meiotic or mitotic behavior is considered one of the most reliable indices to estimate the potency of a mutagen. Hence, investigations on cytological aberrations and their genetic consequences form an integral part for most of the mutation studies and provide a considerable clue to assess the sensitivity of plants for different mutagens. In addition, the cytogenetic information vis-à-vis chromosome deformities provides an exhaustive overview related to the improvement of the desired trait via induced mutagenesis. The present documentation is an attempt to reveal the impact of mutagens on cytological behavior and their overall role in crop improvement.

Keywords Chromosomal abnormalities · Genetic variation · Improved traits · Developmental processes

M. I. Kozgar (✉)
KERB Biotech, Bemina, Srinagar, Jammu and Kashmir 190018, India
e-mail: m.i.kozgar@gmail.com

S. Hussain
Department of Biosciences, Jamia Millia Islamia, New Delhi 110025, India
e-mail: shaheenfatma11@gmail.com

M. R. Wani
Department of Botany, Government Degree College (Boys), Anantnag 192102, Jammu and Kashmir, India
e-mail: botanyrafiq@gmail.com

S. Khan
Mutation Breeding Laboratory, Department of Botany, Aligarh Muslim University, Aligarh, Uttar Pradesh 202002, India
e-mail: khan_drsami@yahoo.co.in

P. Ahmad et al. (eds.), *Improvement of Crops in the Era of Climatic Changes,* Volume 2, DOI 10.1007/978-1-4614-8824-8_11,

1 Introduction

Cytological abnormalities are considered one of the dependable criteria for estimating the potency of mutagenic effects, besides manipulating chromosome segments or whole sets of chromosomes to solve a particular problem in different crop plants (Reddy 1990). The spectrum of chromosomal abnormalities is generally broad at particular stages of cell division (especially diakinesis–metaphase) of meiosis (Azad 2011). Studies on different plant species have revealed that the decline in seed production is correlated with cytological irregularities (La Fleur and Jalal 1972; Smith and Murphy 1986; Consolaro et al. 1996; Kumar and Rai 2007; Kumar and Gupta 2007). It has been reported that chromosomal mutations leading to the formation of nonfunctional gametes are the most common effects of mutagen-induced sterility with reduced reproductive capacity (Swaminathan et al. 1962). In higher plants, cytological abnormalities due to single or combination treatments of various mutagens have been comprehensively reported by several workers in different crop species like *Hordeum vulgare, Pisum sativum, Triticale, Nigella sativa, Vigna* spp., *Physalis, Lycopersicon esculentum, Lens culinaris, Solanum melongena, Cicer arietinum, Capsicum annuum, Lathyrus sativus, Chlorophytum comosum,* and *Catharanthus roseus* (Swaminathan et al. 1962; Jayabalan and Rao 1987; Venkateswarlu et al. 1988; Ignacimuthu and Babu 1989a, b; Edwin and Reddy 1993; Mitra and Bhowmik 1996; Dhamayanthi and Reddy 2000; Pagliarini et al. 2000; Goyal and Khan 2009; Fatma and Khan 2009; Husain et al. 2013).

The application of a varied number of mutagens, having different physicochemical properties, alters the chemical pathways and provides information regarding the molecular actions of chromosomes and the genes involved through molecular assays (Haliem et al. 2013). Cytogenetic tests are regarded as the indicators of genotoxicity, genetic variation, cytotoxicity, and mutagenic potency (Juchimiuk and Maluszynska 2003; Husain et al. 2013). In the light of previously documented works, the main goal of compiling this literature at one place is to evaluate and reveal the cytogenetic paradigm in terms of aberrations caused by mutagenesis for better understanding of mutational processes.

2 Major Types of Cytological Aberrations

In any mutation breeding program, various chromosomal aberrations are noticed and the different types that come across from experimental mutation studies are illustrated in Fig. 11.1 (aberrations of a model crop *Vicia faba* L.).

2.1 Stickiness

Sticky chromosomes were first reported in maize (Beadle 1932) and are mostly seen as intense chromatin clustering at the pachytene stage. The phenotypic

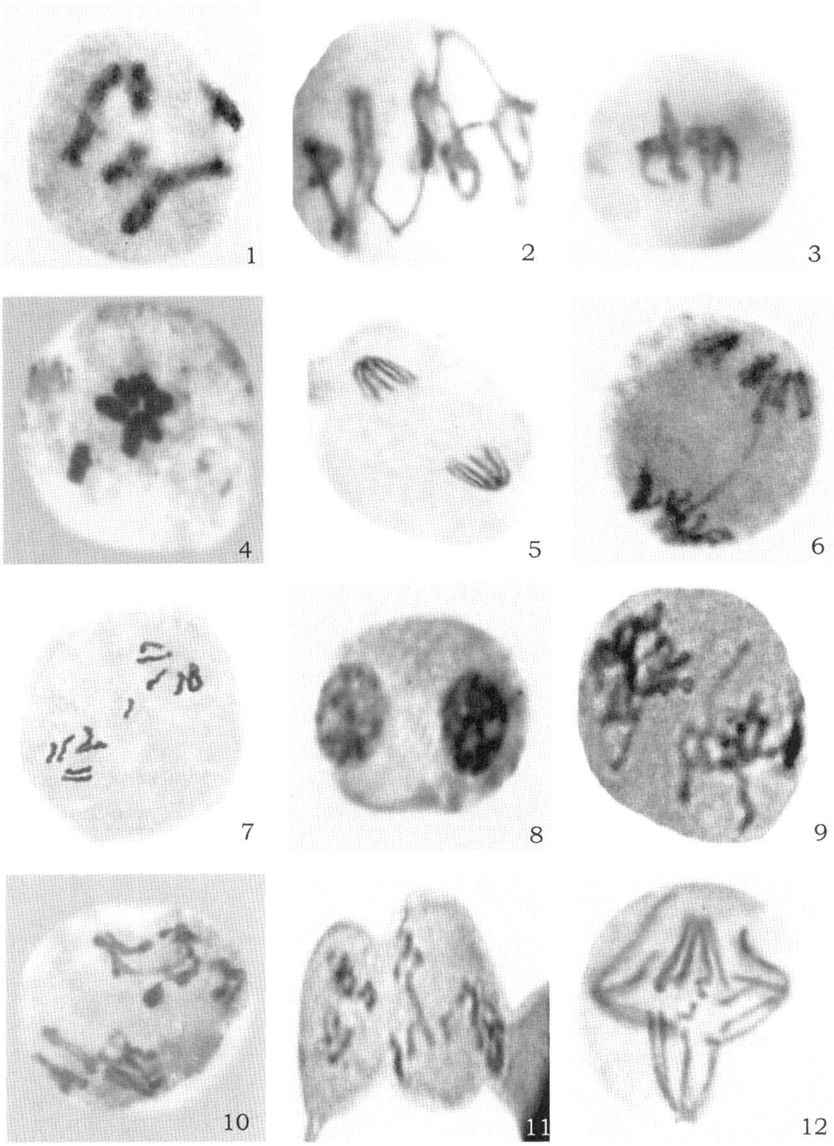

Fig. 11.1 Control diakinesis (*1*), control metaphase (*2*), stickiness at metaphase I (*3*), stray bivalent with sticky chromosomes at metaphase I (*4*), control anaphase I (*5*), bridge at anaphase I (*6*), laggard at anaphase I (*7*), control telophase I (*8*), control metaphase II (*9*), stray chromosomes at metaphase II (*10*), cytomixis at metaphase II (*11*), and laggard at anaphase (*12*). (Source: Husain et al. 2013)

manifestation of stickiness may vary from mild, when only a few chromosomes of genome are involved, to intense with the formation of pycnotic nuclei, that may involve the entire genome culminating in chromatin degeneration. Chromosomal stickiness may be caused by genetic or environmental factors. Genetically controlled stickiness has been described in various cultivated plants such as maize (Golubovaskaya 1989; Caetano-Pereira et al. 1995), pearl millet (Rao et al. 1990), and wheat (Zanella et al. 1991). Different reasons have been put forward for the formation of sticky chromosomes; however, the prime cause and biochemical basis of chromosomal stickiness are still anonymous. Sticky chromosomes might result from depolymerization of nucleic acids caused by mutagenic treatments or due to the partial dissociation of nucleoproteins and alterations in their pattern of organization (Kumar et al. 2003; Kumar and Srivastava 2010). It may also arise due to the improper clustering of chromosomes at any phase of the cell cycle, which causes the chromatids to be connected by subchromatid bridges (Mc Gill et al. 1974).

2.2 Bridges

Bridges may be produced due to subchromatic exchanges or unequal exchange of dicentric chromosomes. The occurrence of breaks at the same locus and their lateral fusion leads to the formation of a dicentric chromosome which is pulled equally to both the poles forming bridges. Another striking reason seems to be the delay or failure of terminalization of chiasmata, as a result of which the ends of separated chromosomes remain midway between the two poles. Sax (1960) and Saylor and Smith (1966) suggested that the formation of the chromatin bridge might be either due to the failure of chiasmata in a bivalent to terminalize or due to the stretching of chromosomes between the poles.

2.3 Cytomixis

Cytomixis generally refers to the migration of chromatin materials from one cell to the neighboring cells through cytoplasmic/intercellular connections. It is considered to be the production source of aneuploid and polyploid gametes (Kaul 1990; Yen et al. 1993). The intensity of chromosome passage from one cell to the other during cytomixis depends upon the number and nature of cytoplasmic connections (de Souza and Pagliarini 1997). The important factors proposed to cause cytomixis are the influence of genes, fixation effects, pathological conditions, herbicides, and temperature (Caetano-Pereira and Pagliarini 1997). Cytomixis may have serious genetic consequences by causing deviation in the chromosome number and may represent an additional mechanism for the origin of aneuploidy and polyploidy (Sarvella 1958).

2.4 Univalents and Multivalents

The univalent is either a chromosome which fails to pair at the zygotene stage or the one which pairs to form a bivalent whose two component chromosomes are, however, separated at the diplotene stage due to non-chiasma formation or precocious anaphase separation of bivalents (Sarbhoy 1977). Multivalent can be attributed to irregular pairing and breakage followed by translocation and inversion. According to Koduru and Rao (1981), the univalents which are formed just before metaphase I lie close to each other with their kinetochores directed towards the spindle axis and the arms towards the outside and also near to the equator. The anaphasic separation of univalents and bivalents depends on their position during metaphase I (Koller 1938). According to him, the bivalents and univalents, which are at or near the equator during metaphase I, segregate normally in anaphase I (Ostergren 1951).

2.5 Laggards

A laggard is a chromosome that did not overlap along the long axis of the spindle with any of the properly segregating chromosome (Janicke et al. 2007); henceforth, it is the cytological aberration, wherein the chromosomes do not attach to the spindle apparatus, and thus randomly segregate in daughter cells. The occurrence of micronuclei may cause variation in the number and size of pollen grains resulting from a mother cell, especially when found in meiotic telophase II stages (Bhattacharjee 1953). The disturbed polarity at anaphase and telophase stages could be due to the spindle disturbances. These abnormal segregations vis-à-vis laggard patterns have not been seen to resolve the balanced products, and hence the random segregation of multiple chromosome types produces mostly aneuploid gametes (Woodhouse et al. 2009).

2.6 Disturbed Polarity

Disturbed polarity is mostly seen at telophase II and could be due to alteration in the genes that control the reactions of biochemical pathways determining the position of spindle poles as reported by Suarez and Bullrich (1990) and Ansari and Ali (2009). Such abnormalities may be attributed to spindle mutations. The cellular abnormalities are mostly studied in pollen mother cells as meiotic stages; however, the studies on root tips of treated plants for mitotic analyzation cannot be avoided. The decline in mitotic index, due to mutagenic treatments, has also been studied earlier (Kaur and Grover 1985; Adam et al. 1990; Ahmad and Yasmin 1992; Kumar and Kumar 2000) in various crop plants. Mitotic inhibition might be due to the mitotic poison which causes metabolic imbalance, interferes with the synthesis and structure of DNA, and results in physiological effects and structural changes in chromosomes during the division (Soni et al. 1982).

3 Interrelation Between Mutagens and Cytological Aberrations

Several agents have been reported to cause chromosomal stickiness, including X-rays, gamma rays, temperature, herbicides, and/or some chemicals present in soil (Caetano-Pereira et al. 1995). Different types of abnormalities like chromatin bridges, laggards, fragments, cytomixis, micronuclei, etc. have been reported in the past following mutagenic treatments of varying nature (Kumar and Gupta 1978). Chromosomal abnormalities and disturbed cytological division patterns are found at different stages like laggards and bridges at anaphase and sticky chromosomes, fragments, and ring chromosomes at metaphase as have been reported by Grover and Tejpual (1982) in *Vigna radiata* using maleic hydrazide (MH), gamma rays, and their combinations. Shah et al. (1992) reported misorientations at metaphase, bridges at anaphase, fragmentation, and multinucleolate conditions in *Vigna mungo* using gamma rays. Kumar and Gupta (1978) have observed an induced asynaptic mutant, which had shown a large number of univalents and irregular anaphase division in black gram.

The mutagenic action alters the pathways of normal cytological divisions which lead to abnormalities. For example, the mutation may alter the functioning of proteins involved in chromosome origination, and thus leads to stickiness (induced stickiness) or can alter the functioning of structural gene coding for them (hereditary stickiness). Mutagen-induced chromosome stickiness has been described in various crops like *Phaseolus vulgaris* (Al-Rubeai and Godward 1981), *Capsicum* spp. (Katiyar 1978), *Turnera ulmifolia* (Tarar and Dnyansagar 1980), *Lycopersicon esculentum* (Jayabalan and Rao 1987), *Nigella sativa* (Mitra and Bhowmik 1996), and *Secale montanum* (Akgün and Tosun 2004). The univalents and multivalents are found in mutagen-treated population, and their frequencies are generally higher at higher mutagenic doses (Gupta and Roy 1985). Kaul and Murthy (1985), in their communication, refer to such genes that control the normal spindle formation leading to proper separation of bivalents and univalents. Mutation in these genes could lead to abnormal spindle formation and function causing uneven segregation of chromosomes to the opposite poles, thus resulting in additional number of microspores or formation of micronuclei (Sjodin 1970). The laggards observed during the experimental mutagenic studies might be due to the delayed terminalization, stickiness of chromosome ends, or failure of chromosomal movements (Jayabalan and Rao 1987). The laggards and micronuclei are regarded as the major factors in promoting pollen sterility in plants (Patil 1992). The percentage of cells with univalents and laggards in mutation breeding programs seems to be higher than the percentage of cells with micronuclei and indicates that some chromosomes get involved in the main nucleus (Dewitte et al. 2010; Gonza'lez-Sa'nchez et al. 2011).

The laggard chromosomes and their presence as univalents and unequal separation may result in the production of aneuploid gametes which may be utilized in breeding programs (Tel-Zur et al. 2005; Miko 2008). Mutagen-induced multivalent formations have been reported in various plants (Dhamayanthi and Reddy 2000;

Khan et al. 2009; Gulfishan et al. 2010, 2011; Sharma and Gohil 2011). In most cases, multivalent formation among mutagen-treated plants has been attributed to reciprocal translocation which results in segmental homology between nonhomologous chromosomes. However, Kleinhofs et al. (1974) reported that sodium azide (SA) induces mostly gene mutations with negligible frequency of chromosomal aberrations. In addition to altering the cytological behavior, a relative account of mutagenic effects on pollen fertility of crop species has been reported by Bhat et al. (2007). There is a close relationship between pollen sterility and cytological (meiosis) abnormalities as has been also proved by Goyal and Khan (2009). In general, the cytological abnormalities have been found to be dose dependent (Zhou et al. 2003; Goyal and Khan 2009; Husain et al. 2013). It is being viewed that physical mutagens produce more cytological abnormalities than chemical ones, but Dhamayanthi and Reddy (2000) contrarily reported more cytological abnormalities from chemical mutagen-treated *Capsicum annuum* L. Most researchers in the past reaffirmed that meiosis and mitosis is a complex and coordinated activity involving several genes, and that mutation in any one of these may lead to irregularities.

4 Case Studies

There have been ample case studies on the effect of mutagens in the past from our coworkers on different plant species; however, the two cases have been mentioned herein. In one of the case studies, the effects of two potent chemical mutagens, viz., ethylmethane sulphonate (EMS) and SA, on the cytological behavior of isolated morphological mutants of *Vigna mungo* have been elucidated (Goyal and Khan 2010), and the summary of the results has been provided in Table 11.1. However, in the second case, the effects of hydrazine hydrate (HZ) and MH on two varieties of *Vicia faba* have been reported by Husain et al. (2013) and accordingly illustrated in Tables 11.2 and 11.3. In both case studies, it has been concluded that certain cytological abnormalities (e.g., stickiness) lead to the reduction of pollen fertility, hence reducing the yield. Some abnormalities like cytomixis are beneficial in inducing the aneuploidy and polyploidy levels as have been reported earlier by Lavia et al. (2011), Kumar and Dwivedi (2013), and Kumar and Naseem (2013).

5 Modern Trends

Cytological tests help to analyze the frequencies of chromosomal aberrations, and in the past various methodologies were used to depict the same. However, with the advent of molecular diagnostics, modern methods are being adopted which lead proximity to perfection. One of the tests is the comet assay, which has been primarily used to test the apoptosis process and is subsequently exploited as a useful tool to investigate the capacity of DNA repair-damage processes in the cells induced

Table 11.1 Meiotic abnormalities induced by chemical mutagens in various morphological mutants. (Source: Goyal and Khan 2010; reproduced by permission)

Treatment/mutant	Total no. of PMCs observed	Total no. of abnormal PMCs	PMCs on metaphase I (%)		PMCs on anaphase I (%)		PMCs on telophase II (%)	Abnormal PMCs (%)	%age pollen fertility
			Univalents	Stickiness	Bridges	Laggards	Cytomixis		
Control	248	–	–	–	–	–	–	–	97.39
0.3% EMS Bushy	238	9	0.84 (2)		1.26 (3)	1.68 (4)	–	3.78	70.00
0.4% EMS Dwarf	249	14	2.81 (7)	0.80 (2)	–	2.00 (5)	–	5.62	48.25
0.02% SA Narrow leaves	225	3	–	–	–	–	1.33 (3)	1.33	96.17

PMCs pollen mother cells

by mutagens as has been advocated by Menke et al. (2001). The comet assay was used to detect the DNA damage of several species after the mutagenic treatments (Navarrete et al. 1997); however, it has not yet been adopted by a large number of researchers who are working in the field of mutation breeding in countries like India. Another method used for cytological assays was the TUNEL (terminal deoxynucleotidyl transferase deoxyuridine triphosphate (dUTP) nick end labeling) test and has been well illustrated to be used in plant species by Havel and Durzan (1996). The TUNEL test has been recommended for the evaluation of genotoxicity and mutagenicity of any new tested agent (Juchimiuk and Maluszynska 2003). The main advantages of the test include the detection of DNA breaks at a single nucleus and a short time requirement with an easy screening (Maluszynska and Juchimiuk 2005). With the advent of molecular markers adopted in different fields, newer assays were introduced to analyze the cells and one of the tests graded as being a useful assay is fluorescent in situ hybridization (FISH) which has provided a new insight into studying the chromosomal aberrations caused by mutagenesis. All the types of assays have been well illustrated and documented by Maluszynska and Juchimiuk (2005).

6 Conclusions and Future Perspective

Developing and disseminating the economically important mutant varieties to the end users has been a primary goal of any mutation breeding program. To fulfill the requirements of this goal, the basic information about the mutagenic sensitivity proved highly beneficial for breeders to chalk out policies vis-à-vis isolating the desired and potent mutants. Induction of cytological disturbances in the mitotic as well as meiotic cells is of great value, as it results in genetic damage that is transmitted to

Table 11.2 Frequency and spectrum of chromosome abnormalities induced by HZ in two varieties of *Vicia faba* L. (Source: Husain et al. 2013; reproduced by permission)

Treatment	Total no. of PMCs observed	PMCs with diakinesis to metaphase I/II					PMCs with anaphase I/II					PMCs with telophase I/II				Total frequency (%)
		Univalents	Multivalents	Stickiness	Stray bivalents	Cytomixis	Misorientation	Bridges	Laggard	Nonsynchronization	Unequal separation	Micronuclei	Disturbed polarity	Multinuclei	Cytomixis	
Var. NDF-1																
Control	255	–	–	–	–	–	–	–	–	–	–	–	–	–	–	–
0.01 % HZ	250	0.40	0.81	0.30	0.45	0.52	1.20	1.25	0.59	0.52	–	1.50	1.22	1.08	1.19	11.03
0.02 % HZ	241	0.80	0.85	0.60	0.55	0.59	1.25	1.26	0.62	0.58	0.98	1.55	1.23	1.20	1.20	13.26
0.03 % HZ	245	0.90	–	2.10	1.45	0.79	1.26	1.28	0.65	0.59	1.19	–	1.50	1.60	1.25	14.56
0.04 % HZ	252	1.20	0.80	2.11	1.50	0.99	1.27	1.30	1.20	0.60	1.20	1.56	–	1.65	1.28	16.66
	Total	*3.30*	*2.46*	*5.11*	*3.95*	*2.89*	*4.98*	*5.09*	*3.06*	*2.29*	*3.37*	*4.61*	*3.95*	*5.53*	*4.92*	*55.51*
Var. HB-405																
Control	255	–	–	–	–	–	–	–	–	–	–	–	–	–	–	–
0.01 % HZ	253	0.30	–	1.00	0.43	0.50	1.10	1.23	0.53	0.50	0.91	1.45	1.20	1.01	1.18	11.34
0.02 % HZ	245	0.75	0.80	–	0.59	0.55	1.17	1.25	0.59	0.58	–	1.52	1.21	1.19	1.19	11.39
0.03 % HZ	248	0.80	0.81	1.50	1.25	–	1.20	1.27	0.59	0.59	0.99	1.53	1.28	1.20	1.21	14.22
0.04 % HZ	254	1.19	0.83	2.08	1.48	1.10	1.21	1.29	–	0.60	1.19	1.50	–	1.64	1.20	15.31
	Total	*3.04*	*2.44*	*4.58*	*3.75*	*2.15*	*4.68*	*5.04*	*1.17*	*2.27*	*3.09*	*6.00*	*3.69*	*5.04*	*4.78*	*52.26*

PMCs pollen mother cells

Table 11.3 Frequency and spectrum of chromosome abnormalities induced by MH in two varieties of *Vicia faba* L. (Source: Husain et al. 2013; reproduced by permission)

Treatment	Total no. of PMCs observed	PMCs with Diakinesis to metaphase I/II					PMCs with anaphase I/II					PMCs with telophase I/II				Total frequency (%)
		Univalents	Multivalent	Stickiness	Stray bivalent	Cytomixis	Misorientation	Bridges	Laggard	Nonsynchronization	Unequal separation	Micronuclei	Disturbed polarity	Multinuclei	Cytomixis	
Var. NDF-1																
Control	225	–	–	–	–	–	–	–	–	–	–	–	–	–	–	–
0.004 % MH	240	0.88	0.89	1.02	1.00	0.79	1.22	1.67	0.78	–	1.50	2.01	1.20	1.19	1.50	15.64
0.006 % MH	241	1.10	0.99	2.15	1.50	0.98	1.29	1.68	0.79	2.20	1.88	–	1.50	2.20	1.70	19.96
0.008 % MH	253	1.25	1.25	2.19	1.51	1.10	1.30	1.80	1.25	2.25	1.99	2.20	1.50	2.25	1.80	23.64
	Total	*3.23*	*3.13*	*5.36*	*4.01*	*2.87*	*3.81*	*5.15*	*2.82*	*4.45*	*5.37*	*4.21*	*4.20*	*5.64*	*5.00*	*59.24*
Var. HB-405																
Control	255	–	–	–	–	–	–	–	–	–	–	–	–	–	–	–
0.004 % MH	245	0.86	0.87	1.00	1.00	0.69	1.10	–	0.60	2.19	1.50	2.00	1.19	1.18	1.40	15.58
0.006 % MH	240	1.09	0.98	2.00	1.45	0.97	1.28	1.99	0.75	–	1.70	2.18	1.45	2.19	1.65	19.68
0.008 % MH	250	1.20	1.23	2.18	1.48	1.09	1.29	2.00	1.19	2.20	1.89	–	1.45	2.20	1.78	21.18
	Total	*3.15*	*3.08*	*5.18*	*3.93*	*2.75*	*3.67*	*3.99*	*2.54*	*4.39*	*5.09*	*4.18*	*4.09*	*5.57*	*4.83*	*56.44*

PMCs pollen mother cells

the subsequent generation. The cytological analysis clarifies the specific response of different genotypes to a specific mutagen and provides significant evidence for selection of a desirable trait (Avijeet et al. 2011). For the success of adopting cytological studies in the right direction, the espousal of techniques like comet assay, TUNEL, FISH, etc. has to be enhanced to a large extent in mutational experiments in order to reveal the hidden treasure of mechanisms which are involved to shape up the plant in the desired direction.

Acknowledgments The lead author highly acknowledges the financial assistance in the form of a fellowship (BSR-UGC) received from UGC, New Delhi, India, during the Ph.D. program (2008–2012) carried out at Aligarh Muslim University, Aligarh, India.

References

Adam ZM, Ebad FA, EL-Shaikh IA (1990) Alterations in nucleic acid protein content and mitotic division of *Vicia faba* root tip cells as affected by malathion and tamron insecticides. Cytologia 55:349–355
Ahmad S, Yasmin R (1992) Effect of methyl parathion and trimiltox on the mitosis of *Allium cepa*. Cytologia 57:155–160
Akgün I, Tosun M (2004) Agricultural and cytological characteristics of M_1 perennial rye (*Secale montanum* Guss.) as affected by the application of different doses of gamma rays. Pak J Biol Sci 7(5):827–833
Al-Rubeai MAF, Godward MBE (1981) Genetic control of radiosensitivity in *Phaseolus vulgaris* L. Environ Exp Bot 21:211–216
Ansari MYK, Ali Z (2009) Cytomorphological variabilities induced by aniline in *Vicia faba* L. J Cytol Genet 11:57–64
Avijeet C, Shukla S, Rastogi A, Mishra BK, Ohri D, Singh SP (2011) Impact of mutagenesis on cytological behavior in relation to specific alkaloids in opium poppy (*Papaver somniferum* L.). Caryologia 64(1):14–24
Azad SA (2011) Mutagen induced meiotic chromosomal abnormalities in variety Asha of mungbean. Indian J Sci Res 2(4):29–31
Beadle GW (1932) A gene for sticky chromosomes in *Zea mays*. Z Indukt Abstamm Verebungle 63:195–217
Bhat TA, Parveen S, Khan AH (2007) Meiotic studies in two varieties of *Vicia faba* L. (Fabaceae) after EMS treatment. Asian J Plant Sci 61:51–55
Bhattacharjee SK (1953) Cytogenetics of *Lens esculenta* Monesch. Caryologia 5:159–166
Caetano-Pereira CM, Pagliarini MS (1997) Cytomixis in maize microsporocytes. Cytologia 65:351–355
Caetano-Pereira CM, Pagliarini MS, Brasil EM, Martins EN (1995) Influence of aluminum in causing chromosome stickiness in maize microsporocytes. Maydica 40:325–330
Consolaro MEL, Pagliarini MS, Chaves LJ (1996) Meiotic behavior, pollen fertility and seed production in Brazilian populations of *Centella asiatica* (L.) Urban (Umbelliferae). Cytologia 61:375–381
Dewitte A, Eeckhaut T, Huylenbroeck JV, Bockstaele EV (2010) Meiotic aberrations during 2n pollen formation in *Begonia*. Heredity 104:215–223
Dhamayanthi KPM, Reddy VRK (2000) Cytogenetic effects of gamma rays and ethylmethane sulphonate in chilli pepper (*Capsicum annuum* L.). Cytologia 65:129–133
Edwin R, Reddy VRK (1993) Cytological effect of gamma rays and EMS in triticale. J Natcon 5(1):9–23

Fatma S, Khan S (2009) Diethyl sulphate induced meiotic abnormalities in *Vicia faba* L. J Phytol Res 22(2):195–198

Golubovskaya IN (1989) Meiosis in maize: *mei* genes and conception of genetic control of meiosis. Adv Genet 26:149–192

Gonza'lez-Sa'nchez M, Rosato M, Chiavarino M, Puertas MJ (2011) Chromosome instabilities and programmed cell death in tapetal cells of maize with B chromosomes and effects on pollen viability. Genetics 166:999–1009

Goyal S, Khan S (2009) A comparative study of chromosomal aberrations in *Vigna mungo* induced by ethylmethane sulphonate and hydrazine hydrate. Thai J Agric Sci 42(3):117–182

Goyal S, Khan S (2010) Cytology of induced morphological mutants in *Vigna mungo* (L.) Hepper. Egypt J Biol 12:81–85

Grover IS, Tejpaul SK (1982) Cytogenetical effects of gamma rays and maleic hydrazide in mungbean. Acta Bot Indica 10:210–216

Gulfishan M, Khan AH, Bhat TA (2010) Studies on cytotoxicity induced by DES and SA in *Vicia faba* var. Major. Turk J Bot 34:31–37

Gulfishan M, Khan AH, Haneef I, Bhat TA (2011) Genotoxic effect of diethyl sulphate in two varieties of *Capsicum annuum* L. Nucleus 54(2):107–111

Gupta SK, Roy SK (1985) Comparison of meotic abnormalities induced by gamma rays between a diploid and a tetraploid species of *Physalis*. Cytologia 50:167–175

Haliem EA, Abdullah H, AL-Huqail AA (2013) Oxidative damage and mutagenic potency of fast neutron and UV-B radiation in pollen mother cells and seed yield of *Vicia faba* L. Biomed Res Int. doi:10.1155/2013/824656

Havel L, Durzan DJ (1996) Apoptosis in plants. Bot Acta 109:268–277

Husain S, Kozgar MI, Jafrey IF, Khan S (2013) Meiotic changes in *Vicia faba* L. subsequent to treatments of hydrazine hydrate and maleic hydrazide. J BioSci Biotechnol 2(1):33–38

Ignacimuthu S, Babu CR (1989a) Induced variation in protein quantity and quality in the wild and cultivated urd and mungbean. Indian J Genet Plant Breed 49(2):173–181

Ignacimuthu S, Babu CR (1989b) Induced chromosomal abnormality and pollen sterility in wild and cultivated urd and mungbean. Cytologia 54:159–167

Janicke MA, Lasko L, Oldenbourg R, LaFountain JR (2007) Chromosome mal-orientations after meiosis II arrest cause non-disjunctions. Mol Biol Cell 18(5):1645–1656

Jayabalan N, Rao GR (1987) Gamma radiation induced cytological abnormalities in *Lycopersicon esculentum* Mill. var. Pusa Ruby. Cytologia 52:1–4

Juchimiuk J, Maluszynska J (2003) Detection of DNA fragmentation caused by chemical mutagens using the TUNEL test. In: Maluszynska J, Plewa M (eds) Bioassays in plant cells for improvement of ecosystem and human health. Wydawnictwo Uniwersytetu Śląskiego, Katowice, pp 133–138

Katiyar RB (1978) Radio-cytogenetical studies on *Capsicum* I. Meiotic anomalies. Cytologia 43:415–420

Kaul KK (1990) Cytomixis in pollen mother cells of *Alopecurus arundinaceus* Poir. Cytologia 55:169–173

Kaul MLH, Murthy TGK (1985) Mutant genes affecting higher plant meiosis. Theor Appl Genet 70:449–466

Kaur P, Grover IS (1985) Cytological effects of some organophosphorous pesticides. Cytologia 50:187–197

Khan Z, Gupta H, Ansari MYK, Chaudhary S (2009) Methylmethane sulphonate induced chromosomal variations in a medicinal plant *Cichorium intybus* L. during microsporogenesis. Biol Med 1(2):66–69

Kleinhofs A, Sander C, Nilan RA, Konzak CF (1974) Azide mutagenecity-mechanism and nature of mutants produced. In: Polyploidy and induced mutations in plant breeding. International Atomic Energy Agency, Vienna

Koduru PRK, Rao MK (1981) Cytogenetics of synaptic mutants in higher plants. Theor Appl Genet 59:197–214

Koller PC (1938) Asynapsis in *Pisum stivum*. J Genet 36:275–306

Kumar G, Dwivedi K (2013) Cytogenetical evidences of abnormal meiosis and 2n pollen formation via colchicine in microsporogenesis of *Brassica campestris* L. Int J Res Plant Sci 3(2):18–24
Kumar S, Gupta PK (1978) An induced sterile mutant in blackgram showing failure of chromosome pairing. Natl Acad Sci Lett 1:51–53
Kumar G, Gupta P (2007) Mutagenic efficiency of lower doses of gamma rays in black cumin (*Nigella sativa* L.). Cytologia 72(4):435–440
Kumar G, Kumar R (2000) Chromotoxic and mito-inhibitory effects of pesticides in *Trigonella foenum-graecum*. Cytol Genet 1:11–15
Kumar G, Naseem S (2013) Gamma-induced intra-specific polyploidy via cytomixis-mediated syncyte formation in *Papaver somniferum* (Papaveraceae). Turk J Bot 37:530–536
Kumar G, Rai PK (2007) EMS induced karyo-morphological variations in maize (*Zea mays* L.) inbreds. Turk J Biol 31:187–195
Kumar G, Srivastava P (2010) Comparative radio-cytological effect of gamma rays and laser rays on safflower. Rom J Biol 55(2):105–111
Kumar G, Kesarwani S, Sharma V (2003) Clastogenic effect of individual and combined treatments of gamma rays and EMS in *Lens culinaris*. J Cytol Genet 4:149–154
La Fleur JD, Jalal SM (1972) Chromosomal aberrations and fertility inter-relationships in prairie *Bromus inermis* Leyrss population. Cytologia 37:747–757
Lavia GI, Ortiz AM, Robledo G, Fernández A, Seijo G (2011) Origin of triploid *Arachis pintoi* (Leguminosae) by autopolyploidy evidenced by FISH and meiotic behavior. Ann Bot 108:103–111
Maluszynska J, Juchimiuk J (2005) Plant genotoxicity: a molecular cytogenetic approach in plant bioassays. Arh Hig Rada Toksikol 56(2):177–184
de Souza AM, Pagliarini MS (1997) Cytomixis in *Brassica napus* var. Oleifera and *Brassica compestris* var. Oleifera (Brassicaceae). Cytologia 62:25–29
McGill M, Pathak S, Hsu TC (1974) Effects of ethidium bromide on mitosis and chromosomes: a possible material basis for chromosome stickiness. Chromosoma 47:157–167
Menke M, Chen IP, Angelis KJ, Schubert I (2001) DNA damage and repair in *Arabidopsis thaliana* as measured by the comet assay after treatment with different classes of genotoxins. Mutat Res 493:87–93
Miko I (2008) Mitosis, meiosis and inheritance. Nat Educ 1:1. www.nature.com/scitable/topicpage/mitosis-meiosis-and-inheritance-476. Accessed 30 Dec 2011
Mitra PK, Bhowmik G (1996) Cytological abnormalities in *Nigella sativa* induced by gamma rays and EMS. J Cytol Genet 31:205–215
Navarrete MH, Carrera P, de Miguel M, de la Torre C (1997) A fast comet assay variant for solid tissue cells. The assessment of DNA damage in higher plants. Mutat Res 389:271–277
Ostergren G (1951) The mechanism of co-orientation in bivalents and multivalent: the theory of orientation by pulling. Heredity 37:85–156
Pagliarini MS, De Freitas PM, Batista LAR (2000) Chromosome stickiness in meiosis of a Brazilian *Paspalun accession*. Cytologia 65:289–294
Patil BC (1992) The induction of tetraploid in *Crotalaria linifolia* Linn. Cytologia 57:247–252
Rao PN, Kanganadhaam P, Nirmala A (1990) Behavior of sticky desynaptic mutant in pearl millet. Genetica 81:221–227
Reddy VRK (1990) Cytology of some morphological mutants in triticale. Acta Bot Indica 18:146–149
Sarbhoy RK (1977) Cytogenetical studies in the genus *Phaseolus* Linn. III evolution in the genus *Phaseolus*. Cytologia 42:401–413
Sarvella P (1958) Cytomixis and the loss of chromosomes in meiotic and somatic cells of *Gossypium*. Cytologia 23:14–24
Sax K (1960) Meiosis in inter-specific pine hybrids. Forest Sci 6:135–138
Saylor LG, Smith BN (1966) Meiotic irregularities in species of inter-specific hybrids in *Pisum*. Am J Bot 53:453–468
Shah D, Singh RR, Kaur A (1992) Gamma rays induced nucleolar behaviour and mitotic anomalies in *Vigna mungo*. Acta Bot Indica 20:300–302

Sharma G, Gohil RN (2011) Occurrence of differential meiotic associations and additional chromosomes in the embryo-sac mother cells of *Allium roylei* Stearn. J Genet 90(1):45–49
Sjodin J (1970) Induced asynaptic mutant in *Vicia faba* L. Hereditas 66:215–232
Smith SE, Murphy RP (1986) Relationships between inbreeding, meiotic irregularity and fertility in alfalfa. Can J Genet Cytol 28:130–137
Soni NK, Oswal P, Yadav SK (1982) Effects of solar eclipse on cell division in the meristematic root tip cells of *Allium sativum*. Cytologia 47:353–358
Suarez E, Bullrich L (1990) Meiotic spindle disturbances in a commercial bread wheat cultivar. Cytologia 55:79–86
Swaminathan MS, Chopra VL, Bhaskaran S (1962) Chromosome aberrations and frequency and spectrum of mutations induced by EMS in barley and wheat. Indian J Genet 22:192–207
Tarar JL, Dnyansagar VR (1980) Comparison of ethylmethane sulphonate and radiation induced meiotic abnormalities in *Turnera ulmifolia* Linn. var. Angustifolia Wild. Cytologia 45:221–231
Tel-Zur N, Abbo S, Mizrahi Y (2005) Cytogenetics of semi-fertile triploid and aneuploid intergeneric vine cacti hybrids. J Hered 96(2):124–131
Venkateswarlu M, Susheelamma BN, Kumar P (1988) Studies on induced mutation frequency in *Catharanthus roseus* (L.) G. don by gamma rays and EMS individually and in combination. Indian J Genet Plant Breed 48(3):313–316
Woodhouse M, Burkart-Waco D, Comai L (2009) Polyploidy. Nat Educ 2(1):1
Yen C, Yang JL, Sun GL (1993) Internal connection and cytomixis in inter-generic hybrid *Roegeneria ciliares* (Trin) Nevski with *Psathyrosatachys luashancia* Keng. Cytologia 58:187–193
Zanella CC, Bodanese-Zanettini MH, Moraes-Fernandes MIB, Zinn DM (1991) Differential effect of soil acidity and lime treatment on the chromosome of two wheat cultivars. Rev Bras Genet 14:1021–1032
Zhou Y, Shi S, Wei Z, Wu J, Nie Q, Li G (2003) A procedure to minimize the linkage drag in backcross breeding of rapeseed (Brassica napus) through irradiation. In: Improvement of new and traditional industrial crops by induced mutations and related biotechnology, IAEA-TEC-DOC-1369. International Atomic Energy Agency, Vienna, pp 125–131

Chapter 12
Wheat Improvement: Historical Perspective and Mutational Approach—A Review

Subzar Ahmad Sheikh, Mohd Rafiq Wani, Mohd Imran Kozgar and Parvaiz Ahmad

Abstract Wheat (*Triticum aestivum* L.) occupies a prominent position among the cereal crops and supplements dietary requirement of nearly one-third of the world's population. The importance of wheat can be gauged by the fact that the Food and Agricultural Organization (FAO) of the United Nations has chosen a wheat spike in its symbol with the description "Let there be bread." In the present scenario, the likings for wheat-based food products have increased, thus widening the scope for industrial production and consequently providing the livelihood to millions of people world over directly or indirectly. In India, the All India Coordinated Wheat Improvement Project established in 1965 by Indian Council of Agricultural Research (ICAR) and later raised to the status of Directorate of Wheat Research is largely involved with the coordination of wheat research in the country. More than 300 varieties of wheat suitable for different agronomical conditions of the country have been released by the organization. Though, at present, India is self-sufficient in wheat production, but the emerging challenges of population explosion, environmental stresses, new pathogenic races, decreased arable lands, depleting water resources, and degraded soils make it imperative to consistently aim at the development of efficient and promising varieties so that the food security in the country can be ensured. It is estimated that, by 2030, India will require some 100 million tonnes of wheat to satisfactorily feed its population. Therefore, there is a pressing need to enhance wheat productivity so as to keep up the pace with the mounting demand and to maintain price stability by making it physically available and economically accessible. The key objectives of wheat breeding are yield enhancement, good nutritional

M. R. Wani (✉) · S. A. Sheikh
Department of Botany, Government Degree College (Boys),
Anantnag 192102, Jammu and Kashmir, India
e-mail: botanyrafiq@gmail.com

S. A. Sheikh
e-mail: sheikhsubzar00@gmail.com

M. I. Kozgar
KERB Biotech, Bemina, Srinagar 190018, Jammu and Kashmir, India
e-mail: m.i.kozgar@gmail.com

P. Ahmad
Department of Botany, S.P. College, Srinagar 190001, Jammu and Kashmir, India
e-mail: parvaizbot@yahoo.com

P. Ahmad et al. (eds.), *Improvement of Crops in the Era of Climatic Changes,* Volume 2,
DOI 10.1007/978-1-4614-8824-8_12,

quality, biotic and/or abiotic stress tolerance, etc. Presently, conventional methods of wheat breeding are not enough to make any considerable breakthrough to cope with the world's rising demand, with the exhausted genetic variability among the existing genotypes being the major constraint in the progression of these varieties and also making them susceptible to various abiotic and biotic stresses. In this scenario, the possibility offered by experimental mutagenesis to induce new genetic variability is of extreme interest, as it has played an enormous role in increasing the world food security by contributing significantly in the improvement of wheat crop production. The present chapter enfolds various historical aspects in addition to contemporary knowledge of wheat crop improvement programs through induced mutagenesis.

Keywords Genetic diversity · Historical perspective · Molecular approaches · Wheat crop improvement · Mutagenesis

1 Introduction

Wheat (*Triticum aestivum* L.), a member of the Poaceae family, is an important cereal crop domesticated by ancient farming communities in the Fertile Crescent region (James 2009). It is mainly cultivated for its edible grain called caryopsis (Fig. 12.1). The other cereal crops which are part of the staple food category alongside wheat are maize, rice, barley, sorghum, oats, triticale, and rye. The production of these major cereal crops around the world is provided in Table 12.1. The total production of cereals in some major cereal producing countries is given in Table 12.2. From the description of Tables 12.1 and 12.2, it is evident that wheat occupies a prominent position among the cereal crops and supplements nearly one-third of the world's population diet (Dhanda et al. 2004). Since wheat is important for human diet, its further improvement is desired to fulfill the requisite demands of the growing population. The goal can be achieved through adoption of various agricultural and breeding strategies, especially induced mutagenesis—the main focus of the present compilation.

2 Background Information of Wheat

Wheat is among the first cereal crops reported to have been cultivated some 10,000 years ago in Pre-Pottery Neolithic near the East Fertile Crescent (Harlan and Zohary 1966). After about 8000 B.C., wheat cultivation began to spread beyond its point of origin—the Fertile Crescent. Cultivation of emmer wheat started in the Fertile Crescent in about 8500 B.C., reached Greece, Cyprus, and India by 6500 B.C., Egypt shortly after 6000 B.C., and Germany and Spain by 5000 B.C. (Diamond 1997). By 3000 B.C., wheat reached England and Scandinavia, while it reached to China a millennium later (Smith 1995). The first identifiable bread wheat with sufficient gluten has been identified by DNA analysis of a sample from a granary dating to

Fig. 12.1 Wheat crop at its maturity. (Source: http://www.freegreatpicture.com)

1350 B.C. at Assiros in Greek Macedonia (Sheffield.ac.uk). Traditionally, cultivated wheat often consists of landraces, also known as farmer's varieties or folk varieties (Belay et al. 1995). These varieties developed by farmers through years of natural and human selection are well adapted to local environmental conditions and management practices (Zeven 1999). Owing to their genetic structure, buffering capacity, and some morphophysiological traits conferring adaptability to stress environments, they (the landraces) are better adapted to stress environments than modern cultivars and prove to be a vital source for broadening the genetic base of cultivated genotypes (Jaradat 2013). The landraces play a significant role in the development of new varieties with improved yield and yield stability particularly under changing environments (Witcomb et al. 1996). Additionally, these landraces also harbor genes and gene complexes for quality traits (Zencirci and Karagoz 2005), stress tolerance, and for wider adaptation to low-input and organic farming systems (Jaradat 2006).

3 Species of Wheat

Wheat (*Triticum* spp.) is a monocot angiospermic plant belonging to the order Poales, family Poaceae, tribe Triticeae, and genus *Triticum*. The number of species (wild as well as cultivated) of the genus *Triticum* varies according to different classifications (Merezhko 1998). The wild species are still a valuable source of useful agronomic traits for continued improvement of cultivated wheat (Sramkova et al. 2009). Based on their ploidy levels, the wheat species are classified into three types: (a) diploid with two sets of chromosomes (14 chromosomes, for example, *T. monococcum*), (b) tetraploids with four sets of chromosomes (28 chromosomes, for example, *T. dicoccoides*), and (c) hexaploids with six sets of chromosomes (42 chromosomes, for example, *T. aestivum*). According to Huang et al. (2002), grasses (family Poaceae) evolved 50–70 million years ago. Wheat, oats, and barley diverged about 20 million years ago (Inda et al. 2008). Tetraploid wild emmer wheat (*T. dicoccoides,* genome AABB) emerged as a result of hybridization between diploid wheat (*T. urartu,* genome AA) and goat grass (*Aegilops speltoides,* genome BB)

Table 12.1 World production of main cereal crops (in tones; Source: FAOSTAT, http://faostat3.fao.org/faostat-gateway/go/to/download/Q/QC/E)

Crops	1970	1980	1990	2000	2008	2009	2010	2011	2012
Maize	265,831,145	396,623,388	483,372,615	592,479,375	830,262,010	820,015,367	849,792,119	885,289,935	875,098,631
Rice, Paddy	316,345,703	396,871,310	518,568,653	599,355,292	688,405,734	684,808,704	701,047,510	722,559,584	718,345,380
Wheat	310,740,954	440,187,901	592,311,011	585,690,886	683,014,574	686,836,963	651,906,870	701,395,334	674,884,372
Barley	119,378,695	156,702,822	178,074,020	133,119,046	154,753,540	151,561,497	123,682,055	133,049,075	132,350,225
Sorghum	55,773,304	57,238,185	56,807,007	55,856,128	66,511,675	56,659,716	59,053,223	58,583,460	58,098,158
Oats	52,411,105	41,433,288	39,917,119	26,098,626	25,862,791	23,336,921	19,693,610	22,676,189	20,974,945
Triticale	0	167,210	4,453,148	9,105,528	14,226,241	15,835,539	13,727,551	13,526,041	13,701,426
Rye	27,677,728	25,377,975	38,193,583	20,116,038	18,101,210	18,293,822	11,942,275	13,162,017	14,544,170

Table 12.2 Total cereal production (tonnes) in some major cereal-producing countries. (Source: FAOSTAT, http://faostat3.fao.org/faostat-gateway/go/to/download/Q/QC/E)

Country	1970	1980	1990	2000	2008	2009	2010	2011	2012
Australia	12,904,533	16,402,312	23,045,332	34,446,938	35,211,060	34,499,849	33,505,761	39,987,043	43,371,698
China	197,635,008	277,190,199	401,934,965	405,224,140	478,474,000	481,563,000	496,891,000	519,403,200	540,830,000
France	31,443,420	48,024,778	55,110,621	65,698,424	70,108,165	70,031,326	68,309,112	65,980,811	70,981,610
Germany	23,753,864	32,712,941	37,579,589	45,271,234	50,104,859	49,808,770	44,314,391	41,938,200	44,942,300
India	113,909,504	140,490,600	193,919,312	234,931,192	266,835,300	250,783,400	267,838,300	287,860,000	286,500,000
Indonesia	22,156,220	33,642,843	51,912,780	61,75,000	76,574,994	82,028,630	84,797,030	833,69,979	88,422,171
Iran	6,448,000	8,582,850	13,683,863	12,873,964	13,475,497	20,835,765	22,246,967	20,104,800	20,833,000
Mexico	15,006,401	20,893,795	25,561,601	27,990,817	36,106,577	31,282,168	34,922,462	28,405,871	32,595,101
Myanmar	8,305,587	13,680,630	14,423,776	22,125,724	34,346,602	34,520,905	34,554,617	31,088,328	35,100,000
Pakistan	12,097,337	17,073,600	20,957,200	30,460,700	35,527,800	38,157,384	34,811,500	39,190,524	36,981,000
Poland	16,262,100	18,336,070	28,013,511	22,340,612	27,664,297	29,571,134	26,652,568	26,172,545	27,667,535
Russian Federation	167,008,000[a]	170,359,000[a]	209,108,512[a]	64,326,238	106,417,890	95,615,930	59,624,036	91,792,153	68,766,551
Thailand	15,857,700	20,612,202	21,169,497	30,529,251	36,192,087	37,039,697	40,764,688	39,716,591	42,36,500
Turkey	15,989,280	24,418,700	30,201,369	32,248,694	29,279,920	33,569,627	32,764,875	35,195,055	33,371,844
United States of America	186,860,751	269,883,982	312,410,604	342,628,297	403,541,004	419,380,398	401,669,955	386,815,964	356,961,850

[a]Data for USSR

Table 12.3 List of different species of wheat. (Source: ITIS 42236 11-09-2013)

S. No.	Scientific name	Common/local name	Ploidy
1.	*Triticum aestivum* L.	Common wheat, wheat	Hexaploid
2.	*Triticum aethiopicum* Jakubz.	Ethiopian wheat	NA[a]
3.	*Triticum araraticum* Jakubz.	Wheat	Tetraploid
4.	*Triticum boeoticum* Boiss.	Wild einkorn	Diploid
5.	*Triticum carthlicum* Nevski	Persian wheat	Tetraploid
6.	*Triticum compactum* Host.	Clubbed wheat	Hexaploid
7.	*Triticum dicoccoides* (Körn.) Körn. ex Schweinf.	Wild emmer	Tetraploid
8.	*Triticum dicoccum* Schrank ex Schübl.	Emmer	Tetraploid
9.	*Triticum durum* Desf.	Durum wheat	Tetraploid
10.	*Triticum ispahanicum* Heslot	Wheat	Tetraploid
11.	*Triticum karamyschevii* Nevski	Karamyschev wheat	Tetraploid
12.	*Triticum monococcum* L.	Einkorn	Diploid
13.	*Triticum polonicum* L.	Polish wheat	Tetraploid
14.	*Triticum spelta* L.	Spelt	Hexaploid
15.	*Triticum thaoudar* (Hausskn.) Jakubz.	NA[a]	NA[a]
16.	*Triticum timopheevii* (Zhuk.) Zhuk.	Timopheev wheat	Tetraploid
17.	*Triticum turanicum* Jakubz.	Oriental wheat	Tetraploid
18.	*Triticum turgidum* L.	Rivet wheat or English wheat	Tetraploid
19.	*Triticum urartu* Thumanian ex Gandilyan	Wheat	Diploid
20.	*Triticum vavilovii* Jakubz.	Vavilov wheat	Hexaploid
21.	*Triticum zhukovskyi* Menabde et Ericzjan	Zhukovsky wheat	Hexaploid

[a]NA: Not available

some 300,000–500,000 years ago (Huang et al. 2002; Dvorak and Akhunov 2005). Hexaploid bread wheat does not have a wild hexaploid progenitor and is the result of spontaneous hybridization between tetraploid wheat and diploid *Aegilops tauschii* (Kihara 1944; McFadden and Sears 1946; Kerber 1964; Kislev 1980; Dvorak et al. 1998; Matsuoka and Nasuda 2004). According to Peng et al. (2011), wild emmer wheat *T. dicoccoides* has the same genome as durum wheat. It is the progenitor of cultivated wheat and has contributed two genomes to bread wheat that contains three genomes. About 10,000 years ago, hunter-gatherers began to cultivate wild emmer, and later from the domesticated emmer, several cultivated tetraploid wheats like *T. turgidum, T. carthlicum,* etc. were derived (Peng et al. 2011). Cultivated emmer (*T. dicoccum,* genome AABB) is the result of slow subconscious selection, and around 9,000 years ago, it naturally hybridized with *Aegilops tauschii* having genome DD to produce early spelt (*T. spelta,* genome AABBDD). Moreover, spontaneous mutation about 8,500 years ago, altered the ears of both emmer and spelt to the easily threshed type that subsequently evolved into free-threshing ears of durum wheat (*T. durum,* genome AABB) and bread wheat (*T. aestivum,* genome AABBDD) as described by Peng et al. (2011). Some cultivated species of wheat are bread wheat, spelt wheat, durum wheat, emmer wheat, and einkorn. A list of different wheat species is indexed in Table 12.3.

Table 12.4 Nutrient content of wheat (per 100 g; Source: Nutrient Data Laboratory, United States Department of Agriculture)

Component	Amount	Component	Amount
Water (g)	11	Copper (mg)	0.55
Energy (kJ)	1419	Manganese (mg)	3.01
Protein (g)	13.7	Selenium (mcg)	89.4
Fat (g)	2.47	Thiamine (mg)	0.42
Carbohydrates (g)	71	Riboflavin (mg)	0.12
Fiber (g)	10.7	Niacin (mg)	6.74
Calcium (mg)	34	Pantothenic acid (mg)	0.94
Iron (mg)	3.52	Vitamin B_6 (mg)	0.42
Magnesium (mg)	144	Foliate total (mcg)	43
Phosphorous (mg)	508	Saturated fatty acid (g)	0.45
Potassium (mg)	431	Monounsaturated fatty acid (g)	0.34
Sodium (mg)	2	Polyunsaturated fatty acid (g)	0.98
Zinc (mg)	4.16		

4 Importance of Wheat

Wheat has been playing an essential role in leveraging the agrarian scenario of India, bringing with it the loads of health benefits. Bread, made from wheat flour, is extremely nourishing and provides strength and vitality to people by inducing high appetite. Wheat helps in curing constipation and is reported to be beneficial for people suffering from cancer. Additionally, it also helps in treating nasal bleeding when consumed with milk and sugar. The whole grain of wheat helps in preventing diabetes by controlling body weight and maintaining the insulin level. Besides, wheat is also a rich source of vitamin B and basic amino acids, including arginine and lysine (www.rajasthanagriculture.com/wheat-grains-india.php). Wheat is the world's most favored staple food crop providing adequate nourishment for humans than any other food source. It is an important source of carbohydrates, proteins, minerals, vitamins, and fats (http://www.nal.usda.gov/fnic/foodcomp/search/). Wheat is used to make rotis, chapatis, biscuits, cookies, cakes, pasta, noodles, etc. and is fermented to make alcoholic beverages like beer (Palmer 2001), besides being used for making biofuel (DAA 1957). The nutrient content of wheat is depicted in Table 12.4.

5 Wheat as per Indian Scenario

India produces about 94,880,000 t of wheat per year or about 14% of world production and is now the second largest producer and consumer of wheat after China. The data regarding area, yield, and production of wheat for 2012 in some major wheat-producing countries and of the world are given in Table 12.5. During the mid-twentieth century, the average yield of wheat was low in India (below 1 t/ha;

Table 12.5 Area, yield, and production of wheat for the world and some major wheat-producing countries (2012; Source: FAOSTAT, http://faostat3.fao.org/faostat-gateway/go/to/download/Q/QC/E)

Country	Area (ha)	Yield (kg/ha)	Production (tonnes)
China	24,139,080	49,952	120,580,000
India	29,900,000	31,732	94,880,000
USA	19,826,170	31,148	61,755,240
France	5,303,300	75,992	40,300,800
Russian Federation	21,277,900	17,727	37,719,640
World	*216,638,762*	*31,153*	*674,884,372*

Swaminathan 2013). India mainly relied on imports to meet the domestic requirements in the past, however, a severe drought in 1965–1966 necessitated India to reform its agricultural policy so as to attain food self-sufficiency and ensure food security. Accordingly, major policy reforms in agriculture aiming at food grain self-sufficiency were made which triggered India's Green Revolution. This yield revolution in wheat in India became possible because of the introduction of high-yielding, disease-resistant, and semidwarf wheat varieties carrying Norin dwarfing genes that were supplied by Dr. Norman E. Borlaug (Swaminathan 1965). As a result, the wheat yields witnessed a quantum jump in 1968 which greatly improved the food security of the country (Swaminathan 2013). In 1963, Borlaug supplied the germplasm of Sonora 63, Sonora 64, Mayo 64, and Lerma Rojo 64A (Mexican semidwarf strains). During 1963 and 1964, after multilocational trials, it was realized that Lerma Rojo 64A and Sonora 64 could help in increasing the yield significantly but had undesired red-colored grains associated with it; however, efficient selection, among the segregating populations, bred amber grain varieties like Kalyan Sona and Sonalika, possessing good chapati-making quality, and proved to be very popular among farmers and consumers (Swaminathan 1965). Since then, hundreds of high-yielding, good quality, and disease-resistant wheat varieties suitable for different agroecological regions of the country have been released. The Punjab state pioneered India's Green Revolution and earned the distinction of being India's bread basket (GOPR 2004). The manifold increase in wheat production made India self-sufficient by 1970s. Presently, India is the exporter of various food grains including rice and wheat. In 2011, India exported about 2 billion kg of wheat and rice to Africa, Nepal, Bangladesh, and other regions of the world (Bloomberg 2011). The all India scenario of area, production, and yield of wheat from 2001 to 2012 is given in Table 12.6.

6 Wheat-Breeding Objectives and Important Research Areas

The major objectives of wheat breeding include high grain yield, good quality, disease resistance, and tolerance to biotic and/or abiotic stresses. In the present scenario, the main thrust areas of wheat breeding in Indian context (DWR Perspective Plan: Vision 2025) can be summarized as below:

Table 12.6 All India area, yield, and production of wheat. (Source: FAOSTAT, http://faostat3.fao.org/faostat-gateway/go/to/download/Q/QC/E)

Year	Area (ha)	Yield (kg/ha)	Production (tonnes)
2001	25,730,600	27,081	69,680,896
2002	26,344,700	27,621	72,766,304
2003	25,195,700	26,100	65,760,800
2004	26,594,700	27,132	72,156,200
2005	26,382,900	26,016	68,636,900
2006	26,483,600	26,188	69,354,500
2007	27,994,500	27,079	75,806,700
2008	28,038,600	28,022	78,570,200
2009	27,752,400	29,071	80,679,400
2010	28,457,400	28,395	80,803,600
2011	29,068,600	29,886	86,874,000
2012	29,900,000	31,732	94,880,000

1. To exploit the heterosis by developing hybrids based on cytoplasmic male sterility systems
2. To broaden the genetic base of varieties by inducing new genetic variability, by spring and winter wheat hybridization and use of alien species, etc
3. To make use of biotechnological interventions for gene pyramiding and marker-aided selection for biotic and abiotic stress tolerance and quality traits
4. To develop tillage-specific varieties and those with water and nutrient use efficiency
5. To develop varieties with resistance to insects, nematodes, rusts, karnal bunt, etc
6. To develop product-specific varieties with desired quality traits
7. To improve protein content, gluten strength, and extensibility for enhancing bread-making quality
8. To improve nutritional quality by manipulating starch composition, lysine content, and biofortification of micronutrients, their enhanced bioavailability and reduced anti-nutritional factors
9. To combat diseases and pests through integrated pest management, durable resistance, etc
10. Survey and scrutiny for new races and elimination of such new races like yellow rust race 78S84 and black rust race Ug99
11. To develop genotypes with inherent capacity of responding well under moisture stress conditions

Genetic variability is a prerequisite for any plant breeding program as it forms the basis for yield enhancement and genetic resistance to biotic, abiotic, or other stresses in addition to architect the crop species in desired direction. Hence, plant breeders always aim at creating genetic variability before exercising selection. Though wheat shows heterosis, but under normal conditions, hybrid seed production is not easy as the wheat flowers are complete and self-pollinating (Bajaj 1990). However, hybrid seeds in wheat have been produced using chemical hybridizing agents, naturally occurring cytoplasmic male sterility, or plant growth regulators that selectively interfere with pollen development (McRae 1985; Duvick 1999). The hybrid wheat production and exploitation of heterosis helps in enhancing the yield. Improvement of nutritional quality is one of the major concerns of wheat breeding programs. Understanding the molecular basis of grain quality will facilitate the improvement of these characters. To ensure nutritional security, efforts are being made globally to enhance the bioavailability and biofortification of micronutrients in wheat. Many high-yielding wheat varieties resistant to different diseases have been developed by different breeding procedures, which in turn have lead to yield enhancement. However, the progress has been hindered in the soils with insufficient irrigation/moisture. For such areas/soils, efforts are made to develop/identify genotypes which are able to perform well even in moisture stress conditions. Heat, salinity, and alkalinity resistance genes could be screened from the wild species and deployed in the otherwise high-yielding genotypes. Moreover, in the improvement of wheat quality, many types of pathological problems like yellow rust, stem rust, leaf blight, karnal bunt, powdery mildew, and head scab are still challenges to wheat breeders. These challenges can be overcome by screening and transferring resistant genes in good agronomical varieties. In the scenario of enhanced food grain requirement and growing environmental stresses, the induced mutagenesis seems to be a viable solution.

7 Mutagenesis

Mutations are sudden heritable changes occurring in the genetic material (DNA) of an organism. Mutations may occur spontaneously in nature or may be induced artificially utilizing various mutagens. Many endogenous factors like free radicals (Wiseman and Halliwell 1996) generated during metabolic activities and exogenous factors such as ultraviolet (UV) radiations or ionizing radiations (Taylor 1990) are known to interfere with genome integrity (Rastogi et al. 2010). The genome of an organism is subject to damages of radiations (UV and ionizing), mutagenic chemicals, bacterial or fungal toxins, and endogenously generated free radicals and/or alkylating agents, etc. (Tuteja et al. 2001). Moreover, the DNA may also be damaged because of replication errors. The damage may be in the form of altered bases, missing bases, mismatch bases, deletions or insertions, linked pyrimidines, strand breaks, and cross-links which could prove genotoxic or cytotoxic for the cell.

Endogenously, mutagenesis may be the result of spontaneous hydrolysis, replication and repair errors, or normal metabolic processes generating reactive oxygen species (ROS) and DNA adducts (Loeb 1989). These adducts may induce conformational changes in DNA, besides resulting in depurination of DNA (Schaaper et al. 1982). The mechanism by which mutation arises depends upon the mutagen involved. Mutagens either act directly or indirectly via mutagenic metabolites, producing lesions; however, some may affect the replication or chromosomal partition mechanisms and other cellular processes. Mutations may occur by one or a combination of methods, viz., spontaneous hydrolysis, base modification, cross-linking, dimerization, intercalation between bases, backbone damage, insertional mutagenesis involving transposons and viruses, or by replication errors. According to Rastogi et al. (2010), DNA damage may result in (1) base mis-incorporation during replication; (2) deamination, depurination, and depyrimidination due to hydrolytic damage (Lindahl 1993); (3) oxidative damage, caused by direct interaction of ionizing radiations with DNA, free radicals, or ROS induced by UV radiations (Valko et al. 2006; Halliwell and Gutteridge 2007); and (4) alkylating agents that may result in base modifications (Dizdaroglu 1992; Lindahl 1993). The exposure of radiations (UV, ionizing) and certain chemicals may also result in single- as well as double-strand breaks. DNA double strand breaks are the most deleterious leading to loss of genetic material (Rastogi et al. 2010).

At high concentrations, oxygen free radicals or ROS can induce damage to cell structure, lipids, proteins as well as DNA (Valko et al. 2007). The hydroxyl radicals can damage all components of DNA molecules such as purines, pyrimidines, and sugar backbone, thus inhibiting the normal functioning of the cell (Valko et al. 2006; Halliwell and Gutteridge 2007; Rastogi et al. 2010). Different hydrolytic reactions target DNA in living cells. The hydrolysis of the glycosidic bonds between bases and the sugar backbone are the most common, resulting in apurinic/apyrimidinic (AP) sites (Britt 1996). Abasic sites are most prevalent endogenous DNA lesions, with an estimate of 10,000 lesions/human cell/day (Lindahl 1993). Nakamura and Swenberg (1999) reported about 50,000–200,000 apurinic (AP) sites/genome in many human and rodent tissues. AP sites are not only produced by spontaneous depurination but also by ROS (Nakamura et al. 2000). In DNA, purines are lost at a rate 20 times higher than pyrimidines (Lindahl and Karlstrom 1973) and are able to prevent normal DNA replication and transcription. It has been estimated that about 10,000 purine sites in DNA are depurinated each day in a cell (Loeb 1989). Despite the overall rate of depurination being slow, still in organisms with a large genome, spontaneous hydrolysis would induce the loss of several thousand purines daily in a cell (Britt 1996).

Although numerous DNA repair pathways exist, nevertheless if an apurinic site fails to repair, mis-incorporation of nucleotide may occur during replication. AP sites also arise as a result of occasional lesion bypass events during DNA replication and are potentially mutagenic (Gentil et al. 1984). In depurination/depyrimidination, there is a complete removal of purine/pyrimidine bases, which may result in breakage of the DNA backbone (Rastogi et al. 2010). The hydrolytic deamination can directly convert one base to another, e.g., deamination of cytosine results in uracil, methylcytosine to thymine, and at a much lower frequency, adenine to hypoxanthine (Rastogi et al. 2010). Uracil and thymine differ in their mutagenic abilities. Uracil is easily recognized and repaired. However, a transition to thymine is less likely to be detected as an error because being the normal DNA component, it is not recognized by proofreading enzymes. 5-methylcytosine deamination is one of the most important causes of point mutations in mammalian cells (Rideout et al. 1990). As the plant genome contains ten times more 5-methylcytosine than the human genome (Shapiro 1976), there is a greater possibility of this source of point mutations in plants.

Base alkylation and arylation may result in replication errors. DNA methylation can occur spontaneously even in the absence of exogenous alkylating agents

(Rebeck and Samson 1991). Majority of the bonds of bases are vulnerable to methylation to varying extents, and if left unrepaired, may prove pre-mutagenic or lethal. Purines are more susceptible to methylation damage. The most frequently generated alkylation product, 7-methyladenine, pairs normally and is not mutagenic. Contrastingly, 3-methyladenine does not act as a template for DNA synthesis, thus acts as a replication block. O^6-methylguanine efficiently pairs with thymine and is a potent pre-mutagenic lesion (Britt 1996). N^7 and O^6 of guanine and N^3 and N^7 of adenine are most susceptible to methylation damage. Methylation of O^6-guanine results in G to A transition, whereas O^4-methylthymine can mispair with guanine. O^6-alkylguanine has been reported to be the major ethylmethane sulfonate (EMS)-induced mutagenic lesion in *Arabidopsis* seeds (Dolferus et al. 1990; Niyogi et al. 1993; Orozco et al. 1993).

Ionizing radiations can engender free radicals that can break DNA. Double-stranded breaks are more damaging and produce chromosomal translocations and deletions. UV radiation, a prominent natural mutagen, is cytotoxic even at low doses. Higher doses of UV radiations may lead to replicative arrest and cell death by apoptosis (Latonen et al. 2001). Longer UV wavelengths (UVB and UVA) induce oxidative stress and protein denaturation (Ravanat et al. 2001). Mutations at CC sites can be due to the indirect influence of UV light mediated by ROS (Ried and Loeb 1993). UV radiations result in dimerization of flanking pyrimidines (Friedberg 2003). Gamma irradiation generally leads to DNA double strand breaks (Puchta 2005). The living cells are regularly exposed to potentially damaging ROS that can modify various biomolecules including DNA. Reactions in chloroplast (Bowler et al. 1992) and mitochondrion (Wallace 1992) frequently misdirect electrons to oxygen, generating reactive superoxides. The extracellular ROS-inducing factors include ozone (Mehlhorn et al. 1990; Kanofsky and Sima 1991), high levels of UVB radiations (Hariharran and Cerutti 1977), heavy metal ions (Kovalchuk et al. 2001), ionizing radiations, etc. These species cause oxidative damage to DNA resulting in the formation of modified bases, strand breaks, and clustered damage sites (Friedberg et al. 1995). The most prevalent base damage to purines is 7, 8-dihydro-8-oxoguanine (also named as 8-oxoguanine) and for pyrimidines is the formation of thymine glycol (Slupphaug et al. 2003). 8-oxoguanine pairs with equal capability to A and C and represents an important source of point mutations (Maki and Sekiguchi 1992). Ionizing radiation can effectively convert 5-methylcytosine to thymine glycol under aerobic conditions or peroxides. Thymine glycol is thought to be primarily a replication-blocking lesion (Gerd and Besaratinia 2009) and an additional source of mutations.

8 Induced Mutagenesis

Induced mutagenesis by Muller (1927) opened a new era in the field of crop improvement and now it has become a reputable gizmo in the field of plant breeding to supplement the existing germplasm and improve the cultivars in certain specific traits. Mutagenesis is of great use in those cases where there is low genetic variability or some inherited defects need to be rectified in an otherwise agronomically superior cultivar (Chopra and Sharma 1985). The main focus of mutation-breeding programs has been the altering of one or few major traits which limit their productivity or enhance their quality (Srivastava et al. 2011; Wani et al. 2012). Mutations have played a significant role in increasing the world food security (Kharakwal and Shu 2009) and enables the induction of desired attributes that either cannot be found in nature or have been lost during evolution (Srivastava et al. 2011). Mutations can be the result of natural DNA replication errors. Though most of these errors are repaired, nevertheless some may get established in the offsprings as spontaneous mutations. Gene mutations without phenotypic expressions are usually not recognized. Accordingly, genetic variations appear quite limited and breeders have to resort to mutation induction (Novak and Brunner 1992). Mutations have been used to develop many improved cultivars in addition to study the genetic and developmental phenomena of plants (Van Den-Bulk et al. 1990; Bertagne-Sagnard et al. 1996).

Mutations are induced through physical mutagens such as X-rays, gamma rays, fast neutrons, thermal neutrons, UV, and beta radiations; besides, a number of chemical mutagens are also used for inducing mutations in crop plants. With the exception of UV rays, all other types of radiations ionize atoms in a tissue by detaching electrons from them (Anonymous 1977). Most of the chemical mutagens like methylmethane sulfonate (MMS), EMS, ethylene imine (EI), diethyl sulfate (DES), N-ethyl-N-nitrosourea (ENU), N-methyl-N-nitrsourea (MNU), etc. are alkylating agents and induce a broad genetic variation for morphological, biochemical, and yield parameters. Mutagenic efficiencies of various mutagens have been compared on different crops by different researchers and their results seem to be entirely specific for particular species. Many researchers have found chemical mutagens to be more effective than physical ones (Dhanayanth and Reddy 2000; Bhat et al. 2005; Wani et al. 2011), while others have found the physical mutagens to be more effective (Zeerak 1991; Laxami et al. 2013). The role of mutation breeding in increasing the genetic variability for quantitative traits in various crops has been proved by a number of workers (Ricardo and Ando 1998; Khan et al. 1999; Rachovska and Dimova 2000; Baloch et al. 2002; Kumar and Mishra 2004; Erdem and Oldacay 2004; Khan and Wani 2006; Addai and Kantanka 2006; Bhat et al. 2007; Senevirantne and Wijesundara 2007; Tai et al. 2007; Adamu and Aliyu 2007; Kozgar et al. 2011; Mostafa 2011; Srivastava et al. 2011; Sheikh et al. 2012; Husain et al. 2013).

In traditional mutation breeding, mutations are induced in crops and the resultant mutants are used either directly or indirectly for developing a new variety. Presently, it is possible to tag mutated genes, pyramid them into a single elite breeding line, and follow them up in subsequent breeding programs (Shu 2009). Although mutations may have some grievous consequences for organisms, they are still essential to life and evolution (Friedberg 2003). The knowledge of the type of mutation caused by a particular mutagen will be helpful in choosing a proper mutagen for inducing a particular mutation. Induced mutagenesis is the method by which DNA mutations are intentionally engineered to produce mutant genes, proteins, or strains of organisms. Improved techniques for site-directed mutagenesis, combinatorial mutagenesis, and insertional mutagenesis have been developed which have increased the significance and contribution of mutagenesis tremendously. Mutation research has witnessed a renewed interest in recent years due to the availability of a variety of new technologies, which have made possible to induce novel and targeted mutations for the study of mechanisms involved in various developmental and physiological processes. Presently, it is possible to either insert new genes or disrupt and alter preexisting genes in a targeted manner (Chen et al. 1998). Insertional mutagenesis approach involving T-DNA and transposable elements (TEs) and gene targeting involving the use of nucleases like zinc finger nucleases (ZFN) and transcription activator like effector (TALE) proteins have become available for mutation research (Balyan et al. 2008; Townsend et al. 2009; Urnov et al. 2010; Miller et al. 2011). The availability of DNA-based molecular markers have also made it possible to identify, map, or clone specific genes (Kuraparthy et al. 2007; Wu et al. 2007). Mutant-assisted gene identification and characterization (MAGIC) and bulked segregant RNA sequencing (BSR-Seq) are some of the recent approaches for mapping quantitative trait loci (QTLs)/genes (Johal et al. 2008; Liu et al. 2012).

The eukaryotic genomes have a large number of repeat sequences, like minisatellites, microsatellites, and expanded simple tandem repeats, collectively called the tandemly repeated DNA loci (TRDLs). They are inherently unstable and also have shown an increased frequency of induced mutations (Dubrova et al. 1996; Ellegren et al. 1997; Kovalchuck et al. 2000; Bridges 2001; Joshi and Gopalakrishnan 2007). The high frequencies of radiation-induced mutations have indicated that these mutations are not directly linked to the site of DNA damage and hence are "untargeted." In addition, the radiation-induced genomic instability has another feature of "delayed mutations," where mutations are caused long after the radiation exposure has ended (Niwa 2006). The mechanism of such a nontargeted and delayed effect of radiations is not clear though, but it is assumed that some radiation-induced events trigger these sequences to become genetically less stable (Bouffler et al. 2006). Several studies have demonstrated the use of tandem repeats as reporters of mutation (Yauk 2004; Armour 2006; Joshi and Gopalakrishnan 2007).

The conventional mutation breeding involving generation of random mutations in chromosomal DNA requires screening or selection from massive numbers of mutants to obtain the desired mutation (Hong and Ames 1971, Botstein and Shortle 1985). However, the site-directed mutations of DNA allow precise changes in discrete and manageable segments of the genome with relatively little efforts (Cosby and Lesley 1997). Induced mutations have been greatly used to improve various crops including cereals. By 2013, a total of 3,218 mutant varieties have been released throughout the world. These include cereals (1,541), ornamentals (709), pulses (432), oilseeds (145), and other crops (391; Joint FAO/IAEA, Vienna Mutant Variety Database (MVD); http://mvgs.iaea.org). In India, the number of mutant varieties released for major cereal crops till 2009 are given in Fig. 12.2.

Fig. 12.2 Number of mutant varieties released in major cereal crops in India. (2009; Source: Kharkwal and Shu 2009)

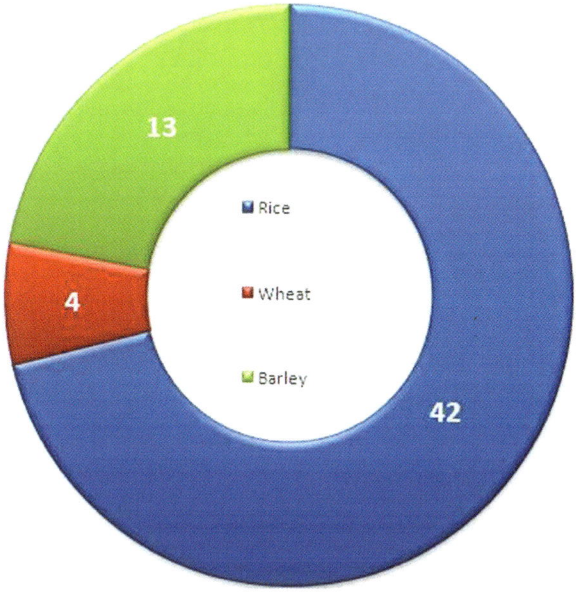

9 Wheat in Relation to Induced Mutagenesis

Wheat breeding by conventional methods has been practiced for centuries, but these methods are now scant to make any further breakthrough with limited genetic variability being the major constraint. The challenge, at present, is to produce high-yielding varieties coupled with good agronomical quality and resistance to a range of biotic and abiotic stresses. Also, owing to critical nutritional status of human population, there is a burning need to develop varieties with improved protein and mineral (e.g., zinc, manganese, iron) contents for meeting the required health demands, and artificial mutagenesis seems to be a viable solution for addressing these challenges. Stadler (1929) pioneered the mutagenesis work in wheat followed by several other workers (Pal and Swaminathan 1960; Swaminathan et al. 1963; Varughese and Swaminathan 1967; Singh 1969; Singh et al. 1973; Chowdhury 1978; Khamankar 1981). Hassan et al. (1988) while working with gamma rays and sodium azide (SA) reported that higher doses of gamma rays delayed the maturity, while higher doses of SA induced earliness in wheat. Moreover, SA caused more reduction in tillers per plant, spikelets per spike, and spike length. The mutagenic potential of SA has been proved in several organisms, including plants and animals, by several screening assays (Rines 1985; Raicu and Mixich 1992; Grant and Salamone 1994). Vasil and Knayazyuk (1988), working with different chemical mutagens on wheat, showed that mutation frequency depends on the genotype, type of mutagens, concentration of mutagens, and the duration of treatment. According to Siddiqui et al. (2007), *Triticum aestivum* offers many opportunities for the exploitation of mutations, recombination, and for increasing the genetic variability of quantitatively inherited traits. Bobryshev et al. (1991) reported increased effectiveness of intervarietal hybridization in wheat using chemo-mutagenesis-generated lines. Ayub et al. (1989), while working with gamma rays, observed that the effect of radiation was depressive and the magnitude of depression varied with the strength of irradiation dose.

Borojevic (1990) induced mutations in wheat and presented a mutational review for morphological characters and disease resistance. Melnik and Yanchenko (1990) studied the morphological and biological features of induced mutations in durum wheat and reported three useful mutants for grain quality and lodging resistance. Ibrahim et al. (1991) reported the development of new Iraqi gamma-ray-derived mutant wheat variety with resistance to *Puccinia recondita,* resistance to lodging, and having enhanced grain weight and protein content. Suganthi et al. (1992) reported the induced variability for quantitative characters, including grain yield, with gamma rays and EMS in wheat. Lapochkina (1998) reported increased yield in wheat hybrids irradiated with 1.5 kR of gamma rays. Eiges (1990) analyzed the mutation spectra and cytogenetics of winter wheat mutants using different concentrations of ethylene imine. Knott (1990) examined the advantages and disadvantages of mutation breeding and concluded that despite the production of directly useful mutants being infrequent, the mutagens can still successfully be used for specific purposes. Dzhelepov (1991) found an association between height, coleoptile length, and harvest index in gamma-irradiation-derived short strawed mutants of wheat. Reddy (1992) studied the mutagenic parameters in single and combined treatments of gamma rays, EMS, and SA in wheat and reported that the combined treatments produced more number of mutations. Chhuneja and Minocha (1993), by using EMS, induced male sterility in three cultivars of wheat.

Wang et al. (1995) irradiated the seeds of two varieties viz., 77-Zhong-2882 and 79-p-17 of wheat with ^{137}Cs and ^{60}Co gamma rays, and reported that both the irradiation sources had similar effects on the heading data and plant height. Khan and Malik (1999) studied the effect of gamma irradiation on some morphological characters of three wheat varieties and reported a gradual increase in the number of tillers and spike length. Jamil and Khan (2002) studied the induced variation for yield components of wheat cultivar Bakhtawar-92 by gamma irradiation and concluded that the most beneficial dose was 20 kR. Further, the impact of this dose was promising on germination, plant height, grains per plant, and grain yield. Srivastava et al. (2011) studied the mutagenic effects of SA on growth and yield characteristics in wheat and reported desirable progenies for earliness, spike length, tillering, and grain yield. They concluded that SA mutagenesis can provide an efficient method for breeding disease-resistant varieties. Sheikh et al. (2012) treated two varieties of wheat (*Triticum aestivum* L.) with SA and concluded that the mutagen induced biological damage and variability for various quantitative traits as well as protein content. Agarwal et al. (2013), by using ENU treatment, induced eight different types of mutations in common wheat which were grouped into stem, reproductive, leaf, and spike mutants. Two of the stem mutants, namely axillary branching and reduced node were novel. After conducting in silico studies using candidate genes producing similar mutant phenotypes in other species, six wheat genomic sequences were identified and considered to be orthologous to the sequences for branching genes from rice and maize. Further, using the expressed sequence tag (EST) database, 11 unigenes were identified which matched a gene responsible for reduction in the number of nodes in maize. They concluded that these sequences involved in axillary branching and reducing the number of nodes may be used as candidates for further studies of above mutants in bread wheat.

Mutagenesis in wheat has been widely employed to modify some protein subunits determining the grain quality (Kiribuchi-Otobe et al. 1998; Yasui et al. 1998; Maluszynski et al. 2001). Behl et al. (1991) found amino acid and protein changes under prolonged heat stress in gamma-ray-induced, high-temperature-tolerant spring wheat mutants. Perez et al. (1999) concluded that radiography is a rapid and nondestructive method which offers the possibility of predicting the field behavior of irradiated material. The impact of mutagens for successful release of mutant wheat varieties in different countries of the world is given in Table 12.7. It is evident from the data that many countries world over though have extensively exploited the potential of induced mutagenesis in wheat and have released a considerable number of mutant varieties, India, despite being one of the largest wheat producers, is yet lagging in adopting the techniques of mutagenesis fully for obtaining the desired mutant varieties embedded with good qualitative and quantitative traits.

10 Conclusions and Future Perspective

Though tremendous enhancement in wheat yields has been witnessed in India, sustaining wheat production and quality with reduced inputs and developing genotypes for specific industrial or nutritional uses and suitable for intensive agriculture continues to be one of the main concerns of wheat breeding. In view of increased demand and high production inputs, the prices of wheat and wheat products have escalated to the extent that it becomes economically inaccessible to a large proportion of the global population. To address this grave concern, it is imperative to produce more wheat involving fewer resources in a sustainable and cost-effective manner. Although, Indian breeders through their consistent and coherent efforts have developed wheat genotypes with higher yield potentials, they do not match with those of China, France, etc. In this scenario, the need of the hour is to boost the yield potential and to develop super wheat genotypes with highly improved agronomical traits. Despite the fact that India has witnessed self-sufficiency in wheat production and is worth of stockpiling adequate buffer stock to thwart the adverse weather and other calamities, the wheat breeders still face innumerable challenges in view of mounting demand, various stresses due to changing environments, depleted water resources, and degraded soils. In the past, many breeding methods have contributed to the yield enhancement, but in the contemporary era, most of them are insignificant to make any further substantial breakthrough. However, modern mutagenesis coupled with genomics continues to be one such promising method in wheat breeding. With the advent of the genomics era, it has become possible to study mutations

Table 12.7 Number and names of mutant wheat varieties released in different countries. (Source: IAEA Mutant Database, http://mvgs.iaea.org)

Country	No. of mutant varieties released	Names of mutant wheat varieties released
Albania	01	Ni 792
Argentina	01	Sinvalocho Gama
Brazil	02	*BR4*, IAS 63
Bulgaria	05	Altimir 67, Fermer (2009), Guinness/1322, Myriana, Zlatostrui
Chile	01	Carolina
China	*163*	092, 1161, 352, 503, 62-10, 62-8, 77L15, 78 A, *Baichun 5*, Changwei 19, Changwei 20, *Changwei 51503*, Chuanfu 1, Chuanfu 2, *Chuanfu 3*, Chuanfu 4, Chuanfu 5, Emai 23, Emai 6, Emai 9, *Fuer*, Fumai 2008, Fuou 1, Fusheabo 1, Ganchun 20, Guifu 12, H6765, *Hangmai 96*, Hangmai 901, Heichun 2, Heima 1, Henong 1, *Hezu 8, Humai 3*, Jiaxuan 1, Jichun 14, *Jienmai 2*, Jihe 02, Jimai 28, Jingdong 23, Jingfen 1, *Jingmai 34*, Jingmai 35, Jinmai 22, Jinmai 23, Kexing 15, Longfu 2, Longfumai 1, Longfumai 10, *Longfumai 11, Longfumai 13, Longfumai 14, Longfumai 15*, Longfumai 16, *Longfumai 17, Longfumai 18, Longfumai 19, Longfumai 2, Longfumai 3, Longfumai 4, Longfumai 5, Longfumai 6, Longfumai 7, Longfumai 8, Longfumai 8, Longfumai 9, Lumai 11, Lumai 16, Lumai 20, Lumai 4, Lumai 5, Lumai 6, Lumai 8,Luten 1, Luyuan 301, Luyuan502, Nanjing 3,Nanyang 75-6, Neimai 5*, Ningmai 3, Qicheng115, Qichun 1, Qinchun 415, Qinghai 570, Qinmai 6, Qunzhong 42, *Shaannong 138, Shannong 12, Shannong 78,* Shannongfu 63, Shenmai 1, *Taifu 1, Taifu 10, Taifu 15, Taifu 22, Taifu 23, Taikong 5, Taikong 6, Wanmai 32, Wanyuan 28-88, Wanyuan 75-6, Wei 9133, Weifu 6757, Weimai 6, Wuchun 3, Xiaoyan 6, Xifu 12, Xifu 3, Xifu 4, Xifu 5, Xifu 6, Xifu 7, Xifu 8, Xinchun 2, Xinchun 3, Xinchun 33, Xinchun No.30, Xingchun 6, Xingchun 7, Xingdong 19, Xinmai 16, Xinong-Mai 2, Xinshukuang 1, Yanfu 188, Yanfuzao, Yangfumai 2, Yangfumai 3, Yangmai 158, Yannong 5158, Yannoun 685, YF188, Yuanchun 7112, Yuandon 2, Yuandong 1, Yuandong 3*, Yuandong 772, *Yuandong 7848, Yuandong 94, Yuanfeng 1, Yuanfeng 2, Yuanfeng 3, Yuanfeng 4, Yuanfeng 5, Yuangnong 53, Yuanyuan 18-37, Yumai 12, Yumai 4, Yumai 43, Yunfu 2, Yunfuzao, Yuyuan 1*, Zhangchun 10, *Zhangchun 12, Zhangchun 13, Zhangchun 14, Zhangchun 17, Zhangchun 18, Zhemai 3, Zhemai 4, Zhemai 5, Zhengliufu, Zhonga 1, Zhonghong 1*
Finland	01	*Hankkija's Taava*
Germany	02	Els, *Sirius*
Hungary	01	*Mv 8*
India	04	NI-5643, *NP 836*, Pusa Lerma, *Sharbati Sonora*
Iraq	06	Intesar, Iratom, *Rabia, Sali, Tammuz-2, Tammuz-3*
Italy	02	Claudia (=Mv 8), *Spinnaker*
Japan	07	*Akebono-mochi, Ibuki-mochi*, Shirowase komugi, Tama-izumi, *Wheat Noh PL 7, Wheat Noh PL 8,* Zenkoji Komugi (Zenkouzi-Komugi)

Table 12.7 (continued)

Country	No. of mutant varieties released	Names of mutant wheat varieties released
Kenya	01	Njoro-BW1
Mexico	02	Bajio Plus, Centauro
Mongolia	04	*Darkhan-106*, Darkhan-35, Darkhan-49, Kharaa 86
Pakistan	06	Bakhtawar-92, Jauhar-78, Kiran-95, Nishte-95, Soghat 90, Tatara,
Russian Federation	36	Albidum 12, Belchanka, Birlik, Deda, Dnestryanka, Erytrospermum 103, Inna, Jubileinaya 75, Kazanskaya 84, *Kharkovskaya 90*, Khersonskaya 86, Kiyanka, Kormovaya 30, Ljubov, *Lutestsens 7*, *Meshenskaya*, *Moskovskaya 70*, *Moskovskaya nizkostebelnaya*, Motsinave 100, *Mriya Khersona*, Nemchinovskaya 52, Nemchinovskaya 86, *Novosibirskaya 67*, *Odesskaja 75*, *Odesskaja Polukarlykovaja*, *Omskaya ozimaya*, *Pitikul*, *Polukarlik 3*, *Polukarlykovaja-49*, Progress, *Schedraja Polesja*, *SGT 17*, *Sibirskaya niva*, Skifyanka, Spartanka, *Yunnat odesskii*
Switzerland	01	Tambo
Ukraine	03	Lybid, Tsarivna, *Yasochka*
USA	04	Above, Lewis, Payne, *Stadler*

at the molecular level using genome sequences. RNA interference (RNAi) and virus-induced gene silencing (VIGS) techniques facilitate the discovery of genes for a variety of traits (Holzberg et al. 2002: McCallum et al. 2000), while those of TILLING and Eco-TILLING allow allele mining at the molecular level (Balyan et al. 2008). Moreover, high-throughput and cost-effective next-generation sequencing (NGS) technologies also enable the use of mutations for gene/allele discovery (Liu et al. 2012). In view of the above techniques, induced mutants are becoming increasingly important for studying the structure–function relationships of genome sequences with unknown functions utilizing the modern approaches of molecular genetics. In future, the potential of these techniques can be effectively exploited in wheat breeding programs.

References

Adamu AK, Aliyu H (2007) Morphological effects of sodium azide on tomato (*Lycopersicon esculentum* Mill). Sci World J 2(4):9–12

Addai IK, Kantanka OS (2006) Effect of ^{60}Co gamma irradiation on storability of soybean seed. Asian J Plant Sci 5(3):221–225

Agarwal P, Kumar S, Mir RR, Balyan HS, Gupta PK (2013) Some ENU induced mutations: a resource for functional genomics in bread wheat. Plant Mutat Rep 3(1):9–17

Anonymous (1977) Manual on mutation breeding. Technical series 199, Joint FAO/IAEA under the auspices of the International Atomic Energy Agency, Vienna, Austria

Armour JAL (2006) Tandemly repeated DNA: why should anyone care? Mut Res 598:6–14

Ayub M, Rehman S, Khan AD (1989) The response of different wheat varieties to gamma irradiation in relation to yield. Gomal Univ J Res 9:77–84
Bajaj YPS (1990) Wheat. In: Biotechnology in agriculture and forestry, vol 13. Springer, pp 161–163, ISBN:3540518096
Baloch AW, Soomro AM, Javed MA, Bughio MS, Mastoi NN (2002) Impact of reduced culm length on yield and yield parameters in rice. Asian J Plant Sci 1:39–40
Balyan HS, Sreenivasulu N, Rieralizarazu O, Azhaguvel P, Kianian SF (2008) Mutagenesis and high-throughput functional genomics in cereal crops: current status. Adv Agron 88:357–417
Behl RK, Moawad AM, Achtmich W (1991) Amino acid and protein profile changes in spring wheat mutant under prolonged heat stress. Ann Biol 7:63–67
Belay G, Tesemma T, Bechere E, Mitiku D (1995) Natural and human selection for purple grain tetraploid wheats in the Ethiopian highlands. Gen Res Crop Evol 42:387–391
Bertagne-Sagnard B, Fouilloux G, Chupeau Y (1996) Induced albino mutations as a tool for genetic analysis and cell biology in flax (*Linum usitatissimum*). J Exp Bot 47:189–194
Bhat TA, Khan AH, Parveen S (2005) Comparative analysis of meiotic abnormalities induced by gamma rays, EMS and MMS in *Vicia faba* L. J Indian Bot Soc 84:45–58
Bhat TA, Sharma M, Anis M (2007) Comparative analysis of meiotic aberrations induced by diethyl sulphate and sodium azide in broad bean (*Vicia faba* L.). Asian J Plant Sci 6:1051–1057
Bloomberg (2011) India allows wheat exports for the first time in four years. Bloomberg 8, September 2011
Bobryshev FI, Kostin VI, Eiges NS, Zubov AE (1991) Increasing the effectiveness of intervarietal hybridization in winter wheat using mutant lines. Khimicheskii Mutagenez I Problemy Seleksii, pp 107–109
Borojevic K (1990) Induced mutations and their genetic aspects in wheat (*Triticum aestivum* vulgare). In: Plant mutation breeding for crop improvement, vol 2. Proc Int Symp on the contribution of plant mutation breeding to crop improvement, Joint FAO/IAEA, Vienna, Austria, pp 317–326
Botstein D, Shortle D (1985) Strategies and applications of *in vitro* mutagenesis. Science 229(4719):1193–1201
Bouffler SD, Bridges BA, Cooper DN, Dubrova Y, McMillan TJ, Thacker J, Wright EG, Waters R (2006) Assessing radiation-associated mutational risk to the germline: repetitive DNA sequence as mutational targets and biomarkers. Rad Res 165:249–268
Bowler C, Van Montagu M, Inzé D (1992) Superoxide dismutase and stress tolerance. Ann Rev Plant Phy Plant Mol Biol 43:83–116
Bridges BA (2001) Radiation and germline mutation at repeat sequences: are we in the middle of paradigm shift? Rad Res 156:631–641
Britt AB (1996) DNA damage and repair in plants. Ann Rev Plant Physiol Plant Mol Biol 47:75–100
Chen L, Marmey P, Taylor NJ, Brizard JP, Espinoza C, D'Cruz P, Huet H, Zhang S, De Kochko A, Beachy RN, Fauquet CM (1998) Expression and inheritance of multiple transgenes in rice plants. Nat Biotechnol 16:1060–1064
Chhuneja P, Minocha JL (1993) Induction of male sterility mutations in three cultivars of *Triticum aestivum* L. heterosis breeding in crop plants theory and application. Short Communications Symposium, Ludhiana, India, 23–24 February, pp 130–131
Chopra VL, Sharma RP (1985) Induced mutations in crop improvement. In: Chopra VL (ed) Genetic manipulations in crop plants. Oxford and IBH Publishing Company, New Delhi, pp 22–48
Chowdhury S (1978) Frequency and spectrum of mutations in bread wheat by physical and chemical mutagens. Indian J Genet 38(1):142–147
Cosby N, Lesley S (1997) Site-directed mutagenesis. Promega Notes Magazine 61, p 12
DAA (1957) Department of Agriculture Appropriations: 84th Congress, 2nd session, United States House Committee on Appropriations, p 242
Dhanayanth KPM, Reddy V (2000) Cytogenetic effects of gamma rays and ethylmethane sulphonate in chilli pepper (*Capsicum annum*). Cytologia 65:129–133

Dhanda SS, Sethi GS, Behl RK (2004) Indices of drought tolerance in wheat genotypes at early stages of plant growth. J Agron Crop Sci 190(1):6–12
Diamond J (1997) Guns, germs and steel: a short history of everybody for the last 13,000 years. Viking UK Random House, ISBN:0-09-930278-0
Dizdaroglu M (1992) Oxidative damage to DNA in mammalian chromatin. Mutat Res 275(3–6):331–342
Dolferus R, van den Bossche D, Jacobs M (1990) Sequence analysis of two null-mutant alleles of the single *Arabidopsis* Adh locus. Mol Gen Genet 224:297–302
Dubrova YE, Nesterov VN, Krouchinsky NG, Ostapenko VA, Neumann R, Neil DL, Jeffreys AJ (1996) Human minisatellite mutation rate after the Chernobyl accident. Nature 380:683–686
Duvick DN (1999) Commercial strategies for exploitation of heterosis. In: Coors JG, Pandey S (eds) The genetics and exploitation of heterosis in crops. American Society of Agronomy, Inc, Crop Science Society of America, Inc, Soil Science Society of America, Inc., Madison, WI, USA, pp 295–304
Dvorak J, Akhunov E (2005) Tempos of gene locus deletions and duplications and their relationship to recombination rate during diploid and polyploid evolution in the *Aegilops-Triticum alliance*. Genetics 17:323–332
Dvorak J, Luo MC, Yang ZL (1998) Genetic evidence on the origin of *Triticum aestivum* L. In: Damania AB, Valkoun J, Willcox G, Qualset CO (eds) The origins of agriculture and crop domestication. Proc Harlan Symp, ICARDA, Aleppo, pp 235–251
DWR Perspective Plan: Vision 2025. Directorate of Wheat Research, Karnal-132 001, Haryana, India
Dzhelepov K (1991) Association between height, coleoptile length and harvest index in short-strawed mutants of wheat. Genetika I Selektsiya 24(2):99–103
Eiges NS (1990) Mutation spectra and the cytogenetics of winter wheat mutations using different concentrations of ethylene imine. Tritogenetika Zernovykh Kul Tur Jallinn, Estonian SSRC, pp 168–175
Ellegren H, Lindgren G, Primmer CR, Moller AP (1997) Fitness loss and germline mutations in barn swallows breeding in Chernobyl. Nature 389:593–596
Erdem G, Oldacay S (2004) Employment of RAPD technique to assess the genetic stability of *Helianthus annus* treated with differrent mutagenic agents. J Appl Sci 5:277–2281
Friedberg EC (2003) DNA damage and repair. Nature 421:436–440
Friedberg EC, Walker GC, Siede W (1995) DNA repair and mutagenesis. ASM, Washington, DC, p 698
Gentil A, Margot A, Sarasin A (1984) Apurinic sites cause mutations in simian virus 40. Mut Res 129:141–147
Gerd PP, Besaratinia A (2009) Mutational spectra of human cancer. Hum Genet 125(5–6):493–506
GOPR (2004) The Government of Punjab Report 2004. Human Development Report 2004, Punjab (Report) "The Green Revolution" pp 17–20
Grant WF, Salamone MF (1994) Comparative mutagenicity of chemicals selected for test in the international program on chemical safety collaborative study on plant systems for the detection of environmental mutagens. Mutat Res 310(2):187–209
Halliwell B, Gutteridge J (2007) Free radicals in biology and medicine, 4th ed. Oxford University Press, Oxford
Hariharan PV, Cerutti PA (1977) Formation of products of the 5, 6-dihydroxydihydrothymine type by ultraviolet light in HeLa cells. Biochemistry 16(12):2791–2795
Harlan JR, Zohary D (1966) Distribution of wild emmer wheat and barley. Science 153:1074–1080
Hassan S, Ali I, Muhammad T, Shah SA (1988) Effect of gamma rays and sodium azide (NaN_3) on morphological characteristics of wheat. Nucleus 25(1–2):19–22
Holzberg S, Brosio P, Gross C, Pogue GP (2002) Barley stripe mosaic virus-induced gene silencing in a monocot plant. Plant J 30:315–327
Hong JS, Ames BN (1971) Localized mutagenesis of any specific small region of the bacterial chromosome. Proc Natl Acad Sci U S A 68(12):3158–3162

Huang S, Sirikhachornkit A, Su X, Faris J, Gill B, Haselkorn R, Gornicki P (2002) Genes encoding plastid acetyl-CoA carboxylase and 3-phosphoglycerate kinase of the *Triticum/Aegilops* complex and the evolutionary history of polyploid wheat. Proc Natl Acad Sci U S A 99:8133–8138
Husain S, Kozgar MI, Jafrey IJ, Khan S (2013) Meiotic changes in *Vicia faba* L. subsequent to treatments of hydrazine hydrate and maleic hydrazide. J Biosci Biotechnol 2(1):33–38
Ibrahim IF, Al-Janabi KK, Al-Maaroof EM, Al-Aubaidi MO, Mahmoud AH, Al-Janabi AA (1991) Development of new Iraqi wheat varieties induced by gamma rays. Isotope Rad Res 23(1):59–66
Inda LA, Segarra-Moragues JG, Müller J, Peterson PM, Catalán P (2008) Dated historical biogeography of the temperate LoHinae (Poaceae, Pooideae) grasses in the northern and southern hemispheres. Mol Phylogenet Evol 46:932–957
James B (2009) Assembly of a complex genome: Defining elements of structure and function. PhD Thesis, Murdoch University, Western Australia
Jamil M, Khan UQ (2002) Study of genetic variation in yield components of wheat cultivar Bukhtwar-92 as induced by gamma radiation. Asia J Plant Sci 1(5):579–580
Jaradat AA (2006) Phenotypic divergence in the meta-population of the Hourani durum wheat landrace. J Food Agric Environ 4(3–4):186–191
Jaradat AA (2013) Wheat landraces: A mini review. Emir J Food Agric 25(1):20–29
Johal GS, Balint-Kurti P, Weil CF (2008) Mining and harnessing natural variation: a little MAGIC. Crop Sci 48:2066–2073
Joshi SA, Gopalakrishna T (2007) Inter simple sequence repeat (ISSR) markers for detecting radiation induced polymorphism and its application as genetic marker system in *Sesbania rostrata* (Bremek. & Obrem). Plant Mutat Rep 1:41–45
Kanofsky JR, Sima P (1991) Singlet oxygen production from the reactions of ozone with biological molecules. J Biol Chem 266:9039–9042
Kerber ER (1964) Wheat: reconstitution of the tetraploid component (AABB) of hexaploids. Science 143:253–255
Khamankar YG (1981) Micromutations induced by physical and chemical mutagens in bread wheat (*Triticum aestivum*). Indian J Genet 41:322–327
Khan AS, Malik MA (1999) Effect of gamma irradiation on some morphological characters of three wheat varieties. Pak J Biol Sci 2:903–905
Khan S, Wani MR (2006) MMS and SA induced genetic variability for quantitative traits in mungbean. Indian J Pulses Res 19:50–52
Khan S, Siddiqui BA, Rehman MU (1999) Mutation genetic studies in mungbean III: screening of high yield mutants. J Cytol Genet 34:75–78
Kharkwal MC, Shu QY (2009) The role of induced mutations in world food security. In: Shu QY(ed) Induced plant mutations in genomics era, Food and Agricultural Organization of the United Nations, Rome, pp 33–38
Kihara H (1944) Discovery of the DD-analyser, one of the ancestors of *Triticum vulgare*. Agric Hortic 19:13–14 (Tokyo)
Kiribuchi-Otobe C, Yanagisawa T, Yamaguchi I, Yoshida H (1998) Wheat mutant with waxy starch showing stable hot paste viscosity. Cereal Chem 75(5):671–672
Kislev ME (1980) *Triticum parvicoccum* sp. nov, the oldest naked wheat. Isr J Bot 28:95–107
Knott DR (1990) What determines the success of mutation breeding? In: Plant mutation breeding for crop improvement. Proc Int Symp on the contribution of plant mutation breeding to crop improvement, Joint FAO/IAEA, Division of Nuclear Techniques in Food and Agriculture, Vienna, Austria, pp 111–118
Kovalchuk I, Kovakchuk O, Barbara H (2001) Biomonitoring the genotoxicity of environmental factors with transgenic plants. Trends Plants Sci 6:306–310
Kozgar MI, Goyal S, Khan S (2011) EMS induced mutational variability in *Vigna radiata* and *Vigna mungo*. Res J Bot 6:31–37
Kumar A, Mishra MN (2004) Gamma rays irradiation under dry, pre and post soaked condition on yield and its attributes in M_2 populations of urdbean (*Vigna mungo* (L.) Hepper). Adv Plant Sci 17:475–478

Kuraparthy V, Sood S, Dhaliwal HS, Chhuneja P, Gill BS (2007) Identification and mapping of a tiller inhibition gene (*tin3*) in wheat. Theor Appl Genet 114:285–294
Lapochkina IF (1998) Cytogenetic and morphological features of bread wheat hybrids obtained with the use of irradiated pollen of *Aegilops triumcialies* L. Genetika-Mosk Va 34:1263–1268
Latonen L, Taya Y, Laiho M (2001) UV-radiation induces dose-dependent regulation of p53 response and modulates p53-HDM2 in human fibroblasts. Oncogene 20(46):6784–6793
Laxami S, Desai SA, Lohithaswa HC, Hanchinal RR (2013) The utility of chemical and radiation induced mutations in tetraploid wheat. Indian J Genet 73(3):252–258
Lindahl T (1993) Instability and decay of the primary structure of DNA. Nature 362(6422):709–715
Lindahl T, Karlström O (1973) Heat-induced depyrimidination of deoxyribonucleic acid in neutral solution. BioChemistry 11:5151–5154
Liu S, Yeh C-T, Tang HM, Nettleton D, Schnable PS (2012) Gene mapping via bulked segregant RNA-Seq (BSR-Seq). PLoS One 7:e36406
Loeb LA (1989) Endogenous carcinogenesis: molecular oncology into the twenty first century presidential address. Cancer Res 49(20):5489–5496
Maki H, Sekiguchi M (1992) Mut T protein specifically hydrolyses a potent mutagenic substrate for DNA synthesis. Nature 355:273–275
Maluszynski M, Szareiko I, Maluszinska J (2001) Induced mutations in wheat. The world wheat book: a history of wheat breeding. Lavoisier, Londres-Paris, New York, pp 939–977
Matsuoka Y, Nasuda S (2004) Durum wheat as a candidate for the unknown female progenitor of bread wheat: an empirical study with a highly fertile F_1 hybrid with *Aegilops tauschii* Coss. Theor Appl Genet 109:1710–1717
McCallum CM, Comai L, Greene EA, Henikoff S (2000) Targeting induced local lesions in genomes (TILLING) for plant functional genomics. Plant Physiol 123:439–442
McFadden ES, Sears ER (1946) The origin of *Triticum spelta* and its free-threshing hexaploid relatives. J Hered 37(81–89):107–116
McRae DH (1985) Advances in chemical hybridization. Plant Breed Rev 3:169–191
Mehlhorn H, Tabner B, Wellburn AR (1990) Electron spin resonance evidence for the formation of free radicals in plants exposed to ozone. Phys Plant 79:377–383
Melnik VM, Yanchenko VI (1990) Morphological/biological features of induced mutants of durum wheat. Selektsiya i genetika sel'skokhozyaistvennykh kul'tur na Altae, pp 39–43
Merezhko AF (1998) Impact of plant genetic resources on wheat breeding. Euphytica 100(1–3):295–303
Miller JC, Tan S, Qiao G, Barlow KA, Wang J, Xia DF, Meng X, Paschon DE, Leung E, Hinkley SJ, Dulay GP, Hua KL, Ankoudinova I, Cost GJ, Urnov FD, Zhang HS, Holmes MC, Zhang L, Gregory PD, Rebar EJ (2011) A TALE nuclease architecture for efficient genome editing. Nat Biotechnol 29(2):143–148
Mostafa GG (2011) Effect of sodium azide on the growth and variability induction in *Helianthus annuus* L. Int J Plant Breed Genet 5(1):76–85
Muller HJ (1927) Artificial transmutation of the gene. Science 66:84–871
Nakamura J, Swenberg JA (1999) Endogenous apurinic/apyrimidinic sites in genomic DNA of mammalian tissues. Cancer Res 59:2522–2526
Nakamura J, La DK, Swenber JA (2000) 5′-nicked apurinic/apyrimidinic sites are resistant to β-elimination by β-polymerase and are persistent in human cultured cells after oxidative stress. J Biol Chem 275:5323–5328
Niwa O (2006) Indirect mechanism of genomic instability and the biological significance of mutations at tandem repeat loci. Mut Res 598:61–72
Niyogi KK, Last RL, Fink GR, Keith B (1993) Suppressors of trp1 fluorescence identify a new *Arabidopsis* gene TRP4 encoding the anthranilate synthase b subunit. Plant Cell 5:1011–1027
Novak FJ, Brunner H (1992) Plant breeding: induced mutation technology for crop Improvement. Int Atomic Energy Agency Bull 4:25–32

Orozco BM, McClung CR, Werneke JM, Ogren WL (1993) Molecular basis of the ribulose-1, 5-bisphosphate carboxylase/oxygenase activase mutation in *Arabidopsis thaliana* is a guanine-to-adenine transition at the 5'-splice junction of intron 3. Plant Physiol 102:227–232

Pal BP, Swaminathan MS (1960) Induced mutations, evolution and wheat breeding. Proc Nat Inst Sci India 26(Suppl):109–119

Palmer JJ (2001) How to brew. Defenestrative Publishing Company, ISBN:0-9710579-0, p 233

Peng J, Sun D, Nevo E (2011) Wild emmer wheat, *Triticum dicoccoides*, occupies a pivotal position in wheat domestication process. Aust J Crop Sci 5(9):1127–1143

Perez TS, Caballero TI, Mico IM, Perez LA, Guerra IM (1999) Radiation stress on M_1 wheat seeds studies by densitometry of X-ray radiographic images. Alimentaria 302:113–115

Puchta H (2005) The repair of double-strand breaks in plants: mechanisms and consequences for genome evolution. J Exp Bot 56:1–14

Rachovska G, Dimova D (2000) Effect of sodium azide and gamma rays on M_1 quantitative characteristics of the productivity and their connection with M_2 mutation changes in winter common wheat. Rasteniev dni Nauki 37:413–419

Raicu P, Mixich F (1992) Cytogenetic effects of sodium azide encapsulated in liposomes on heteroploides cell cultures. Mutat Res 283:215–219

Rastogi RP, Richa, Kumar A, Tyagi MB, Sinha RP (2010) Molecular mechanisms of ultraviolet radiation induced DNA Damage and repair. J Nucleic Acids. http://dx.doi.org/10.4061/2010/592980

Ravanat JL, Douki T, Cadet J (2001) Direct and indirect effects of UV radiation on DNA and its components. J Photochem Photobiol 63:88–102

Rebeck GW, Samson L (1991) Increased spontaneous mutation and alkylation sensitivity of *E. coli* strains lacking the ogt O6- methylguanine methyltransferase. J Bacteriol 173:2068–2076

Reddy VRK (1992) Mutagenic parameters in single and combined treatments of gamma rays, EMS and sodium azide in triticale, barley and wheat. Ad Plant Sci 5(II):542–553

Ricardo M, Ando A (1998) Effects of gamma radiation and sodium azide on quantitative characters in rice (*Oryza sativa* L.). Genet Mol Biol 21(1):244–251

Rideout WM, Coetzee GA, Olumi AF, Jones PA (1990) 5-methylcytosine as an endogenous mutagen in the human LDL receptor and p53 genes. Science 249:1288–1290

Ried TM, Loeb LA (1993) Tandem double CC → TT mutations are produced by reactive oxygen species. Proc Natl Acad Sci USA 90:3904–3907

Rines HW (1985) Sodium azide mutagenesis in diploid and hexaploid oats and comparison with ethylmethane sulfonate treatments. Env Exp Bot 25:7–16

Schaaper RM, Glickman BW, Loeb LA (1982) Role of depurination in mutagenesis by chemical carcinogens. Cancer Res 42(9):3480–3485

Seneviratne KACN, Wijesundara DSA (2007) First African violets (*Saintpaulia ionantha* (H.) Wendl.) with a changing colour pattern induced by mutation. Am J Plant Physiol 2:233–236

Shapiro HS (1976) Distribution of purines and pyrimidines in deoxyribonucleic acids. In: Fasma GD (ed) CRC handbook of Biochemistry and Molecular Biology. CRC, Cleveland, pp 241–262

Sheffield.ac.uk. (2011) The science in detail-wheats DNA-Research-Archaeology-The University of Sheffield. Sheffield.ac.uk.2011-07-19

Sheikh SA, Wani MR, Lone MA, Tak MA, Malla NA (2012) Sodium azide induced biological damage and variability for quantitative traits and protein content in wheat (*Triticum aestivum* L.). J Plant Genomics 2(1):34–38

Shu QY (2009) Turning plant mutation breeding into a new era: molecular mutation breeding. In: Shu QY (ed) Induced plant mutations in the genomics era. Food and Agriculture Organization of the United Nations, Rome, pp 425–427

Siddiqui S, Meghvansi MK, Zia-ul H (2007) Cytogenetic changes induced by sodium azide (NaN_3) on *Trigonella foenum-graecum* seeds. South Afr J Bot 73:632–635

Singh MP (1969) Some radiation induced changes at Q locus in bread wheat. Caryologia 22:119–126

Singh MOP, Dubey SD, Singhal NC (1973) Trends of mutational studies in bread wheat (*T. aestivum*). In: Adv Frontier in Cytogenetics, Hindustan Publishing Corporation, India, pp 21–27

Slupphaug G, Kavli B, Krokan HE (2003) The interacting pathways for prevention and repair of oxidative DNA damage. Mut Res 531(1–2):231–251

Smith CW (1995) Crop production, evolution and history. John Wiley and Sons, ISBN:0471079723

Sramkova Z, Gregova E, Sturdik E (2009) Genetic improvement of wheat-a review. Nova Biotechnol 9(1):27–51

Srivastava P, Marker S, Pandey P, Tiwari DK (2011) Mutagenic effects of sodium azide on the growth and yield characteristics in wheat (*Triticum aestivum* (L.) em. Thell). Asia J Plant Sci 10:190–201

Stadler LJ (1929) Chromosome number and the mutation rate in *Avena* and *Triticum*. Proc Natl Acad Sci U S A 15:876–881

Suganthi CP, Nalini R, Reddy VRA, Edwin R, Saikia A (1992) Mutation breeding in some cereals V: quantitative variability. Biosci Res Bull 8:73–79

Swaminathan MS (1965) The impact of dwarfing genes on wheat production. J Postgrad Sch Indian Agric Res Inst 3:57–62

Swaminathan MS (2013) Genesis and growth of the yield revolution in wheat in India: lessons for shaping our agricultural destiny. Agric Res 2(3):183–188

Swaminathan MS, Jaganthesen D, Chopra VL (1963) Induced spherococoid mutations in *Triticum aestivum* and their phylogenetic and breeding significance. Curr Sci 32:539–540

Tai YS, Bragg J, Meinhardt SW (2007) Functional characterization of ToxA and molecular identification of its intracellular targeting protein in wheat. Am J Plant Physiol 2:76–89

Taylor JS (1990) DNA, sunlight and skin cancer. J Chem Educ 67(10):833–841

Townsend JA, Wright DA, Winfrey RJ, Fu F, Maeder ML, Joung JK, Voytas DF (2009) High frequency modification of plant genes using engineered zinc finger nucleases. Nature 459:442–445

Tuteja N, Singh MB, Misra MK, Bhalla PL, Tuteja R (2001) Molecular mechanisms of DNA damage and repair: progress in plants. Crit Rev Biochem Mol Biol 36(4):337–397

Urnov FD, Rebar EJ, Holmes MC, Zhang HS, Gregory PD (2010) Genome editing with engineered zinc finger nucleases. Nat Rev Genet 11:636–646

Valko M, Rhodes CJ, Moncol J, Izakovic M, Mazur M (2006) Free radicals, metals and antioxidants in oxidative stress-induced cancer. Chem Biol Interact 160(1):1–40

Valko M, Leibfritz D, Moncol J, Cronin MTD, Mazur M, Telser J (2007) Free radicals and antioxidants in normal physiological functions and human disease. Int J Biochem Cell Biol 39(1):44–84

Van Den-BulkRW, Loffer HJM, Lindhout WH, Koornneef M (1990) Somaclonal variation in tomato: effect of explants source and a comparison with chemical mutagenesis. Theor Appl Genet 80:817–825

Varughese G, Swaminathan MS (1967) Sharbati Sonora a symbol of the age of algeny. Ind Farm 17(5):8–9

Vasil SP, Knayazyuk VI (1988) Effect of chemical mutagens and the mutability of winter wheat varieties. Progress Sctektsil Ozimai Pshenitsy Factor Intensifikasil Proizyodstva, pp 18–20

Wallace DC (1992) Mitochondrial genetics: a paradigm for aging and degenerative diseases? Science 256:628–632

Wang G, Zhu X, Wang GX, Zhu XD (1995) Genetic effects of ^{137}Cs gamma rays and ^{60}Co gamma rays irradiation on heading data and plant height of wheat. Acta Agric Nucleatae Sinica 9:20–24

Wani MR, Khan S, Kozgar MI (2011) Induced chlorophyll mutations I. mutagenic effectiveness and efficiency of EMS, HZ and SA in mungbean. Front Agr China 5(4):514–518

Wani MR, Lone MA, Sheikh SA, Dar MS, Tak MA, Ahmad P, Khan S (2012) Induction and assessment of genetic variability for yield and yield contributing traits of chickpea (*Cicer arietinum* L). J Plant Genom 2(1):28–33

Wiseman H, Halliwell B (1996) Damage to DNA by reactive oxygen and nitrogen species: role in inflammatory disease and progression to cancer. Biochem J 313(1):17–29

Witcomb JR, Joshi A, Joshi KD, Sthapti BR (1996) Farmer participatory crop improvement I. varietal selection and breeding methods and their impact on biodiversity. Exp Agric 32:445–460

Wu Z, Zhang X, He B, Diao L, Sheng S, Wang J, Guo X, Su N, Wang L, Jiang L, Wang C, Zhai H, Wan J (2007) A chlorophyll-deficient rice mutant with impaired chlorophyllide esterification in chlorophyll biosynthesis. Plant Physiol 145:29–40

Yasui T, Sasaki J, Matsuki J (1998) Waxy bread wheat mutants, K107W × 1 and K107W × 2, have a new null allele on Wx-D1 locus. Breed Sci 48:405–407

Yauk CL (2004) Advances in the application of germline tandem repeat instability for *in situ* monitoring. Mut Res 566:169–182

Zeerak NA (1991) Cytogenetical effect of gamma rays and ethylmethane sulphonate in brinjal (*Solanum melongena* L.). Cytologia 56:639–643

Zencirci N, Karagoz A (2005) Effect of developmental stages length on yield and some quality traits of Turkish durum wheat (*Triticum durum* L. convar. durum (Desf.) MacKey) landraces: influence of developmental stages length on yield and quality of durum wheat. Gen Res Crop Evol 52:765–774

Zeven AC (1999) The traditional inexplicable replacement of seed and seed ware of landraces and cultivars: a review. Euphytica 110:181–191

Chapter 13
Cotton Leaf Curl Virus Disease Predictive Model Based on Environmental Variables

Safdar Ali, M. Aslam Khan, Shahbaz Talib Sahi, M. Atiq and A. Hannan

Abstract Cotton leaf curl virus disease (CLCuVD), caused by a geminivirus, is a major threat to the cotton industry particularly in the province of Punjab. It has devastated cotton production during the past couple of decades or so causing serious problems in its management. This study, therefore, was initiated to develop a disease predictive model to characterize epidemiological factors conducive for disease spread/severity. Five years' data of CLCuVD severity, whitefly population density, and environmental conditions were collected for the development of a predictive model from different districts of cotton-growing areas of Punjab. A close relationship was observed between CLCuVD severity and whitefly population. The disease severity was high (54.06–55.31 %) when whitefly population attack was at its maximum (8.48–10.97/leaf) and low (18.37–36.30) when its infestation fell down (2.97–3.78/leaf). The maximum temperature conducive for the development of disease ranged from 37.16 to 37.78 °C contributing 58.3–61.4 % influence on the development of CLCuVD. The role of minimum temperature remained relatively low (33.6–30.3 %). The range of 58.35–60.39 % of relative humidity was found to be more conducive for the development of disease. The contribution remained poor so far as the role of rainfall and wind speeds was concerned. A predictive model based on 5 years' data (2002–2006) of CLCuV disease severity, whitefly population density, and environmental conditions was developed ($Y = 145 + 4.47x_1 -$

A. Hannan (✉)
Irrigation Department, Directorate of Land Reclamation, Sargodha-40100, Pakistan
e-mail: ahannan70@gmail.com

A. Hannan · S. Ali · M. A. Khan · S. Talib Sahi · M. Atiq
Department of Plant Pathology, University of Agriculture, Faisalabad-38000 , Pakistan

S. Ali
e-mail: safdar147m@uaf.edu.pk

M. A. Khan
e-mail: aslamkhanuaf@uaf.edu.pk

S. Talib Sahi
e-mail: shahbazsahi@uaf.edu.pk

M. Atiq
e-mail: dratiq1@yahoo.com

P. Ahmad et al. (eds.), *Improvement of Crops in the Era of Climatic Changes,* Volume 2,
DOI 10.1007/978-1-4614-8824-8_13,

$0.151x_2 - 0.490x_3 - 1.83x_4 + 1.58x_5 - 4.84x_6$; $R^2 = 0.79$). The model statistically justified ($R^2 = 0.79$) at $P < 0.05$ was calibrated and then validated using 2 years' data (2007–08) collected from the experiment in the research area of the Department of Plant Pathology, University of Agriculture, Faisalabad. A close relationship was observed between the two models. It was envisaged that the model would be helpful in forecasting the disease to decide the correct timing of pesticide application.

Keywords CLCuVD · Relative humidity · Temperature · Rainfall · Wind speed · Whitefly · Punjab · Epidemics

1 Introduction

Cotton is known as the backbone of Pakistan's economy due to its wide production and diverse uses. Pakistan ranks fourth in area and production of cotton in the world (Anonymous 2010). Cotton plants are naturally susceptible to a number of diseases; about 75 cotton diseases have been reported to be destructive (Watkins 1981). Cotton leaf curl virus disease (CLCuVD) is a serious threat to successful cotton production, and the average cotton yield of Pakistan dropped to 30% between 1992 and 1997 resulting in losses of US$ 5 billion due to CLCuVD (Briddon and Markham 2000; Anonymous 2003–2004). From 1998 to 2002, losses of 7.7 million bales of cotton have been reported by Ahmad et al. (2002). In 1999–2000, the cotton production was 11.24 million bales which declined to 10.73 million bales in 2000–2001. It further decreased to 10.61 million bales in 2001–2002 and still further dropped to 10.21 million bales in 2002–2003. The downslide continued until 2003–2004 reaching 10.00 million bales with 1.6% lower yield than the last year. For the previous 2 decades, epidemics of CLCuVD have induced considerable losses to this crop and continue to be a major threat to future cotton production (Briddon and Markham 2000). During 2012–2013, 10.3% decline was observed in cotton production, the major cause being CLCuVD (Anonymous 2012–2013).

In Pakistan, this disease was first reported on few plants near Multan in 1967 (Hussain and Ali 1975) and remained endemic until 1986 (Hussain and Mahmood 1988). In 1988, the disease appeared in an epidemic form and damaged the crop in about 60 hectares near Multan. Since then, the disease has been gradually spreading and causing huge losses in yield (Mahmood 1999). CLCuVD is the most important constraint to cotton production in Pakistan and India. A circular single-stranded DNA monopartite begomo virus from family *Geminiviridae* (transmitted by *Bemisia tabaci*) is the cause of the disease. Among many strains of CLCuV, *cotton leaf curl Burewala virus* (CLCuBuV) along a specific satellite (beta satellite) is responsible for the expression of severe symptoms (Brown et al. 2012; Ali et al. 2013; Kumar et al. 2013; Sattar et al. 2013). Out of many factors, environmental conditions create a favorable situation for the disease to spread in epidemic form as these play a significant role in the insect vector population buildup (Khan and Khan 2000). In view of the heavy losses caused by CLCuVD, it is necessary to determine environmental conditions conducive for the development of CLCuVD in major cotton-growing areas of Punjab, Pakistan, and to develop a disease predictive model to establish a forecasting system in future which will provide a base to take preemptive decision against the insect vector and CLCuVD under a given set of environmental conditions for better management practices.

2 Materials and Methods

Five years' data (2002–2006) of CLCuV disease incidence, whitefly population density, and environmental conditions were collected from different districts of cotton-growing areas of Punjab (Pakistan) to develop a disease predictive model. To validate/verify the model so developed, 2 years' data of disease incidence and whitefly population were collected from the disease-screening nursery established in the research area of Department of Plant Pathology, University of Agriculture, Faisalabad, from June 2007 to October 2008. During the same period, maximum and minimum air temperature, rainfall, relative humidity, and wind velocity were recorded on a daily basis at the University of Agriculture, Faisalabad. The detection of virus was confirmed through whitefly transmission and graft inoculation in the net house (Akhtar et al. 2001). The influence of each environmental variable (maximum and minimum air temperature, rainfall, relative humidity, and wind velocity) on CLCuVD was determined by regression analysis.

3 Results and Discussion

3.1 Conducive Environmental Conditions for the Development of clcuvd

In the present study, maximum whitefly population (10.97/leaf) was recorded in 2005 (Table 13.1), and the corresponding value for the CLCuVD incidence was found at its peak of 55.31 % (Fig. 13.1). A similar pattern of disease development was observed during 2006 (CLCuVD = 54.06 %, whitefly population = 8.48). From 2002 to 2004, a significantly low whitefly population (2.97–3.78/leaf) was recorded as compared to 2005 and 2006 (8.48–10.97/leaf). The CLCuVD incidence during the same period, i.e., from 2002 to 2004, was found to be lower (18.37–36.30 %) as compared to 2005 and 2006 (54.06–55.31 %). One of the most important factors in the development of CLCuVD was whitefly—the vector of this disease (Mansoor et al. 1993; Hameed et al. 1994). The maximum temperature conducive to CLCuVD recorded during 2005 was 37.16 °C contributing 61.4 % influence on the

Table 13.1 Comparison of environmental conditions for CLCuVD during 2002–2006

Environmental parameters	2002	2003	2004	2005	2006	LSD
CLCuVD	18.37 c	35.24 b	36.30 b	55.31 a	54.06 a	14.73
Bemisia tabaci	2.97 b	3.78 b	3.18 b	10.97 a	8.48a	2.818
Maximum temperature (°C)	40.63 a[a]	39.94 ab	39.63 a	37.16 ab	37.78 ab	1.943
Relative humidity (%)	62.43 a	65.49 a	65.19 a	58.35ab	60.39a	7.824
Minimum temperature (°C)	26.84 a	27.33 a	27.41 a	27.02 a	26.84 a	1.250
Rainfall (mm)	13.44 a	17.28 b	12.68 a	10.06 a	10.54a	6.706
Wind velocity (km/hr)	4.61 a	4.93 a	5.40 a	5.04a	4.75 a	0.864

[a]Mean values within rows not sharing the same letter differ significantly at 0.05 level of probability

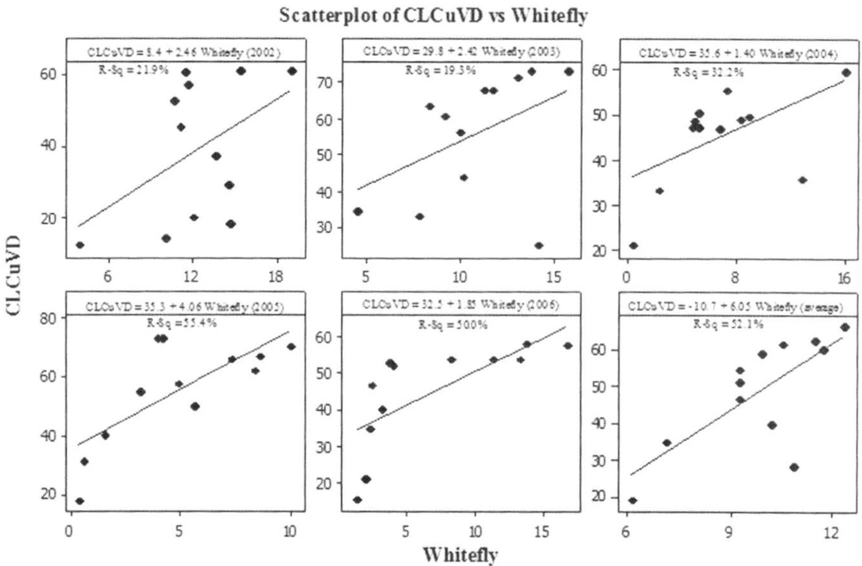

Fig. 13.1 Effect of whitefly in the development of CLCuVD

development of disease (Table 13.1; Fig. 13.2). The role of this variable in disease development was to the tune of 58.3% in 2006 when the temperature was 37.78 °C. The maximum temperature on an average contributed maximum influence as compared to the other variables. The ideal relative humidity conducive for maximum disease development was found to be 58.35 and 60.39% in 2005 and 2006, respectively (Table 13.1). The contribution of this variable remained 34.8 and 28.4% during 2005 and 2006, respectively (Fig. 13.3). The value of minimum tem-

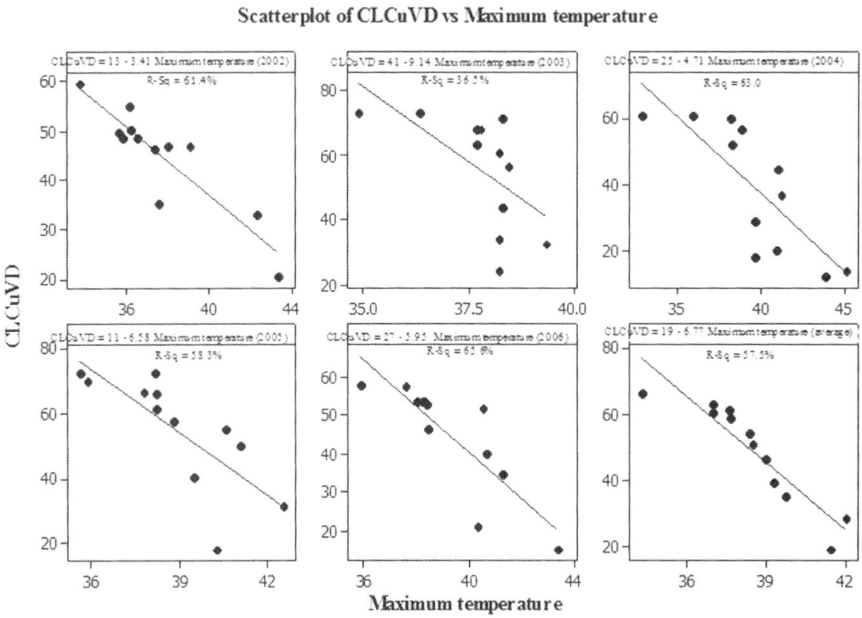

Fig. 13.2 Effect of maximum temperature in the development of CLCuVD

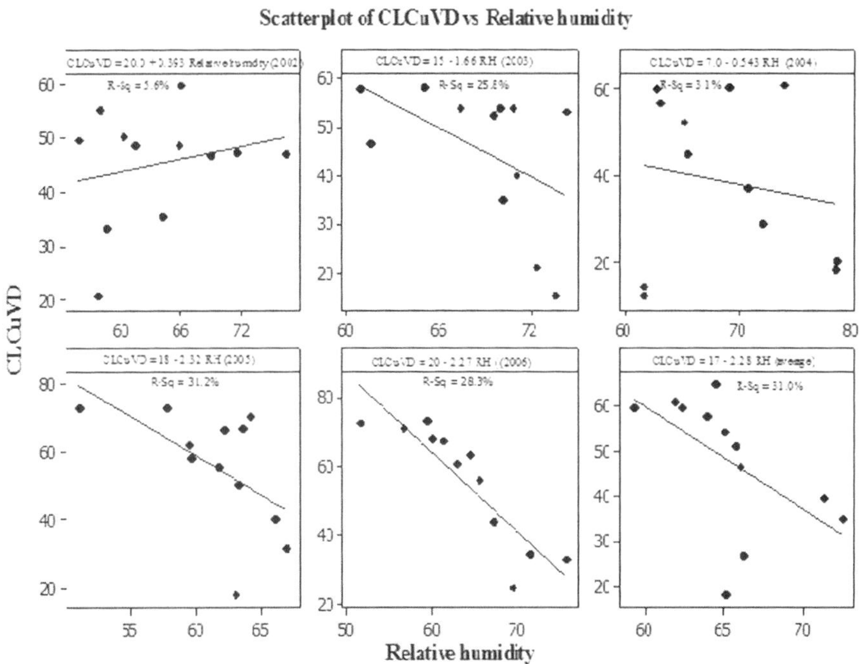

Fig. 13.3 Effect of relative humidity in the development of CLCuVD

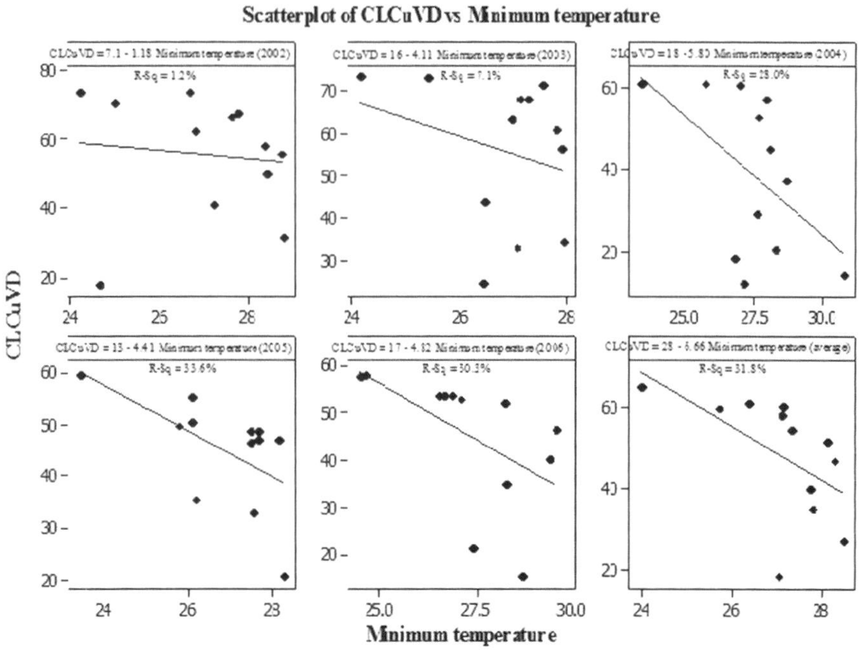

Fig. 13.4 Effect of minimum temperature in the development of CLCuVD

perature towards CLCuVD development was 27.43 °C averaged across 2005–2006 (Fig. 13.4). The contribution of minimum temperature towards the disease development remained poor but significant at $P=0.05$. The overall relationship between CLCuVD and rainfall was found to be negative (Fig. 13.5). The contribution of this variable to the development of CLCuVD was 24.8 and 28.4 % during 2005 and 2006, respectively. The wind velocity remained statistically nonsignificant from 2002 to 2006 (Table 13.1) ranging from 4.61 to 5.40 km/hr. On an overall basis, the individual contribution of this variable remained low ranging from 5.12 to 20.4 % in 2006 and 2005, respectively (Fig. 13.6).

3.2 Development of a CLCuVD Predictive Model

In research, there is a great thrust to know how and to what extent a certain variable (response variable) is related to a set of other variables (explanatory variables). In a statistical relationship, the relation between variables is not known exactly and one has to approximate the relationship and develop models that characterize their main features. Regression analysis is concerned with developing such "approximating" models (Graham 2003). The objective of this investigation was to understand and explain the nature of relationship of different environmental variables with

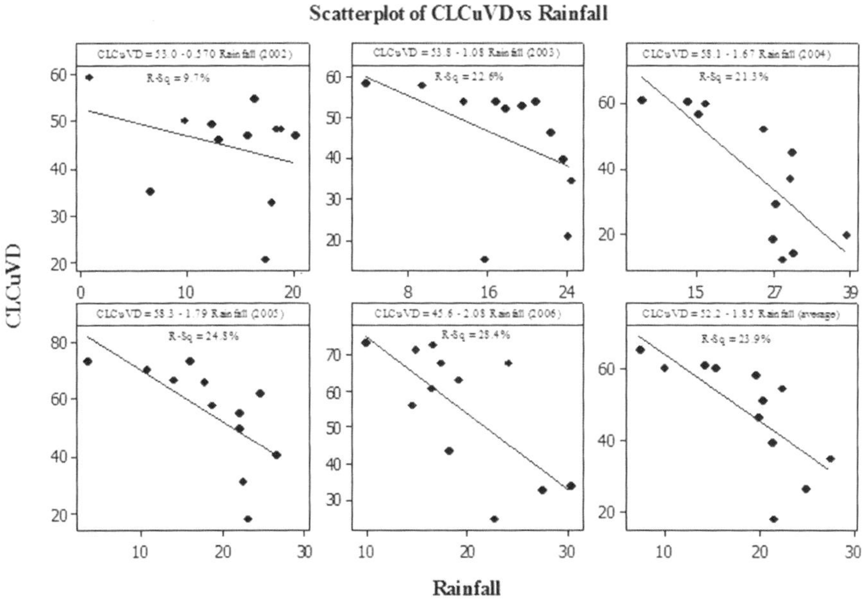

Fig. 13.5 Effect of rainfall and wind velocity in the development of CLCuVD

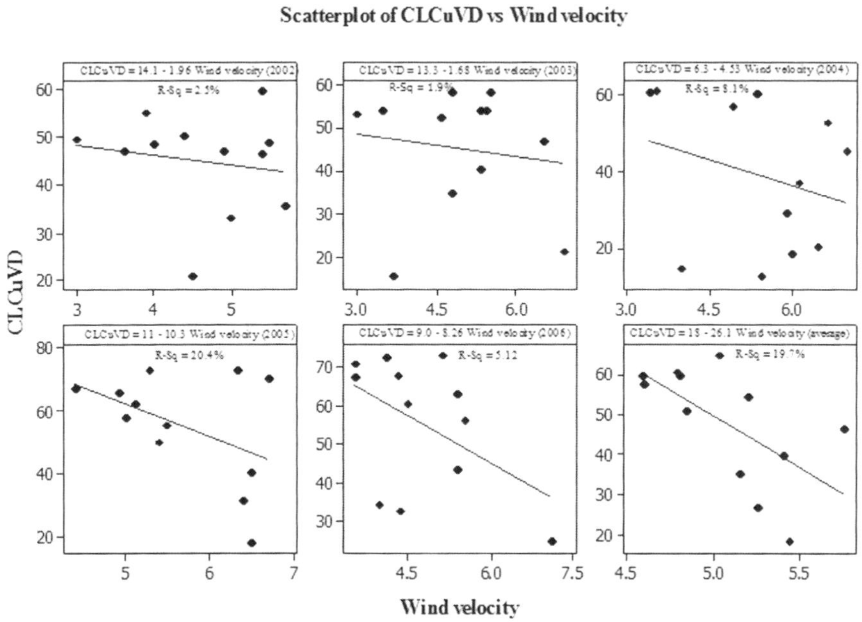

Fig. 13.6 Effect of wind velocity in the development of CLCuVD

Table 13.2 Regression statistics of the predictive model for CLCuVD based on 5 years (2002–2006)

Regression statistics	
R-square	0.79
Adj-R square	0.76
Standard error	8.185
Observations	60

CLCuVD and its vector—whitefly. By identifying such disease-contributing variables, it would be rather easy for the farmers to have better control over the disease. The model was developed in four phases.

3.2.1 clcuvd Predictive Model Based on 5 Years' Data (2002–2006)

In the present study, a multiple regression model was developed. The statistically justified model ($R^2=0.79$) at $P<0.05$ developed to predict the probable attack of CLCuVD under a given set of environmental variables is: $Y=145+4.47x_1-0.151x_2-0.490x_3-1.83x_4+1.58x_5-4.84x_6$ with $R^2=0.79$, where Y=CLCuVD, x_1= minimum temperature, x_2= rainfall, x_3= relative humidity, x_4= wind velocity, x_5= whitefly, and x_6= maximum temperature. It is evident from the model that major factors responsible for the attack of CLCuVD were temperature and the extent/intensity of whitefly population prevalent at that time. It indicated that with one unit change in minimum temperature, there would be a probable change of 4.47 units in CLCuVD. The change would be 4.84 units in the case of maximum temperature. Whitefly, considered as a vector in the onset of CLCuVD, has a significant contributing role in the spread of this viral disease, i.e., linear expansion in a single part of whitefly would cause an equivalent swell in 1.58 units of CLCuVD.

3.2.2 Model Assessment

The above-cited model was assessed according to the procedures described by Chattefuee and Hadi (2006) and Snee (1977). One of the most important parameters to check the model reliability is the value of coefficient of determination, i.e., R^2. In the present study, it was 0.79 which is considered fairly good particularly under field conditions (Table 13.2). Similarly, the standard error of estimate was not so high (8.185). The F-distribution of the regression model was significant at $P<0.05$ (Table 13.3). The relative contribution of maximum temperature, minimum temperature, relative humidity, and whitefly towards the development of model was significant at <0.05, except wind velocity and rainfall. It may be concluded that the model is good for prediction/forecasting purpose from a set of unknown variables based on the physical theory. The model assessment phase was completed by comparing observed and predicted data. There was a small difference between the slope of observed and predictive model values (0.411 and 0.287) with R^2 of 0.18 and 0.11, respectively (Fig. 13.7). It was obvious from the graph displayed in Fig. 13.8 that

Table 13.3 Analysis of variance of the predictive model for CLCuVD based on 5 years (2002–2006)

Source	Df	SS	MS	*F*	*P* value	Significance *F*
Regression	6	13,450.1	2,241.7	33.46	0.000	***
Residual error	53	3,550.7	67			
Total	59	17,000.8				

***$P<0.10$

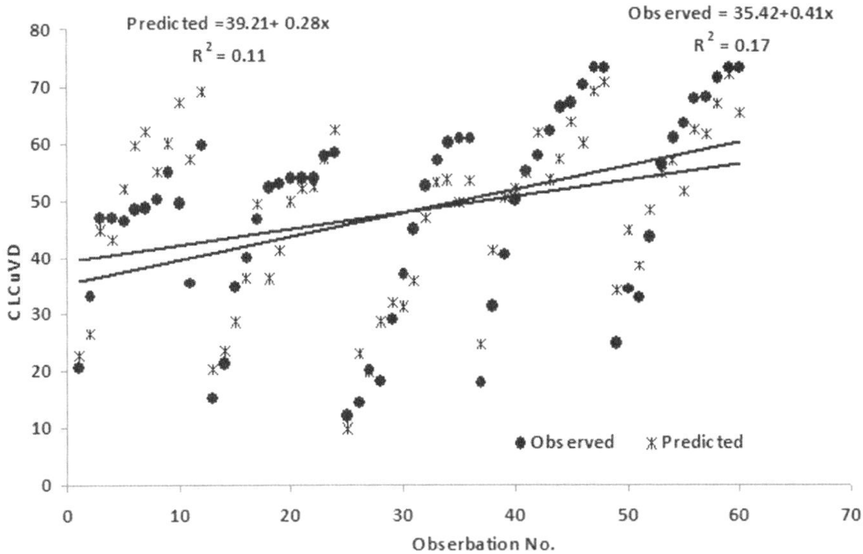

Fig. 13.7 A comparison of predicted and observed data points based on 5 years' data

the majority of the predictions (out of 60 data points, 47 data points have difference less than 10) were centralized between 95 % confidence interval (CI) and 95 % predictive interval (PI) indicating a fairly good fit between observed and predicted data.

3.2.3 CLCuVD Predictive Model Based on 2 Years' Data (2007–2008)

Similarly, a predictive model based on 2 years' data (2007–2008) was developed (significant at $P=0.05$; Table 13.4) on the same lines as was done for 5 years' data based on a given set of environmental variables. $Y=145+2.78x_1-0.998x_2-0.400x_3-1.02x_4+3.85x_5-2.25x_6$ with $R^2=0.75$ and SE=6.30, where Y=CLCuVD,

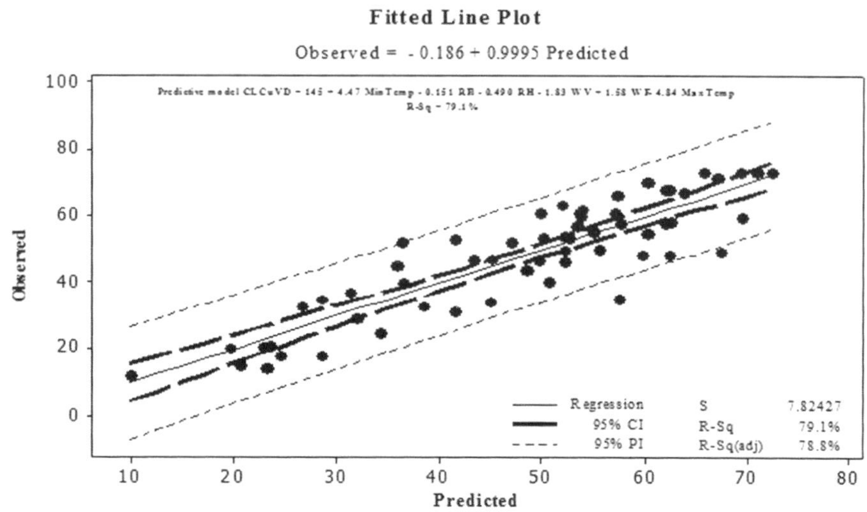

Fig. 13.8 A fitted line plot for CLCuVD with predicted and observed data points at 95% confidence and predictive interval

Table 13.4 Analysis of variance of the predictive model for CLCuVD based on 2 years (2007–2008)

Source	Df	SS	MS	*F*	*P* value	Significance *F*
Regression	6	2,628.20	438.03	10.83	0.000	***
Residual error	17	687.81	40.46			
Total	23	3,316.01				

***$P<0.10$

x_1= minimum temperature, x_2= rainfall, x_3=relative humidity, x_4= wind velocity, x_5= whitefly, and x_6= maximum temperature. The model was significant at $P=0.05$ (Table 13.4). It may be noticed from the above model that the relative contribution of maximum temperature, minimum temperature, relative humidity, wind velocity, rainfall, and whitefly towards the model development was similar to that observed in the previous model based on 5 years' data.

3.2.4 CLCuVD Model Validation

The most reliable option in validating the model is making a prediction by collecting data from some other source/location, i.e., any other data not used in the model development. Therefore, the model so developed on 5 years' data (2002–2006) was validated on 2 years (2007–2008). There was a small difference between the slope of observed and predicted values (0.80 and 1.03) with R^2 of 0.22 and 0.21, respec-

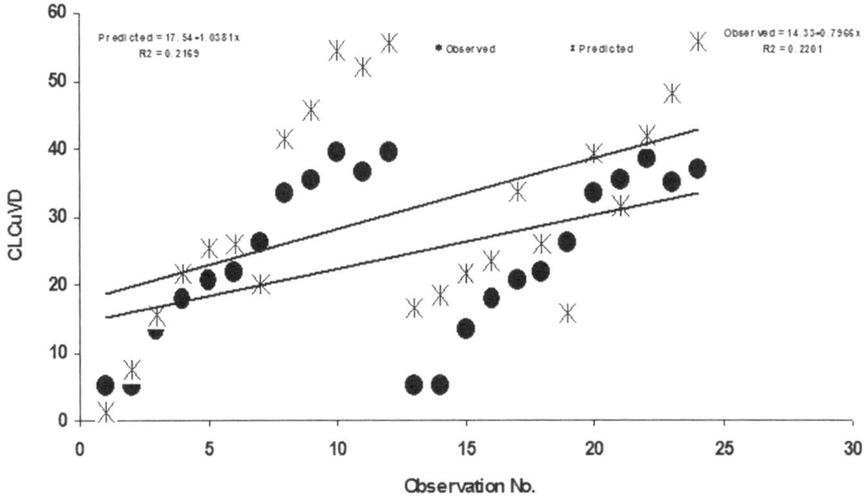

Fig. 13.9 A comparison of predicted and observed data points based on 2 years' data using model developed on 5 years' data

tively (Fig. 13.9). The forecasting accuracy was considerably better, $R^2=0.78$, significant at $P<0.05$ with a low standard error of 5.67 (Fig. 13.10). The P value for model comparison was 0.972 which was nonsignificant indicating that two models have close association with one another (Table 13.5).

4 Conclusions and Future Perspective

The maximum temperature and relative humidity were identified as main environmental variables contributing to the development of disease. The influence of rainfall and wind speeds was either nil or meager with respect to disease development. A predictive model ($Y=145+4.47x_1-0.151x_2-0.490x_3-1.83x_4+1.58x_5-4.84x_6$; $R^2=0.79$) statistically justified at $P<0.05$ ($R^2=0.79$) based on 5 years' data (2002–2006) was developed which was then calibrated and validated using 2 years' data (2007–08). The forecasting accuracy of the predictive model was considerably better, $R^2=0.78$, significant at $P<0.05$ with a low standard error of 5.67. The model indicated a good fit between observed and simulated data. However, the model needs to be expanded at the country level based on historic data comprising the past 30–40 years. The most important soil properties (e.g., electrical conductivity (EC), sodium adsorption ratio (SAR), NPK status, texture, organic matter, etc.) need to be evaluated in such a model. The nutritional status of affected plants also may be explored in the model. Research on management practices in controlling some particular environmental variable will certainly go a long way in reducing its epidemic.

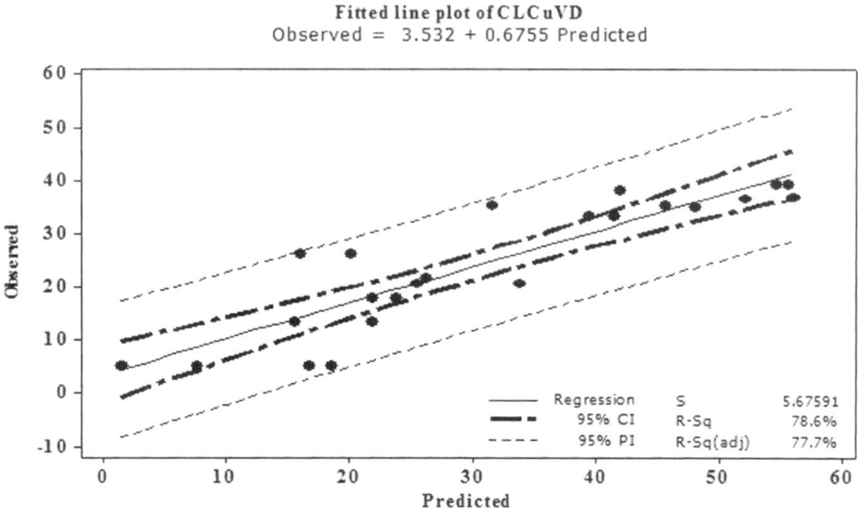

Fig. 13.10 A fitted line plot for CLCuVD with predicted and observed data points at 95 % confidence and predictive interval

Table 13.5 Comparison of the two models

Model	Regression equations	*F* value	*P* value
Model (I)	$Y=145+4.47x_1-0.151x_2-0.490x_3-1.83x_4+1.58x_5-4.84x_6$		
vs.		0.243	0.972
Model (II)	$Y=145+2.78x_1-0.998x_2-0.400x_3-1.02x_4+3.85x_5-2.25x_6$		

Model (I) Model based on 5 years' data
Model (II) Model based on 2 years' data

References

Ahmad G, Malik SA, Mahmood Z, Iqbal MZ, Ahmad S (2002) Effect of cotton leaf curl virus disease severity on morphology, yield and fibre characteristics of susceptible lines/cultivars of cotton (*G. hirsutum* L.). Asian J Plant Sci 1:705–707

Akhtar KP, Khan AI, Khan MSI (2001) Response of some cotton varieties to leaf curl virus through grafting. Pak J Phytopathol 13:91–95

Ali I, Amin I, Briddon RW, Mansoor S (2013) Artificial microRNA-mediated resistance against the monopartite begomovirus cotton leaf curl Burewala virus. Virol J 10:231

Anonymous (2003–2004) Economic survey 2003–04. Ministry of Finance, Govt. of Pakistan, Islamabad, pp 132

Anonymous (11 March 2010) World cotton production, US cotton market monthly economic letter

Anonymous (2012–2013) Pakistan economic survey. Ministry of Finance, Govt. of Pakistan, Islamabad, pp 19

Briddon RW, Markham PG (2000) Cotton leaf curl virus disease. Virus Res 71:151–159

Brown JK, Fauquet CM, Briddon RW, Zerbini M, Moriones E, Navas-Castillo J (2012) Geminiviridae. In: King AMQ, Adams MJ, Carstens EB, Lefkowitz EJ (eds) Virus taxonomy-ninth report of the international committee on taxonomy of viruses. Elsevier, London, pp 351–373

Chattefuee S, Hadi AS (2006) Regression analysis by example. Wiley, Hoboken

Graham MH (2003) Confronting multicollinearity in ecological multiple regression. Ecology 84:2809–2815
Hameed S, Khalid S, Ehsan-ul-Haq, Hashmi AA (1994) Cotton leaf curl disease in Pakistan caused by whitefly transmitted geminivirus. Plant Dis 78:528–529
Hussain T, Ali M (1975) A Review of cotton diseases of Pakistan. Pak Cottons 19:71–86
Hussain T, Mahmood T (1988) A note on the leaf curl disease of cotton. Pak Cottons 32:248–251
Khan MA, Khan HA (2000) Cotton leaf curl virus disease severity in relation to environmental conditions. Pak J Biol Sci 3(10):1688–1690
Kumar J, Gunapati S, Singh SP, Kumar A, Lalit A, Sharma NC, Puranik R, Tuli R (2013) A new betasatellite associated with cotton leaf curl Burewala virus infecting tomato in India: influence on symptoms and viral accumulation. Arch Virol 158(6):1349–1353
Mahmood T (1999) Cotton leaf curl virus and its status in Pakistan. Proceedings of the ICAC-CCRI, regional consultation insecticide resistance management in Cotton, Multan, Pakistan, 28 June–1 July, pp 234–244
Mansoor S, Bedford ID, Pinner MS, Stanley J, Markham PG (1993) A whitefly-transmitted geminivirus associated with cotton leaf curl disease in Pakistan. Pak J Bot 25:105–107
Sattar MN, Kvarnheden A, Saeed M, Briddon RW (2013) Cotton leaf curl disease-An emerging threat to cotton production worldwide. J Gen Virol 94:695–710
Snee RD (1977) Validation of regression models: methods and examples. Technometrics 19:415–428
Watkins GM (1981) Compendium of cotton diseases. The American Phytopathological Society 3340. Pilot Knob Road, Minnesota, USA, pp 87

Chapter 14
Transcription Factors in Abiotic Stress Responses: Their Potentials in Crop Improvement

Xuan Lan Thi Hoang, Nguyen Binh Anh Thu, Nguyen Phuong Thao and Lam-Son Phan Tran

Abstract Abiotic stresses, especially drought, high salinity, flooding, and extreme temperatures, have become a big concern due to their high frequency of occurrence and usually beyond human control capacity, as well as their severe impacts on agricultural crop productivities. Under the pressures of climate change and reduction in total cultivated land worldwide for other purposes, sustaining food security to feed an increasing human population while coping with these environmental constraints is a greater challenge than ever. Generating new varieties with better traits based on gene exchange from available sources via conventional breeding methods currently no longer provides an adequate solution in coping with abiotic stresses. Therefore, another research theme attracting the scientists over the past 20 years has been to elucidate molecular mechanisms that the plants employ to defend and adapt to stress conditions. The final aims are to identify and characterize the function of important genes involved in plant responses to stress that can be used for genetic manipulation. Thanks to advances in molecular biotechnology, including gene transfer techniques such as particle bombardment, microinjection, and *Agrobacterium*-mediated transformation, new varieties with better stress tolerance and yield production could be made by this strategy; thus, in combination with traditional approaches, development of new lines with improved traits has become more practical. According to our current knowledge, transcription factors (TFs) have been recognized to play essential roles in regulating plant responses against adverse abiotic factors. Many TFs belonging to families AP2/EREBP, bZIP, MYB, WRKY, and NAC have been

Xuan Lan Thi Hoang and Nguyen Binh Anh Thu contributed equally to this work.

N. P. Thao (✉) · X. L. T. Hoang · N. B. A. Thu
School of Biotechnology, International University, Vietnam National University HCMC, Quarter 6, Linh Trung Ward, Thu Duc District, Ho Chi Minh City, Vietnam
e-mail: npthao@hcmiu.edu.vn

L.-S. P. Tran (✉)
Signaling Pathway Research Unit, RIKEN Center for Sustainable Resource Science, 1-7-22, Suehiro-cho, Tsurumi, Yokohama 230-0045 Japan
e-mail: son.tran@riken.jp

P. Ahmad et al. (eds.), *Improvement of Crops in the Era of Climatic Changes,* Volume 2,
DOI 10.1007/978-1-4614-8824-8_14,

reported to participate in plant responses to various stressors. A number of TFs whose encoding genes are appropriately altered in expression level have shown enhanced tolerance capacity toward drought, salt, and suboptimal temperatures in transgenic model and crop plants. In this chapter, we summarize our current understanding about TF activities in plants under adverse stress conditions and their use in crop improvement.

Keywords Abiotic stress · Transcriptional factor · Stress tolerance · Genetic engineering · Crop improvement

1 Introduction

Abiotic stress is defined as a "non-living environmental factor that can negatively or even harmfully affect the growth and productivity of plant" (Ku et al. 2013). Among the abiotic stressors, drought is the biggest threat to plants in general and to crops in particular. Shortage of water supply leads to reduction of photosynthesis due to lack of intracellular CO_2 availability because of stress-induced stomatal closing and reduction of root capacity to absorb nutrients in soil, and thus total energy production of the plant is decreased (Lawlor and Cornic 2002; Lawlor and Tezara 2009; Aroca et al. 2012). Excessive light can exacerbate stress in plants by affecting its photosynthesis. When chloroplasts are exposed to excessive light, photooxidation is induced, which results in the increased production of reactive oxygen species (ROS), such as hydrogen peroxide, superoxide anion, and hydroxyl radicals (Grassi and Magnani 2005; Aguado-Santacruz 2006; Zhou et al. 2007; Miller et al. 2010; Osakabe et al. 2014). These biological molecules can cause negative effects on plant productivity. Moreover, water deficit triggers the stomata closure and a decrease in leaf water potential as well as the downregulation of photosynthesis-related genes and slow diffusion of CO_2 (Osakabe et al. 2014). More severe drought exposures can finally lead to plant death.

The invasion of salinity into mainland and higher frequency of flooding as a consequence of ice melt or urbanization process make these the second- and third-ranked concerns to sustainable agriculture. Under high salinity, susceptible plants suffer ion toxicity due to excessive entry of ions, particularly Na^+, into the cells, resulting in homeostatic disruption, water loss, and oxidative stress, and thereof inhibition of photosynthesis and enzyme activities, cell division, and plant growth (Munns et al. 2006; Munns and Tester 2008; Parvaiz and Satyawati 2008; Shabala and Munns 2012). It is projected that by 2050, more than half of arable land would encounter salinity problem (Wang et al. 2003; Setia et al. 2011). Opposite to water deficit, waterlogging is another type of water stress whereby it prevents plants, or at least their root part, from accessing oxygen supplied from environment (Mahajan and Tuteja 2005; Calvo-Polanco et al. 2012; Kreuzwieser and Rennenberg 2014). The consequences of this are negative effects on root functions in respiration and taking up nutrients (Mahajan and Tuteja 2005). Meanwhile, exposure to nonoptimal temperature conditions, such as cold (chilling and frost) and heat (high temperature), may cause injuries to cellular structure and functions, and thus significantly affecting plant metabolism, development, and reproduction (Yadav 2010; Theocharis et al. 2012; Hasanuzzaman et al. 2013). For example, cellular ice formation, dehydration, and fluidity movement reduction in the phospholipid membrane are among early effects caused by low temperatures (Mahajan and Tuteja 2005; Wang et al. 2006).

It has been known that chilling, freezing, heat, drought, and salinity all lead to hyperosmotic condition as a secondary stress in plants. Depletion in cellular water is the common consequence caused by each of these stress factors. Hyperosmosis is explained as a condition in which there is an increase in concentration of extracellular solutes due to water deficit, and thus triggering the movement of water out of the cells into the environment (Aguado-Santacruz 2006). As a result, cellular turgor cannot be maintained and intracellular concentration of solutes is increased (Aguado-Santacruz 2006; Solanke and Sharma 2008; Chaves et al. 2009). It is important to mention that although some of the molecular responses to these types of stressors overlap as they produce similar effects in the water status of cells, stress-specific responses are also present depending on which particular stress the plant is facing with (Aguado-Santacruz 2006).

In addition to the well-mentioned abiotic stressors, such as drought, salinity, heat, cold, and flooding, there are other abiotic stressors appearing with less frequency, including UV radiation, heavy metal toxicity, and inorganic nutrient deficiency (Clemens 2006; Hossain et al. 2012; Hideg et al. 2013; Liang et al. 2013). Nowadays, abiotic stresses are not purely considered as natural causes since their occurrence and severity are under the influence of human activities such as not only inappropriate practices in agriculture like intensive land use without break and excessive organic fertilization but also in other areas, such as deforestation and industrial gas release into the atmosphere. As a consequence of these human errors, climate change has been a hot topic for discussion, and of course this exacerbates unpredictability of abiotic stresses, making them more serious threats to the food security of humankind.

To reduce agricultural loss by the impact of abiotic stresses, scientists have been trying to widen our current understanding of the mechanisms of plant tolerance to various abiotic stressors by using different methodologies. Although how expression of genes alters under stress conditions can be easily determined, working out their roles in the stress responses and identifying their relationship with other members in the whole picture of stress acclimation are not easy tasks (Chaves et al. 2009). That may be the reason why the number of successes in the development of crop plants coping with the major stressors, including water scarcity, salinity, and temperature stresses, is still limited. To accelerate this process, a combination of modern techniques in molecular biology, advances in phenotyping, quantitative trait locus (QTL) mapping, and conventional breeding strategies has been used. The purposes of this chapter are to summarize our current knowledge about stress-adaptive mechanisms in plants, mainly focusing on major abiotic stresses (drought, salinity, and extreme temperatures) and transcription factors (TFs)—important members in the stress-adaptive pathways—and then review the progress in applications of TFs in improving abiotic stress tolerance in crop plants.

2 Mechanisms in Plant Responses to Abiotic Stresses

It is predicted that abiotic stresses will continue to be major constraints affecting global crop yields (Sharma and Lavanya 2002; Osakabe et al. 2013b), and thus, through evolution plants have developed various mechanisms to cope with them. These mechanisms have been classified into three groups: (1) stress escape by adjusting plant development and reproduction in accordance with the environmental conditions; (2) stress avoidance by developing morphologically and physiologically advantageous traits; and (3) stress tolerance by regulating morphological, physiological, biochemical, and molecular activities to minimize the stress impact (Turner et al. 2001; Mittler 2002; Chaves et al. 2003). The last defending strategy attracts the largest research attention and hence will be discussed in detail later in this section.

2.1 Physiological and Biochemical Responses

Adaptation to abiotic stress is undoubtedly one of the most complex biological processes in plants since it involves multigenic traits. Reponses that are usually seen at the physiological level are closure of stomata, shedding of leaves, decrease in light absorbance and photosynthetic activities, adjustment in relative water content, increase in root growth, and reduction in shoot growth (Manavalan et al. 2009; Shanker and Venkateswarlu 2011; Thu et al. 2014; Ha et al. 2013). For example, under high CO_2 and abiotic stress conditions, excess C may favor root growth over that of photosynthetic organs (Hachiya et al. 2014). Shoot-growth restriction could be considered as an advantage to help plants preserve limited carbohydrate resource for maintaining the most important metabolism activities to ensure their survival over the stress period and their capacity to recover quickly after the demission of the stress (Aguado-Santacruz 2006). In contrast, continuation of root growth is favorably promoted to search for water from deeper soil layers (Alsina et al. 2011; Aroca et al. 2012). Examination of various root characteristics, such as dry weight, total length, and number of lateral roots, enables us to quickly assess the tolerant capacity of a particular plant variety under stress conditions (Liu et al. 2005; Gowda et al. 2011; Kumar et al. 2012). Other physiological symptoms can be observed under abiotic stress, including decrease in leaf expansion, wilting, discoloration, chlorosis, surface lesion and/or sterility at reproduction stage, senescence, stunted shoots, and lower biomass production (Wen et al. 2002; Yadav 2010; Jaarsma et al. 2013). Hence, many common parameters, such as survival rate, productivity, and relative growth rate, are used for the evaluation of plant tolerance to various stresses (Ashraf and Harris 2004; Parvaiz and Satyawati 2008).

The most important, commonly seen response of plants to major abiotic stresses at biochemical level is perhaps the osmotic adjustment. A wide variety of solutes are involved in this regulation, including amino acids and amino acid-associated compounds (proline, aspartate, glutamate, and glycine betaine etc.), sugars (sucrose, fructose, fructans, and trehalose etc.), and polyols (mannitol, and sorbitol etc.) (Chaves et al. 2003; Koyro et al. 2012). These metabolites are also called osmoprotective substances that contribute to maintaining the cell turgor and prevention of water loss. Additionally, increase of other products, such as antioxidant substances (e.g., ascorbate, glutathione, carotenoids, anthocyanins, and α-tocopherol), detoxifying enzymes (e.g., catalase, superoxide dismutase, peroxidase, ascorbate peroxidase, glutathione peroxidase, and glutathione reductase), heat shock proteins (HSPs), and late embryogenesis abundant (LEA) proteins, is also observed under stresses (Goyal et al. 2005; Chakrabortee et al. 2010; Aguado-Santacruz 2006). Moreover, under stress conditions, such as drought and salinity, transportation activities of anions and cations, including Cl^-, Na^+, and K^+, as well as the participation of water transport systems play important roles to maintain an ion balance between tissues and cells in stress adaptation (Osakabe et al. 2013a; Osakabe et al. 2014).

2.2 Signal Perception, Transduction, and Regulating Pathways Involved in Plant Responses to Abiotic Stresses

Regulation of molecular activities such as gene expression is considered the most critical factor determining the response degree of a plant as a whole toward a specific stress condition. In this process, being able to sense changes in the surrounding environment is the first important step for plants to respond, survive, and adapt to living conditions. Disruption of normal intracellular homeostasis in plants caused by stress is regarded as early alerting signals that include alterations of cellular size, solute concentration, and cellular turgidity and/or increase in ROS levels (Loutfy et al. 2012; Suzuki et al. 2012). A number of stress detectors have been identified in plants and most of them are located on the cell membrane (Mahajan and Tuteja 2005). Under cold, drought, or salinity stress conditions, Ca^{2+} ions have been shown to be shuttled into the cell cytoplasm, indicating participation of Ca^{2+} permeable channels in the stress-responsive networks as stress signal receptors (Boudsocq and Sheen 2013). Calcium ions have been indicated to interact with kinase enzymes, including Ca-dependent protein kinases (CDPKs), calmodulin-binding protein kinases, and Ca^{2+}/calmodulin-dependent protein kinases that are known to participate in signaling pathways via phosphorylation activities. Taking a strategy used by plants to cope with salt stress as an example, the salt overly sensitive (SOS)-responsive pathway, which aids the reestablishment of intracellular ionic balance, includes a sensor Ca-binding protein SOS3 that is able to mediate the activity of kinase SOS2 (Boudsocq and Laurière 2005; Thapa et al. 2011). These activated kinases will in turn directly regulate activities of TFs in the downstream transduction pathway (Chinnusamy et al. 2004). It was reported that an increase in ROS concentration during stress also led to the enhancement of activity of these Ca channels (Mori and Schroeder 2004).

Another type of membrane protein sensors involved in the perception of drought, high salinity, and low temperature, which deserves to be mentioned, is histidine kinases (HKs) that are members of two-component system (TCS). In *Arabidopsis thaliana,* a number of HKs (AHKs) have been identified (Schaller et al. 2008). HKs functioning as osmosensors, such as Sln1 and Sho1, are also present in yeast in which both of them are membrane spanning but have no similarity in structure. In *Arabidopsis,* a homolog of Sln1, AtHK1/AHK1, has been identified which is able to suppress the salt-sensitive phenotype of the yeast double-mutant sln1Δ sho1Δ that lacked both osmosensors (Urao et al. 1999). The interaction between AHK1 and the downstream *Arabidopsis* histidine phosphotransfers (AHPs) in the multistep phosphorelay has been identified (Urao et al. 2000; Kumar et al. 2013). In addition to these above-described sensors, G protein-associated receptors may function in the perception of a secondary signal derived from these stresses (Perfus-Barbeoch et al. 2004; Misra et al. 2007). They might also serve as a kind of membrane-bound receptors for abscisic acid (ABA; Pandey et al. 2009).

It is important to emphasize that a specific abiotic stress can generate more than one type of signals that can be detected by one or more sensors (Gong et al. 2013). For example, drought causes both osmotic and oxidative stresses to plants. Oxidative stress and appearance of photorespiration as secondary stress effects from drought can lead to increased production of ROS, possibly deriving from organelles such as chloroplasts, mitochondria, glyoxysomes, and peroxisomes, and from activities of β-oxidation of lipids and several different oxidases (Apel and Hirt 2004; Miller et al. 2010; Voss et al. 2013). ROS inhibits the function of many enzymes, increases protein and DNA susceptibility to degradation due to structural modifications, reduces permeability of plasma membrane, and stimulates the activity of Ca-dependent proteases and nucleases (Sharma et al. 2012).

The participators working as second messengers can be divided into three groups, including: (1) diacylglycerol, inositol triphosphate, such as IP3, and phosphate idylinositols (group of hydrophobic molecules) which are located on the membrane and are able to pass the signal to membrane-associated effector proteins; (2) cAMP, cGMP, sugars, such as sucrose, glucose, and fructose, and Ca^{2+} (group of hydrophilic molecules) which are presented within the cytosol; and (3) nitric oxide and carbon monoxide (group of gas molecules) which can diffuse through cytosol and cellular membranes (Bhargava and Sawant 2013; Aguado-Santacruz 2006; Chaves et al. 2009). Additionally, ROS can also be assigned as another member of second messenger since its damage effects to cellular components during stress switch on specific genes involved in the activation of mitogen-activated protein kinase (MAPKs) as well as the production of antioxidants, HSPs, and ROS-scavenging enzymes.

Several second messengers, such as IP3, can regulate intracellular Ca^{2+} levels, while the Ca^{2+} ion itself is also regarded as a second messenger. Ca^{2+} has been known to be involved in the closure of stomatal aperture (Zhang et al. 2011a), and often initiates a protein phosphorylation cascade that regulates expression of downstream regulatory and functional genes, contributing to stress tolerance (Gong et al. 2013). It is noticed that in order to successfully transfer the signal message through the whole signal transduction pathway, extra assistants that are in charge of modification, delivery, or assembly of signaling components are required as well. Examples of these molecules are protein modifiers involved in protein lipidation, methylation, glycosylation, and ubiquitination; scaffolds; and adaptors (Xiong et al. 2002).

A great variety of other elements are also recruited in the networks, leading to acclimative reactions to stress, such as plant growth regulators, including phytohormones, polyamines, and Ca, and various proteins, including CDPKs, MAPKs (including MAPKKKs, MAP kinase kinase kinases; MAPKKs, MAP kinase kinases), and TFs (Ohnishi et al. 2008; Huang et al. 2012). Particularly, involvement of ABA and its induced biosynthesis in response to various abiotic stresses, such as salinity, drought, and cold, have been well reported in literature (Fujita et al. 2011; Danquah et al. 2014). Many stress-responsive functional genes, such as *RD22* (responsive to dehydration 22), *RD29A, COR15* (cold responsive 15), *COR47, and P5CS* (-Δ-1-pyrroline-5-carboxylate synthase), and regulatory genes, including dehydration-responsive element-binding protein (DREB), ABA-responsive element (AREB), myeloblastosis (MYB), and NAC (NAM, no apical meristem; ATAF, *Arabidopsis* transcription activation factor; and CUC, cup-shaped cotyledon), whose expression is regulated by ABA, have been identified to be involved in regulating plant responses to stresses (Xiong et al. 2001; Dalal et al. 2009; Fujita et al. 2011; Nakashima et al. 2009; Tran et al. 2010).

Protein phosphorylation is an important and effective mechanism that plants use in the signal transduction process to trigger stress responses as quickly as possible. In plants, in addition to the TCSs, activities of the two classes of stress-activated protein kinases, MAPKs and CDPKs, are also performed via phosphorylation at specific amino acid residues present in the structure (Schaller et al. 2008; Huang et al. 2012). In the working module of MAPK members, an MAPKKK is activated by phosphorylation which in turn phosphorylates the activity of MAPKK (Bhargava and Sawant 2013). At the end of the phosphorylation cascade, activation of a cytoplasmic MAPK often results in its translocation into the nucleus to regulate gene expression via controlling TFs also by phosphorylation (Danquah et al. 2014). Alternatively, these terminal MAPKs remain in the cytoplasma where they phosphorylate enzymes or cytoskeleton components (Danquah et al. 2014).

In further progression of the signaling process, TFs should be highlighted as the key participators. They work as final transducers in the transduction module and directly mediate gene expression since they can bind to regulatory regions of specific promoters (e.g., *cis*-acting elements). There is a fact that expression of a gene can be regulated by several mechanisms of which some are still unknown (Weake and Workman 2010). Nonetheless, the regulation of gene expression via the interaction of TFs and promoters are well documented (Yamaguchi-Shinozaki and Shinozaki 2006). Detailed information about TFs will be discussed in the next section of this chapter.

3 Plant TFs and their Roles in Abiotic Stress Responses

In plant genomes, approximately 7% of the coding sequences are assigned to TFs (Udvardi et al. 2007), which are divided into families based on their distinct signatures in structure. Major families are well known to be involved in various abiotic stresses either in ABA-dependent (MYB/MYC, bZIP) (Zhang et al. 2009a; Fujita et al. 2011), ABA-independent pathway (WRKY) (Chen et al. 2012; Rushton et al. 2008; Umezawa et al. 2006), or in both pathways (NAC, AP2/EREBP) (Souer et al. 1996; Olsen et al. 2005; Nakashima et al. 2012; Puranik et al. 2012). With the aim to generate new varieties, which are able to cope with abiotic stress(es) more efficiently, genetic engineering has been considered as a powerful approach in addition to conventional breeding methods. Thanks to recent advances in technologies, including real-time quantitative polymerase chain reaction (RT-qPCR), omic technologies, cloning, transformation, gene overexpressing and silencing, and other biochemical, molecular, and physiological methods employed in stress-response analyses at physiological and molecular levels, essential candidate genes contributing to plant adaptation to stress have been identified. These genes include those encoding members functioning in various stages of stress signal transduction cascade. Among these, TFs have drawn particular attention due to their important function in stress regulation and their potential in genetic engineering. Therefore, numerous attempts have been made all around the world, mainly by overexpressing the TF-encoding genes, and several promising results have been reported. Table 14.1 summarizes a number of studies using TFs to create transgenic model and crop plants with improved abiotic stress tolerance within the past 5 years.

3.1 The AP2/EREBP Family

In plants, AP2/EREBP (APETALA2/ethylene-responsive element-binding protein) has been known as a large family of TF genes which contains the highly conserved AP2/ethylene-responsive element-binding factor (ERF) DNA-binding domain (Riechmann and Meyerowitz 1998), and has been identified in various species, such as *Arabidopsis,* tobacco (*Nicotiana tabacum*), and tomato (*Solanum lycopersicum*) (Fischer and Dröge-Laser 2004; Oñate-Sánchez and Singh 2002; Tournier et al. 2003). Based on the number of AP2/ERF domains and the gene structure, the AP2/EREBP gene family can be divided into four subfamilies AP2, RAV (related to ABI3/VP1), DREB, and ERF (Sakuma et al. 2002; Sharoni et al. 2011). Among these, the ERF and DREB TF subfamilies were discovered in various plant species, including rice (*Oryza sativa*) (Quan et al. 2010), *Arabidopsis* (Sakuma et al. 2006), and tobacco (Agarwal et al. 2010), and their functions in the plant responses to biotic and abiotic stresses have been extensively studied (Agarwal et al. 2006; Agarwal et al. 2010).

The DREB-type TFs contain a conserved DNA-binding domain of 58–60 amino acids, and thus are able to bind to the PyCCGACAT *cis*-elements named

Table 14.1 Studies on abiotic stress tolerance improvement in crop plants using transcription factors over the past 5 years

Family	Gene	Source of isolation	Transgenic host	Gene manipulation	Stress tolerance	References
AP2/EREBP	*AtDREB1A*	*Arabidopsis thaliana*	Rice	Stress-inducible overexpression	Drought↑	(Ravikumar et al. 2014)
	GmERF3	*Glycine max*	Tobacco	Constitutive overexpression	Drought↑, salinity↑	(Zhang et al. 2009b)
	GmERF7	*Glycine max*	Tobacco	Constitutive overexpression	Salinity↑	(Zhai et al. 2013)
	HARDY	*Arabidopsis thaliana*	Clover	Constitutive overexpression	Drought↑, salinity↑	(Abogadallah et al. 2011)
	JERF3	*Solanum lycopersicum*	Rice	Constitutive overexpression	Drought↑	(Zhang et al. 2010)
	LcDREB3a	*Leymus chinensis*	*Arabidopsis*	Constitutive overexpression	Drought↑, salinity↑	(Peng et al. 2011)
	LcDREB2	*Leymus chinensis*	*Arabidopsis*	Constitutive overexpression	Salinity↑	(Peng et al. 2013)
	OsAP23	*Oryza sativa*	*Arabidopsis*	Constitutive overexpression	Salinity↓	(Zhuang et al. 2013)
	OsDREB2A	*Oryza sativa*	Rice	Stress-inducible overexpression	Drought↑, salinity↑	(Mallikarjuna et al. 2011)
	OsDREB2A	*Oryza sativa*	Soybean	Constitutive overexpression	Salinity↑	(Zhang et al. 2013)
	OsERF4a	*Oryza sativa*	Rice	Constitutive and ABA-inducible overexpression	Drought↑	(Joo et al. 2013)
	StDREB1	*Solanum tuberosum*	Potato	Constitutive overexpression	Salinity↑	(Bouaziz et al. 2013)
	TaERF3	*Triticum aestivum*	Wheat	Constitutive overexpression Constitutive knockdown	Drought↑, salinity↑ Drought↓, salinity↓	(Rong et al. 2014)
	TaPIE1	*Triticum aestivum*	Wheat	Constitutive overexpression	Cold↑	(Zhu et al. 2014)
	TaPIE1	*Triticum aestivum*	Wheat	Constitutive knockdown	Cold↓	(Zhu et al. 2014)
	TSRF1	*Solanum lycopersicum*	Rice	Constitutive overexpression	Drought↑	(Quan et al. 2010)

Table 14.1 (continued)

Family	Gene	Source of isolation	Transgenic host	Gene manipulation	Stress tolerance	References
bZIP	*ABP9*	*Zea may*	*Arabidopsis*	Constitutive overexpression	Drought↑, salinity↑, cold↑	(Zhang et al. 2011b)
	AtbZIP1	*Arabidopsis thaliana*	*Arabidopsis*	Knockout	Drought↓, salinity↓	(Sun et al. 2012)
	GmbZIP1	*Glycine max*	*Arabidopsis*	Constitutive and ABA-inducible overexpression	Drought↑, salinity↑, cold↑	(Gao et al. 2011)
	GmbZIP1	*Glycine max*	Wheat	Constitutive overexpression	Drought↑, salinity↑, cold↑	(Gao et al. 2011)
	GmbZIP1	*Glycine max*	Tobacco	Constitutive and ABA-inducible overexpression	Drought↑, salinity↑, cold↑	(Gao et al. 2011)
	LrbZIP	*Nelumbo nucifera*	Tobacco	Constitutive overexpression	Salinity↑	(Cheng et al. 2013a)
	OsbZIP72	*Oryza sativa*	Rice	Constitutive overexpression	Drought↑	(Lu et al. 2009)
	OsbZIP71	*Oryza sativa*	Rice	Constitutive overexpression Constitutive knockdown	Drought↑, salinity↑ Salinity↓	(Liu et al. 2014a)
	PtrABF	*Poncirus trifoliata*	Tobacco	Constitutive overexpression	Drought↑	(Huang et al. 2010)
	ZmbZIP72	*Zea may*	*Arabidopsis*	Constitutive overexpression	Drought↑, salinity↑	(Ying et al. 2012)
NAC	*AhNAC3*	*Arachis hypogaea*	Tobacco	Constitutive overexpression	Drought↑	(Liu et al. 2013)
	BnNAC5	*Brassica napus*	*Arabidopsis*	Constitutive overexpression	Salinity↑	(Zhong et al. 2012)
	EcNAC	*Eleusine coracana*	Tobacco	Constitutive overexpression	Salinity↑	(Ramegowda et al. 2012)
	JcNAC1	*Jatropha curcas*	Model woody	Constitutive overexpression	Drought↑	(Qin et al. 2014)
	ONAC063	*Oryza sativa*	*Arabidopsis*	Constitutive overexpression	Salinity↑	(Yokotani et al. 2009)
	OsNAC5	*Oryza sativa*	Rice	Constitutive overexpression Constitutive knockdown	Drought↑, salinity↑, cold↑ Drought↓, salinity↓, cold↓	(Song et al. 2011)
	ONAC045	*Oryza sativa*	Rice	Constitutive overexpression	Drought↑, salinity↑	(Zheng et al. 2009)
	OsNAC52	*Oryza sativa*	*Arabidopsis*	Constitutive overexpression	Drought↑	(Gao et al. 2010)
	OsNAP	*Oryza sativa*	Rice	Constitutive overexpression	Drought↑, salinity↑, cold↑	(Chen et al. 2014)
	SNAC1	*Oryza sativa*	Cotton	Constitutive overexpression	Drought↑, salinity↑	(Liu et al. 2014b)
	SNAC1	*Oryza sativa*	Wheat	Constitutive overexpression	Drought↑, salinity↑	(Saad et al. 2013)
	TaNAC2a	*Triticum aestivum*	Tobacco	Constitutive overexpression	Drought↑	(Tang et al. 2012)
	TaNAC67	*Triticum aestivum*	*Arabidopsis*	Constitutive overexpression	Drought↑, salinity↑, cold↑	(Mao et al. 2014)
	TaNAC69	*Triticum aestivum*	Wheat	Drought-inducible overexpression	Drought↑	(Xue et al. 2011)
	ZmSNAC1	*Zea may*	*Arabidopsis*	Constitutive overexpression	Drought↑	(Lu et al. 2012)

Table 14.1 (continued)

Family	Gene	Source of isolation	Transgenic host	Gene manipulation	Stress tolerance	References
MYB	*AtMYB14*	*Arabidopsis thaliana*	*Arabidopsis*	Constitutive knockdown	Cold↑	(Chen et al. 2013a)
	AtMYB44	*Arabidopsis thaliana*	Soybean	Constitutive overexpression	Drought↑, salinity↑	(Seo et al. 2012)
	AtMYBR1	*Arabidopsis thaliana*	*Arabidopsis*	Constitutive overexpression	Drought↓	(Jaradat et al. 2013)
	GmMYBJ1	*Glycine max*	*Arabidopsis*	Constitutive overexpression	Drought↑, cold↑	(Su et al. 2014)
	LcMYB1	*Leymus chinensis*	*Arabidopsis*	Constitutive overexpression	Salinity↑	(Cheng et al. 2013b)
	MYB15	*Arabidopsis thaliana*	*Arabidopsis*	Constitutive overexpression	Drought↑, salinity↑	(Ding et al. 2009)
	TaPIMP1	*Triticum aestivum*	Tobacco	Constitutive overexpression	Drought↑, salinity↑	(Liu et al. 2011b)
WRKY	*AtWRKY28*	*Arabidopsis thaliana*	*Arabidopsis*	Constitutive overexpression	Salinity↑	(Babitha et al. 2013)
	AtWRKY30	*Arabidopsis thaliana*	*Arabidopsis*	Constitutive overexpression	Salinity↑	(Scarpeci et al. 2013)
	GhWRKY39	*Gossypium hirsutum*	Tobacco	Constitutive overexpression	Salinity↑	(Shi et al. 2014)
	MusaWRKY71	*Musa* spp.	Banana	Constitutive overexpression	Salinity↑	(Shekhawat et al. 2011)
	OsWRKY11	*Oryza sativa*	Rice	Heat-inducible overexpression	Drought↑, heat↑	(Wu et al. 2009)
	OsWRKY45	*Oryza sativa*	*Arabidopsis*	Constitutive overexpression	Drought↑, salinity↑	(Qiu and Yu 2009)
	TaWRKY10	*Triticum aestivum*	Tobacco	Constitutive overexpression	Drought↑, salinity↑	(Wang et al. 2013)
	TaWRKY79	*Triticum aestivum*	*Arabidopsis*	Constitutive overexpression	Drought↑, salinity↑	(Qin et al. 2013)
	VvWRKY11	*Vitis vinifera*	*Arabidopsis*	Constitutive overexpression	Drought↑	(Liu et al. 2011a)
	ZmWRKY33	*Zea may*	*Arabidopsis*	Constitutive overexpression	Salinity↑	(Li et al. 2013)

as DRE/C-repeat (CRT) motifs located in the promoter regions of target genes to activate or suppress their transcription for achieving stress adaptation (Park et al. 2001; Ito et al. 2006; Jaglo et al. 2001; Kasuga et al. 1999; Liu et al. 1998; Agarwal et al. 2006; Sakuma et al. 2002). According to Sakuma et al. (2002), DREB subfamily has been further classified into six subgroups termed A-1 to A-6, of which A-1 and A-2 are the two largest groups. DREB1/C-repeat-binding factor (CBF; A-1) subgroup of *Arabidopsis* with three major regulator factors, DREB1A/CBF3, DREB1B/CBF1, and DREB1C/CBF2, has shown its importance in plant responses to cold stress. Overexpression of *RD29A:DREB1* or *35S:CBF* genes showed significantly improved tolerance to freezing, drought, and high salinity in *Arabidopsis* (Jaglo-Ottosen et al. 1998; Liu et al. 1998; Mizoi et al. 2012). Furthermore, the DREB1s can be used to produce transgenic crops with higher tolerance to drought, high salt, high temperature, and cold stress. Bouaziz et al. (2013) reported that transgenic potato plants overexpressing *35S:StDREB1,* which had been isolated from potato (*S. tuberosum*) and classified in the A-4 group of DREB subfamily based on multiple sequence alignments and phylogenetic characterization, had enhanced salt and drought tolerance. Moreover, *StDREB1* was significantly induced by NaCl, drought, low temperature, and ABA conditions in potato. In *Populus trichocarpa,* the development of *Populus* varieties with a greater tolerance to many adverse environments has been aided by understanding the characterization of the *DREB* subfamily (Chen et al. 2013b). Based on the expression analysis of 15 selected *PtrDREB* genes under abiotic stress conditions in *Populus,* Chen et al. (2013b) claimed that there were 14 genes with induced expression under different abiotic stresses. *GmDREB2* isolated from soybean is a novel *DREB* gene of A-5 subgroup which was found to have induced expression by drought, high salt, and low temperature stresses (Chen et al. 2007). Overexpression of *GmDREB2* by constitutive *35S* promoter or stress-inducible *RD29A* promoter resulted in upregulation of downstream genes in transgenic *Arabidopsis* (Chen et al. 2007). As a result, the transgenic plants showed enhanced tolerance to drought and high salinity without any growth retardation even with constitutive overexpression (Chen et al. 2007), as was observed with constitutive overexpression of *OsDREB1A* (Dubouzet et al. 2003). *LeDREB2* was discovered from tomato genome and classified into an A-2 group member of the *DREB* family. The expression of this gene was induced by high salinity, drought, and cold (Islam and Wang 2012). In another independent research on expression analysis in *Pisum sativum, PsDREB2A* was reported to be induced in pea roots and leaves under water deficit (Jovanovic et al. 2013). Mallikarjuna et al. (2011) developed transgenic rice lines overexpressing *OsDREB2A* under control of stress-inducible *RD29A* promoter, which resulted in the enhanced growth performance and significant tolerance to osmotic, salt, and dehydration stresses during simulated stress conditions as compared with the wild type. Later on, Zhang et al. (2013) used this gene to enhance salt tolerance of soybeans. The authors reported that salt tolerance was enhanced in the *35S:OsDREB2A* transgenic soybean plants due to accumulation of osmolytes, such as soluble sugars and free proline, as well as induced expression of several stress-responsive TFs and key genes (Zhang et al. 2013). *LcDREB3a* from the drought-tolerant forage grass

Leymus chinensis was shown to function in the improvement of drought and salt tolerance in *Arabidopsis* overexpressing *35S:LcDREB3a* without causing growth retardation by inducing expression of stress tolerance genes when compared to control (Peng et al. 2011). Besides, transgenic expression of another *LcDREB* member (*LcDREB2*) in combination with its downstream gene (S-adenosyl-methionine decarboxylase, LcSAMDC2,-encoding gene) obtained from *L. chinensis* under the control of the *35S* promoter could improve the salt tolerance in *Arabidopsis* (Peng et al. 2013). In *Malus domestica,* RT-qPCR analysis showed significant upregulation of some putative *MdDREB* genes under various abiotic stress treatments, which proved their vital roles during stress adaptation (Zhao et al. 2012).

In another subfamily of AP2/EREBP TFs, ERFs have been indicative of their participation in plant responses to biotic and abiotic stresses by recognizing the *cis*-acting element AGCCGCC, known as the GCC box (Hao et al. 1998; Ohme-Takagi and Shinshi 1995; Fujimoto et al. 2000). Based on the phylogenetic analyses of 125 *AP2/ERF* members in *Arabidopsis,* the ERF subfamily could be divided into six subgroups, from ERF-B1 to ERF-B6 (Sakuma et al. 2002). In wheat (*Triticum aestivum*), 47 ERF-encoding genes have already been identified (Zhuang et al. 2011). Among these, constitutively overexpressing *TaPIE1* controlled by maize *ubiquitin* promoter in wheat exhibited significantly enhanced resistance to both pathogen (triggered by *Rhizoctonia cerealis*) and freezing stress, whereas constitutive knockdown wheat plants by a recombinant construct between barley stripe mosaic virus (BSMV) and *TaPIE1* were more susceptible to both stresses relative to control plants (Zhu et al. 2014). Functional analysis of *TSRF1,* a member of tomato ERF TFs, demonstrated that overexpression of *35S:TSRF1* improved the osmotic and drought tolerance of rice seedlings without growth retardation as indicated by physiological analyses of root and leaf growth, leaf water loss, and survival rate under stress conditions compared to control (Quan et al. 2010). In another study in rice, Joo et al. (2013) reported that *ERF* genes, including *OsERF4a* and *OsERF10a,* had an important contribution in conferring drought stress tolerance. Both constitutive and ABA-inducible expression of the ERF-associated amphiphilic repression (EAR) domain-containing protein-encoding *OsERF4a* showed increased drought tolerance as a consequence of suppression of a putative repressor *Silent information regulator 2* (*Sir2*) involved in response to drought. By using a yeast-one hybrid system, *OsAP23* belonging to the B3 group of the *ERF* subfamily was isolated from rice (Zhuang et al. 2013). When exposed to high salt concentrations, several stress-responsive genes were induced significantly in the wild-type lines compared to *Arabidopsis* overexpressing *35S:OsAP23,* suggesting a negative regulatory role of OsAP23 in salt stress response (Zhuang et al. 2013). Besides, characterization of an *ERF* gene from soybean, *GmERF3,* showed its inducible expression in soybean by high salinity, drought, ABA, salicylic acid (SA), jasmonic acid (JA), ethylene (ET), and soybean mosaic virus (SMV), whereas *GmERF3* mRNA was not significantly accumulated under cold stress treatment (Zhang et al. 2009b). Transgenic tobacco plants with overexpressed *35S:GmERF3* displayed not only enhanced resistance against infection by *Ralstonia solanacearum, Alternaria alternata,* and tobacco mosaic virus (TMV) but also higher tolerance to high salinity and dehydration.

Functional analyses of a tomato *ERF* (*JERF3*-jasmonate and ethylene responsive factor 3) in transgenic tobacco overexpressing *35S:JERF3* indicated that expression of this gene could enhance the tolerance to salt, drought, and freezing (Wang et al. 2004; Wu et al. 2008). Also in further report of this group, transgenic rice plants overexpressing *35S:JERF3* exhibited better tolerance to drought and osmotic stress in comparison with non-transgenic rice seedlings (Zhang et al. 2010).

3.2 The bZIP Family

The bZIP (basic leucine zipper) family is another large group of TFs in plants. At present, most reports on stress responses have shown that bZIP TFs regulate stress response in ABA-dependent manner through interaction with specific ABA-responsive *cis*-acting elements (ABRE) to promote transcription of downstream target genes (Kobayashi et al. 2005; Kim 2006; Zou et al. 2008; Shinozaki and Yamaguchi-Shinozaki 2007; Uno et al. 2000). In *Arabidopsis, AtbZIP1*-knockout mutants showed a decrease in salt and osmotic stress tolerance, suggesting its positive regulation of plant response to these stresses (Sun et al. 2012). Transgenic *Arabidopsis* overexpressing the maize *ABP9* gene under the control of *35S* promoter, which encodes a bZIP TF, enhanced tolerance to multiple stresses (Zhang et al. 2011b). Another *bZIP* from maize was identified as *35S:ZmbZIP72* whose overexpression improved drought and partial salt tolerance of transgenic *Arabidopsis* plants (Ying et al. 2012). Meanwhile, expression analysis of *LrbZIP* in tips of lotus roots (*Nelumbo nucifera*) showed strong upregulation by low temperature, salt, and ABA treatments (Cheng et al. 2013a). Transgenic tobacco transformed with *35S:LrbZIP* exhibited higher salt tolerance comparing to the control under salt stress (Cheng et al. 2013a). Lu et al. (2009) identified OsbZIP72 as a positive regulator whose constitutive overexpression increased hypersensitivity to ABA and transcript level of ABA-responsive genes to improve drought tolerance in transgenic rice. Several soybean bZIPs, including GmbZIP44, GmbZIP62, and GmbZIP78, displayed notable roles in stress acclimation. These TFs functioned as negative regulators of ABA signaling and plant responses to salt and freezing tolerance (Liao et al. 2008b). Gao et al. (2011) also indicated the positive role of GmbZIP1 in the enhancement of multiple abiotic stress tolerance, including drought, salinity, and cold stresses in transgenic *Arabidopsis* by stimulating the expression of ABA- or stress-related genes. Overexpression of this gene by *ubiquitin* promoter also resulted in enhanced drought, salt, and freezing tolerance in transgenic wheat (Gao et al. 2011). The authors indicated that *Arabidopsis* and tobacco overexpressing *GmbZIP1* by stress-inducible *RD29A* or *35S* promoter also showed increased tolerance under similar stresses. *ScAREB1, SpAREB1,* and *SlAREB1* belonging to the *bZIP* family of *S. chilense, S. peruvianum,* and *S. lycopersicum,* respectively, were upregulated by salt stress (Yáñez et al. 2009). Moreover, expression of *SlAREB1* was induced by other stresses, such as drought and cold, and ABA in *S. lycopersicum*, and its encoded TF upregulated stress-responsive genes in *35S:SlAREB1*-overexpressing transgenic tobacco and tomato plants (Yáñez et al. 2009).

3.3 The MYB Family

The MYB TFs have also been known to form one of the largest TF families, which interact with one or more of the two stress-inducible *cis*-elements known as MYB-binding sites (MBS) that contain consensus sequences CNGTT(A/G) (MBSI) or C(G/T)T(A/T)GTT(A/G) (MBSII) to activate their downstream genes (Abe et al. 1997; Pabo and Sauer 1992; Riechmann and Ratcliffe 2000). The MYB TFs possess a MYB domain containing 1–4 imperfect tandem repeats (MYB repeat) located near the N-terminus and thus showing their distinctive characteristics (Ambawat et al. 2013). Based on the number of adjacent repeat(s) in the MYB domain, the MYB family is divided into different types, including 4R-MYB (four repeats), 3R-MYB (R1R2R3-MYB) (three consecutive repeats), R2R3-MYB (two repeats), and the MYB-related type with just a single repeat (Rosinski and Atchley 1998; Jin and Martin 1999; Dubos et al. 2010). Members belonging to this family have been found in different plant species, such as 204 members in *Arabidopsis,* 218 members in rice, 279 members in grapevine, and 197 members in *Populus* (Wilkins et al. 2009; Velasco et al. 2007; Chen et al. 2006). The roles of MYB proteins have been indicated in many physiological and biochemical processes which include regulation of primary and secondary metabolism, control of cell development and cell cycle, hormone synthesis, and signal transduction. The MYBs are also involved in plant responses to various biotic and abiotic stresses (Dubos et al. 2010; Feller et al. 2011; Stracke et al. 2001). Mochida et al. (2009) also found approximately 160 *GmMYBs* in soybean, of which 43 out of 48 analyzed genes showed expression changes at least by one of the following treatments: ABA, high salinity, drought, and cold.

A number of studies have indicated the potential of MYB TFs in genetic engineering for improved stress tolerance. When the rice *OsMYB3R-2* and the wheat *TaPIMP1* genes were overexpressed in *Arabidopsis* and tobacco using *35S* promoter, respectively, the transgenic lines displayed increased drought tolerance (Liu et al. 2011b; Dai et al. 2007). Besides, salt and freezing tolerance was elevated significantly in *Arabidopsis* overexpressing either *35S:GmMYB76* or *35S:GmMYB177* (Liao et al. 2008a). In addition, while *GmMYB177* was upregulated by both drought and NaCl treatments, *GmMYB76* was induced by NaCl treatment only. Jaradat et al. (2013) characterized the function of AtMYBR1 by using *mybr1*-mutant and *35S:AtMYBR1*-overexpressing *Arabidopsis* lines. Its negative regulatory functions in drought response and senescence, as well as in the downregulation of many ABA-responsive genes involved in abiotic stresses, were revealed in their study. Moreover, expression of *Arabidopsis AtMYB44* gene has been shown to improve salt and drought stress tolerance in soybean and *Arabidopsis* by preventing excessive ROS accumulation (Persak and Pitzschke 2014; Seo et al. 2012). Expression analysis of the *OsMyb4* in rice suggested that this gene might be involved in rice response to dehydration and cold stress (Baldoni et al. 2013). Moreover, the authors found induction of *OsMyb4*-like genes in wheat and *Arabidopsis* under similar stress treatments. Overexpression of *35S:OsMyb4* in apple could improve adaptive responses to drought and cold stresses (Pasquali et al. 2008).

3.4 The WRKY Family

The WRKY TFs were first reported in the study of Ishiguro and Nakamura (1994) and the name WRKY (pronounced "worky") was coined with the identification of WRKY1, WRKY2, and WRKY3 from parsley (*Petroselinum crispum*) (Rushton et al. 1996). The WRKY family is among the largest families of TFs in higher plants (Rushton et al. 2010). The WRKY domain is about 60 residues in length, and based on the number of WRKY domains, WRKY TFs were divided into three groups, group I (two domains), group II (one domain), and group III (one domain with structure of zinc fingers C2HC) (Eulgem et al. 2000). Till date, the functions of WRKY TFs have been intensively studied in not only biotic stress responses but also abiotic stress responses, as well as in seed germination, flower development, and senescence (Tripathi et al. 2014; Rushton et al. 2012; Thao and Tran 2012).

Overexpression of the heat- and drought-inducible rice *OsWRKY11* gene mediated using the heat-inducible *HSP101* promoter showed significant heat and drought tolerances in transgenic rice plants (Wu et al. 2009). Mochida et al. (2009) identified more than 210 putative WRKY TF-encoding genes in soybean. Under various abiotic stresses, 24 of 64 examined *GmWRKY* genes were found to be induced by drought (Zhou et al. 2008). Zhou et al. (2008) reported that *35S:GmWRKY21*-overexpressing *Arabidopsis* plants exhibited improved tolerance to cold stress in comparison with wild type. The same authors also demonstrated that transgenic *Arabidopsis* plants overexpressing *35S:GmWRKY54* were more tolerant to drought and salt stress, whereas those overexpressing *35S:GmWRKY13* were more sensitive to salt stress (Zhou et al. 2008), suggesting these two WRKY TFs have opposite functions in plant responses to salt stress. Overexpression of the rice *35S:OsWRKY45* enhanced salt and drought tolerance of *Arabidopsis* transgenic plants in addition to increased disease resistance (Qiu and Yu 2009). The grapevine VvWRKY11 TF played a role in osmotic stress tolerance as improved tolerance of *35S:VvWRKY11*-overexpressing transgenic *Arabidopsis* seedlings to mannitol-induced osmotic stress was observed in comparison with wild-type plants (Liu et al. 2011a). Recently, overexpression of the *35S:AtWRKY30* in *Arabidopsis* showed enhanced abiotic stress tolerance during early growth stages due to the binding of the TF to W-boxes in the promoter region of many stress/development-related genes, leading to the activation of their expression (Scarpeci et al. 2013). Results of RT-qPCR analyses showed that *ZmWRKY33* of maize belonging to the group I subfamily was induced by high salt, dehydration, cold, and ABA treatments. Overexpression of this gene under control of *35S* promoter in *Arabidopsis* activated stress-responsive genes, such as *RD29A*, under both normal growth and stress conditions, thereby improving tolerance of transgenic plants to salt stress (Li et al. 2013). Babitha et al. (2013) overexpressed *35S:AtWRKY28* in *Arabidopsis* and observed enhanced tolerance of transgenic plants to high NaCl, high mannitol, and oxidative stress. Additionally, higher root growth was observed in transgenic lines under mannitol-induced stress conditions. These transgenic plants showed their capacity in growth recovery to normal level after an 8-day drought exposure period followed by 6 days of rewatering.

3.5 The NAC Family

The plant-specific NAC TF family was initially described in *Petunia* and *Arabidopsis* more than 15 years ago (Aida et al. 1997; Souer et al. 1996). In plants, the NAC TFs have been reported to regulate diverse biological processes, such as flowering (Yoo et al. 2007), regulation of secondary cell wall synthesis and cell division (Zhong et al. 2007), embryo development (Duval et al. 2002), auxin signaling and lateral root formation (Xie et al. 2000), senescence (Kjaersgaard et al. 2011), as well as biotic and abiotic stress responses (Olsen et al. 2005; Puranik et al. 2012). The typical features of an NAC TF include an N-terminal conserved DNA-binding domain involving nucleus-oriented localization and a variable domain located at the C-terminal end which is essential for transcriptional activation (Fang et al. 2008; Riechmann and Ratcliffe 2000; Hao et al. 2011). Alignments of *Arabidopsis* and rice NAC domains suggested eight NAC subfamilies (from NAC-a to NAC-h), mainly distinguished by their unique structures in motif at the C-terminal of NAC domain (Fujita et al. 2004; Fang et al. 2008; Shen et al. 2009). Till date, NAC TFs have been systematically identified in many plant species thanks to the availability of their sequenced genomes. For instance, at least 117, 151, 163, 152, and 200 *NAC* genes have been identified in *Arabidopsis,* rice (Nuruzzaman et al. 2010), poplar (Hu et al. 2010), tobacco (Rushton et al. 2008), and soybean (Mochida et al. 2009), respectively. In drought signaling, NAC TFs were reported to function in both ABA-dependent and ABA-independent pathways (Shinozaki and Yamaguchi-Shinozaki 2007). The role of NACs in relation to drought response was initially proposed in a study of overexpression of either *ANAC019, ANAC055,* or *ANAC072* in *Arabidopsis* which led to considerable increase in drought tolerance of transgenic plants (Tran et al. 2004). Thereafter, stress-related *NAC* genes have been detected in other plant species, such as rice (*OsNAC5, OsNAC6, SNAC1,* and *ONAC45*) (Puranik et al. 2012; Nakashima et al. 2007; Hu et al. 2006; Zheng et al. 2009; Song et al. 2011; Takasaki et al. 2010), wheat (*TaNAC4, TaNAC69,* and *TaNAC2a*) (Xue et al. 2011; Tang et al. 2012; Xia et al. 2010), oilseed rape (*Brassica napus*) (*BnNAC2* and *BnNAC5*) (Zhong et al. 2012), and peanut (*Arachis hypogaea*) (*AhNAC3*) (Liu et al. 2013), which showed strong potential for genetic engineering of improved biotic and/or abiotic stress-tolerant crops.

Transgenic *Arabidopsis* displayed enhanced tolerance to drought stress without growth retardation when overexpressing the rice *OsNAC52* using *35S* promoter (Gao et al. 2010). In another independent study of rice, *ONAC063* expression was highly induced in roots by high salinity as well as by high osmotic pressure and ROS levels (Yokotani et al. 2009). *35S:ONAC063*-overexpressing transgenic *Arabidopsis* also displayed enhanced tolerance to high salinity and osmotic pressure (Yokotani et al. 2009). In a study of Lu et al. (2012), a maize *NAC* gene, *ZmSNAC1,* was cloned and functionally characterized. Low temperature, high salinity, drought stress, and ABA treatment strongly induced the expression of this gene. Overexpression of *35S:ZmSNAC1* in *Arabidopsis* resulted in hypersensitivity of transgenic plants to ABA and osmotic stress at the germination stage, but enhanced dehydration tolerance at the seedling stage as compared with non-transgenic control (Lu et al. 2012). Recently, a novel TF named *JcNAC1* from the new model woody plant, *Jatropha curcas,* was reported to have function in responses to abiotic stresses and pathogen infection as overexpression of this gene under control of *35S* promoter not only changed the expression of stress-related genes but also increased tolerance of transgenic *J. curcas* to drought (Qin et al. 2014). Another novel NAC TF, *EcNAC* from finger millet (*Eleusine coracana*), was overexpressed under control of either *35S* promoter or synthetic *4xABRE* stress-inducible promoter in tobacco, and both transgenic lines led to enhanced tolerance to various abiotic stresses, including stresses induced by polyethylene glycol (PEG) and mannitol, as well as high salinity (Ramegowda et al. 2012). In alfalfa (*Medicago sativa*), a NAC TF involved in response to abiotic stress was identified, and the expression of this gene was shown to be induced by drought, high salinity, and ABA (Wang 2013). Results revealed that transgenic *Arabidopsis* overexpressing this TF using *35S* promoter had better drought tolerance than the wild type (Wang 2013). Baloglu et al. (2012) performed expression analysis of *TaNAC69-1* and *TtNAMB-2* in durum wheat (*T. turgidum*) under different abiotic stress conditions. Specifically, *TaNAC69-1* was upregulated after 3 h of salt treatment, and the highest level of expression was observed at 24 and 48 h of post-treatment with heat and salinity respectively. On the other hand, *TtNAMB-2* was significantly induced by salt and low temperature stresses. In soybean, the first *GmNAC* genes identified were *GmNAC1–6* in a study conducted by Meng et al. (2007). Subsequently, expression of these genes in response to various stress and hormone treatments, including ABA, JA, high salinity, and PEG-induced osmotic stress, was analyzed in detail by another group (Pinheiro et al. 2009). Later on, Tran et al. (2009) initiated a study of the *GmNAC* family at a wider scale, covering the expression analysis of 31 *GmNAC* genes at seedling stage and under different abiotic stress conditions, including dehydration, salinity, cold, and ABA treatment, as well as examination of their transcriptional activity. According to the results, nine genes were shown to be upregulated by at least one of the tested treatments. Except *GmNAC028,* all remaining genes (*GmNAC002, 003, 004, 010, 012, 013, 015,* and *020*) also had transcriptional activation activity as shown by a yeast one-hybrid assay. More recently, Tran's laboratory studied 152 *GmNAC* genes, which could be detected in soybean genome with full-length cDNA, and proposed a comprehensive nomenclature for the *GmNAC* members (Le et al. 2011). Furthermore, the authors reported that 31 genes displayed significantly altered expression upon dehydration, with 25 up- and 6 downregulated genes. Additionally, the same research group demonstrated the complexity in the dynamics of drought-responsive expression of the *GmNAC* genes as they indicated that expression of several *GmNAC* genes were tissue- and/or development stage-dependent (Le et al. 2012). More recently, Thao et al. (2013) found differential expression of a subset of drought-responsive *GmNAC*s in soybean cultivars differing in drought tolerance, and identified positive correlation between *GmNAC* expression levels in these cultivars and their drought-tolerant degree. On the basis of their results, the authors also suggested a number of promising candidate *GmNAC* genes with potential application in genetic engineering of improved drought-tolerant soybean varieties (Thao et al. 2013).

4 Conclusions and Future Perspectives

Given the fact that using conventional breeding methods to create better stress-tolerant cultivars is not really effective since it is applied for varieties with close relationship only, creating transgenic plants by genetic technologies is a much more powerful approach. Thanks to advanced development in molecular cloning and plant transformation methods, barriers in gene transfer across species can be easily overcome. Therefore, the main challenge for scientists is to gain more and more in-depth understanding of mechanisms that plants employ to respond to stresses, especially to conditions similarly to the field environment. By doing this, important genes involved in plant adaptation to environmental stresses can be identified and used for crop improvement. Till date, many genes have been assigned as crucial contributors and uncountable attempts have been made to evaluate their roles in transgenic plants regarding the tolerance capacity toward abiotic stresses. The results of applications of TF-encoding genes so far have been quite promising. Accordingly, TFs bear a high potential for crop improvement using genetic engineering, and thus their characterization should deserve even more attention from the research community in the coming years.

Acknowledgments This work is funded by Vietnam National Foundation for Science and Technology Development (NAFOSTED) under grant number 106.16-2011.37 to Nguyen Phuong Thao.

References

Abe H, Yamaguchi-Shinozaki K, Urao T, Iwasaki T, Hosokawa D, Shinozaki K (1997) Role of Arabidopsis MYC and MYB homologs in drought-and abscisic acid-regulated gene expression. Plant Cell 9(10):1859–1868

Abogadallah GM, Nada RM, Malinowski R, Quick P (2011) Overexpression of HARDY, an AP2/ERF gene from *Arabidopsis*, improves drought and salt tolerance by reducing transpiration and sodium uptake in transgenic *Trifolium alexandrinum* L. Planta 233(6):1265–1276

Agarwal PK, Agarwal P, Reddy M, Sopory SK (2006) Role of DREB transcription factors in abiotic and biotic stress tolerance in plants. Plant Cell Rep 25(12):1263–1274

Agarwal P, Agarwal PK, Joshi AJ, Sopory SK, Reddy MK (2010) Overexpression of PgDREB2A transcription factor enhances abiotic stress tolerance and activates downstream stress-responsive genes. Mol Biol Rep 37(2):1125–1135

Aguado-Santacruz GA (2006) Genetic manipulation of plants for increased drought tolerance. Adv Agr Food Biotechnol 2006:71–98

Aida M, Ishida T, Fukaki H, Fujisawa H, Tasaka M (1997) Genes involved in organ separation in Arabidopsis: an analysis of the cup-shaped cotyledon mutant. Plant Cell 9(6):841–857

Alsina MM, Smart DR, Bauerle T, De Herralde F, Biel C, Stockert C, Negron C, Save R (2011) Seasonal changes of whole root system conductance by a drought-tolerant grape root system. J Exp Bot 62(1):99–109

Ambawat S, Sharma P, Yadav NR, Yadav RC (2013) MYB transcription factor genes as regulators for plant responses: an overview. Physiol Mol Biol Plants 19(3):307–321

Apel K, Hirt H (2004) Reactive oxygen species: metabolism, oxidative stress, and signal transduction. Annu Rev Plant Biol 55:373–399

Aroca R, Porcel R, Ruiz-Lozano JM (2012) Regulation of root water uptake under abiotic stress conditions. J Exp Bot 63(1):43–57
Ashraf M, Harris P (2004) Potential biochemical indicators of salinity tolerance in plants. Plant Sci 166(1):3–16
Babitha K, Ramu S, Pruthvi V, Mahesh P, Nataraja KN, Udayakumar M (2013) Co-expression of *AtbHLH17* and *AtWRKY28* confers resistance to abiotic stress in *Arabidopsis*. Transgenic Res 22(2):327–341
Baldoni E, Genga A, Medici A, Coraggio I, Locatelli F (2013) The *OsMyb4* gene family: stress response and transcriptional auto-regulation mechanisms. Biol Plant 57(4):691–700
Baloglu MC, Oz MT, Oktem HA, Yucel M (2012) Expression analysis of *TaNAC69-1* and *TtNAMB-2*, wheat NAC family transcription factor genes under abiotic stress conditions in durum wheat (*Triticum turgidum*). Plant Mol Biol Rep 30(5):1246–1252
Bhargava S, Sawant K (2013) Drought stress adaptation: metabolic adjustment and regulation of gene expression. Plant Breed 132(1):21–32
Bouaziz D, Pirrello J, Charfeddine M, Hammami A, Jbir R, Dhieb A, Bouzayen M, Gargouri-Bouzid R (2013) Overexpression of *StDREB1* transcription factor increases tolerance to salt in transgenic potato plants. Mol Biotechnol 54(3):803–817
Boudsocq M, Laurière C (2005) Osmotic signaling in plants. Multiple pathways mediated by emerging kinase families. Plant Physiol 138(3):1185–1194
Boudsocq M, Sheen J (2013) CDPKs in immune and stress signaling. Trends Plant Sci 18(1):30–40
Calvo-Polanco M, Señorans J, Zwiazek JJ (2012) Role of adventitious roots in water relations of tamarack (Larix laricina) seedlings exposed to flooding. BMC Plant Biol 12(1):99–108
Chakrabortee S, Meersman F, Schierle GSK, Bertoncini CW, McGee B, Kaminski CF, Tunnacliffe A (2010) Catalytic and chaperone-like functions in an intrinsically disordered protein associated with desiccation tolerance. Proc Natl Acad Sci U S A 107(37):16084–16089
Chaves MM, Maroco JP, Pereira JS (2003) Understanding plant responses to drought-from genes to the whole plant. Funct Plant Biol 30(3):239–264
Chaves M, Flexas J, Pinheiro C (2009) Photosynthesis under drought and salt stress: regulation mechanisms from whole plant to cell. Ann Bot 103(4):551–560
Chen Y, Yang X, He K, Liu M, Li J, Gao Z, Lin Z, Zhang Y, Wang X, Qiu X (2006) The MYB transcription factor superfamily of Arabidopsis: expression analysis and phylogenetic comparison with the rice MYB family. Plant Mol Biol 60(1):107–124
Chen M, Wang Q-Y, Cheng X-G, Xu Z-S, Li L-C, Ye X-G, Xia L-Q, Ma Y-Z (2007) *GmDREB2*, a soybean DRE-binding transcription factor, conferred drought and high-salt tolerance in transgenic plants. Biochem Biophys Res Commun 353(2):299–305
Chen L, Song Y, Li S, Zhang L, Zou C, Yu D (2012) The role of WRKY transcription factors in plant abiotic stresses. Biochim Biophys Acta 1819(2):120–128
Chen Y, Chen Z, Kang J, Kang D, Gu H, Qin G (2013a) *AtMYB14* regulates cold tolerance in *Arabidopsis*. Plant Mol Biol Rep 31(1):87–97
Chen Y, Yang J, Wang Z, Zhang H, Mao X, Li C (2013b) Gene structures, classification, and expression models of the *DREB* transcription factor subfamily in *Populus trichocarpa*. Scientific World Journal 2013(2013):1–12
Chen X, Wang Y, Lv B, Li J, Luo L, Lu S, Zhang X, Ma H, Ming F (2014) The NAC family transcription factor OsNAP confers abiotic stress response through the ABA pathway. Plant Cell Physiol. 55(3):604–619
Cheng L, Li S, Hussain J, Xu X, Yin J, Zhang Y, Chen X, Li L (2013a) Isolation and functional characterization of a salt responsive transcriptional factor, LrbZIP from lotus root (*Nelumbo nucifera* Gaertn). Mol Biol Rep 40(6):4033–4045
Cheng L, Li X, Huang X, Ma T, Liang Y, Ma X, Peng X, Jia J, Chen S, Chen Y (2013b) Overexpression of sheepgrass R1-MYB transcription factor *LcMYB1* confers salt tolerance in transgenic *Arabidopsis*. Plant Physiol Biochem 70:252–260
Chinnusamy V, Schumaker K, Zhu J-K (2004) Molecular genetic perspectives on cross-talk and specificity in abiotic stress signalling in plants. J Exp Bot 55(395):225–236

Clemens S (2006) Toxic metal accumulation, responses to exposure and mechanisms of tolerance in plants. Biochimie 88(11):1707–1719

Dai X, Xu Y, Ma Q, Xu W, Wang T, Xue Y, Chong K (2007) Overexpression of an R1R2R3 MYB gene, *OsMYB3R-2*, increases tolerance to freezing, drought, and salt stress in transgenic *Arabidopsis*. Plant Physiol 143(4):1739–1751

Dalal M, Tayal D, Chinnusamy V, Bansal KC (2009) Abiotic stress and ABA-inducible Group 4 *LEA* from *Brassica napus* plays a key role in salt and drought tolerance. J Biotechnol 139(2):137–145

Danquah A, de Zelicourt A, Colcombet J, Hirt H (2014) The role of ABA and MAPK signaling pathways in plant abiotic stress responses. Biotechnol Adv 32(1):40–52

Ding Z, Li S, An X, Liu X, Qin H, Wang D (2009) Transgenic expression of MYB15 confers enhanced sensitivity to abscisic acid and improved drought tolerance in *Arabidopsis thaliana*. J Genet Genomics 36(1):17–29

Dubos C, Stracke R, Grotewold E, Weisshaar B, Martin C, Lepiniec L (2010) *MYB* transcription factors in *Arabidopsis*. Trends Plant Sci 15(10):573–581

Dubouzet JG, Sakuma Y, Ito Y, Kasuga M, Dubouzet EG, Miura S, Seki M, Shinozaki K, Yamaguchi Shinozaki K (2003) *OsDREB* genes in rice, *Oryza sativa* L., encode transcription activators that function in drought-, high-salt- and cold-responsive gene expression. Plant J 33(4):751–763

Duval M, Hsieh T-F, Kim SY, Thomas TL (2002) Molecular characterization of *AtNAM*: a member of the *Arabidopsis* NAC domain superfamily. Plant Mol Biol 50(2):237–248

Eulgem T, Rushton PJ, Robatzek S, Somssich IE (2000) The WRKY superfamily of plant transcription factors. Trends Plant Sci 5(5):199–206

Fang Y, You J, Xie K, Xie W, Xiong L (2008) Systematic sequence analysis and identification of tissue-specific or stress-responsive genes of NAC transcription factor family in rice. Mol Genet Genomics 280(6):547–563

Feller A, Machemer K, Braun EL, Grotewold E (2011) Evolutionary and comparative analysis of MYB and bHLH plant transcription factors. Plant J 66(1):94–116

Fischer U, Dröge-Laser W (2004) Overexpression of *NtERF5*, a new member of the tobacco ethylene response transcription factor family enhances resistance to tobacco mosaic virus. Mol Plant Microbe Interact 17(10):1162–1171

Fujimoto SY, Ohta M, Usui A, Shinshi H, Ohme-Takagi M (2000) *Arabidopsis* ethylene-responsive element binding factors act as transcriptional activators or repressors of GCC box-mediated gene expression. Plant Cell 12(3):393–404

Fujita M, Fujita Y, Maruyama K, Seki M, Hiratsu K, Ohme-Takagi M, Tran L-SP, Yamaguchi-Shinozaki K, Shinozaki K (2004) A dehydration-induced NAC protein, RD26, is involved in a novel ABA-dependent stress-signaling pathway. Plant J 39(6):863–876

Fujita Y, Fujita M, Shinozaki K, Yamaguchi-Shinozaki K (2011) ABA-mediated transcriptional regulation in response to osmotic stress in plants. J Plant Res 124(4):509–525

Gao F, Xiong A, Peng R, Jin X, Xu J, Zhu B, Chen J, Yao Q (2010) *OsNAC52*, a rice NAC transcription factor, potentially responds to ABA and confers drought tolerance in transgenic plants. Plant Cell Tiss Org 100(3):255–262

Gao S-Q, Chen M, Xu Z-S, Zhao C-P, Li L, Xu H-j, Tang Y-m, Zhao X, Ma Y-Z (2011) The soybean GmbZIP1 transcription factor enhances multiple abiotic stress tolerances in transgenic plants. Plant Mol Biol 75(6):537–553

Gong Y, Rao L, Yu D (2013) Abiotic stress in plants. In: Stoytcheva M, Zlatev R (eds) Agricultural chemistry. InTech, Rijeka. doi:10.5772/55163

Gowda VR, Henry A, Yamauchi A, Shashidhar H, Serraj R (2011) Root biology and genetic improvement for drought avoidance in rice. Field Crop Res 122(1):1–13

Goyal K, Walton L, Tunnacliffe A (2005) LEA proteins prevent protein aggregation due to water stress. Biochem J 388:151–157

Grassi G, Magnani F (2005) Stomatal, mesophyll conductance and biochemical limitations to photosynthesis as affected by drought and leaf ontogeny in ash and oak trees. Plant Cell Environ 28(7):834–849

Ha CV, Le DT, Nishiyama R, Watanabe Y, Tran UT, Dong NV, Tran L-SP (2013) Characterization of the newly developed soybean cultivar DT2008 in relation to the model variety W82 reveals a new genetic resource for comparative and functional genomics for improved drought tolerance. Biomed Res Int 2013:1–8

Hachiya T, Sugiura D, Kojima M, Sato S, Yanagisawa S, Sakakibara H, Terashima I, Noguchi K (2014) High CO_2 triggers preferential root growth of *Arabidopsis thaliana* via two distinct systems at low pH and low N stresses. Plant Cell Physiol 55(2):269–280

Hao D, Ohme-Takagi M, Sarai A (1998) Unique mode of GCC box recognition by the DNA-binding domain of ethylene-responsive element-binding factor (ERF domain) in plant. J Biol Chem 273(41):26857–26861

Hao YJ, Wei W, Song QX, Chen HW, Zhang YQ, Wang F, Zou HF, Lei G, Tian AG, Zhang WK (2011) Soybean NAC transcription factors promote abiotic stress tolerance and lateral root formation in transgenic plants. Plant J 68(2):302–313

Hasanuzzaman M, Nahar K, Alam MM, Roychowdhury R, Fujita M (2013) Physiological, biochemical, and molecular mechanisms of heat stress tolerance in plants. Int J Mol Sci 14(5):9643–9684

Hideg É, Jansen MA, Strid Å (2013) UV-B exposure, ROS, and stress: inseparable companions or loosely linked associates? Trends Plant Sci 18(2):107–115

Hossain MA, Piyatida P, da Silva JAT, Fujita M (2012) Molecular mechanism of heavy metal toxicity and tolerance in plants: central role of glutathione in detoxification of reactive oxygen species and methylglyoxal and in heavy metal chelation. J Bot 2012(2012):1–37

Hu H, Dai M, Yao J, Xiao B, Li X, Zhang Q, Xiong L (2006) Overexpressing a NAM, ATAF, and CUC (NAC) transcription factor enhances drought resistance and salt tolerance in rice. Proc Natl Acad Sci U S A 103(35):12987–12992

Hu R, Qi G, Kong Y, Kong D, Gao Q, Zhou G (2010) Comprehensive analysis of NAC domain transcription factor gene family in Populus trichocarpa. BMC Plant Biol 10(1):145

Huang X-S, Liu J-H, Chen X-J (2010) Overexpression of *PtrABF* gene, a bZIP transcription factor isolated from *Poncirus trifoliata*, enhances dehydration and drought tolerance in tobacco via scavenging ROS and modulating expression of stress-responsive genes. BMC Plant Biol 10(1):230

Huang G-T, Ma S-L, Bai L-P, Zhang L, Ma H, Jia P, Liu J, Zhong M, Guo Z-F (2012) Signal transduction during cold, salt, and drought stresses in plants. Mol Biol Rep 39(2):969–987

Ishiguro S, Nakamura K (1994) Characterization of a cDNA encoding a novel DNA-binding protein, SPF1, that recognizes SP8 sequences in the 5′ upstream regions of genes coding for sporamin and β-amylase from sweet potato. Mol Gen Genet 244(6):563–571

Islam M, Wang M (2012) Expression patterns of an abiotic stress-inducible dehydration responsive element binding protein-2 (DREB2) gene in tomato. Bangladesh J Plant Breed Genet 25(1):01–09

Ito Y, Katsura K, Maruyama K, Taji T, Kobayashi M, Seki M, Shinozaki K, Yamaguchi-Shinozaki K (2006) Functional analysis of rice DREB1/CBF-type transcription factors involved in cold-responsive gene expression in transgenic rice. Plant Cell Physiol 47(1):141–153

Jaarsma R, de Vries RS, de Boer AH (2013) Effect of salt stress on growth, Na^+ accumulation and proline metabolism in potato (*Solanum tuberosum*) cultivars. PloS ONE 8(3):e60183

Jaglo-Ottosen KR, Gilmour SJ, Zarka DG, Schabenberger O, Thomashow MF (1998) *Arabidopsis CBF1* overexpression induces *COR* genes and enhances freezing tolerance. Science 280(5360):104–106

Jaglo KR, Kleff S, Amundsen KL, Zhang X, Haake V, Zhang JZ, Deits T, Thomashow MF (2001) Components of the Arabidopsis C-repeat/dehydration-responsive element binding factor cold-response pathway are conserved in *Brassica napus* and other plant species. Plant Physiol 127(3):910–917

Jaradat MR, Feurtado JA, Huang D, Lu Y, Cutler AJ (2013) Multiple roles of the transcription factor AtMYBR1/AtMYB44 in ABA signaling, stress responses, and leaf senescence. BMC Plant Biol 13(1):192

Jin H, Martin C (1999) Multifunctionality and diversity within the plant *MYB*-gene family. Plant Mol Biol 41(5):577–585

Joo J, Choi HJ, Lee YH, Kim Y-K, Song SI (2013) A transcriptional repressor of the ERF family confers drought tolerance to rice and regulates genes preferentially located on chromosome 11. Planta 238(1):155–170

Jovanovic Z, Stanisavljevic N, Mikic A, Radovic S, Maksimovic V (2013) The expression of drought responsive element binding protein (DREB2 A) related gene from pea (*Pisum sativum* L.) as affected by water stress. Aust J Crop Sci 7(10):1590–1596

Kasuga M, Liu Q, Miura S, Yamaguchi-Shinozaki K, Shinozaki K (1999) Improving plant drought, salt, and freezing tolerance by gene transfer of a single stress-inducible transcription factor. Nat Biotechnol 17(3):287–291

Kim SY (2006) The role of ABF family bZIP class transcription factors in stress response. Physiol Plant 126(4):519–527

Kjaersgaard T, Jensen MK, Christiansen MW, Gregersen P, Kragelund BB, Skriver K (2011) Senescence-associated barley NAC (NAM, ATAF1, 2, CUC) transcription factor interacts with radical-induced cell death 1 through a disordered regulatory domain. J Biol Chem 286(41):35418–35429

Kobayashi Y, Murata M, Minami H, Yamamoto S, Kagaya Y, Hobo T, Yamamoto A, Hattori T (2005) Abscisic acid-activated SNRK2 protein kinases function in the gene-regulation pathway of ABA signal transduction by phosphorylating ABA response element-binding factors. Plant J 44(6):939–949

Koyro H-W, Ahmad P, Geissler N (2012) Abiotic stress responses in plants: an overview. In: Ahmad P, Prasad MNV (eds) Environmental adaptations and stress tolerance of plants in the era of climate change. Springer, New York, pp 1–28

Kreuzwieser J, Rennenberg H (2014) Molecular and physiological responses of trees to waterlogging stress. Plant Cell Environ. doi:10.1111/pce.12310

Ku Y-S, Au-Yeung W-K, Yung Y-L, Li M-W, Wen C-Q, Liu X, Lam H-M (2013) Drought stress and tolerance in soybean. In: Board JE (ed) A comprehensive survey of internaitonal soybean research—genetics, physiology, agronomy and nitrogen relationships. InTech, New York, pp 209–237

Kumar N, Nandwal AS, Waldia RS, Singh S, Devi S, Sharma KD, Kumar A (2012) Drought tolerance in chickpea as evaluated by root characteristics, plant water status, membrane integrity and chlorophyll fluorescence techniques. Exp Agric 48(03):378–387

Kumar MN, Jane W-N, Verslues PE (2013) Role of the putative osmosensor arabidopsis histidine kinase1 in dehydration avoidance and low-water-potential response. Plant Physiol 161(2):942–953

Lawlor D, Cornic G (2002) Photosynthetic carbon assimilation and associated metabolism in relation to water deficits in higher plants. Plant Cell Environ 25(2):275–294

Lawlor DW, Tezara W (2009) Causes of decreased photosynthetic rate and metabolic capacity in water-deficient leaf cells: a critical evaluation of mechanisms and integration of processes. Ann Bot 103(4):561–579

Le DT, Nishiyama R, Watanabe Y, Mochida K, Yamaguchi-Shinozaki K, Shinozaki K, Tran LSP (2011) Genome-wide survey and expression analysis of the plant-specific NAC transcription factor family in soybean during development and dehydration stress. DNA Res 18(4):263–276

Le DT, Nishiyama R, Watanabe Y, Tanaka M, Seki M, Yamaguchi-Shinozaki K, Shinozaki K, Tran L-SP (2012) Differential gene expression in soybean leaf tissues at late developmental stages under drought stress revealed by genome-wide transcriptome analysis. PloS ONE 7(11):e49522

Li H, Gao Y, Xu H, Dai Y, Deng D, Chen J (2013) *ZmWRKY33*, a WRKY maize transcription factor conferring enhanced salt stress tolerances in *Arabidopsis*. Plant Growth Regul 70(3):207–216

Liang C, Tian J, Liao H (2013) Proteomics dissection of plant responses to mineral nutrient deficiency. Proteomics 13(3-4):624–636

Liao Y, Zou H-F, Wang H-W, Zhang W-K, Ma B, Zhang J-S, Chen S-Y (2008a) Soybean *GmMYB76*, *GmMYB92*, and *GmMYB177* genes confer stress tolerance in transgenic *Arabidopsis* plants. Cell Res 18(10):1047–1060

Liao Y, Zou H-F, Wei W, Hao Y-J, Tian A-G, Huang J, Liu Y-F, Zhang J-S, Chen S-Y (2008b) Soybean *GmbZIP44*, *GmbZIP62* and *GmbZIP78* genes function as negative regulator of ABA signaling and confer salt and freezing tolerance in transgenic *Arabidopsis*. Planta 228(2):225–240

Liu Q, Kasuga M, Sakuma Y, Abe H, Miura S, Yamaguchi-Shinozaki K, Shinozaki K (1998) Two transcription factors, DREB1 and DREB2, with an EREBP/AP2 DNA binding domain separate two cellular signal transduction pathways in drought-and low-temperature-responsive gene expression, respectively, in *Arabidopsis*. Plant Cell 10(8):1391–1406

Liu F, Liu Q, Liang X, Huang H, Zhang S (2005) Morphological, anatomical, and physiological assessment of ramie [*Boehmeria nivea* (L.) Gaud.] tolerance to soil drought. Genet Resour Crop Evol 52(5):497–506

Liu H, Yang W, Liu D, Han Y, Zhang A, Li S (2011a) Ectopic expression of a grapevine transcription factor *VvWRKY11* contributes to osmotic stress tolerance in *Arabidopsis*. Mol Biol Rep 38(1):417–427

Liu H, Zhou X, Dong N, Liu X, Zhang H, Zhang Z (2011b) Expression of a wheat MYB gene in transgenic tobacco enhances resistance to *Ralstonia solanacearum*, and to drought and salt stresses. Funct Integr Genomics 11(3):431–443

Liu X, Liu S, Wu J, Zhang B, Li X, Yan Y, Li L (2013) Overexpression of *Arachis hypogaea NAC3* in tobacco enhances dehydration and drought tolerance by increasing superoxide scavenging. Plant Physiol Biochem 70:354–359

Liu C, Mao B, Ou S, Wang W, Liu L, Wu Y, Chu C, Wang X (2014a) OsbZIP71, a bZIP transcription factor, confers salinity and drought tolerance in rice. Plant Mol Biol 84(1-2):19–36

Liu G, Li X, Jin S, Liu X, Zhu L, Nie Y, Zhang X (2014b) Overexpression of rice NAC gene *SNAC1* improves drought and salt tolerance by enhancing root development and reducing transpiration rate in transgenic cotton. PloS ONE 9(1):e86895

Loutfy N, El-Tayeb MA, Hassanen AM, Moustafa MF, Sakuma Y, Inouhe M (2012) Changes in the water status and osmotic solute contents in response to drought and salicylic acid treatments in four different cultivars of wheat (*Triticum aestivum*). J Plant Res 125(1):173–184

Lu G, Gao C, Zheng X, Han B (2009) Identification of OsbZIP72 as a positive regulator of ABA response and drought tolerance in rice. Planta 229(3):605–615

Lu M, Ying S, Zhang D-F, Shi Y-S, Song Y-C, Wang T-Y, Li Y (2012) A maize stress-responsive NAC transcription factor, ZmSNAC1, confers enhanced tolerance to dehydration in transgenic *Arabidopsis*. Plant Cell Rep 31(9):1701–1711

Mahajan S, Tuteja N (2005) Cold, salinity and drought stresses: an overview. Arch Biochem Biophys 444(2):139–158

Mallikarjuna G, Mallikarjuna K, Reddy M, Kaul T (2011) Expression of *OsDREB2A* transcription factor confers enhanced dehydration and salt stress tolerance in rice (*Oryza sativa* L.). Biotechnol Lett 33(8):1689–1697

Manavalan LP, Guttikonda SK, Tran LSP, Nguyen HT (2009) Physiological and molecular approaches to improve drought resistance in soybean. Plant Cell Physiol 50(7):1260–1276

Mao X, Chen S, Li A, Zhai C, Jing R (2014) Novel NAC transcription factor TaNAC67 confers enhanced multi abiotic stress tolerances in *Arabidopsis*. PloS ONE 9(1):e84359

Meng Q, Zhang C, Gai J, Yu D (2007) Molecular cloning, sequence characterization and tissue-specific expression of six NAC-like genes in soybean (*Glycine max* (L.) Merr.). J Plant Physiol 164(8):1002–1012

Miller G, Suzuki N, Ciftci-yilmaz S, Mittler R (2010) Reactive oxygen species homeostasis and signalling during drought and salinity stresses. Plant Cell Environ 33(4):453–467

Misra S, Wu Y, Venkataraman G, Sopory SK, Tuteja N (2007) Heterotrimeric G-protein complex and G-protein-coupled receptor from a legume (*Pisum sativum*): role in salinity and heat stress and cross-talk with phospholipase C. Plant J 51(4):656–669

Mittler R (2002) Oxidative stress, antioxidants and stress tolerance. Trends Plant Sci 7(9):405–410

Mizoi J, Shinozaki K, Yamaguchi-Shinozaki K (2012) AP2/ERF family transcription factors in plant abiotic stress responses. Biochim Biophys Acta 1819(2):86–96

Mochida K, Yoshida T, Sakurai T, Yamaguchi-Shinozaki K, Shinozaki K, Tran LSP (2009) In silico analysis of transcription factor repertoire and prediction of stress responsive transcription factors in soybean. DNA Res 16(6):353–369

Mori IC, Schroeder JI (2004) Reactive oxygen species activation of plant Ca^{2+} channels. A signaling mechanism in polar growth, hormone transduction, stress signaling, and hypothetically mechanotransduction. Plant Physiol 135(2):702–708

Munns R, James RA, Läuchli A (2006) Approaches to increasing the salt tolerance of wheat and other cereals. J Exp Bot 57(5):1025–1043

Munns R, Tester M (2008) Mechanisms of salinity tolerance. Annu Rev Plant Biol 59:651–681

Nakashima K, Tran LSP, Van Nguyen D, Fujita M, Maruyama K, Todaka D, Ito Y, Hayashi N, Shinozaki K, Yamaguchi-Shinozaki K (2007) Functional analysis of a NAC-type transcription factor OsNAC6 involved in abiotic and biotic stress-responsive gene expression in rice. Plant J 51(4):617–630

Nakashima K, Ito Y, Yamaguchi-Shinozaki K (2009) Transcriptional regulatory networks in response to abiotic stresses in Arabidopsis and grasses. Plant Physiol 149(1):88–95

Nakashima K, Takasaki H, Mizoi J, Shinozaki K, Yamaguchi-Shinozaki K (2012) NAC transcription factors in plant abiotic stress responses. Biochim Biophys Acta 1819(2):97–103

Nuruzzaman M, Manimekalai R, Sharoni AM, Satoh K, Kondoh H, Ooka H, Kikuchi S (2010) Genome-wide analysis of NAC transcription factor family in rice. Gene 465(1):30–44

Ohme-Takagi M, Shinshi H (1995) Ethylene-inducible DNA binding proteins that interact with an ethylene-responsive element. Plant Cell 7(2):173–182

Ohnishi T, Nakazono M, Tsutsumi N (2008) Abiotic stress. In: Hirano H-Y, Sano Y, Hirai A, Sasaki T (eds) Rice biology in the genomics era. Springer, Berlin, pp 337–355

Olsen AN, Ernst HA, Leggio LL, Skriver K (2005) NAC transcription factors: structurally distinct, functionally diverse. Trends Plant Sci 10(2):79–87

Oñate-Sánchez L, Singh KB (2002) Identification of *Arabidopsis* ethylene-responsive element binding factors with distinct induction kinetics after pathogen infection. Plant Physiol 128(4):1313–1322

Osakabe Y, Yamaguchi-Shinozaki K, Shinozaki K, Tran L-SP (2013a) ABA control of plant macroelement membrane transport systems in response to water deficit and high salinity. New Phytol 202(1):35–49

Osakabe Y, Yamaguchi-Shinozaki K, Shinozaki K, Tran L-SP (2013b) Sensing the environment: key roles of membrane-localized kinases in plant perception and response to abiotic stress. J Exp Bot 64(2):445–458

Osakabe Y, Osakabe K, Shinozaki K, Tran L-SP (2014) Response of plants to water stress. Front Plant Sci 5. doi:10.3389/fpls.2014.00086

Pabo CO, Sauer RT (1992) Transcription factors: structural families and principles of DNA recognition. Annu Rev Biochem 61(1):1053–1095

Pandey S, Nelson DC, Assmann SM (2009) Two novel GPCR-type G proteins are abscisic acid receptors in *Arabidopsis*. Cell 136(1):136–148

Park JM, Park C-J, Lee S-B, Ham B-K, Shin R, Paek K-H (2001) Overexpression of the tobacco *Tsi1* gene encoding an EREBP/AP2-type transcription factor enhances resistance against pathogen attack and osmotic stress in tobacco. Plant Cell 13(5):1035–1046

Parvaiz A, Satyawati S (2008) Salt stress and phyto-biochemical responses of plants-a review. Plant Soil Environ 54(3):89–99

Pasquali G, Biricolti S, Locatelli F, Baldoni E, Mattana M (2008) *Osmyb4* expression improves adaptive responses to drought and cold stress in transgenic apples. Plant Cell Rep 27(10):1677–1686

Peng X, Ma X, Fan W, Su M, Cheng L, Iftekhar A, Lee B-H, Qi D, Shen S, Liu G (2011) Improved drought and salt tolerance of *Arabidopsis thaliana* by transgenic expression of a novel DREB gene from *Leymus chinensis*. Plant Cell Rep 30(8):1493–1502

Peng X, Zhang L, Zhang L, Liu Z, Cheng L, Yang Y, Shen S, Chen S, Liu G (2013) The transcriptional factor *LcDREB2* cooperates with *LcSAMDC2* to contribute to salt tolerance in *Leymus chinensis*. Plant Cell Tissue Org 113(2):245–256

Perfus-Barbeoch L, Jones AM, Assmann SM (2004) Plant heterotrimeric G protein function: insights from *Arabidopsis* and rice mutants. Curr Opin Plant Biol 7(6):719–731

Persak H, Pitzschke A (2014) Dominant repression by *Arabidopsis* transcription factor MYB44 causes oxidative damage and hypersensitivity to abiotic stress. Int J Mol Sci 15(2):2517–2537

Pinheiro GL, Marques CS, Costa MDBL, Reis PAB, Alves MS, Carvalho CM, Fietto LG, Fontes EPB (2009) Complete inventory of soybean NAC transcription factors: sequence conservation and expression analysis uncover their distinct roles in stress response. Gene 444(1):10–23

Puranik S, Sahu PP, Srivastava PS, Prasad M (2012) NAC proteins: regulation and role in stress tolerance. Trends Plant Sci 17(6):369–381

Qin Y, Tian Y, Han L, Yang X (2013) Constitutive expression of a salinity-induced wheat WRKY transcription factor enhances salinity and ionic stress tolerance in transgenic *Arabidopsis thaliana*. Biochem Biophys Res Commun 441(2):476–481

Qin X, Zheng X, Huang X, Lii Y, Shao C, Xu Y, Chen F (2014) A novel transcription factor JcNAC1 response to stress in new model woody plant *Jatropha curcas*. Planta 239(2):511–520

Qiu Y, Yu D (2009) Over-expression of the stress-induced *OsWRKY45* enhances disease resistance and drought tolerance in *Arabidopsis*. Environ Exp Bot 65(1):35–47

Quan R, Hu S, Zhang Z, Zhang H, Zhang Z, Huang R (2010) Overexpression of an ERF transcription factor *TSRF1* improves rice drought tolerance. Plant Biotechnol J 8(4):476–488

Ramegowda V, Senthil-Kumar M, Nataraja KN, Reddy MK, Mysore KS, Udayakumar M (2012) Expression of a finger millet transcription factor, *EcNAC1*, in tobacco confers abiotic stress-tolerance. PloS ONE 7(7):e40397

Ravikumar G, Manimaran P, Voleti S, Subrahmanyam D, Sundaram R, Bansal K, Viraktamath B, Balachandran S (2014) Stress-inducible expression of *AtDREB1* A transcription factor greatly improves drought stress tolerance in transgenic *indica* rice. Transgenic Res:1–19

Riechmann JL, Meyerowitz EM (1998) The AP2/EREBP family of plant transcription factors. Biol Chem 379:633–646

Riechmann JL, Ratcliffe OJ (2000) A genomic perspective on plant transcription factors. Curr Opin Plant Biol 3(5):423–434

Rong W, Qi L, Wang A, Ye X, Du L, Liang H, Xin Z, Zhang Z (2014) The ERF transcription factor TaERF3 promotes tolerance to salt and drought stresses in wheat. Plant Biotechnol J. 12(4): 468–479

Rosinski JA, Atchley WR (1998) Molecular evolution of the Myb family of transcription factors: evidence for polyphyletic origin. J Mol Evol 46(1):74–83

Rushton PJ, Torres JT, Parniske M, Wernert P, Hahlbrock K, Somssich I (1996) Interaction of elicitor-induced DNA-binding proteins with elicitor response elements in the promoters of parsley PR1 genes. EMBO J 15(20):5690

Rushton PJ, Bokowiec MT, Han S, Zhang H, Brannock JF, Chen X, Laudeman TW, Timko MP (2008) Tobacco transcription factors: novel insights into transcriptional regulation in the Solanaceae. Plant Physiol 147(1):280–295

Rushton PJ, Somssich IE, Ringler P, Shen QJ (2010) WRKY transcription factors. Trends Plant Sci 15(5):247–258

Rushton DL, Tripathi P, Rabara RC, Lin J, Ringler P, Boken AK, Langum TJ, Smidt L, Boomsma DD, Emme NJ (2012) WRKY transcription factors: key components in abscisic acid signalling. Plant Biotechnol J 10(1):2–11

Saad ASI, Li X, Li H-P, Huang T, Gao C-S, Guo M-W, Cheng W, Zhao G-Y, Liao Y-C (2013) A rice stress-responsive *NAC* gene enhances tolerance of transgenic wheat to drought and salt stresses. Plant Sci 203:33–40

Sakuma Y, Liu Q, Dubouzet JG, Abe H, Shinozaki K, Yamaguchi-Shinozaki K (2002) DNA-binding specificity of the ERF/AP2 domain of *Arabidopsis* DREBs, transcription factors involved in dehydration-and cold-inducible gene expression. Biochem Biophys Res Commun 290(3):998–1009

Sakuma Y, Maruyama K, Osakabe Y, Qin F, Seki M, Shinozaki K, Yamaguchi-Shinozaki K (2006) Functional analysis of an *Arabidopsis* transcription factor, DREB2A involved in drought-responsive gene expression. Plant Cell 18(5):1292–1309
Scarpeci TE, Zanor MI, Mueller-Roeber B, Valle EM (2013) Overexpression of *AtWRKY30* enhances abiotic stress tolerance during early growth stages in *Arabidopsis thaliana*. Plant Mol Biol 83(3):265–277
Schaller GE, Kieber JJ, Shiu S-H (2008) Two-component signaling elements and histidyl-aspartyl phosphorelays. In: The Arabidopsis book, vol 6. p e0112
Seo JS, Sohn HB, Noh K, Jung C, An JH, Donovan CM, Somers DA, Kim DI, Jeong S-C, Kim C-G (2012) Expression of the *Arabidopsis AtMYB44* gene confers drought/salt-stress tolerance in transgenic soybean. Mol Breed 29(3):601–608
Setia R, Lewis M, Marschner P, Raja Segaran R, Summers D, Chittleborough D (2011) Severity of salinity accurately detected and classified on a paddock scale with high resolution multispectral satellite imagery. Land Degrad Dev 24(4):375–384
Shabala S, Munns R (2012) Salinity stress: physiological constraints and adaptive mechanisms. In: Shabala S (ed) Plant stress physiology. CAB International, Oxford, pp 59–93
Shanker A, Venkateswarlu B (2011) Abiotic stress response in plants-physiological, biochemical and genetic perspectives. InTech, Rijeka. doi:10.5772/1762
Sharma KK, Lavanya M (2002) Recent developments in transgenics for abiotic stress in legumes of the semi-arid tropics. JIRCAS Working Report 23:61–73
Sharma P, Jha AB, Dubey RS, Pessarakli M (2012) Reactive oxygen species, oxidative damage, and antioxidative defense mechanism in plants under stressful conditions. J Bot 2012 (2012):217037
Sharoni AM, Nuruzzaman M, Satoh K, Shimizu T, Kondoh H, Sasaya T, Choi I-R, Omura T, Kikuchi S (2011) Gene structures, classification and expression models of the AP2/EREBP transcription factor family in rice. Plant Cell Physiol 52(2):344–360
Shekhawat UKS, Ganapathi TR, Srinivas L (2011) Cloning and characterization of a novel stress-responsive WRKY transcription factor gene (*MusaWRKY71*) from *Musa* spp. cv. *Karibale Monthan* (ABB group) using transformed banana cells. Mol Biol Rep 38(6):4023–4035
Shen H, Yin Y, Chen F, Xu Y, Dixon RA (2009) A bioinformatic analysis of *NAC* genes for plant cell wall development in relation to lignocellulosic bioenergy production. Bioenerg Res 2(4):217–232
Shi W, Liu D, Hao L, Wu C-a, Guo X, Li H (2014) *GhWRKY39*, a member of the WRKY transcription factor family in cotton, has a positive role in disease resistance and salt stress tolerance. Plant Cell Tiss Org 1–16
Shinozaki K, Yamaguchi-Shinozaki K (2007) Gene networks involved in drought stress response and tolerance. J Exp Bot 58(2):221–227
Solanke AU, Sharma AK (2008) Signal transduction during cold stress in plants. Physiol Mol Biol Plants 14(1-2):69–79
Song S-Y, Chen Y, Chen J, Dai X-Y, Zhang W-H (2011) Physiological mechanisms underlying OsNAC5-dependent tolerance of rice plants to abiotic stress. Planta 234(2):331–345
Souer E, van Houwelingen A, Kloos D, Mol J, Koes R (1996) The no apical meristem gene of petunia is required for pattern formation in embryos and flowers and is expressed at meristem and primordia boundaries. Cell 85(2):159–170
Stracke R, Werber M, Weisshaar B (2001) The *R2R3-MYB* gene family in *Arabidopsis thaliana*. Curr Opin Plant Biol 4(5):447–456
Su L, Li J, Liu D, Zhai Y, Zhang H, Li X, Zhang Q, Wang Y, Wang Q (2014) A novel MYB transcription factor, GmMYBJ1, from soybean confers drought and cold tolerance in *Arabidopsis thaliana*. Gene. doi:10.1016/j.gene.2014.01.024
Sun X, Li Y, Cai H, Bai X, Ji W, Ding X, Zhu Y (2012) The *Arabidopsis AtbZIP1* transcription factor is a positive regulator of plant tolerance to salt, osmotic and drought stresses. J Plant Res 125(3):429–438
Suzuki N, Koussevitzky S, Mittler R, Miller G (2012) ROS and redox signalling in the response of plants to abiotic stress. Plant Cell Environ 35(2):259–270

Takasaki H, Maruyama K, Kidokoro S, Ito Y, Fujita Y, Shinozaki K, Yamaguchi-Shinozaki K, Nakashima K (2010) The abiotic stress-responsive NAC-type transcription factor OsNAC5 regulates stress-inducible genes and stress tolerance in rice. Mol Genet Genomics 284(3):173–183

Tang Y, Liu M, Gao S, Zhang Z, Zhao X, Zhao C, Zhang F, Chen X (2012) Molecular characterization of novel *TaNAC* genes in wheat and overexpression of *TaNAC2a* confers drought tolerance in tobacco. Physiol Plant 144(3):210–224

Thao NP, Tran L-SP (2012) Potentials toward genetic engineering of drought-tolerant soybean. Crit Rev Biotechnol 32(4):349–362

Thao NP, Thu NBA, Hoang XLT, Ha VC, Tran LSP (2013) Differential expression analysis of a subset of drought-responsive *GmNAC* genes in two soybean cultivars differing in drought tolerance. Int J Mol Sci 14(12):23828–23841

Thapa G, Dey M, Sahoo L, Panda S (2011) An insight into the drought stress induced alterations in plants. Biol Plant 55(4):603–613

Theocharis A, Clément C, Barka EA (2012) Physiological and molecular changes in plants grown at low temperatures. Planta 235(6):1091–1105

Thu NBA, Nguyen QT, Hoang XLT, Thao NP, Tran LSP (2014) Evaluation of drought tolerance of the Vietnamese soybean cultivars provides potential resources for soybean production and genetic engineering. Biomed Res Int 2014(2014):809736

Tournier B, Sanchez-Ballesta MT, Jones B, Pesquet E, Regad F, Latché A, Pech J-C, Bouzayen M (2003) New members of the tomato ERF family show specific expression pattern and diverse DNA-binding capacity to the GCC box element. FEBS Lett 550(1):149–154

Tran L-SP, Nakashima K, Sakuma Y, Simpson SD, Fujita Y, Maruyama K, Fujita M, Seki M, Shinozaki K, Yamaguchi-Shinozaki K (2004) Isolation and functional analysis of Arabidopsis stress-inducible NAC transcription factors that bind to a drought-responsive cis-element in the early responsive to dehydration stress 1 promoter. Plant Cell 16(9):2481–2498

Tran L-SP, Quach TN, Guttikonda SK, Aldrich DL, Kumar R, Neelakandan A, Valliyodan B, Nguyen HT (2009) Molecular characterization of stress-inducible *GmNAC* genes in soybean. Mol Genet Genomics 281(6):647–664

Tran L-SP, Nishiyama R, Yamaguchi-Shinozaki K, Shinozaki K (2010) Potential utilization of NAC transcription factors to enhance abiotic stress tolerance in plants by biotechnological approach. GM Crops 1(1):32–39

Tripathi P, Rabara RC, Rushton PJ (2014) A systems biology perspective on the role of WRKY transcription factors in drought responses in plants. Planta 239(2):255–266

Turner NC, Wright GC, Siddique K (2001) Adaptation of grain legumes (pulses) to water-limited environments. Adv Agron 71:193–231

Udvardi MK, Kakar K, Wandrey M, Montanari O, Murray J, Andriankaja A, Zhang J-Y, Benedito V, Hofer JM, Chueng F (2007) Legume transcription factors: global regulators of plant development and response to the environment. Plant Physiol 144(2):538–549

Umezawa T, Fujita M, Fujita Y, Yamaguchi-Shinozaki K, Shinozaki K (2006) Engineering drought tolerance in plants: discovering and tailoring genes to unlock the future. Curr Opin Biotechnol 17(2):113–122

Uno Y, Furihata T, Abe H, Yoshida R, Shinozaki K, Yamaguchi-Shinozaki K (2000) *Arabidopsis* basic leucine zipper transcription factors involved in an abscisic acid-dependent signal transduction pathway under drought and high-salinity conditions. Proc Natl Acad Sci U S A 97(21):11632–11637

Urao T, Yakubov B, Satoh R, Yamaguchi-Shinozaki K, Seki M, Hirayama T, Shinozaki K (1999) A transmembrane hybrid-type histidine kinase in *Arabidopsis* functions as an osmosensor. Plant Cell 11(9):1743–1754

Urao T, Miyata S, Yamaguchi-Shinozaki K, Shinozaki K (2000) Possible His to Asp phosphorelay signaling in an *Arabidopsis* two-component system. FEBS Lett 478(3):227–232

Velasco R, Zharkikh A, Troggio M, Cartwright DA, Cestaro A, Pruss D, Pindo M, FitzGerald LM, Vezzulli S, Reid J (2007) A high quality draft consensus sequence of the genome of a heterozygous grapevine variety. PloS ONE 2(12):e1326

Voss I, Sunil B, Scheibe R, Raghavendra A (2013) Emerging concept for the role of photorespiration as an important part of abiotic stress response. Plant Biol 15(4):713–722
Wang W, Vinocur B, Altman A (2003) Plant responses to drought, salinity and extreme temperatures: towards genetic engineering for stress tolerance. Planta 218(1):1–14
Wang H, Huang Z, Chen Q, Zhang Z, Zhang H, Wu Y, Huang D, Huang R (2004) Ectopic overexpression of tomato *JERF3* in tobacco activates downstream gene expression and enhances salt tolerance. Plant Mol Biol 55(2):183–192
Wang X, Li W, Li M, Welti R (2006) Profiling lipid changes in plant response to low temperatures. Physiol Plant 126(1):90–96
Wang YX (2013) Characterization of a novel *Medicago sativa NAC* transcription factor gene involved in response to drought stress. Mol Biol Rep 40(11):6451–6458
Wang C, Deng P, Chen L, Wang X, Ma H, Hu W, Yao N, Feng Y, Chai R, Yang G (2013) A wheat WRKY transcription factor TaWRKY10 confers tolerance to multiple abiotic stresses in transgenic tobacco. PloS ONE 8(6):e65120
Weake VM, Workman JL (2010) Inducible gene expression: diverse regulatory mechanisms. Nat Rev Genet 11(6):426–437
Wen J-Q, Oono K, Imai R (2002) Two novel mitogen-activated protein signaling components, OsMEK1 and OsMAP1, are involved in a moderate low-temperature signaling pathway in rice. Plant Physiol 129(4):1880–1891
Wilkins O, Nahal H, Foong J, Provart NJ, Campbell MM (2009) Expansion and diversification of the *Populus* R2R3-MYB family of transcription factors. Plant Physiol 149(2):981–993
Wu L, Zhang Z, Zhang H, Wang X-C, Huang R (2008) Transcriptional modulation of ethylene response factor protein JERF3 in the oxidative stress response enhances tolerance of tobacco seedlings to salt, drought, and freezing. Plant Physiol 148(4):1953–1963
Wu X, Shiroto Y, Kishitani S, Ito Y, Toriyama K (2009) Enhanced heat and drought tolerance in transgenic rice seedlings overexpressing *OsWRKY11* under the control of *HSP101* promoter. Plant Cell Rep 28(1):21–30
Xia N, Zhang G, Liu X-Y, Deng L, Cai G-L, Zhang Y, Wang X-J, Zhao J, Huang L-L, Kang Z-S (2010) Characterization of a novel wheat *NAC* transcription factor gene involved in defense response against stripe rust pathogen infection and abiotic stresses. Mol Biol Rep 37(8):3703–3712
Xie Q, Frugis G, Colgan D, Chua N-H (2000) Arabidopsis NAC1 transduces auxin signal downstream of TIR1 to promote lateral root development. Genes Dev 14(23):3024–3036
Xiong L, Ishitani M, Lee H, Zhu J-K (2001) The *Arabidopsis LOS5/ABA3* locus encodes a molybdenum cofactor sulfurase and modulates cold stress-and osmotic stress-responsive gene expression. Plant Cell 13(9):2063–2083
Xiong L, Lee H, Ishitani M, Zhu J-K (2002) Regulation of osmotic stress-responsive gene expression by the *LOS6/ABA1* locus in *Arabidopsis*. J Biol Chem 277(10):8588–8596
Xue G-P, Way HM, Richardson T, Drenth J, Joyce PA, McIntyre CL (2011) Overexpression of *TaNAC69* leads to enhanced transcript levels of stress up-regulated genes and dehydration tolerance in bread wheat. Mol Plant 4(4):697–712
Yadav SK (2010) Cold stress tolerance mechanisms in plants. A review. Agron Sustain Dev 30(3):515–527
Yamaguchi-Shinozaki K, Shinozaki K (2006) Transcriptional regulatory networks in cellular responses and tolerance to dehydration and cold stresses. Annu Rev Plant Biol 57:781–803
Yáñez M, Cáceres S, Orellana S, Bastías A, Verdugo I, Ruiz-Lara S, Casaretto JA (2009) An abiotic stress-responsive bZIP transcription factor from wild and cultivated tomatoes regulates stress-related genes. Plant Cell Rep 28(10):1497–1507
Ying S, Zhang D-F, Fu J, Shi Y-S, Song Y-C, Wang T-Y, Li Y (2012) Cloning and characterization of a maize bZIP transcription factor, ZmbZIP72, confers drought and salt tolerance in transgenic *Arabidopsis*. Planta 235(2):253–266
Yokotani N, Ichikawa T, Kondou Y, Matsui M, Hirochika H, Iwabuchi M, Oda K (2009) Tolerance to various environmental stresses conferred by the salt-responsive rice gene *ONAC063* in transgenic *Arabidopsis*. Planta 229(5):1065–1075

Yoo SY, Kim Y, Kim SY, Lee JS, Ahn JH (2007) Control of flowering time and cold response by a NAC-domain protein in *Arabidopsis*. PloS ONE 2(7):1–10

Zhai Y, Wang Y, Li Y, Lei T, Yan F, Su L, Li X, Zhao Y, Sun X, Li J (2013) Isolation and molecular characterization of *GmERF7*, a soybean ethylene-response factor that increases salt stress tolerance in tobacco. Gene 513(1):174–183

Zhang C-L, He X-Y, He Z-H, Wang L-H, Xia X-C (2009a) Cloning of *TaCYP707A1* gene that encodes ABA 8′-hydroxylase in common wheat (*Triticum aestivum* L.). Agr Sci China 8(8):902–909

Zhang G, Chen M, Li L, Xu Z, Chen X, Guo J, Ma Y (2009b) Overexpression of the soybean *GmERF3* gene, an AP2/ERF type transcription factor for increased tolerances to salt, drought, and diseases in transgenic tobacco. J Exp Bot 60(13):3781–3796

Zhang H, Liu W, Wan L, Li F, Dai L, Li D, Zhang Z, Huang R (2010) Functional analyses of ethylene response factor JERF3 with the aim of improving tolerance to drought and osmotic stress in transgenic rice. Transgenic Res 19(5):809–818

Zhang W, Jeon BW, Assmann SM (2011a) Heterotrimeric G-protein regulation of ROS signalling and calcium currents in *Arabidopsis* guard cells. J Exp Bot 62(7):2371–2379

Zhang X, Wang L, Meng H, Wen H, Fan Y, Zhao J (2011b) Maize ABP9 enhances tolerance to multiple stresses in transgenic *Arabidopsis* by modulating ABA signaling and cellular levels of reactive oxygen species. Plant Mol Biol 75(4-5):365–378

Zhang XX, Tang YJ, Ma QB, Yang CY, Mu YH, Suo HC, Luo LH, Nian H (2013) *OsDREB2A*, a rice transcription factor, significantly affects salt tolerance in transgenic soybean. PloS ONE 8(12):e83011

Zhao T, Liang D, Wang P, Liu J, Ma F (2012) Genome-wide analysis and expression profiling of the *DREB* transcription factor gene family in *Malus* under abiotic stress. Mol Genet Genomics 287(5):423–436

Zheng X, Chen B, Lu G, Han B (2009) Overexpression of a *NAC* transcription factor enhances rice drought and salt tolerance. Biochem Biophys Res Commun 379(4):985–989

Zhong R, Richardson EA, Ye Z-H (2007) Two NAC domain transcription factors, SND1 and NST1, function redundantly in regulation of secondary wall synthesis in fibers of *Arabidopsis*. Planta 225(6):1603–1611

Zhong H, Guo Q-Q, Chen L, Ren F, Wang Q-Q, Zheng Y, Li X-B (2012) Two *Brassica napus* genes encoding NAC transcription factors are involved in response to high-salinity stress. Plant Cell Rep 31(11):1991–2003

Zhou Y, Lam HM, Zhang J (2007) Inhibition of photosynthesis and energy dissipation induced by water and high light stresses in rice. J Exp Bot 58(5):1207–1217

Zhou QY, Tian AG, Zou HF, Xie ZM, Lei G, Huang J, Wang CM, Wang HW, Zhang JS, Chen SY (2008) Soybean WRKY-type transcription factor genes, *GmWRKY13*, *GmWRKY21*, and *GmWRKY54*, confer differential tolerance to abiotic stresses in transgenic *Arabidopsis* plants. Plant Biotechnol J 6(5):486–503

Zhu XL, Qi L, Liu X, Cai SB, Xu HJ, Huang RF, Li JR, Wei XN, Zhang ZY (2014) The wheat ERF transcription factor TaPIE1 mediates host responses to both the necrotrophic pathogen *Rhizoctonia cerealis* and freezing stresses. Plant Physiol. 164(3):1499–514

Zhuang J, Chen JM, Yao QH, Xiong F, Sun CC, Zhou XR, Zhang J, Xiong AS (2011) Discovery and expression profile analysis of AP2/ERF family genes from *Triticum aestivum*. Mol Biol Rep 38(2):745–753

Zhuang J, Jiang HH, Wang F, Peng RH, Yao QH, Xiong AS (2013) A rice OsAP23, functioning as an AP2/ERF transcription factor, reduces salt tolerance in transgenic *Arabidopsis*. Plant Mol Biol Rep 31(6):1336–1345

Zou M, Guan Y, Ren H, Zhang F, Chen F (2008) A bZIP transcription factor, OsABI5, is involved in rice fertility and stress tolerance. Plant Mol Biol 66(6):675–683

Index

P. Ahmad et al. (eds.), *Improvement of Crops in the Era of Climatic Changes,* Volume 2,
DOI 10.1007/978-1-4614-8824-8,

Printed by Printforce, the Netherlands